Hollow Fiber Membrane Contactors

Hollow Fiber Membrane Contactors

Module Fabrication, Design and Operation, and Potential Applications

Edited by

Anil K. Pabby, S. Ranil Wickramasinghe, Kamalesh K. Sirkar,
Ana-Maria Sastre

CRC Press
Taylor & Francis Group
Boca Raton London New York

CRC Press is an imprint of the
Taylor & Francis Group, an **informa** business

First edition published 2020
by CRC Press
6000 Broken Sound Parkway NW, Suite 300, Boca Raton, FL 33487-2742

and by CRC Press
2 Park Square, Milton Park, Abingdon, Oxon, OX14 4RN

© 2021 Taylor & Francis Group, LLC

CRC Press is an imprint of Taylor & Francis Group, LLC

Library of Congress Cataloging-in-Publication Data

Names: Pabby, Anil Kumar, editor.
Title: Hollow fiber membrane contactors : module fabrication, design and operation, and potential applications / edited by Anil Kumar Pabby, Sumith Ranil Wickramasinghe, Kamalesh K. Sirkar, Ana Maria Sastre Raquena.
Description: First edition. | Boca Raton : Taylor and Francis, 2020. | Includes bibliographical references and index.
Identifiers: LCCN 2020024677 (print) | LCCN 2020024678 (ebook) | ISBN 9780367025786 (hardback) | ISBN 9780429398889 (ebook)
Subjects: LCSH: Membranes (Technology) | Filters and filtration. | Engineering--Case studies.
Classification: LCC TP159.M4 H65 2020 (print) | LCC TP159.M4 (ebook) | DDC 660/.28424--dc23
LC record available at https://lccn.loc.gov/2020024677
LC ebook record available at https://lccn.loc.gov/2020024678

ISBN: 9780367025786 (hbk)
ISBN: 9780429398889 (ebk)

Typeset in Times
by Deanta Global Publishing Services, Chennai, India

Visit the [companion website/eResources]: [insert comp website/eResources URL]

Contents

PART I   Introductory Chapters

PART II   Chapters on Gas–Liquid Contacting

PART III   Chapters on Liquid–Liquid Contacting

PART IV   Chapters on Supported Liquid Membranes

PART V   Chapters on Supported Gas Membranes

PART VI   Chapters on Fluid–Fluid Contacting

Foreword

Hollow Fiber Membrane Contactors: Module Fabrication, Design and Operation, and Potential Applications by Anil K. Pabby, S. Ranil Wickramasinghe, Kamalesh K. Sirkar, Ana-Maria Sastre, Editors

The literature covering hollow fiber membrane contactors and their applications is highly fragmented and contained in a large number of different scientific journals, book chapters, or special issues. This makes it difficult to gain a reasonably complete overview of hollow fiber contactor-based technology and its applications. It is certainly not an easy task to provide a complete overview of hollow fiber contactors and their applications in a single monograph. However, these were the objectives of the current book. This book is also a first attempt to bridge the gap between theory and practice.

Since the discovery of hollow fiber membrane contactors by W.S. Winston Ho, Lester T.C. Lee, and K.J. Fred Liu in the 1970s (U.S. Patents 3,957,504, 3,956,112 and 3,951,789) for membrane solvent extraction and hydrometallurgical extraction, the contactors have found many applications. The applications have included gas absorption and stripping, metal removal and recovery from wastewaters, acid gas removal and capture, solvent extraction operations, membrane distillation for juice concentration, air dehumidification, antibiotic recovery from fermentation broths, and separations in the food, pharmaceutical, and biotechnological industries. CO_2 carbonization of beverages, stripping of O_2 from beverages and boiler waters, blood oxygenation, and metal (e.g., indium) removal or recovery from wastewaters are in commercial operation. The deoxygenation of water to sub-ppb level to produce ultrapure water for semiconductor industries is one of the largest industrial applications of membrane contactors. Commercial units are in operation for membrane distillation in smaller scale. Membrane solvent extraction is being commercially used to extract organics, e.g., phenol. In addition, a great deal of research has been done both in fundamental understanding and in practical engineering. The research work has included the development of supported liquid membranes (SLMs) with strip dispersion that has solved the long-standing SLM stability problem as well as many aforementioned applications. Hence, there has been a tremendous amount of publications in the literature.

This book has nicely organized the key publications of membrane contactor technology in an effective manner for ease of understanding for the reader. It has organized on the basis of fluid contacting in (1) gas absorption and stripping, such as CO_2 carbonization of beverages, stripping of O_2 from beverages and boiler waters, and blood oxygenation through gas-liquid contacting, (2) solvent extraction, treatment of metal-bearing aqueous solutions, process intensification in integrated use of liquid membranes, non-dispersive solvent extraction and strip dispersion, and uranium-containing solution treatment, as well as emulsification and membrane nanoprecipitation and crystallization, via liquid-liquid contacting, (3) nuclear fuel cycle application and analytical techniques using supported liquid membranes, (4) air dehumidification, ammonia liquid fertilizers from urban wastewater, membrane distillation, and membrane condenser using supported gas membranes, and (5) fermentation, enzymatic transformation, chiral separation, and separations in the food, pharmaceutical, and biotechnological industries.

This book may benefit several groups and is intended to be a single and unique source of underlying principles, membrane contactor configurations, process design, and applications for hollow fiber membrane contactors. It can serve as educational material for industrial personnel contemplating the use of the membrane contactor technology. For scientists and engineers active in membrane contactors, it will provide a single source of reference in this field. Engineers evaluating separation processes will find this book to be a helpful guide to allow the membrane contactor approach to be compared with other separation processes. For students examining separation processes, membrane processes, and membrane contactors, it should be an invaluable sourcebook.

W.S. Winston Ho, Ph.D.
Distinguished Professor of Engineering
The Ohio State University
Columbus, Ohio

William G. Lowrie
Department of Chemical and Biomolecular Engineering
Department of Materials Science and Engineering
The Ohio State University
Columbus, Ohio

Preface

This book on hollow fiber contractors presents an up to date compilation of the latest developments and milestones in membrane technology. Membrane contactors, wherein phase contacting is implemented or facilitated by the porous membrane structure and form, provide a new dimension to the growth of membrane science and technology; further they satisfy the requirements of process intensification. In addition, these membrane contactors represent significant progress beyond the first success in membrane contactors: blood oxygenators. Their integration with other membrane systems including membrane reactors could lead to redesigning membrane-based integrated production lines.

The chapters in this book have been classified using the following, based on different ways of contacting fluids with each other:

- Gas–liquid contacting
- Liquid–liquid contacting
- Supported liquid membrane
- Supported gas membrane
- Fluid–fluid contacting.

A variety of original contributions in the area of membrane contactors has been collected. Each section based on the above categories is divided into chapters that deal with the subject matter in depth and focus on cutting-edge advancements in the field. Several authors were commissioned to write chapters under the supervision of the editors, and each chapter was peer-reviewed for content and style before it was accepted for publication. The aim has been to maintain the perspective of an informative book rather than merely a collection of review chapters.

The editors would like to acknowledge the contributions of a number of authors and institutions that have played a major role in drafting the book, from conception to publication. The book would not have been possible without their input. These contributors are leading experts in their fields and bring a great wealth of experience to this book. The editors would also like to acknowledge the efforts of the reviewers who devoted their valuable time in critically evaluating the chapters before the set deadlines and suggested improvements to maintain the high standard of the book. Finally, we would like to acknowledge the support of our home institutions at every stage in the book's conception: the Bhabha Atomic Research Centre, Mumbai, India, University of Arkansas, Fayetteville, AR, New Jersey Institute of Technology, Newark, NJ and the Technical University of Catalonia, Barcelona, Spain.

Anil K. Pabby
S. Ranil Wickramasinghe
Kamalesh K. Sirkar
Ana-Maria Sastre

Acknowledgment

Having an idea and turning it into a book is as hard as it sounds. The experience is both internally challenging and rewarding. I am thankful to my colleagues and research group working on membranes in Bhabha Atomic Research Centre and Homi Bhabha National Institute who inspired me to initiate this project and helped make this happen. I especially want to thank my family members, my loving wife, Anju, my son Anubhav, daughter Akanksha and son-in-law, Arjun and my cute grandson, Avyan. It is well understood that this book would not have been possible without their unconditional support and love during the different stages of this document.

Anil K. Pabby

I am very appreciative of my research group for their dedication to the pursuit of knowledge. Finally, this book would not have been possible without the love, dedication and support of my wife Xianghong and son Aroshe, which has enabled me to pursue my scientific interests.

S. Ranil Wickramasinghe

I would like to acknowledge my wife Keka Sirkar for all the help, sustenance, and understanding she has provided during this effort. I would also like to acknowledge many of our former graduate students and post-doctoral fellows whose work has been crucial to the progress in membrane contactor development. Contributions over the years by Amit Sengupta, Gregory Frank, Ravi Prasad, Sudip Majumdar, Asim Guha, Rahul Basu, Uttam Shambhag, Gordana Obuskovic, Atsawin Thongsukmak, and Tripura Mulukutla form the basic building blocks of our work in membrane contactors.

Kamalesh K. Sirkar

I would like to acknowledge my research group and colleagues in my department for their valuable discussions and help. I would like to acknowledge, as well, my husband Antonio and my sons Carles and Jordi for their support and their patience during all these times. Without their support this project would not have been possible.

Ana-Maria Sastre

Editors

Dr. Anil K. Pabby is presently serving with one of the pioneer research centers of India, Bhabha Atomic Research Centre (BARC), Tarapur, Mumbai, Maharashtra, as Scientific Officer (Senior Scientist) and is associated with Department of Atomic Energy activities including research and development work under the DAE programme. He did his Ph.D. from the University of Mumbai, India and subsequently carried out his postdoctoral work at the Technical University of Catalunya, Barcelona, Spain. Dr. Pabby has more than 190 publications to his credit including 20 chapters and two patents on non-dispersive membrane technology. He was invited to join as an associate editor of the international journal, *Journal of Radioanalytical and Nuclear Chemistry* during 2002–2005. He has also served as a consultant to IAEA, Vienna, Austria, developing a technical book volume on "Application of membrane technologies for liquid radioactive waste processing". Dr. Pabby has taken a leading role in publishing *Handbook of membrane separation: Chemical, pharmaceutical, food and biotechnological applications* in July 2008 by CRC Press, New York, USA. The second edition of this book was published in April 2015.

Dr. Pabby has been a regular reviewer for several national and international journals and has also served on the editorial board of a few reputed international journals. His research interest includes membrane based solvent extraction, liquid membrane and its modeling aspects, pressure driven membrane processes, macrocyclic compounds, etc. Dr Pabby was elected as a Fellow of the Maharashtra Academy of Sciences (FMASc) in 2003 for his outstanding contribution to membrane science and technology. Also, he has been awarded the prestigious Tarun Datta Memorial award (instituted by Indian Association Nuclear Chemist and Allied Scientist) for his contribution to Nuclear and Radiochemistry for the year 2005. He was appointed as Associate Professor and Ph.D. guide and also M.Tech guide, Homi Bhabha National Institute, Mumbai. He has served as Secretary to the Indian Association Nuclear Chemist and Allied Scientist, Tarapur Chapter since 2008.

He has delivered several keynote/plenary/invited talks in national and international conferences. He has also served as chairman of the different sessions in conference in India and abroad.

Dr. Ana-Maria Sastre is a Professor of Chemical Engineering at the Universitat Politècnica de Catalunya (Barcelona, Spain), where she has been teaching Chemistry for more than 40 years. She received a Ph.D. from the Autonomous University of Barcelona in 1982 and has been working for many years in the field of solvent extraction, solvent impregnated resins, and membrane technology and adsorption processes.

She was a visiting fellow at the Department of Inorganic Chemistry, The Royal Institute of Technology, Sweden, during 1980–1981 and carried out postdoctoral research work during October 1986–April 1987 at Laboratoire de Chimie Minerale, de l'Ecole Europeenne des Hautes Etudes des Industries Chimiques d'Estrasbourg, France.

During 2015 she was Fulbright visiting scholar at the Chemical and Biomolecular Engineering Department of the University of California, Berkeley.

Prof. Sastre has published more than 200 journal publications and more than 80 papers in international conferences. She holds seven patent applications. She has advised 13 Ph.D. students and 16 master theses, and is a reviewer of many international journals. She was awarded the "Narcis Monturiol medal for scientific and technological merits" given by the Generalitat de Catalunya for outstanding contribution in science and technology in 2003.

Prof. Sastre was the head of the Chemical Engineering Department from 1999 to 2005, and from 2006 to 2013 was Vice President for Academic Policy at the Universitat Politècnica de Catalunya.

Dr. S. Ranil Wickramasinghe has published over 200 peer-reviewed journal articles, several book chapters and is co-editor of a book on responsive membrane and materials. He is active in the American Institute of Chemical Engineers and the North American Membrane Society. He is the executive editor of Separation Science and Technology. Prof Wickramasinghe's research interests are in membrane science and technology. His research focuses on synthetic membrane-based separation processes for purification of pharmaceuticals and biopharmaceuticals, treatment and re-use of water and for the production of biofuels. Typical unit operations include: microfiltration, ultrafiltration, virus filtration, nanofiltration, membrane extraction, etc. A current research focus is surface modification of membranes in order to impart unique surface properties. His group is actively developing responsive membranes. These membranes change their physical properties in response to changed environmental conditions. A second research focus is the development of catalytic membranes for biomass hydrolysis by grafting catalytic groups to the membrane surface. Prof Wickramasinghe obtained his Bachelor's and Master's degrees from the University of Melbourne, Australia in Chemical Engineering. He obtained his Ph.D. from the University of Minnesota, also in Chemical Engineering. He worked for five years in the biotechnology/biomedical industry in the Boston area before joining the faculty of Chemical Engineering at Colorado State University. He joined the Department of Chemical Engineering at the University of Arkansas in 2011 where he holds the Ross E. Martin Chair in Emerging Technologies and is an Arkansas Research Alliance Scholar.

 **Dr. Kamalesh K. Sirkar** is a Distinguished Professor of Chemical Engineering and Foundation Professor of Membrane Separations at New Jersey Institute of Technology (NJIT). He was a Professor of Chemical Engineering at Stevens Institute of Technology and Indian Institute of Technology, Kanpur. He received B. Tech (Hons) from IIT, Kharagpur and a Ph.D. from the University of Illinois, Urbana. His research involves synthetic membranes and membrane separation processes and techniques. He pioneered the notion of microporous membranes as membrane contactors of immiscible phase-pairs. He invented the commercialized membrane-based solvent extraction technology for which Hoechst Celanese, Inc. received

Honorable Mention in the 1991 Kirkpatrick Award. He also invented the hollow-fiber contained liquid membrane technology. His other innovations include: novel hollow-fiber membrane technology for membrane distillation; multilayer ultrafiltration membranes for protein purification; membrane-based ozonation; dendrimers for CO_2 separation; integrated filtration-chromatography technique for direct protein purification from a fermentation broth; solid hollow fiber cooling crystallization; and porous hollow fiber-based anti-solvent crystallization. His advisees include: 28 post-doctoral fellows, 36 Ph.D. students, and 47 MS students. He has authored 207 refereed journal papers and 22 book chapters. He has delivered 350 conference presentations, 10 Plenary Lectures and 14 Keynote Lectures. He is Editor-in-Chief of "Current Opinion in Chemical Engineering" since its inception. He is coeditor of *Membrane Handbook* which won the Professional and Scholarly Publishing Award for the Most Outstanding Engineering Work, 1992. He was/is on the Editorial Boards of J. Membrane Science, I&EC Research, Separation Science and Technology and Indian Chemical Engineer (IIChE). His awards and honors are: M. Eng. (Hon), Stevens Institute (1987); Harlan Perlis Award for Research (NJIT) (1997); Honorary Fellow, IIChE (2001); AIChE's Institute Award for Excellence in Industrial Gases Technology (2005); Clarence Gerhold Award, Separations Division (AIChE) (2008); AAAS Fellow (2008); 2009 NJIT Board of Overseers Excellence in Research Prize and Medal; 2016 Fellow, National Academy of Inventors; 2017 Alan S. Michaels Award for Innovation in Membrane Science and Technology by North American Membrane Society (NAMS); 2017 Fellow, NAMS; 2020 Fellow, AIChE. He has received 34 US patents, three Canada patents and one Indian patent. He was a Director of Separations Division (AIChE) (1994–1996). He was President of NAMS for 1998–1999 and on its Board of Directors during 1996–2001. His book entitled *Separation of Molecules, Macromolecules and Particles: Principles, Phenomena and Processes* was published in 2014 by Cambridge University Press in the Cambridge Series in Chemical Engineering.

Contributors

M. S. Abdullah
Advanced Membrane Technology Research Centre
Universiti Teknologi Malaysia
Skudai, Malaysia

F. J. Alguacil
Centro Nacional de Investigaciones Metalúrgicas (CSIC)
Madrid, Spain

N. de Arespacochaga
CETaqua, Carretera d'Esplugues
Barcelona, Spain.

F. Bazzarelli
National Research Council of Italy
University of Calabria Campus
Rende, Italy

R. B. Bhatt
Nuclear Recycle Board
Bhabha Atomic Research Centre
Tarapur, India

J. Chau
Chemical and Materials Engineering
New Jersey Institute of Technology
Newark, New Jersey

J. L. Cortina
Chemical Engineering Department
Universitat Politècnica de Catalunya
Barcelona, Spain

P. S. Dhami
Nuclear Recycle Group
Bhabha Atomic Research Centre
Mumbai, India

E. Drioli
Institute on Membrane Technology
National Research Council of Italy (CNR-ITM)
Rende, Italy

M. Fallanza
Department of Chemical and Biomolecular Engineering
Universidad de Cantabria
Santander, Spain

R. Farnood
Department of Chemical Engineering and Applied
    Chemistry
University of Toronto
Toronto, Canada

M. Frappa
Institute on Membrane Technology
National Research Council of Italy (CNR-ITM)
Rende, Italy

L. Giorno
National Research Council of Italy
University of Calabria Campus
Rende, Italy

P. S. Goh
Advanced Membrane Technology Research Centre
Universiti Teknologi Malaysia
Skudai, Malaysia

D. Gorri
Department of Chemical and Biomolecular Engineering
Universidad de Cantabria
3Santander, Spain

A. M. Isloor
Membrane Technology Laboratory
National Institute of Technology Karnataka
Surathkal, India

A. F. Ismail
Advanced Membrane Technology Research Centre
Universiti Teknologi Malaysia
Skudai, Malaysia

V. A. Juvekar
Department of Chemical Engineering
Indian Institute of Technology
Mumbai, India

C. P. Kaushik
Nuclear Recycle Group
Bhabha Atomic Research Centre
Mumbai, India

F. Macedonio
Institute on Membrane Technology
National Research Council of Italy (CNR-ITM)
Rende, Italy

M. Madhumala
Membrane Separations Group
CSIR-Indian Institute of Chemical Technology
Hyderabad, India

A. Mayor
Chemical Engineering Department
Universitat Politècnica de Catalunya
Barcelona, Spain.

S. Mishra
Chemical Engineering Group
Bhabha Atomic Research Centre
Mumbai, India

V. K. Mittal
Nuclear Recycle Board
Bhabha Atomic Research Centre
Tarapur, India

S. Mukhopadhyay
Chemical Engineering Group
Bhabha Atomic Research Centre
Mumbai, India

T. Nagamani
Membrane Separations Group
CSIR-Indian Institute of Chemical Technology
Telangana, India

R. Naim
Faculty of Chemical Engineering Technology
    and Process
Universiti Malaysia Pahang
Kuantan, Malaysia

B. C. Ng
Advanced Membrane Technology Research Centre
Universiti Teknologi Malaysia
Skudai, Malaysia

I. Ortiz
Department of Chemical and Biomolecular Engineering
Universidad de Cantabria
Santander, Spain

A. Ortiz
Department of Chemical and Biomolecular Engineering
Universidad de Cantabria
Santander, Spain

A. K. Pabby
Nuclear Recycle Board
Bhabha Atomic Research Centre
Tarapur, India
and
Homi Bhabha National Institute,
Mumbai, India

S. Panja
Nuclear Recycle Group
Bhabha Atomic Research Centre
Mumbai, India

T. Patra
Department of Biomedical Engineering
University of Arkansas
Fayetteville, Arkansas

P. A. Peterson
3M Separation and Purification Sciences Division
Charlotte, North Carolina

E. Piacentini
National Research Council of Italy)
University of Calabria Campus
Rende, Italy

M. Reig
Chemical Engineering Department
Universitat Politècnica de Catalunya
Barcelona, Spain.

A. M. Sastre
Chemical Engineering Department
Universitat Politècnica de Catalunya
Barcelona, Spain

D. B. Sathe
Nuclear Recycle Board
Bhabha Atomic Research Centre
Maharashtra, India

A. Sengupta
Ralph E. Martin Department of Chemical Engineering
University of Arkansas
Fayetteville, Arkansas

and

Radiochemistry Division
Bhabha Atomic Research Centre
Mumbai, India

A. Sengupta
3M Separation and Purification Sciences Division
Charlotte, North Carolina

K. T. Shenoy
Chemical Engineering Group
Bhabha Atomic Research Centre
Mumbai, India

K. K. Sirkar
Chemical and Materials Engineering
New Jersey Institute of Technology
Newark, New Jersey

N. L. Sonar
Nuclear Recycle Board
Bhabha Atomic Research Centre
Tarapur, India

S. Sridhar
Membrane Separations Group
CSIR-Indian Institute of Chemical Technology,
Telangana, India

B. Swain
Nuclear Recycle Board
Bhabha Atomic Research Centre
Tarapur, India
and
Homi Bhabha National Institute
Mumbai, India

G. P. Syed Ibrahim
Membrane Technology Laboratory
National Institute of Technology
Surathkal, India

G. P. Taylor
3M Separation and Purification
Sciences Division
Charlotte, North Carolina

C. Valderrama
Chemical Engineering Department
Universitat Politècnica de Catalunya
Barcelona, Spain.

T. P Valsala
Nuclear Recycle Board
Bhabha Atomic Research Centre
Tarapur, India

X. Vecino
Chemical Engineering Department
Universitat Politècnica de Catalunya
Barcelona, Spain.

S. R. Wickramasinghe
Ralph E. Martin Department of Chemical Engineering
University of Arkansas
Fayetteville, Arkansas

J. S. Yadav
Nuclear Recycle Group
Bhabha Atomic Research Centre
Mumbai, India

Part I

Introductory Chapters

1 Hollow Fiber Membrane Contactors
Introduction and Perspectives

Anil K. Pabby, S. Ranil Wickramasinghe,
Kamalesh K. Sirkar, and Ana-Maria Sastre

CONTENTS

1.1 INTRODUCTION

In recent years, the focus of research in membrane contactors has shifted to applications of a hollow fiber membrane contactor (HFMC), which achieves gas/liquid or liquid/liquid mass transfer without dispersion of one phase in another. This is accomplished by passing the fluids on opposite sides of a microporous membrane. By careful control of the pressure difference between the fluids, one of the fluids is immobilized in the pores of the membrane so that the fluid/fluid interface is located at the mouth of each pore. The principle of separation through HFMC is based on inter-phase mass transfer. HFMC-based technology provides non-dispersive contact along with larger interfacial area per unit volume than conventional contactors [1–4]. The operation is free from loading and flooding, and it involves minimal pressure drop and independent flow control of each phase. Diffusional mass transfer occurs at high rates across the interface. Further, the membrane contactor system is advantageous due to its compact operational setup.

In membrane contactors, the membrane functions as a barrier between two phases that avoids mixing but does not control the transport rate of different components between the phases. Membrane contactors typically utilize porous capillary membranes in a shell-and-tube device. The hollow fiber is a solid, microporous polymeric matrix that may be either hydrophobic or hydrophilic. Hydrophobic membranes are generally used for various techniques such as non-dispersive solvent extraction, membrane distillation, removal/absorption of dissolved gases or volatile species from/into water, and so on. Polypropylene (PP), polytetrafluoroethylene (PTFE), and polyvinylidene fluoride (PVDF) are reasonably good hydrophobic materials. These polymeric materials allow only gases to permeate through the pores in the solid membrane phase [5] whereas in the case of non-dispersive solvent extraction (NDSX) or hollow fiber strip dispersion (HFSD) technique, organic extractant gets impregnated into hydrophobic membrane pores [6, 7].

In the case of two aqueous phases flowing on either side of a hydrophobic HFM, the pores are not wetted by the liquid and the phases form an interface at both sides of the pore openings, with the gas trapped inside the pores. In NDSX technique, the aqueous-organic mass transfer interface is located at the pore mouth of the HF membrane [8, 9].

A new technique for solvent extraction with immobilized interfaces in a hydrophobic microporous membrane was first described by Kiani et al. [10]. Under this technique, the extraction of acetic acid from aqueous solutions of different concentrations into either methyl isobutyl ketone or xylene through a flat thin Celgard®-2400 microporous polypropylene film was studied in a flow-type test cell primarily at an aqueous phase pressure 40 psi (2.75×10^5 Pa) greater than an essentially atmospheric organic phase pressure. Studies over a range of 20 to 60 psi (1.378×10^5 to 4.134×10^5 Pa) did not indicate any significant pressure effect for extraction with methyl isobutyl ketone. Extraction rates reported as an overall organic phase based mass transfer coefficient are influenced by the boundary layer resistances on the organic and aqueous sides. The potential of this new technique was described and compared with conventional extraction techniques where hollow microporous hydrophobic fibers are utilized.

Important applications utilizing HFMC technology have been demonstrated in fermentation, pharmaceuticals, wastewater treatment, chiral separations, semiconductor manufacturing, carbonation of beverages, metal ion extraction, protein extraction, VOC removal from waste gas, and osmotic distillation. Some of the specific examples in literature are as follows: gas–liquid applications including oxygen desorption for industrial scale boilers [11], from ultrapure water [12], for absorption applications [13, 14], and at smaller scale, for blood oxygenation in open heart surgery [15, 16]. Their commercial advantage can be ascribed to the phase separation facilitated by the membrane as compared to conventional mass transfer technologies [17]; further, they satisfy the requirements of process intensification. Their integration with other membrane systems, including membrane reactors, could lead to the redesigning of membrane-based integrated production lines.

The chapters in this book have been classified using the following categories based on different ways of contacting fluids with each other [18]:

- Gas–liquid contacting
- Liquid–liquid contacting
- Supported liquid membrane
- Supported gas membrane
- Fluid–fluid contacting.

1.2 FUTURE PERSPECTIVES AND DETAILS OF THE CHAPTERS IN THIS BOOK

Over the last two decades, hollow fiber contactor-based technology has grown into an accepted unit operation for a wide variety of separations. The increase in the use of this technology, owing to strict environmental regulations and legislation, together with a growing preference for it over conventional separation processes, has led to a spectacular advance in membrane development, module configurations, and various applications. Future perspectives on the success of HFMC technology research and its impact on membrane separations are key directions for the researchers and experts working in this field. As considerable progress has been made in a number of key areas, with some important and unanticipated developments, this type of technology assessment can play an important role in setting the directions for future research and development, both for individual scientists and engineers and for the broader membrane community.

This chapter (**Chapter 1**) presents an introduction to and perspectives of HFMCs and also provides an outline of all of the chapters listed in this book. **Chapter 2** presents a comprehensive review, which identifies a variety of advances in hollow fiber membrane contactors beyond their original configurations. These advances have involved novel processing configurations with membrane contactors, novel membrane-contactor structures, novel membranes and membrane structures. For gas–liquid (G–L) systems, the authors will briefly discuss the following: novel ways of operating a HFMC, different configurations of a HFMC, pressure-swing membrane absorption, temperature-swing membrane absorption, and novel membrane structures in HFMCs. **Chapter 3** focuses on updated information on the advancement in design and fabrication of HFMCs dealing with the manufacturing of modules, the type of membrane used, and flow distribution management in modules. A novel concept involving gas absorption and stripping using HFMCs is described in **Chapter 4.** This chapter looks into the key challenges faced by current membrane contactor development. The strategies and breakthroughs made in addressing these challenges are presented. In brief, the advances made in the fabrication and modification of the hollow fiber membranes include

the incorporation of nanomaterials, silanization and amination. **Chapter 5** presents applications of HFMCs in microelectronics, and in the oil, gas, and beverage industries. **Chapter 6** presents CFD based mass transfer modeling in HFMCs for extraction-separation processes. In **Chapter 7**, recent investigations on the treatment of metal-bearing aqueous solutions *via* a HFMC are summarized in order to understand where the hollow fiber technology is positioned in the investigations pursued in *academia* on this particular environmental area of interest. **Chapter 8** deals with the role of process intensification in integrated use of liquid membranes, non-dispersive solvent extraction, and strip dispersion membrane. New metrics introduced under process intensification are estimated, while describing some of the case studies. **Chapter 9** presents the challenges involved and the methodology adopted by the authors for the scaling up of the HFMC for uranium recovery from lean effluents. Results of continuous bench scale demonstration runs with a real stream of uranium production plant will be discussed on a large scale application of the contactor in industry. **Chapter 10** focuses on advances in membrane emulsification using membrane contactors. Important applications and current perspectives with respect to production of nanoparticles are also highlighted. **Chapter 11** deals with advances in the field of hollow fiber liquid membranes and their application at the back end of a nuclear fuel cycle. The present review is a compilation of literature reports on liquid membrane-based separation studies dealing mainly with those radionuclides that are a long-term environmental hazard. Developments in hollow fiber based liquid-phase microextraction in analytical applications are presented in **Chapter 12.** The new concept introduced recently, i.e., air dehumidification by HFMC, is described in **Chapter 13. Chapter 14** deals with an important phenomenon, gas filled membrane pores using HFMC. Ammonium valorization from urban waste water as liquid fertilizers by using HFMC is described in a case study of industrial importance in **Chapter 15.** The authors have also covered the fundamentals and applications of this technique. **Chapter 16** focuses on hollow fiber contactors in facilitated transport based separations dealing with fundamentals and applications (hemoglobin and industrial separations). Applications to liquid phase separations and gas phase separations are also covered by the authors. **Chapter 17** addresses the working principles, the fundamental concepts, and the transport phenomena through microporous hydrophobic membranes of interest in membrane distillation, membrane crystallization, and membrane condenser operations. **Chapter 18** presents the challenges and evolving applications of hollow fiber contactors in fermentation and enzymatic transformation in industry and in chiral separation. The authors also cover the problems associated with hollow fiber contactors, and the advantages and disadvantages of the used system. **Chapter 19** describes advances in hollow fiber contactors in food, pharmaceutical, and biotechnological separations.

REFERENCES

1. K. K. Sirkar, W. S. W. Ho. *Membrane Handbook*. New York, NY: Van Nostrand Reinhold, 1992. Reprinted Kluwer Academic Publishers, Boston, MA, 2001; Chaps. 41, 46.
2. E. Drioli, A. Criscuoli, E. Curcio. *Membrane Contactors: Fundamentals, Applications and Potentialities* (1st ed.). Amsterdam, The Netherlands: Elsevier, 2006.
3. A. K. Pabby, S. S. H. Rizvi, A. M. Sastre. *Handbook of Membrane Separations: Chemical, Pharmaceutical, Food and Biotechnological Application* (1st ed.). New York, NY: CRC Press, 2008 and 2nd edition, 2015.
4. A. K. Pabby, A. M. Sastre. Hollow fiber membrane based separation technology: performance and design perspectives. In M. Aguilar, J. L. Cortina (Eds.), *Solvent Extraction and Liquid Membranes: Fundamental and Applications in New Materials*. New York, NY: Marcel Dekker, 2008.
5. E. Drioli, E. Curcio, G. D. Profio. State of the art and recent progresses in membrane contactors. *Chem. Eng. Res. Des.* 83(A3) (2005) 223–233.
6. A. K. Pabby, A. M. Sastre. State-of-the-art review on hollow fiber contactor technology and membrane-based extraction processes. *J. Membr. Sci.* 430 (2013) 263–303.
7. A. K. Pabby, B. Swain, A. M. Sastre. Recent advances in smart integrated membrane assisted liquid extraction technology. *Chem. Eng. Process. Proc. Inten.* 120 (2017) 27–56.
8. R. Prasad, K. K. Sirkar. Membrane-based solvent extraction. In W. S. W. Ho, K. K. Sirkar (Eds.), *Membrane Handbook*. New York, NY: Van Nostrand Reinhold, 1992, pp. 727–741.
9. A. K. Pabby, A. M. Sastre. Developments in non-dispersive membrane extraction-separation processes. In Y. Marcus, A. Sen Gupta (Eds.), *Ion Exchange and Solvent Extraction*. New York, NY: Marcel Dekker, 2001, Chap. 8, Vol. 15, pp. 331–456.
10. A. Kiani, R. R. Bhave, K. K. Sirkar. Solvent extraction with immobilized interfaces in a microporous hydrophobic membrane. *J. Membr. Sci.* 20 (1984) 125–145.
11. A. Gabelman, S.-T. Hwang. Hollow fiber membrane contactors. *J. Membr. Sci.* 159 (1999) 61–106.
12. A. Sengupta, P. A. Peterson, B. D. Miller, J. Schneider, C. W. Fulk Jr. Large-scale application of membrane contactors for gas transfer from or to ultrapure water. *Sep. Purif. Technol.* 14 (1998)189–200.
13. E. Chabanon, B. Belaissaoui, E. Favre. Gas-liquid separation processes based on physical solvents: opportunities for membranes. *J. Membr. Sci.* 459 (2014) 52–61.
14. S. Zhao, P. H. M. Feron, L. Deng, E. Favre, E. Chabanon, S. Yan, J. Hou, V. Chen, H. Qi. Status and progress of membrane contactors in post-combustion carbon capture: A state-of-the-art review of new developments. *J. Membr. Sci.* 511 (2016) 180–206.
15. K. Esato, B. Eiseman. Experimental evaluation of Gore-Tex membrane oxygenator. *J. Thorac. Cardiovasc. Surg.* 69 (1975) 690–697.
16. S. R. Wickramasinghe, B. Han. Microporous membrane blood oxygenators. In A. K. Pabby, S. S. H. Rizvi, A. M. Sastre (Eds.), *Handbook of Membrane Separations: Chemical, Pharmaceutical, Food and Biotechnological Application*. New York, NY: CRC Press, 2008, Chap. 23, pp. 671–692.
17. B. Belaissaoui, J. Claveria-Baro, A. Lorenzo-Hernando, D. Albarracin Zaidiza, E. Chabanon, C. Castel, S. Rode, D. Roizard, E. Favre. Potentialities of a dense skin hollow fiber membrane contactor for biogas purification by pressurised water absorption. *J. Membr. Sci.* 513 (2016) 236–249.
18. K. K. Sirkar. Membranes, phase interfaces, and separations: Novel techniques and membranes—An overview. *Ind. Eng. Chem. Res.* 47 (2008) 5250–5266.

2 Advanced Concepts in Membrane Contactors

Kamalesh K. Sirkar and J. Chau

CONTENTS

2.1 INTRODUCTION

Three general types of processes and techniques are employed to separate mixtures and purify solutes, namely, selective partitioning between two immiscible fluid phases, membrane separation processes, and external force-field–based processes [1]. We are concerned here primarily with processes involving selective partitioning between two immiscible fluid phases. Conventionally such processes are carried out in devices where one phase is dispersed in another phase in a variety of ways, as drops, bubbles, or thin films. Membrane contactors are employed primarily for selective partitioning-based separation or purification processes involving non-dispersive contact between two immiscible fluid phases. Two types of system are of considerable interest: gas–liquid (also vapor–liquid); and liquid–liquid. Fluid-solid contacting is also of interest. This chapter will focus on a variety of advances that have taken place with such membrane contactors. Table 2.1 provides an elementary identification of various membrane contactor-based processes where there is only one interface between two immiscible phases (Figure 2.1). These membrane

contactor configurations are also used for purely reactive processes, as for example membrane ozonation in gas–liquid systems. Advances in such processes will be briefly identified for various phase-contacting systems. A few acronyms are encountered often in membrane contactor literature on liquid–liquid (L–L) systems: MSX (membrane solvent extraction); and NDSX (non-dispersive solvent extraction).

The pores in a porous membrane are crucial for immobilization of fluid-fluid interfaces. However, in the case of solid–fluid systems, the membrane has no such function. Instead, it is a facilitator for fluid flow with a low pressure drop; it is also a facilitator for efficient accessing of porous solid surface locations to reduce drastically the diffusional resistances of larger molecules.

There is a class of porous/microporous membrane-based processes where two different bulk phases are present on two sides of the membrane whose pores contain a fluid phase which is immiscible with the bulk phases on both sides of the membrane (Figure 2.2). Thus there are two bulk phase interfaces between two immiscible phases, one on each side of the porous membrane. When the pores contain

TABLE 2.1

Membrane Contactors with One Bulk Phase Interface for Phase Equilibrium Processes

Phase Interface	Phase Details	Separation Processes
*Gas–Liquid	Gas phase–Liquid phase	Gas absorption/stripping
*Liquid–Gas	Gas phase may have vapor to be absorbed; liquid phase may have vapor to be stripped	Vapor absorption/stripping
	Hot liquid–gas phase (under vacuum or otherwise)	Vacuum membrane distillation (VMD) of volatile species; sweep gas membrane distillation (SGMD)
*Vapor–Liquid	Vapor stream going up	Distillation
	Liquid stream coming down	
*Liquid–Liquid	Aqueous–Organic	Solvent extraction
	Polar organic–Nonpolar organic	Solvent back extraction
	Aqueous–Aqueous	
*Supercritical Fluid–Liquid	Supercritical fluid–Liquid phase	Extraction
Gas–Solid	Solid adsorbent particles in packed bed	Adsorption
	Packed bed has porous hollow fiber membranes through which gas flows	Desorption
Liquid–Solid	Liquid phase flows through membrane pores	Membrane adsorption
	Membrane pore surfaces have ligands for adsorption	Membrane chromatography
		Elution/Stripping

* In these processes, the interface between the two bulk phases is immobilized at membrane pore mouths

a gas/air immiscible with the liquid on both sides, we have a supported gas membrane (SGM). When the pores contain a liquid immiscible with the liquid phases on both sides of the membrane, the configuration is identified as a supported liquid membrane (SLM) or an immobilized liquid membrane (ILM). One can also have, instead of two liquid phases, two gas phases or one liquid phase and one gas phase; the term ILM is usually used with two gas phases on two sides. To the extent that the phase interface on each side of such a membrane involves a non-dispersive contact

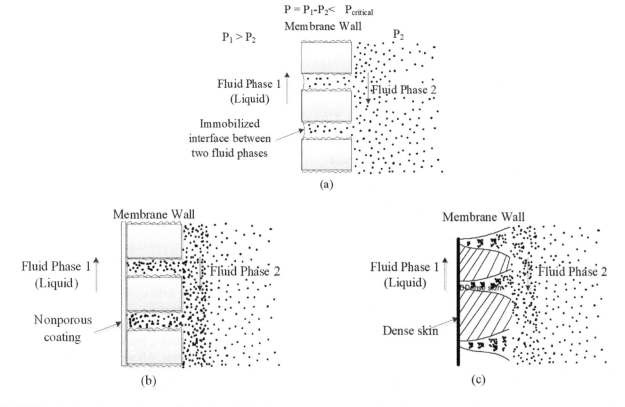

FIGURE 2.1 Membrane contactor allowing two fluid phases to contact each other: (a) porous membrane; (b) porous membrane having a nonporous coating; (c) asymmetric porous membrane with a dense skin.

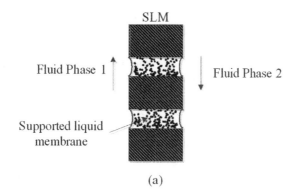

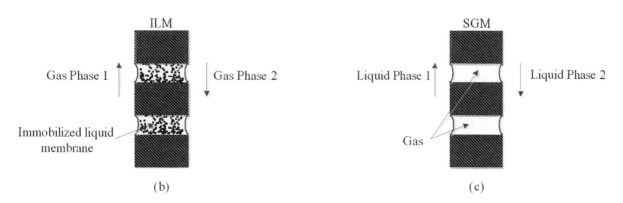

FIGURE 2.2 (a) Supported liquid membrane (SLM) with two fluid phases immiscible with SLM liquid on two sides; (b) immobilized liquid membrane (ILM) with two gas phases on two sides; (c) supported gas membrane (SGM) with two liquid phases on two sides.

between two immiscible fluid phases, these processes are included here under membrane contactors. Table 2.2 identifies a variety of such two-phase-interface-based processes.

Since their inception, membrane contactors were involved in the contacting of two fluid phases via porous membranes. However, contacting between a fluid phase and a solid phase involving adsorption, for example, has also been studied. There have been additional developments using such configurations; these will be briefly described here. Whereas membrane contactors are used in processes where non-dispersive contacting between two different bulk phases is implemented via a porous membrane, there are processes where a porous membrane is used to create contact between two bulk phases in a controlled fashion such

TABLE 2.2

Membrane Contactors with Two Phase Interfaces (1) and (2) for Membrane Processes

Two Phase Interfaces	Membrane Nomenclature	Separation Process
Liquid–Liquid (1) and Liquid–Liquid (2)	Supported Liquid Membrane (SLM) (Liquid in Membrane pores)	Solvent extraction and back extraction combined
Gas–Liquid (1) and Gas–Liquid (2)	Immobilized Liquid Membrane (ILM) (Liquid in Membrane pores)	a) Gas absorption and stripping combined b) Pervaporation with vacuum at interface (2)
Liquid–Liquid (1) and Gas–Liquid (2)	Supported Liquid Membrane in pores (SLM)	Extraction at liquid–liquid interface and stripping via vacuum at gas–liquid interface (2)
Liquid–Gas (1) and Liquid–Gas (2)	Supported Gas Membrane (SGM) Gas-filled pores	a) Gas/vapor stripping at interface (1) and chemisorption at interface (2): Trans-membrane chemisorption b) Membrane distillation: heated liquid feed at interface (1) and cold liquid at interface (2) c) Osmotic distillation: stripping of vapor at interface (1) of a dilute solution with low π_p* and absorption of vapor at interface (2) of a concentrated solution with higher π_p

* π_p, Osmotic pressure

that a dispersion of one phase, e.g., an emulsion, is created in another immiscible phase. This is covered in **Chapter 10**. When the two phases are miscible, one can also achieve crystallization/precipitation in a controlled fashion. The membrane here facilitates the process of the contacting of two phases via highly efficient mixing.

There are three types of fluid-fluid systems of relevance: gas–liquid; vapor–liquid; and liquid–liquid. Supercritical fluid (SCF) liquid systems have been covered briefly by Sirkar [1] and will not be treated here. The order of coverage will be as follows. Gas–liquid systems will be covered first. Subsequent sections will include consideration of contactors for a variety of systems: two fluid-fluid interfaces of ILMs for gas separation; one fluid–fluid interface-based vapor–liquid and then liquid–liquid contacting; and SLM systems for L–L systems. Next, SGM systems will be briefly mentioned. At the end, developments in fluid-solid systems will be identified. A very brief identification of the earliest investigations of and reviews for various membrane contactors will be provided at the beginning of each section that follows; then recent major developments will be identified.

2.2 MEMBRANE CONTACTORS FOR GAS–LIQUID SYSTEMS

One important characteristic of membrane contactor processes for gas–liquid systems involves the relative values of the pressures of the two bulk phases at any location in the contactor. In general, the membranes are hydrophobic. The two bulk phases, namely, the gas phase and the liquid phase, are on two sides of the porous membrane. For operations with gas-filled membrane pores, the liquid phase is maintained at a pressure higher than that of the gas phase (Figure 2.1a). However, the excess liquid phase pressure must not exceed a breakthrough pressure, $\Delta P_{critical}$, which depends among other things on the surface tension of the liquid γ, the contact angle θ and the pore diameter $2r_{pore}$ via the relation,

$$\Delta P_{critical} = \left(2\gamma \cos\theta / r_{pore}\right) \tag{2.1}$$

For cases where the liquid phase wets and fills the membrane pore (wetted mode), the gas phase pressure should be higher than that of the liquid for non-dispersive operation [1–3].

Blood oxygenation systems witnessed the first application of gas–liquid membrane contactors with blood on one side and oxygen/air on the other side: Esato and Eiseman[4] employed microporous hydrophobic Gore-Tex® membranes. Tsuji et al. [5] employed porous hydrophobic hollow fiber membranes (HFMs) of polypropylene (PP). Frank and Sirkar [6] incorporated hydrophobic PP hollow fiber membranes in a bioreactor for alcohol fermentation by yeast; the hollow fibers were used to supply oxygen to the medium and remove CO_2. Qi and Cussler [7] showed the utility of the PP HFMs for industrial applications of CO_2 absorption.

Karoor and Sirkar [3] first showed the effects of pore wetting in G–L membrane contacting with hydrophobic hollow fiber membranes. The earliest review of membrane contactors for G–L systems is available in Sirkar[2]. Additional earlier reviews have been provided by Reed et al. [8], Gableman and Hwang [9], Klaassen et al. [10], Curcio and Drioli [11], Sirkar [1], and Mansourizadeh and Ismail[12] (for acid gases only). The variety of membrane contactor applications in G–L systems has been recently reviewed by Bazhenov et al. [13].

In conventional membrane contactors for gas–liquid systems, the feed gas phase flows on one side of the membrane as it is being scrubbed; the liquid generally flows countercurrently on the other side of the membrane at least for a gas absorber. The absorbent liquid is then taken to a different device with or without membranes, and heated often or subjected to vacuum or steam stripping so that the absorbent is regenerated for recycle back to the membrane contactor acting as the absorber. The system operates continuously; the only difference consists of replacing the conventional gas–liquid contactor with a membrane contactor. In commercial practice, the liquid flow on the shell side is often configured normal to the hollow fibers to enhance the liquid phase mass transfer coefficient.

Significant advancements have taken place in the following areas: (1) Membrane development and structures; (2) PSMAB; (3) TSMAB; (4) Dual membrane devices; (5) Novel solvents; (6) Membrane ozonation. Further, the bulk of the developments reflects an extraordinary emphasis on membrane contactors for CO_2 scrubbing from flue gas (much less on synthesis gas). After considering these sections, we will briefly point out some developments on immobilized liquid membranes (ILMs) in the next section for gas separation.

2.2.1 MEMBRANE DEVELOPMENT AND STRUCTURES

One of the largest volume industrial applications of gas–liquid membrane contactors involves deoxygenation. Very large membrane modules, based on porous hydrophobic symmetric polypropylene hollow fibers, employ a shell-side flow of water to be deoxygenated in cross-flow over the hollow fibers. On the tube-side, either a high vacuum is applied or a sweep stream of nitrogen is provided to remove the dissolved oxygen. Sometimes both are used in what is called the combo operation. The deoxygenation behavior of such a module has been modeled in so far as the water velocity/flow rate dependence is concerned [14]. Zheng et al. [15] have provided a more detailed model of shell-side mass transfer in this type of large transverse flow hollow fiber membrane contactor using the free surface model originally introduced in membrane contactors [3]. For oxygen absorption from air and subsequent stripping applications, see Section 2.5 on novel solvents.

Membrane contactors for gas–liquid systems have been studied for absorption of CO_2, H_2S, and SO_2, from various gaseous streams. Two reviews [16, 17] provide a critical window into the membrane contactor literature on CO_2 scrubbing. For most applications focused on CO_2 scrubbing from flue gas, the solvents are usually aqueous solutions of various amines. Physical solvents have not been explored as much due to the requirement of higher pressures and lower temperatures during absorption followed by lower pressures and higher temperatures for stripping; the conditions during absorption can easily lead to wetting of porous membranes due to the lower surface tensions of the solvents and higher pressure suggesting a need for dense skins [17]. The earliest membrane contactors employed porous polypropylene (PP) hollow fibers; these were also found to be susceptible to wetting by aqueous solvents within a few days. Although the surface tensions of the mostly aqueous solutions employed were much larger than the critical surface tension of PP, the amines being volatile will get adsorbed on the hydrophobic pore surfaces with their hydrophilic groups sticking out and making the pore surface hydrophilic over a longer period of time.

Non-volatile amino acid salts in aqueous solutions were successfully used to absorb CO_2 from gas streams [18]. However at high CO_2 loadings, these amino acid salts (e.g., potassium salt of taurine, potassium glycinate) are known to crystallize. Are there other nonvolatile amines which do not suffer from such deficiency? Using a nonvolatile amine such as polyamidoamine (PAMAM) dendrimer Gen 0 in water, Kosaraju et al. [19] demonstrated no pore wetting of conventional porous PP hollow fiber membranes in a 55-day long run. Other sources of wetting will be accidental pressure fluctuations in the absorbent liquid exceeding the liquid entry pressure for the porous membrane.

The new more hydrophobic materials for porous hollow fiber membranes are polytetrafluoroethylene (PTFE) and polyetheretherketone (PEEK) (2 nm pore size; Li et al. [20]). Additional material modifications have involved porous fluorosiloxane coating on porous PP hollow fibers [21]; longer-term study is needed to verify the utility of such a structure for wetting prevention. Ceramic tubules having a pore size of 5nm and completely hydrophobized via silane coating were used for non-dispersive membrane contacting at high pressures and higher temperatures (100–125°C) [22]. Table 2.2 in the G–L contactor review [13] identifies various publications on H_2S absorption studies. The pressures of operation were as high as 50 bar since natural gas scrubbing is at issue. A new hollow fiber membrane material was identified:poly(tetrafluoroethylene-co-perfluorinated alkylvinyl ether) (PFA). Further there are a number of examples of polyvinylidene fluoride (PVDF) hollow fiber being used in H_2S scrubbing.

An approach being investigated for some time is to employ dense membranes to prevent pore wetting altogether resulting from entry of the absorbent solution (Figure 2.1b). This has involved two strategies so that the resistance of the dense coating is not high as was observed in [23].

(1) Use an asymmetric porous hydrophobic hollow fiber with a dense skin of a polymer (Figure 2.1c) with high free volume polymers such as poly (4-methyl-1-pentene) (PMP) facing the absorbent liquid. Kosaraju et al. [19] employed two such hollow fiber–based membrane contactors for CO_2 scrubbing, one for absorption and the other for stripping, using an aqueous solution of monoethanolamine (MEA) and demonstrated no wetting over a run lasting for more than 55 days. However, there was a very slow loss of MEA through the free volume openings of the dense PMP skin exposed to the aqueous solution.

(2) Use a dense coating of a high free volume polymer on porous hydrophobic hollow fiber substrates facing the absorbent liquid (Figure 2.1b). Chabanon et al. [24] employed this strategy using thin dense coatings of two different polymers, PMP and Teflon-AF perfluoropolymer, on porous PP hollow fibers and found essentially no deterioration in performance over 1,000 hours using a 30% aqueous MEA solution. They did not investigate the loss of MEA through PMP as was done in Kosaraju et al. [19] who had already demonstrated the utility of a dense PMP skin for prevention of pore wetting. On the other hand, a similar study by Scholes et al. [25] using thin dense coatings of poly (1-trimethylsilyl-1-propyne) (PTMSP), a polymer of intrinsic microporosity (PIM-1) and Teflon AF1600 on a porous 0.2 μm PP hollow fiber support showed that after sometime during CO_2 absorption studies, the PP support pores of PTMSP and PIM-1-based membranes got filled up by water but those for AF 1600 did not get wetted. They have suggested that the dense coating layer must have a high CO_2 permeability as well as being resistant to water sorption which was not valid for PTMSP and PIM-1. Ansaloni et al. [26] observed that at room temperature, the Teflon AF polymer coatings had a selectivity of 300–500 for CO_2 over a blend of the amines, diethyl-ethanolamine/3-methylamino-propylamine. The contribution of this enhanced CO_2 selectivity by the dense coating, if any, on the contacting and stripping process has not been adequately investigated.

The demands on a membrane contactor during CO_2 desorption are higher since, among other demands, the temperature is higher. Scholes et al. [27] employed thin dense coatings of PTMSP, PIM-1 and Teflon AF1600 on a porous 0.2 μm PP hollow fiber support (as in [25]) to desorb CO_2 from aqueous MEA solutions at temperatures in the range of 70–105°C. They have shown that it could be carried out successfully. Results from extended-term stability experiments would have been even more useful. A higher temperature of 80°C and a pressure of 20 atm was also used for CO_2 absorption using a simulated syngas mixture of CO_2-He in a contactor of porous PP hollow fibers having a coating of dense Teflon AF 2400 polymer; the absorbent was an ionic liquid [28]. Hollow fibers of hydrophobic PVDF-PEG400

3M™ Liqui-Cel™ Membrane Contactor

FIGURE 2.3 Cutaway version of Liqui-Cel™ Membrane Contactor shown on the left; shown on the right are two 14 × 28 Liqui-Cel™ Membrane Contactor modules vertically mounted (Reproduced with permission. © 3M 2020. All rights reserved.)

showed deterioration in CO_2-stripping performance at 80°C within 10 hr potentially due to interactions between amines in solution and the polymer [29].

In terms of large-scale membrane modules for membrane contactor processes, currently very large PP hollow fiber–based modules of at least a 220 m² surface area are commercially available [30] (see Figure 2(b) for an assembly of such modules for degassing in this reference; also see Figure 2.3 here). The largest PEEK hollow fiber–based module successfully tested as a contactor had a 32 m² surface area; 8 inch modules were prepared. One such module has also been tested for stripping an aqueous amine solution at 40–55 psig and 104–122°C (Li et al. [20]). Porous hollow fibers of PTFE have also been commercially available for some time [31] and have been studied for absorption of CO_2 and NO_2 from a model flue gas [32]; the module had a membrane area of 0.33 m². At the other end of the spectrum, G–L contacting has been studied in a microfluidic contactor for CO_2 removal from anesthesia gas using ionic liquids [33].

2.2.2 PSMAB

Using pressure swing adsorption (PSA) processes, it is a common industrial practice to produce a highly purified stream of the less adsorbed gas species, e.g., production of high purity H_2 gas. In syngas purification processes to produce a purified H_2 stream, one needs to simultaneously recover a purified CO_2 stream for carbon sequestration. Recovering simultaneously a purified CO_2 stream via PSA involves modification of the conventional PSA process operating cycle. For syngas purification, two conditions stand out, high pressure and higher temperature, conditions that are not generally encountered in conventional membrane contactor-based absorption processes suggesting the use of absorbents suitable for high temperatures. The PSMAB process recently developed responds to these needs [34, 35]; a much simpler version was proposed many years back [36].

It uses microporous hydrophobic PEEK hollow fibers. The shell side of the hollow fiber module is filled with a nonvolatile absorbent which can withstand higher temperatures, e.g., ionic liquid, PEG 200/PEG 400. The CO_2 solubility of the absorbent is drastically enhanced by having a nonvolatile amine in solution, e.g., polyamidoamine dendrimer Gen 0. The feed gas is introduced into the tube side of the hollow fibers. Figure 2.4a illustrates the process schematic using a single membrane contactor; Figure 2.4b illustrates the pressure and other conditions in the tube-side and other locations. The system studied used a 60–40 He–CO_2 gas mixture with He acting as a surrogate for H_2.

All PSA processes are cyclic. So is the PSMAB process. Further, the cycle is of short duration. The cycle begins with the high pressure-high temperature 60–40 He–CO_2 feed gas entering the tube-side of the hollow fibers of the cylindrical module for a short period (e.g., 10 s) through the open valve 1 with all other valves in the 5-valve system remaining closed. During this brief period, there is selective absorption of CO_2 into the absorbent liquid in the shell side. Then valve 1 is closed for 30–60 s and the stagnant tube-side gas undergoes considerable selective CO_2 absorption into the stagnant absorbent liquid on the shell side. Then valve 3 is opened at the top for a very short period (e.g., 2 s) to withdraw from the top section a highly He-enriched stream. Next, valve 3 is closed, and valve 4 is opened for a short period (e.g., 2 s) to withdraw a small amount of the gas in the middle section of the module and store it in an outside vessel (Middle part gas). Next, valve 4 is closed and valve 2 is opened for, say, 60 s to a much lower pressure to desorb CO_2 selectively absorbed earlier into the shell-side absorbent and produce a highly CO_2-enriched product stream useful for sequestration. Then valve 2 is closed and valve 5 is opened for a short period, e.g., 2 s, for the middle part gas to enter the tube side of the module and raise its pressure somewhat. The cycle is now complete with valve 5 being closed. Valve 1 opens, high pressure feed

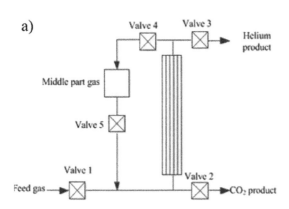
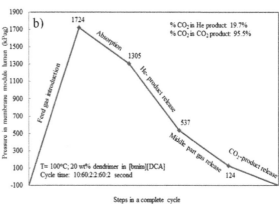

FIGURE 2.4 (a) Cyclic PSMAB process with a hollow fiber membrane contactor for syngas separation with 5 valves and a 6-step cycle (Reprinted from *Ind. Eng. Chem. Res*, 53(8) Jie, X., J. Chau, G. Obuskovic et al., 2014, Enhanced pressure swing membrane absorption process for CO_2 removal from shifted syngas with dendrimer-ionic liquid mixture as absorbent, 3305–3320, Copyright (2014) permission from American Chemical Society)[34].) (b) Pressure profile in the hollow fiber bore for a 6-step 5-valve cycle for He(60)-CO_2(40) syngas purification with 20 wt% PAMAM dendrimer in [bmim][DCA]; 6th step not shown (Reprinted from *Chem. Eng. J.*, 305, Chau, J., X. Jie, K. K. Sirkar, 2016, Polyamidoamine-facilitated poly (ethylene glycol)/ionic liquid based pressure swing membrane absorption process for CO_2 removal from shifted syngas, 212–220, Copyright (2016) permission from Elsevier [35].)

gas comes in, starting a new 6-step cycle in a 5-valve system. The six steps are identified below[34]:

Step 1. Valve 1 is opened and all other valves are closed; fresh feed gas is introduced into the tube side until the desired pressure is established.

Step 2. All valves are closed; absorption happens between the gas and the absorbent.

Step 3. Valve 3 is opened and all others are closed; He-rich product is withdrawn.

Step 4. Valve 4 is opened and all others are closed; middle part gas is withdrawn.

Step 5. Valve 2 is opened and all others are closed; CO_2-rich product is withdrawn via a vacuum.

Step 6. Valve 5 is opened and all others are closed; middle part gas is recycled into the membrane tube side as initial feed gas. This step completes one cycle.

Figure 2.4b shows a cycle duration of 136 sec. For 100°C feed gas of 60–40 He–CO_2coming in at 1724 kPag, using two small hollow fiber modules in a series yielded a 95.5% CO_2-containingstream from valve 2 (Chau et al.[35]). Earlier studies have shown the capacity to produce a 5% CO_2-containing He stream from valve 3 at the top [22]. It requires a significant module length, which can be achieved by having three small modules in a series. An additional important requirement for achieving the proper quality of two gaseous product streams is a drastic reduction of dead space before and after the hollow fibers, especially in the headers of the hollow fiber modules.

For high temperature processes where viscous absorbents need to be used with microporous substrates, mobile absorbents will encounter significant pressure drops in narrow gaps between hollow fiber membranes. To avoid uncertainties about the extent of the pressure drop in the flowing

absorbent which could lead to absorbent breakthrough (since the absorbent pressure has to be higher than the gas pressure for phase interface immobilization), stagnant absorbent is used in the shell side in the PSMAB process. However, since the contact time t is short, the rate of physical gas absorption in a stagnant liquid film(which is proportional to $\sqrt{(D_{CO_2}/\pi t)}$) can be quite high.

2.2.3 TSMAB

In temperature swing adsorption (TSA), a common method of operating a gas adsorption-desorption process, desorption of the adsorbed species is carried out by heating the adsorbent bed. The mass of the absorbent bed is an important parameter in the desorption energy requirement. Considerable research has been/is being carried out in CO_2 sequestration using TSA. Porous solid supports or adsorbents having beneficial properties are frequently used wherein polymeric amines of various types have been grafted, impregnated, or immobilized [37]. The absolute CO_2 adsorption capacity of various supported basic groups/g of the adsorbent material is considerably reduced by the porous support mass and is rarely above 2–2.5 mmol CO_2/g of adsorbent. On the other hand, nonvolatile amine liquids such as PAMAM dendrimer Gen 0 can easily allow for the achievement of values of around 6–7 mmol of CO_2/g of absorbent [38]. Therefore porous polymeric hollow fiber membrane-based temperature swing absorption using nonvolatile/volatile amine liquids is likely to be much less energy intensive [38] since the mass of material used in a thin-walled polymeric hollow fiber is quite small and the corresponding heat consumption for heating is quite low. The temperature of desorption is also important: the lower, the better. Further, the manner of heat supply is also important. A membrane contactor structure and operating procedure to achieve such a goal is described below (see Figure 2.5) using CO_2 absorption and stripping from flue gas as an example.

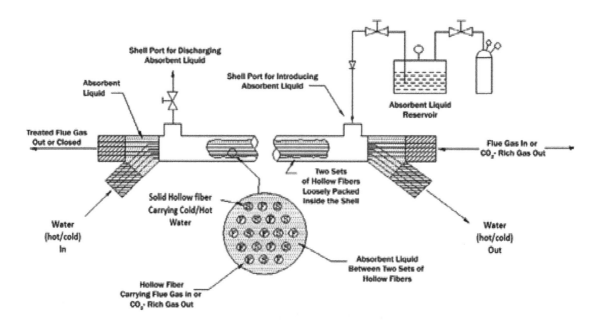

FIGURE 2.5 Schematic of two hollow fiber set-based cylindrical temperature swing membraneabsorption (TSMAB) device (Reprinted from *J. Membr. Sci.*, 493, Mulukutla, T., D. Singh, J. Chau et al., Novel membrane contactor for CO_2 removal from flue gas by temperature swing absorption, 321–328, Copyright (2015) permission from Elsevier **[38]**.)

In the TSMAB process (Figure 2.5), there are two sets of hollow fibers and a cyclic absorption-desorption process. In the first part of the cycle, there is gas absorption. In the second part of the cycle, the bed is heated up. In the third part of the cycle, with heating going on, desorption takes place. In the fourth part of the cycle, the bed is cooled down and readied for absorption in the next cycle. At the start of a cycle, through the bore of one set of hollow fibers whose walls are usually microporous, CO_2-containing feed gas flows for some time. The shell-side is filled with a CO_2-absorbing liquid. Therefore, CO_2 absorption in the absorbent quickly reduces the CO_2 concentration in the exiting gas to zero. It continues for a while; then after sometime there is a breakthrough of CO_2 in the exiting gas signifying increasing saturation of the absorption capacity in the shell-side absorbent liquid. At this time, the feed gas flow is stopped. The contactor device has a second set of solid wall polymeric hollow fibers commingled with the first set of hollow fibers. This fiber wall should be essentially impervious to water vapor and other species under consideration (e.g., solid wall PP hollow fibers). Hot water should be sent through the bores of these hollow fibers to heat up the shell-side absorbent liquid, which leads to desorption of the absorbed CO_2. After sometime, the exiting end of the first set of hollow fibers is opened; it produces a gas stream highly enriched in the CO_2 desorbed from the absorbent liquid. There may be a need at the end for an inert gas stream of, say, N_2, to push the desorbed CO_2 out; in the laboratory He was used. Once the flow rate of this desorbed stream is considerably reduced, this outlet is shut off. Cooler water is introduced into the bore of the solid hollow fiber set to cool down the bed/liquid absorbent and get it ready for absorbing CO_2 again from flue gas in the next cycle. Cold water should be there in the bore of the second set of hollow fibers

to absorb the heat of absorption during the beginning of the next cycle of absorption.

A potential advantage of such a membrane contactor device vis-à-vis a conventional TSA device is that the CO_2 absorption capacity per g of liquid absorbent can easily exceed that of solid adsorbents by 2–4 times plus[38]. This publication used larger diameter porous hydrophobic PVDF hollow fibers for scrubbing. The diameters need to be reduced considerably; further, there needs to be an anti-wetting coating on the outside diameter. By having thin liquid films in between hollow fibers with thin walls and smaller diameters, rapid heating as well faster CO_2 saturation of the absorbent liquid are expected. Improved design requires maximizing absorption per unit volume of the absorbent and reducing the corresponding feed gas volume.

2.2.4 DUAL MEMBRANE DEVICES

The above-mentioned type of membrane contactor configuration, where there is an additional set of solid wall hollow fibers through which heat exchange functions may be carried out, was developed and studied earlier for air gap membrane distillation (AGMD) processes[39]. When the second hollow fiber membrane functions as a stripper of the absorbed species in the shell-side absorbent, then we have a dual membrane function device, the first membrane for absorption and the second for stripping. One such structure was developed by Obuskovic et al. [40] who had used a silicone rubber coated PP hollow fiber membrane for VOC absorption into silicone oil absorbent present in the shell side. The siloxane coating faced the shell-side absorbent liquid to prevent wetting of the hollow fiber substrate pores. A similar type of coated hollow fiber was used as the second set of hollow fibers with vacuum being pulled in its lumen to desorb VOCs absorbed in the oil-absorbent on the shell side. To improve the purity of the treated feed gas, the system on the feed side was operated in a cyclic fashion.

Application of a somewhat similar concept for CO_2 absorption and removal but operating on a steady state has been illustrated by Cai et al. [41]; they had used a porous PP membrane for CO_2 absorption and a silicone rubber tubule for stripping. The absorbents used were water and propylene carbonate and were mobile. These structures are very similar to the hollow fiber contained liquid membrane (HFCLM) structures developed in the late eighties where two porous hydrophobic PP hollow fiber membranes were used to contain a liquid in the shell-side acting as a liquid membrane. A review of such developments until 1996 is available in [42].

2.2.5 NOVEL SOLVENTS

Although solvents are integral to contacting processes for gas–liquid systems, we will not deliberate on them here except to point out the general categories of solvents used. For CO_2 scrubbing, aqueous solution–based systems employ various types of amines or combinations of amines. These include primary (e.g., MEA), secondary, tertiary, and hindered amines as well as dendrimers having primary and tertiary amines. Ionic liquids have also been studied for CO_2 absorption. A few examples are:1-ethyl-3 methylimidazolium ethyl sulfate [Emim][EtSO_4] [43],1-butyl-3-methylimidazolium dicyanamide ([bmim][DCA]) with or without PAMAM dendrimer Gen0 [34], and 1-Butyl-3-methylimidazolium tricyanomethanide [Bmim][TCM] [28]. The use of such solvents requires studies on membrane compatibility and stability with the ionic liquid. Dai et al.[28] investigated the behavior of six different porous and nonporous polymeric hollow fiber membranes exposed to [Bmim][TCM] at 80ºC and elevated pressures. Yang et al. [44] added aqueous MEA to an ionic liquid (30 wt % MEA + 40 wt % [bmim][BF_4] + 30 wt % water), used a PP hollow fiber contactor, and determined that the energy consumption of the mixed ionic liquid solution for absorbent regeneration was 37.2% lower. Sirkar [45] has suggested the use of methyldiethanolamine (MDEA) containing a small amount of water in the TSMAB process.

An additional development involves the production of high purity oxygen from air using an aqueous solution of poly(ethyleneimine)–cobalt (PEI–Co) complex with high O_2 absorption capacity (as high as 1.5 L O_2 (STP)/L solution). Conventional porous PP hollow fiber–based membranes were used; a similar and separate membrane desorber exposed to vacuum was used to generate a high-purity oxygen stream [46].

2.2.6 MEMBRANE OZONATION AND OTHER REACTORS

The earliest G–L membrane contactor-based ozonation studies are described in Shanbhag et al. [47–49] and Guha et al. [50]. It is now being called bubble-less ozonation.

Shanbhag et al. [47]employed silicone capillaries for supplying ozone to the water to be treated: water flowed on the shell side and ozonated oxygen came in through the tube-side into the water and destroyed/degraded organic pollutants. Silicone capillaries were used since the porous PTFE tubules available had too large a pore size and silicone is selective for ozone over oxygen [51]. Additional innovations included multiphase reactor structures where VOCs present in air and candidates for destruction came in through the bore of silicone capillaries into a shell-side fluorocarbon (FC) phase. This FC phase, which had a much higher solubility of ozone than water, was separately supplied with ozone from another set of silicone capillaries[48] and the nonvolatile reaction products were extracted out non-dispersively through a porous hydrophobic PTFE hollow tubule through which water was flowing. For pollutants present in aqueous phase coming though the bore of a porous PTFE hollow tubule, the multiphase ozonation reactor included a set of silicone capillaries which supplied ozone in oxygen to the shell side which had a fluorocarbon phase into which the pollutants were extracted from water and were destroyed by ozone [50]; the reaction products were removed through either capillaries depending on their volatility, removability, and contiguity to the nearest flowing phase. It was found that silicone capillaries get degraded with time. The current availability of smaller pore–based porous PTFE hollow fibers [31] will avoid the degradation encountered by silicone capillaries.

In addition to studying selective ozone permeation through silicone rubber [51], the authors also found earlier that a perfluorodioxole-based amorphous copolymer membrane CMS-3 from Compact Membrane Systems was significantly selective for ozone over oxygen. This was utilized very soon in the late nineties to commercialize a membrane ozonator (by Pall Corp.) called Infuzor® which had a perfluorodioxole membrane-based coating on a porous PTFE substrate. During the nineties, W. L. Gore introduced also a membrane ozonator using larger diameter porous hollow fibers based on expanded PTFE membranes verging on tubules.

Later, Steiner et al. [52] used porous hydrophilic PVDF membranes for ozonation and indicated that the hydrophilic PVDF membranes had to be dried after every day's tests to remove pore water or moisture. Hydrophobic PVDF membranes are likely to be more useful for efficient ozonation provided long-term stability at high ozone concentrations is ensured. Stylianouet al. [53] hydrophobized an α-alumina-based ceramic membrane tubule of 1.4 cm OD and studied bubble-less ozonation. Stylianou et al. [54] also utilized hydrophobizedα-alumina-based ceramic tubules of 1000kDa MWCO for extensive ozonation studies.

Deoxygenation of water flowing on the shell side of a membrane contactor can be achieved by adding hydrogen gas through the tube-side in the presence of palladium nanocatalyst particles embedded on the outer surface of porous PP hollow fibers (Volkov et al. [55]); oxygen is catalytically converted to water. For ultrapure water applications, presence of any dissolved gaseous species is to be avoided although conventional membrane deoxygenation processes do have significant dissolved N_2.

2.3 IMMOBILIZED LIQUID MEMBRANES (ILMs) FOR GAS–LIQUID SYSTEMS

A microporous membrane with a liquid occupying the membrane pores can act as a membrane to separate a gas mixture (Figure 2.2b). Several conditions have to be satisfied for such a structure to be useful: (1) liquid membrane should be nonvolatile; (2) liquid membrane should be able to withstand any pressure difference between the two sides; (3) any condensation of liquid or water vapor from the gas phases should not flood the membrane; and (4) the liquid membrane constituents should not get poisoned by any species from the gas phases. For polar liquid membranes, porous hydrophilic membranes are needed. For nonpolar liquid membranes, porous hydrophobic membranes are needed. These conditions lead to non-dispersive gas–liquid contact on two sides of the microporous membrane.

Obuskovic et al. [56] employed hydrophilic microporous polyethersulfone (PES) hollow fiber membranes immobilized either with 3M-Na/K-glycinate–glycerol or 2M-Na-glycinate in a 50/50glycerol/PEG 400 solution. An ultrathin hydrophobic porous coating on the outside surface of the PES hollow fibers of 100 μm wall thickness (Figure 2.6) prevented flooding of the liquid membrane by moisture condensation or occasional/accidental flowing water on the outside of the hollow fiber. A high CO_2 permeance of 1600 GPU was obtained at 37°C from a feed gas containing a 2% CO_2 and 70% N_2O/O_2 balance; the membranes were highly selective for CO_2 for the purpose of removing CO_2 from a breathing gas mixture due to facilitated CO_2 transport and salting out of N_2. The breathing gas mixture contained halogenated hydrocarbons, e.g., sevoflurane, desflurane, enflurane, and halothane used for anesthesia; the membrane was considerably more selective for CO_2 over such constituents. Thinner wall (40μm thickness) porous polysulfone hollow fibers (hydrophilized as received) containing an immobilized liquid membrane of 3.0M glycine-Na–glycerol

yielded a CO_2/N_2 selectivity in the range of 1100–2700 from a humidified feed gas containing 0.5% CO_2/balance N_2. Extended runs lasting for 300 h showed that the hollow fiber–based ILM permeation performances were stable[57].

Whereas hydrophilic microporous membranes and polar liquids were used for selective and facilitated CO_2 transport, hydrophobic microporous membranes containing hydrophobic liquids are used for selective transport of VOCs. In one configuration [58], essentially nonvolatile silicone oil was immobilized in part of the micropores of a porous polypropylene hollow fiber substrate beneath the plasma-polymerized polydimethylsiloxane (PDMS)-based dense skin on the outer surface of the hollow fiber. Such an immobilized liquid layer in the hollow fiber membrane (Figure 2.7) yielded 2 to 5 times more VOC-enriched permeate and separation factors that were 5 to 20 times compared to what was obtained with the PDMS denseskin only; the exact value depended on the VOC system under consideration: toluene–N_2, methanol–N_2, acetone–N_2. The system was run stably for more than six months with a vacuum being pulled on the permeate side to recover a highly VOC-enriched permeate.

All three examples involved selective removal of a gas/vapor species from a gas stream into another gas stream/vacuum on the other side of the membrane. If a VOC has to be selectively stripped from a liquid stream, e.g., biofuels from an aqueous solution, the strategy is quite different. The liquid membrane should not contact the aqueous solution as this will not only contaminate the aqueous solution but also lead to liquid membrane loss. Thongsukmak et al. [59] developed a composite structure on a porous PP hydrophobic hollow fiber by having a nanoporous fluorosiloxane coating plasma polymerized on the fiber outside diameter (Figure 2.8). The pores of the PP substrate were filled with an 80 to 20 hexane-trioctyl amine (TOA) mixture; this mixture could not wet the flourosiloxane coated pores on the outside surface which remained gas-filled. By pulling a vacuum, hexane was removed and TOA was left to occupy only about one-fifth of the pore length of the substrate PP hollow fiber. The dilute biofuel-containing solution contacted the fibers on the outside surface, the biofuels were evaporated

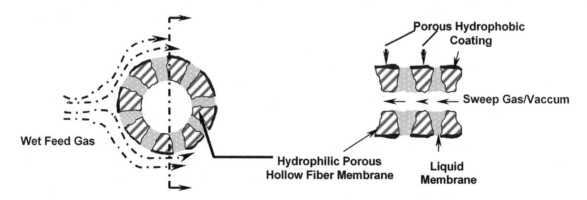

FIGURE 2.6 Polar immobilized liquid membrane (ILM) in the pores of a hydrophilic hollow fiber having a porous hydrophobic coating on the fiber OD exposed to cross flow of moist feed gas with possibilities of moisture condensation on the membrane (Reprinted from *J. Membr. Sci.*, 389,Obuskovic, G., K. K. Sirkar, Liquid membrane-based CO_2 reduction in a breathing apparatus, 424–434, Copyright (2012) permission from Elsevier **[56].**)

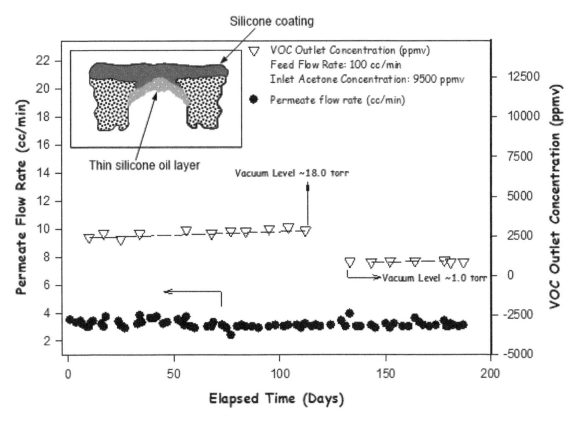

FIGURE 2.7 Performance of VOC permeation from air flowing in fiber bore with vacuum on the shell side; the inset shows the cross section of a hollow fiber pore having a thin ILM of silicone oil (Reprinted from *J. Membr. Sci.*,217,Obuskovic, G., S. Majumdar, K.K. Sirkar, Highly VOC-selective hollow fiber membranes for separation by vapor permeation, 99–116,Copyright (2003) with permission from Elsevier [58].)

into the gas-filled pores, were next absorbed into the immobilized TOA-based liquid membrane and were removed by vacuum pulled on the fiber internal diameter and condensed. The condensate was as high as 85–88% biofuels for a feed temperature of 54°C due to the very high selectivity of the ILM for a feed containing acetone, ethanol, and butanol at the level of 0.8, 0.5, and 1.5% respectively. The membrane system was stable[60] since TOA is essentially nonvolatile; the fluxes of various species such as, butanol,

acetone, and ethanol were enhanced by reducing the thickness of the ILM in the hollow fiber substrate pores.

2.4 MEMBRANE CONTACTORS IN VAPOR–LIQUID SYSTEMS

Several different types of systems are encountered depending on the nature of the feed gas/vapor system. If water vapor is to be removed from an inert gas/air into an absorbent, a

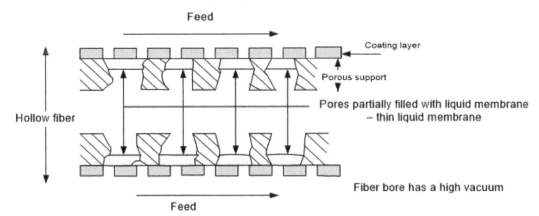

FIGURE 2.8 Microporous hydrophobic hollow fibers having a nanoporousfluorosiloxane coating on the outside surface facing the feed liquid; the hollow fiber substrate pores have a thin immobilized liquid membrane of trioctylamine (Reprinted from *J. Membr. Sci.*, 302, Thongsukmak, A., K.K. Sirkar, Pervaporation membranes highly selective for solvents present in fermentation broths, 45–58, Copyright (2007) with permission from Elsevier[59].)

conventional microporous hydrophobic hollow fiber contactor may be used with an absorbent liquid that does not wet the pores of the membrane. Examples of such absorption liquids for humidification/ dehumidification of air are: aqueous solutions of desiccants such as, LiCl, $MgCl_2$, $CaCl_2$; and triethylene glycol. The contactor hollow fibers studied are frequently made of PVDF since the absorbents are highly polar liquids. The stripping of moisture from the absorbent may be carried out in a separate contactor using dry air on the other side.

If the vapor happens to be organic in nature, e.g., a volatile organic compound (VOC) and is present in for example a N_2/air stream, and the absorbent liquid will wet the pores of conventional porous hydrophobic membranes, then a porous membrane may be used; however, the gas phase pressure has to be higher than that of the liquid phase which is present in the pores since it wets the pores. Alternately, there needs to be an ultrathin dense VOC-permeable silicone rubber skin (e.g., plasma polymerized on the outside surface) on the hollow fiber surface contacting the absorbent so that the membrane pores are not wetted. Such systems have been studied by Poddar et al. [61, 62] using absorbent liquids, e.g., silicone oil 200 fluid (Dow Corning), Paratherm NF, for VOCs such as toluene, methanol, dichloromethane, etc. The VOC stripping from the absorbent is carried out using similar coated hollow fiber modules and a vacuum on the other side.

An olefin-paraffin system containing 74% ethylene and 26% ethane was passed over one side of a porous polysulfone hollow fiber membrane (made more hydrophilic) at a high pressure up to 200 psig; on the other side an aqueous solution of $AgNO_3$ was circulated at a lower pressure and was sent to a flash stripper at atmospheric pressure to strip the highly ethylene-enriched stream. The liquid was present in the membrane pore; the pressure conditions were consistent with the wetted mode of operation [3]. Since silver nitrate is known to complex with ethylene and not ethane, highly selective ethylene absorption took place [63]. Nymeijer et al. [64] employed a composite hollow-fiber membrane with a top layer of a highly permeable block copolymer of poly(ethylene oxide) and poly(butylene terephthalate) on an Accurel PP hollow fiber support for ethylene-ethane system and obtained high performance. There have been a number of other studies in selective olefin/paraffin separation using membrane contactors (see [13] for additional references).

An altogether different approach involved carrying out vapor–liquid distillation using a column made of hollow fiber membranes. A system of great interest involves use of a porous hollow fiber membrane contactor to carry out distillation-based separation of light hydrocarbon mixtures such as, propane-propylene, over a temperature range of 10 and 20°C and iso/n-butane over a temperature range of 15–30°C. A distillation-based separation was carried out using vertical modules of porous PP hollow fiber membranes with liquid introduced into the bores of the hollow fibers at the

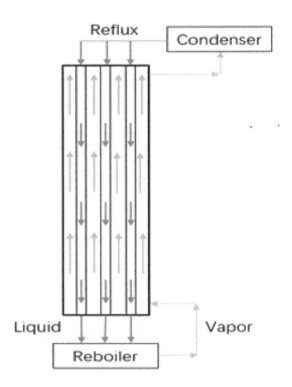

FIGURE 2.9 A microporous hollow fiber membrane-based vertical membrane distillation column for organic solvent mixture separation: total reflux mode of operation (Adapted from *J. Membr.Sci.*, 279, Yang, D., R. S. Barbero, D. J. Devlin et al., Hollow fibers as structured packing for olefin/paraffin separations, 61–69, 2006 **[65]**.)

top and the vapor flowing up in the shell side from the bottom (see Figure 2.9 for operation at total reflux). The devices were pressure tested to withstand a high pressure of 300psig. Significant separation was achieved; the HTU values were on the low side as was expected in high surface area/volume hollow fiber contactor systems. The results and column performance analysis are provided in [65–67]. Earlier, Chung et al. [68] employed a slightly different approach involving either a silicone rubber coated hollow fiber membrane or a nanoporous hollow fiber membrane for distillation separation of methanol–ethanol. No flooding issue was observed since the liquid pressure drop for transfer across the pore wall (as well as the highly permeable silicone rubber coating) was found to be higher than that through the bore of the hollow fiber as the liquid flowed downwards.

For removal and recovery of a volatile species from a liquid, membrane contactors have been investigated for quite some time primarily using vacuum-driven systems. The process is identified as Vacuum Membrane Distillation (VMD) when the feed solution is hot and the other side of the membrane is subjected to a vacuum. The vapor removed is condensed in an external condenser. When a non-condensable sweep gas flows on the other side of the membrane and sweeps away the volatile species, the process is called Sweep Gas Membrane Distillation (SGMD). The sweep gas is introduced into an external condenser to condense the volatile species.

2.5 MEMBRANE CONTACTORS FOR LIQUID–LIQUID SYSTEMS

There are quite a few reviews and extended treatments of membrane contactors for L–L systems, Prasad and Sirkar [69] being the earliest and Song et al. [70] being the latest. Additional reviews and extended treatments include: Reed et al. [8]; Gableman and Hwang [9]; Pabby and Sastre [71]; Klaassen et al. [10]; Schlosser et al. [72] for organic acid mixture separation; Aguilar and Cortina (2008) [73]; Sirkar [1]; Pabby et al. [74]; and Giorno [75]. Membrane extraction in analytical chemistry has been reviewed[76]; it appears that these authors were unaware of the earliest research in NDSX/MSX. Membrane-based non-dispersive liquid–liquid contacting has also been studied in microfluidic devices [77] (this thesis identifies six recent publications with a porous membrane in microfluidic devices).

The first successful non-dispersive operation of a membrane contactor for a liquid–liquid (aqueous-organic) system was reported by Kiani et al. [78] for flat microporous hydrophobic polypropylene (PP) membranes with the organic phase in the membrane pores. It was demonstrated that the aqueous phase pressure needed to be equal to or greater than that of the organic phase present in the membrane pores to carry out non-dispersive solvent extraction; however, the higher pressure cannot become too high lest it should push out the organic solvent from the pores leading to dispersion of the aqueous phase into the organic phase as drops (Figure 2.1a).

Kiani et al. [78] had also speculated in the following fashion about the utility of hollow fibers for such systems: "If a very low value of a of 1000 ft²/ft³ (3274 m²/m³) is used, then one finds for this new technique $K_o a = 47$. Thus the overall rate of extraction by the present technique for a given equipment volume is likely to be several times larger than that in conventional systems since countercurrent extraction can be easily carried out with a hollow fiber system or otherwise. Such a result is possible primarily due to the very large membrane surface area per unit equipment volume available in hollow fiber systems."These projections turned out to be prophetic. Frank and Sirkar [6] illustrated hollow fiber solvent extraction of ethanol from a bioreactor environment with immobilized yeast acting as the biocatalyst this being the first step towards perfusion bioreactors and membrane bioreactors for waste treatment. Extensive quantitative characterizations of mass transport in hollow fiber systems are available in Prasad and Sirkar [79].

Hoechst–Celanese commercialized this licensed technology [80] using PP-based hollow fiber modules called Liqui-Cel® in the 1990s. Larger-scale commercial systems for membrane-based solvent extraction systems using hydrophobic PP-based hollow fiber membrane modules are mentioned in [30]. A few examples follow: Klaassen and Jansen [81] employed the feedstock of a reactor as an organic extractant for an aromatic compound appearing as a pollutant in the waste water from the reactor (see Figure 3b in [30]); Porebski et al. [82] extracted phenol from a hydrocarbon fraction using similar PP hollow fiber modules (see Figure 3c in [30]). Polypropylene hollow fiber and the tube-sheets do show significant swelling with aggressive organic solvents suggesting the need for solvent-resistant hollow fibers.

Flat porous hydrophilic membranes were also studied for solvent extraction in an aqueous–organic system with the aqueous phase in pores and the organic phase at a pressure equal to or higher than the aqueous phase pressure (Prasad and Sirkar [83]). A large-scale commercial application of such a technique being utilized along with a chemical reaction is provided by Lopez and Matson [84]: polyacrylonitrile-based hydrophilic asymmetric hollow fiber membrane was used to extract a product generated by an enzymatic reaction in a multiphase/extractive enzyme membrane reactor for the industrial production of diltiazem chiral intermediate.

In most solvent extraction processes, the solvent has to be regenerated via a back extraction or stripping step. Metal extractions are important examples. Transport analysis for membrane solvent extraction [83] suggests that the back extraction step should be carried out preferably with a hydrophilic membrane having aqueous solution in the membrane pores since the membrane resistance can be reduced by preferentially filling the membrane pores with the phase preferred by the solute. However, in the earlier days, solvent-resistant hydrophilic hollow fiber membranes were difficult to come by. Many metal extraction and back extraction studies [85] continued to use hydrophobic PP membranes for both tasks, which increases the membrane area requirement for back extraction.

Hydrophilic hollow fiber membranes of nylon having some degree of solvent resistance were studied by Basu et al.[86] for back extraction of phenol into a caustic solution. Since these membranes had larger pores and therefore posed problems of interface stability in longer hollow fiber systems with significant flow pressure drop, the nylon hollow fiber pore diameters on the inner fiber diameter were reduced via insolubilization of polyethylenimine (PEI) by crosslinking [87]. A short while earlier, sulphonated poly(ether ether ketone) (SPEEK) was applied as a hydrophilic coating layer on PP-based Accurel hollow fiber membranes by He et al. [88]. Two contactors were used, one for extraction of copper from an aqueous solution and the other for back extraction of copper for 75 days in a stable fashion with reasonable flux. However, it was found that the PP substrate was becoming brittle for longer periods of time in the particular organic solvent environment [89]. For aqueous-organic interface immobilization, the special advantages of a composite solvent extraction membrane where one side is hydrophobic and the other side is hydrophilic were experimentally demonstrated by Prasad and Sirkar [83].

It should be pointed out that a porous hydrophilic substrate, if solvent-resistant, can be used also in the hydrophobic mode, i.e., an organic solvent can be in the pore with the aqueous solution on the outside at a higher pressure [90]. Therefore a solvent-resistant hydrophilic hollow fiber may be used with the organic phase in the pore for solvent extraction and then in another module the same fibers may be used with the aqueous phase in the pores for back extraction.

Non-dispersive solvent extraction was also demonstrated on a lab-scale using other biphasic systems. Prasad and Sirkar [91] demonstrated preferential toluene extraction from a 50–50 n-heptane-toluene mixture into a polar organic solvent, n-methyl pyrrolidone, using a porous hydrophobic membrane whose pores were wetted by the nonpolar organic solvent mixture. Dahuron and Cussler [92] illustrated the extraction of protein in an aqueous biphasic system of polyethylene glycol (PEG) and potassium phosphate with PEG in the membrane pores and at a lower pressure. There are very few membrane contactor studies of polar organic-nonpolar organic and aqueous–aqueous biphasic systems.

On the other hand, solvent extractions of heavy metals such as Cu^{2+}, Cr^{6+}, etc., were investigated extensively from the beginning, an example being the study by Yun et al. [93]. Interfacial reaction resistance can become important in such cases. There are quite a few studies on rare earth extraction and actinide extraction in so far as MSX/NDSX is concerned. In general, polypropylene hollow fibers have been used to study specific separations since the earliest days when neodymium extraction was studied [94]. A more recent example is that in Ambare et al. [95] who studied the extraction of Nd (III) from nitric acid feed solutions using a mixture of di-nonylphenyl phosphoric acid and tri-n-octylphosphine oxide in petrofin as the carrier extractant. The long term-stability of the membrane/module exposed to such an environment is of interest.

A recent study with nanoporous solvent resistant poly(ethylene-co-vinyl alcohol) (EVAL) hollow fiber membranes having a two-layered structure claimed a level of success in terms of having a solvent-resistant membrane (Song et al. [70]). This polymer was found to be stable in the presence of quite a few solvents. Experiments succeeded in showing that lithium ions present in aqueous solutions could be extracted into the organic extract phase stably over a period of 1,000 hr. Four-inch diameter hollow fiber modules were found to be stable in carrying out solvent extraction for over two years.

Modules that have designs other than the hollow fiber–based design have also been studied. Bayer et al. [96] investigated a spiral-wound module for liquid–liquid extraction; unlike conventional spiral-wound modules that are used in various pressure-driven membrane processes, these modules have four openings. The design has reduced the propensity for dead zones and maldistribution; the reactive extraction of phenol was studied.

After the non-dispersive solvent extraction process, NDSX/MSX, was first proposed, published (Kiani et al. [78]) and patented [80] from 1984 onwards, the NDSX/MSX was also termed as "pertraction" (see [10]). The prefix "per" is added to characterize it in general as some type of a membrane process. On the basis that a membrane process has at least two separate bulk phase interfaces, a feed-membrane interface and a permeate-membrane interface, the NDSX process is not a membrane process; it is a membrane-assisted phase-contacting process since there is only one bulk phase interface.

2.6 SUPPORTED LIQUID MEMBRANE (SLM) SYSTEMS FOR LIQUID–LIQUID PROCESSES

One of the most important applications of supported liquid membranes (SLMs) has been in the area of metal extractions where two aqueous liquid phases on two sides of the membrane are separated by an organic liquid membrane in the pores of usually a hydrophobic porous membrane (Figure 2.2a). From the feed aqueous solution, a metal is extracted via complexation through an aqueous–organic interface into the liquid membrane; from the other aqueous–organic interface on the other side of the polymeric support membrane, the metal is back extracted and concentrated into an aqueous stripping solution. An apparently insoluble problem is the slow loss of organic solvent/diluent and complexing agents via solubilization in the surrounding two aqueous phases. Usually the osmotic pressures of the two aqueous solutions on the two sides of the membrane are different. That will cause water to flow from one side to the other primarily by diffusion through the stationary organic liquid membrane. As the amount of organic solvent is depleted via solubilization, the amount of water being transported becomes high and the organic phase in the pore is finally replaced by water; the liquid membrane is lost [97].

Therefore, there has been a search for ways to stem this loss of organic solvent/extractant from the liquid membrane into the surrounding aqueous phases. Brief overviews are provided in [89] and [70] among others. These steps have included gel formation [98], interfacial polymerization at the pore mouths [99], a laminated ion exchange layer of sulfonated poly (etheretherketone) (SPEEK) bonded to a porous hydrophobic membrane on two sides of the hydrophobic support membrane having the organic extractant in the pores [88], etc. All such steps were deficient in preventing organic solvent loss in one fashion or another.

The results of a large-scale pilot test of an SLM for copper extraction from an ammoniacal etching solution has been provided in [100]. A large PP hollow fiber module having a surface area of 130m² was employed successfully for an extended period; a smaller module performed satisfactorily for a month. However one has to be ready to shut down the process for SLM renewal or alternately have a spare SLM-based module ready.

2.6.1 SLM WITH STRIP DISPERSION

There is a process where membrane solvent extraction with a hydrophobic porous membrane is utilized with an organic extracting solvent; however, the extracting organic solvent with dissolved complexing extractants has larger aqueous droplets of the strip liquid dispersed in it. These droplets are not inside the solvent in the membrane pores. After the organic extractant stream goes out of the membrane device, the dispersion is separated in a separate vessel and the aqueous strip stream is recovered. The organic solvent obtained from this separate vessel is next subjected to dispersion of a fresh aqueous strip and reintroduced into the membrane device. This process has been known by two different names: emulsion pertraction [101]; and SLM with strip dispersion (Ho and Poddar [102]). A larger demonstration by Commodore Separation Technologies Inc. is identified in Sirkar [30].

2.6.2 SLM-Based Pervaporation

Consider an SLM process where the feed phase is a liquid but the permeate phase is subjected to a vacuum or has a sweep gas. In it, the feed-membrane phase interface is L–L but the permeate phase interface is L–G, and the process is known as supported liquid membrane–based pervaporation (SLMPV). One such example is provided by [103] wherein an aqueous feed solution containing acetic acid contacted an SLM of a long chain alkyl amine which can complex with acetic acid and extract it efficiently. A vacuum pulled at the other interface and the condensation of the gaseous stream led to a high selectivity of 33 for acetic acid over water from a feed of 1 M at 60°C. "An on-line regeneration technique was used successfully to continuously provide the liquid membrane material to the substrate membrane surface from the permeate side, to make up for the loss of the LM during operation" [103].

2.6.3 Contained Liquid Membrane

Consider two flat porous hydrophobic membranes. If an organic extractant phase is introduced between the two membranes, it will wet the pores of each hydrophobic membrane. If an aqueous solution flows on the other side of each hydrophobic membrane at pressures higher than that of the organic extractant phase, then one can have an aqueous–organic phase interface immobilized on the aqueous side of each membrane. The organic extractant phase now functions as a liquid membrane, the contained liquid membrane (CLM). If instead of two flat membranes, we have two hollow fibers with two different aqueous solutions flowing through the bore of each hollow fiber, then we have a hollow fiber contained liquid membrane (HFCLM) (Sengupta et al. [104]). One of the aqueous solutions will be the feed solution while the other will be the strip aqueous solution. By connecting the shell side of such a module with a reservoir containing the organic liquid, any loss of the organic liquid is automatically taken care of. In a hollow fiber module, the two sets of hollow fibers are commingled together to reduce the gap between the feed solution carrying hollow fiber and the strip solution carrying hollow fiber.

A number of very useful separations were implemented using such a device: Citric acid [105]; Cu++ [106]; diltiazem [107]; 2-Propanol/n-Heptane [108]. A review up until 1996 is available [42]. The diltiazem study [107] employed a Liqui-Cel™ module with a 1.29 m² mass transfer area; modules having a 46.5 m² surface area in each fiber set have been built at Membrana.

2.7 SUPPORTED GAS MEMBRANE (SGM) SYSTEMS FOR LIQUID– LIQUID PROCESSES

There are a number of techniques where two different liquid streams are used on two sides of a porous membrane whose pores are gas-filled (Figure 2.2c). In all such processes, the species being transferred from one liquid stream to another through the gas-filled pore acting as a membrane of sorts is either a gas or a vapor. Although the process of diffusion through the gas-filled pore is unlikely to have much selectivity, this process qualifies as a membrane process since the partition coefficients on the two sides of the membrane are quite different. But the processes also are characteristic of membrane contactors since on each side we have dispersion-free contacting of two immiscible fluid phases.

There are a couple of distinct types of SGM systems depending on the condition of the liquid solutions on the two sides. If the feed liquid solution has a high temperature and the strip side liquid has a lower temperature with the vapor species being transferred from the high temperature liquid to the low temperature liquid, then the process is called Direct Contact Membrane Distillation (DCMD). This is being investigated extensively for desalination, with water vapor being transferred from the hot brine to a cold distillate stream. **Chapter 17** is devoted to this and related techniques.

If, however, the temperatures on both sides of the membrane are identical but the osmotic pressures of the two solutions are different, then for aqueous solutions, water vapor will be transferred to the solution having the higher osmotic pressure; the process is called Osmotic Distillation (OD). If on the other hand, the chemical nature of the two solutions is quite different facilitating the reabsorption of the gas/vapor species stripped from the feed solution into the second solution via some chemical reaction, then one potential designation is Transmembrane Chemisorption (TMCS). **Chapter 15** considers this topic in great detail.

2.8 MEMBRANE CONTACTORS FOR FLUID–SOLID SYSTEMS

Two types of phase-contacting systems are relevant here: gas–solid and liquid–solid.

2.8.1 Gas–Solid Contactors

In conventional gas–solid contactors, gas flows through a bed of adsorbent particles. If the adsorbent particles are small in size, one encounters a significant pressure drop in flow through the packed bed. An approach that was developed sometime back involves using porous hollow fibers along the length of the bed while the shell side is filled with the adsorbent particles. The gas stream can flow through the bore/lumen of the hollow fibers along the bed length. However, as it moves, it can rapidly diffuse through the pores in the membrane wall and contact the adsorbent particles on the shell side. An exchange between the moving gas and the porous adsorbent particles is achieved without incurring a high pressure drop which could happen in the absence of the hollow fibers. Feng and Ivory [109] provide an earlier review of this approach in gas–solid contacting. Note that the membrane is no longer involved in developing an immobilized bulk phase interface here. It is merely facilitating the process by reducing the pressure drop without affecting the gas diffusion process too drastically.

There is an alternate configuration where the adsorbents are integrated into the porous structure of the hollow fiber wall. These are called fiber sorbents and are hollow fiber polymer/zeolite composites prepared by a spinning process employing wet phase inversion at a very high sorbent loading (Lively et al. [110]). These structures allow a rapid diffusion of gases to be separated from the shell side into the polymer/sorbent porous structure. The tubeside of this hollow fiber structure is coated with a gas/vapor-impervious polymer coating of, say, polyvinylidene dichloride to allow water to flow through the fiber lumen for the heating and

cooling of the hollow fiber wall without any gas or water vapor being transported in either direction. An example is provided by silica-supported poly(ethylenimine) hollow fiber sorbents for CO_2 sorption from flue gas (Fan et al. [111]) in a process called Rapid Temperature Swing Adsorption (RTSA). A description of such a cyclic gas–solid contacting process for flue gas CO_2 removal and recovery follows.

Four beds are shown in Figure 2.10 representing different parts of the cycle that a sorbent bed goes through:(1) adsorption; (2) heating; (3)sweeping; and (4) cooling, respectively. At the beginning, flue gas flows through the shell side of the hollow fiber module; CO_2 gets adsorbed during this adsorption step. To make this efficient, cooling water is introduced through the fiber lumen to remove the heat of adsorption, which could potentially be reused in the heating step. Next, the heating step is initiated by introducing hot water into the hollow fiber lumen, which heats the fiber adsorbent leading to the release of adsorbed CO_2. Further, by closing the inlet before this step starts no more feed gas is introduced to the shellside. In the next sweeping step, N_2 gas is introduced to the shell side to push out the desorbed CO_2. This step is followed by the cooling step, where cooling water is made to flow through the fiber lumen to lower the fiber temperature and to get the bed ready for the next cycle of gas adsorption. This process has significant heat integration in order to reduce the total amount of heat needed to carry out the process. For

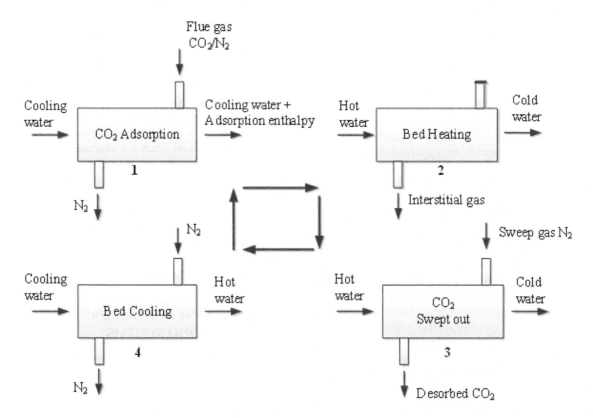

FIGURE 2.10 Overview of an RTSA process with heat integration (Adapted from *AIChE J.*, 60(11), Fan, Y., Y. Labreche, R. P. Lively et al., Dynamic CO_2 adsorption performance of internally cooled silica-supported poly(ethylenimine) hollow fiber sorbents, 3878–3887, 2014; it uses a gas–solid membrane contactor[**111**].)

example, the heating that is undergone by cooling water in the adsorption step reduces the amount of heat needed to heat the water up for the heating step.

2.8.2 LIQUID–SOLID CONTACTORS: MEMBRANE ADSORBERS

In chromatographic processes for the separation of bio-macromolecules that are carried out using porous resin beads, the feed liquid solution flows past a resin bead. Bio-macromolecules diffuse into the porous bead for adsorption and diffuse out during elution. The slow diffusional processes of large bio-macromolecules inside the porous resin lead to a poor utilization of the costly ligands in the resin beads. If, however, the contacting is carried out in convective flow mode with the solution flowing through membrane pores whose surfaces have the ligands dangling, the contacting of bio-macromolecules with the ligands becomes highly efficient; the fractional utilization efficiency of such ligands becomes very high and the process becomes quite rapid. Brandt et al. [112] provided one of the earliest introductions to this technique. One uses microfiltration membranes and a convective flow of the feed liquid to be purified through the membrane pores. Earlier reviews by Roper and Lightfoot [113] and Thommes and Kula [114] are useful.

A major shortcoming of this technique is related to membrane thicknesses that are quite small. Therefore, in practice, one uses a stack of flat membranes to enhance the sorption capacity. Various designs are being developed. A more detailed treatment of this is available in **Chapter 19** of this book.

This technique was adapted for adsorption of heavy metals from contaminated waters via appropriate functionalization of pore surfaces with ligands which interact with heavy metals [115]. The latest example of such a contacting process for removal of a variety of heavy metals is illustrated in [116].

2.9 MEMBRANE CONTACTORS FOR LIQUID–LIQUID–SOLID SYSTEMS IN CRYSTALLIZATION

When an aqueous solution of barium chloride is contacted with an aqueous solution of K_2SO_4, a precipitate of $BaSO_4$ particles is obtained. Kieffer et al. [117, 118] employed a porous hollow fiber membrane-based contacting system in which a solution of K_2SO_4 flowing on the shell-side permeated into the tube side where a solution of $BaCl_2$ was flowing; the contacting led to precipitation of $BaSO_4$ particles in the tube-side of the hollow fiber device. A porous hollow fiber membrane contactor may also be employed to implement anti-solvent crystallization. Zarkadas [119] and Zarkadas and Sirkar [120] successfully demonstrated the crystallization of L-asparagine monohydrate in an aqueous solution with 2-propanol as the miscible anti-solvent. Their feed solution flowing through the hollow fiber bore was pushed to the shell side where the anti-solvent was flowing. The shell-side is a better location for flow of the solution containing crystals unlike the precipitates flowing through the fiber bore in the studies by Kieffer et al. [117, 118].

Chen et al. [121] showed that an even better arrangement is to let the anti-solvent water flow through the hollow fiber bore and come out to the shell-side and contact the crystallizing solution of the drug griseofulvin in acetone flowing on the shell-side at numerous pore mouths. This arrangement reduces the possibility of the plugging of the membrane pores. Chen et al. [122] also showed that if the feed solution of a polymer in acetone flowing on the shell-side had a suspended submicrometer or nanoparticles of silica, precipitation of the polymer by the anti-solvent water coming in from the tube-side can lead to the continuous production of polymer-coated silica particles.

These are examples where porous hollow fiber membrane devices are facilitating efficient contact between miscible phases leading to the production of drug crystals, polymer-coated drug crystals, etc.

2.10 CONCLUDING REMARKS

Research and development in membrane contactors is taking place at a rapid rate and in many directions with significant progress being made in their commercialization. Recent progresses include a selection of better contactor membrane materials and/or modification of existing membranes via coating/surface modification or the development of an altogether new material or membrane structure. High free volume polymer-based dense coatings with high gas permeabilities can eliminate membrane pore wetting caused by various absorbents. Modes of operation of a G–L membrane contactor have been expanded to include pressure-swing and temperature-swing-based processes. Although most studies in G–L systems focus on gas absorption, successful stripping studies have also been made. A number of examples of ILMs for gas separation show extended stability. Membranes with greater solvent resistance have appeared for membrane contactors for L–L processes. The prevention of loss of organic solvent-extractant acting as an SLM continues to be challenging. The introduction of fiber sorbents has facilitated applications of gas–solid membrane contactors. Liquid–solid membrane adsorbers are useful in aqueous separation applications involving bio-macromolecules as well as heavy metal removal. Porous hollow fiber membranes facilitate the high efficiency mixing of anti-solvent with a crystallizing solution to enable continuous processes for crystallization as well as forming a polymer coating on suspended crystals and nanoparticles.

REFERENCES

1. Sirkar, K. K. 2008. Membrane contactors. In N. Li, A. G. Fane, W. S. W. Ho and T. Matsuura (Eds.), *Advanced Membrane Technology and Applications* (Chap. 26, pp. 687–702). Hoboken, NJ: Wiley.

2. Sirkar, K. K. 1992. Other new membrane processes. In W. S. W. Ho and K. K. Sirkar (Eds.), *Membrane Handbook* (Chap. 46, pp. 885–912). New York, NY: Van Nostrand Reinhold. Reprinted (2001), Kluwer Academic Publishers, Boston, MA.

3. Karoor, S. and K. K. Sirkar. 1993. Gas absorption studies in microporous hollow fiber membrane modules. *I&EC Res.* 32: 674–684.

4. Esato, K. and B. Eiseman. 1975. Experimental evaluation of Gore-Tex membrane oxygenator. *J. Thorac. Cardiovasc. Surg.* 69(5): 690–697.

5. Tsuji, T., K. Suma, K. Tanishita et al. 1981. Development and clinical evaluation of hollow fiber membrane oxygenators. *Trans. Am. Soc. Artif. Intern. Organs.* 27: 280–284.

6. Frank, G. T. and K. K. Sirkar. 1985. Alcohol production by yeast fermentation and membrane extraction. Biotech. Bioeng. Symp. *Ser.* 15: 621–631.

7. Qi, Z. and E. L. Cussler. 1985. Microporous hollow fibers for gas absorption, I. Mass transfer in Liquid II. Mass transfer across the membrane. *J. Membr. Sci.* 23: 321–340.

8. Reed, B. W., M. J. Semmens and E. L. Cussler. 1995. Membrane contactors. In R. D. Noble and S. A. Stern (Eds.), *Membrane Separations Technology, Principles and Applications.* New York, NY: Elsevier Science, p. 468.

9. Gableman, A. and Hwang, S-T. 1999. Hollow fiber membrane contactors. *J. Membr. Sci.* 159: 61–106.

10. Klaassen, R., P. H. M. Feron and A. E. Jansen. 2005. Membrane contactors in industrial applications. *Chem. Eng. Res. Des.* 83(3): 234–236.

11. Curcio, E. and E. Drioli. 2005. Membrane distillation and related operations – A Review. *Sep. Purif. Rev.* 34: 35–86.

12. Mansourizadeh, A. and A. F. Ismail. 2009. Hollow fiber gas-liquid membrane contactors for acid gas capture: A review. *J. Hazard. Mater.* 171: 38–53.

13. Bazhenov, S. D., A. V. Bildyukevich and A. V. Volkov. 2018. Gas-liquid hollow fiber membrane contactors for different applications. *Fibers.* 6(4): 76.

14. Sengupta, A., P. A. Peterson, B. D. Miller et al.1998. Large-scale application of membrane contactors for gas transfer from or to ultrapure water. *Sep. Purif. Technol.* 14: 189–200.

15. Zheng, J.-M., Z.-W. Dai, F.-S., Wong et al. 2005. Shell side mass transfer in a transverse flow hollow fiber membrane contactor. *J. Membr. Sci.* 261: 114–12.

16. Favre, E. and H. F. Svendsen. 2012. Membrane contactors for intensified post-combustion carbon dioxide capture by gas–liquid absorption processes. *J. Membr. Sci.*, 407–408: 1–7.

17. Chabanon, E., B. Belaissaoui and E. Favre. 2014. Gas–liquid separation processes based on physical solvents: Opportunities for membranes. *J. Membr. Sci.* 459: 52–61.

18. Kumar, P. S., J. A. Hogendoorn, P. H. M. Feron et al. 2002. New absorption liquids for the removal of CO_2 from dilute gas streams using membrane contactors. *Chem. Eng. Sci.* 57: 1639–1651.

19. Kosaraju, P., A. Korikov, A. Kovvali et al. 2005. Hollow fiber membrane contactor-based CO_2 absorption-stripping using novel solvents and membranes. *I&EC Res.* 44(5): 1250–1258.

20. Li, S., T. J. Pyrzynski, N. B. Klinghoffer et al. 2017. Scale-up of PEEK hollow fiber membrane contactor for post-combustion CO_2 capture. *J. Membr. Sci.* 527: 92–101.

21. Mulukutla, T., G. Obuskovic and K. K. Sirkar. 2014. Novel scrubbing system for post-combustion CO_2 capture and recovery: Experimental studies. *J. Membr. Sci.* 471: 16–26.

22. Jie, X., J. Chau, G. Obuskovic et al. 2013. Preliminary studies of CO_2 removal from precombustion syngas through pressure swing membrane absorption process with ionic liquid as absorbent. *I&EC Res.* 52: 8783–8799.

23. Li, K. and Teo, W. K. 1998. Use of permeation and absorption methods for CO_2 removal in hollow fiber membrane modules. *Sep. Purif. Technol.* 13: 79–88.

24. Chabanon, E., D. Roizard and E. Favre. 2011. Membrane contactors for post-combustion carbon dioxide capture: A comparative study of wetting resistance on long time scales. *I&EC Res,* 50: 8237–8244.

25. Scholes, C. A., S. E. Kentish, G. W. Stevens et al. 2015. Comparison of thin film composite and microporous membrane contactors for CO_2 absorption into monoethanolamine. *International J. Greenhouse Gas Control,* 42: 66–74.

26. Ansaloni, L., A. Arif, A. F. Ciftja et al. 2016. Development of membrane contactors using phase change solvents for CO_2 capture: Material compatibility study. *I&EC Res.* 55: 13102–13113.

27. Scholes, C. A., S. E. Kentish, G. W. Stevens et al. 2015. Thin-film composite membrane contactors for desorption of CO_2 from monoethanolamine at elevated temperatures. *Sep. Purif. Technol.* 156: 841–847.

28. Dai, Z., L. Ansaloni and L. Deng. 2016. Precombustion CO_2 capture in polymeric hollow fiber membrane contactors using ionic liquids: Porous membrane versus nonporous composite membrane. *I&EC Res.* 55: 5983–5992.

29. Naim, R., A. F. Ismail and A. Mansourizadeh. 2012. Effect of non-solvent additives on the structure and performance of PVDF hollow fiber membrane contactor for CO_2 stripping. *J. Membr. Sci.* 423–424: 503–513.

30. Sirkar, K. K. 2008. Membranes, phase interfaces and separations: Novel techniques and membranes-An overview. *I&E Chem. Res.* 47: 5250–5266.

31. Singh, D. and K. K. Sirkar. 2014. High temperature direct contact membrane distillation based desalination using PTFE hollow fibers. *Chem. Eng. Sci.* 116: 824–833.

32. Ghobadi, J., D. Ramirez, S. Khoramfar et al. 2018. Simultaneous absorption of carbon dioxide and nitrogen dioxide from simulated flue gas stream using gas-liquid membrane contacting system. *Int. J. Greenhouse Gas Control,* 77: 37–45.

33. Malankowska, M., C. F. Martins, H. S. Rho et al. 2018. Microfluidic devices as gas–ionic liquid membrane contactors for CO_2 removal from anesthesia gases. *J. Membr. Sci.* 545: 107–115.

34. Jie, X., J. Chau, G. Obuskovic et al. 2014. Enhanced pressure swing membrane absorption process for CO_2 removal from shifted syngas with dendrimer-ionic liquid mixture as absorbent. *I&EC Res.* 53(8): 3305–3320.

35. Chau, J., X. Jie and K. K. Sirkar. 2016. Polyamidoamine-facilitated poly (ethylene glycol)/ionic liquid based pressure swing membrane absorption process for CO_2 removal from shifted syngas. *Chem. Eng. J.* 305: 212–220.

36. Bhaumik, S., S. Majumdar and K. K. Sirkar. 1996. Hollow fiber membrane-based rapid pressure swing absorption. *AIChE J.* 42(2): 409–421.

37. Samanta, A. K., A. Zhao, G. K. H. Shimizu et al. 2012. Post-combustion CO_2 capture using solid sorbents: A review. *Ind. Eng. Chem. Res.* 51: 1438–1463.

38. Mulukutla, T., D. Singh, J. Chau et al. 2015. Novel membrane contactor for CO_2 removal from flue gas by temperature swing absorption. *J. Membr. Sci.* 493: 321–328.

39. Singh, D. and K. K. Sirkar. 2012. Desalination by airgap membrane distillation using a two hollow-fiber-set membrane module. *J. Membr. Sci.* 421–422: 172–179.

40. Obuskovic, G., T. K. Poddar and K. K. Sirkar. 1998. Flow swing membrane absorption-permeation. *I&EC Res.* 37: 212–220.

41. Cai, J., K. Hawboldt and M. A. Abdi. 2016. Improving gas absorption efficiency using a novel dual membrane contactor. *J. Membr. Sci.* 510:249–258.

42. Sirkar, K. K. 1996. Hollow fiber contained liquid membranes for separations: An overview. In R. A. Bartsch and J. D. Way (Eds.), *Chemical Separations with Liquid Membranes* (Chap. 16, pp. 222–238). Washington, DC: ACS Symp. Ser., 642.

43. Albo, J., P. Luis and A. Irabien. 2010. Carbon dioxide capture from flue gases using a cross-flow membrane contactor and the ionic liquid 1-ethyl-3-methylimidazolium ethyl sulfate. *I&EC Res.* 49 (21): 11045–11051.

44. Yang, J., X. Yu, J. Yan et al. 2014. CO_2 capture using amine solution mixed with ionic liquid. *I&EC Res.* 53: 2790–2799.

45. Sirkar, K. K. 2018. Systems and methods for CO_2 removal from flue gas by temperature swing absorption, US Patent10, 005, 022B2, June 26, 2018.

46. Fathizadeh, M., K. Khivantsev, T. J. Pyrzynski et al. 2018. Bio-mimetic oxygen separation via a hollow fiber membrane contactor with O_2 carrier solutions. *Chem Commun.* 54: 9454–9457.

47. Shanbhag, P. V., A. K. Guha and K. K. Sirkar. 1995. Single-phase membrane ozonation of hazardous organic compounds in aqueous streams. *J. Hazardous Mater.* 41(1): 95–104.

48. Shanbhag, P. V., A. K. Guha and K. K. Sirkar. 1996. A membrane-based integrated absorption-oxidation reactor for destroying VOCs in air. *Environ. Sci. Technol.* 30(12): 3435–3440.

49. Shanbhag, P. V., A. K. Guha and K. K. Sirkar. 1998. Membrane-based ozonation of organic compounds. *I&EC Res.* 37(11): 4388–4398.

50. Guha, A. K., P. V. Shanbhag, K. K. Sirkar et al.1995. Multiphase ozonolysis of organics in wastewater by a novel membrane reactor. *AIChE J.* 41(8): 1998–2012.

51. Shanbhag, P. V. and K. K. Sirkar. 1998. Ozone and oxygen permeation behavior of silicone capillary membranes employed in membrane ozonators. *J. Appl. Polym. Sci.* 69(7): 1263–1273.

52. Steiner, M., W. Pronk, U. V. Gunten et al. 2010. *Use of Membrane Contactors for the Diffusion of Ozone (Water Research Foundation, Denver, Colorado: Web Report #2885).* Swiss Federal Institute of Aquatic Science and Technology.

53. Stylianou, S. K., S. D. Sklari, D. Zamboulis et al. 2015. Development of bubble-less ozonation and membrane filtration process for the treatment of contaminated water. *J. Membr. Sci.* 492: 40–47.

54. Stylianou, S. K., I. A. Katsoyiannis, M. Mitrakas et al. 2018. Application of a ceramic membrane contacting process for ozone and peroxone treatment of micropollutant contaminated surface water. *J. Hazard. Mater.* 358: 129–135.

55. Volkov, V. V., V. I. Lebedeva and I. V. Petrova. 2011. Adlayers of palladium particles and their aggregates on porous polypropylene hollow fiber membranes as hydrogenization contractors/reactors. *Adv. Colloid Interface Sci.* 164(1–2): 144–155.

56. Obuskovic, G. and K. K. Sirkar. 2012. Liquid membrane-based CO_2 reduction in a breathing apparatus. *J. Membr. Sci.* 389: 424–434.

57. Chen, H., G. Obuskovic, S. Majumdar et al. 2001. Immobilized glycerol-based liquid membranes in hollow fibers for selective CO_2 separation from CO_2–N_2 mixtures. *J. Membr. Sci.* 183: 75–88.

58. Obuskovic, G., S. Majumdar and K. K. Sirkar. 2003. Highly VOC-selective hollow fiber membranes for separation by vapor permeation. *J. Membr. Sci.* 217:99–116.

59. Thongsukmak, A. and K. K. Sirkar. 2007. Pervaporation membranes highly selective for solvents present in fermentation broths. *J. Membr. Sci.* 302: 45–58.

60. Thongsukmak, A. and K. K. Sirkar. 2009. Extractive pervaporation to separate ethanol from its dilute aqueous solutions characteristic of ethanol-producing fermentation processes. *J. Membr. Sci.* 329: 119–129.

61. Poddar, T. K., S. Majumdar and K. K. Sirkar. 1996. Membrane-based absorption of VOCs from a gas stream. *AIChE J.* 42(11): 3267–3282.

62. Poddar, T. K., S. Majumdar and K. K. Sirkar. 1996. Removal of VOCs from air by membrane-based absorption and stripping. *J. Membr. Sci.* 120(2): 221–237.

63. Tsou, D. T., M. W. Blachman and J. C. Davis. 1994. Silver-facilitated olefin/paraffin separation in a liquid membrane contactor system. *Ind. Eng. Chem. Res.*33: 3209–3216.

64. Nymeijer, K., T. Visser, R. Assen et al. 2004. Olefin-selective membranes in gas-liquid membrane contactors for olefin/paraffin separation. *I&EC Res.*43: 720–727.

65. Yang, D., R. S. Barbero, D. J. Devlin et al. 2006. Hollow fibers as structured packing for olefin/paraffin separations. *J. Membr. Sci.* 279: 61–69.

66. Yang, D., D. J. Devlin and R. S. Barbero. 2007. Effect of hollow fiber morphology and compatibility on propane/propylene separation. *J. Membr. Sci.* 304: 88–101.

67. Yang, D., R. Martinez, B. Fayyaz-Najafi et al. 2010. Light hydrocarbon distillation using hollow fibers as structured packings. *J. Membr. Sci.* 362:86–96.

68. Chung, J. B., J. P. DeRocher and E. L. Cussler. 2005. Distillation with nanoporous or coated hollow fibers. *J. Membr. Sci.* 257: 3–10.

69. Prasad, R. and K. K. Sirkar. 1992. Membrane-based solvent extraction. In W. S. W. Ho and K. K. Sirkar (Eds.), *Membrane Handbook* (pp. 727–763). New York, NY: Van Nostrand Reinhold.

70. Song, J., T. Huang, H. Qiu et al. 2018. A critical review on membrane extraction with improved stability: Potential application for recycling metals from city mine. *Desalination.* 440:18–38.

71. Pabby, A. K. and A. M. Sastre. 2002. Developments in dispersion-free membrane-based extraction –separation processes. In Y. Marcus and A. K. Sengupta (Eds.), *Ion Exchange and Solvent Extraction*, pp. 15. New York, NY: Marcel Dekker.

72. Schlosser, S., R. Kertész and J. Marták 2005. Recovery and separation of organic acids by membrane-based solvent extraction and pertraction: An overview with a case study on recovery of MPCA. *Sep. Purif. Technol.* 41: 237–266.

73. Aguilar, M. and J. L. Cortina (Eds.). 2008. *Solvent Extraction and Liquid Membranes Fundamentals and Applications in New Materials.* Boca Raton, FL: CRC Press.

74. Pabby, A. K., S. S. H. Rizvi, A. M. Sastre (Eds.). 2009. *Handbook of Membrane Separations: Chemical, Pharmaceutical, Food and Biotechnological Applications.* Boca Raton, FL: CRC Press.

75. Giorno, L. 2016. Membrane based solvent extraction. In E. Drioli and L. Giorno (Eds.), *Encyclopedia of Membranes.* Heidelberg, Germany: Springer.

76. Jönsson, J. A. and L. Mathiasson. 2001. Membrane extraction in analytical chemistry. *J. Sep. Sci.* 24: 495–507.

77. Hereijgers, J. 2016. *Solvent extraction in membrane microcontactors: Modeling, spacer structuring and applications.* Thesis for Doctor in Engineering, Vrije Universiteit, Brussel, Belgium.

78. Kiani, A., R. R. Bhave and K. K. Sirkar. 1984. Solvent extraction with immobilized interfaces in a microporous hydrophobic membrane, *J. Membr. Sci.* 20(2): 125–145.

79. Prasad, R. and K. K. Sirkar. 1988. Dispersion-free solvent extraction with microporous hollow fiber modules. *AIChE J.* 34(2): 177–188.

80. Sirkar, K. K. 1988. Immobilized Interface Solute Transfer Apparatus, U.S. Patent 4,789,468, December 6, 1988. Reissued: US Patent Re. 34,828, January 17, 1995.

81. Klaassen, R. and A. E. Jansen. 2001. The membrane contactor: Environmental applications and possibilities. *Environ. Prog.* 20: 37–43.

82. Porebski, T., S. Tomzik, W. Ratajczak et al. 2003. Industrial applications of hollow fiber modules in the process of phenol extraction from the hydrocarbon fraction. In *Proceedings of Conference PERMEA 2003. Abstract in CD ROM, Tatranske´Matliare (SK), Sept 7-11*, 2003; 2 pp.

83. Prasad, R. and K. K. Sirkar. 1987. Solvent extraction with microporous hydrophilic and composite membranes. *AIChE J.* 33(7): 1057–1066.

84. Lopez, J. L. and S. L. Matson. 1997. A multi-phase/extractive enzyme membrane reactor for production of diltiazem chiral intermediate. *J. Membr. Sci.* 125: 189.

85. Ortiz, I., B. Galan and A. Irabien. 1996. Kinetic analysis of the simultaneous non-dispersive extraction and back-extraction of chromium (VI). I *&EC Res.* 35: 1369–1377.

86. Basu, R., R. Prasad and K. K. Sirkar. 1990. Reactive back extraction of phenol using non-dispersive membrane solvent extraction. *AIChE J.* 36(3): 450–460.

87. Kosaraju, P. and K. K. Sirkar. 2007. Novel solvent-resistant hydrophilic hollow fiber membranes for efficient membrane solvent back extraction. *J. Membr. Sci.* 288: 41–50.

88. He, T., L. A. M. Versteeg, M. H. V. Mulder et al. 2004. Composite hollow fiber membranes for organic solvent-based liquid-liquid extraction, *J. Membr. Sci.* 234: 1–10.

89. He, T. 2001. *Composite Hollow Fiber Membranes for Ion Separation and Removal*, Ph. D. Thesis, U. of Twente, Enschede, The Netherlands.

90. Prasad, R., S. Khare, A. Sengupta et al. 1990. Novel liquid-in-membrane pore configurations in membrane solvent extraction. *AIChE J.* 36: 1592–1596.

91. Prasad, R. and K. K. Sirkar. 1987. Microporous membrane solvent extraction. *Sep. Sci. Technol.* 22(2–3): 619–640.

92. Dahuron, L. and E. L. Cussler. 1988. Protein extraction with hollow fibers. *AIChE J.* 34(1): 130–136.

93. Yun, C. H., R. Prasad, A. K. Guha et al. 1993. Hollow fiber solvent extraction removal of toxic heavy metals from aqueous waste streams. *I&EC Res.* 32: 1186–1195.

94. Kathios, O. J., G. D. Jarvinen, S. L. Yarbro et al. 1994. A preliminary evaluation of microporous hollow fiber membrane modules for the liquid–liquid extraction of actinides. *J. Membr. Sci.* 97:251–261.

95. Ambare, D. N., S. A. Ansari, M. Anitha et al. 2013. Non-dispersive solvent extraction of neodymium using a hollow fiber contactor: Mass transfer and modeling studies. *J. Membr. Sci.* 446: 106–112.

96. Bayer, C., M. Follmann, H. Breisig et al. 2013. On the design of a 4-end spiral-wound L/L extraction membrane module. *I&EC Res.*5 2: 1004–1014.

97. Danesi, P. R., L. Reichley-Yinger and P. G. Rickert. 1987. Lifetime of supported liquid membranes: The influence of interfacial properties, chemical composition and water transport on the long-term stability of the membranes. *J. Membr. Sci.* 31:117–145.

98. Kemperman, A. J. B., B. Damink, T. van den Boomgaard et al. 1997. Stabilization of supported liquid membranes by gelation with PVC. *J. Appl. Polym. Sci.* 65: 1205–1215.

99. Kemperman, A. J. B., H. H. M. Rolevink, T. van den Boomgaard et al. 1998, Stabilization of supported liquid membranes by interfacial polymerization top layers. *J. Membr. Sci.* 138: 43–55.

100. Yang, Q. and N. M. Kocherginsky. 2006. Copper recovery and spent ammoniacal etchant regeneration based on hollow fiber supported liquid membrane technology: From bench-scale to pilot-scale tests. *J. Membr. Sci.* 286: 301–309.

101. Klaassen, R. and A. E. Jansen. 1996. Selective removal of heavy metals with emulsion pertraction. In *Minerals, Metals and the Environment II* (pp. 207–216). London, UK: The Institution of Mining and Metallurgy.

102. Ho, W. S. W. and T. K. Poddar. 2001. New membrane technology for removal and recovery of chromium from waste waters. *Environ Prog.* 20(1): 44–52.

103. Qin, Y., J. P. Sheth and K. K. Sirkar. 2003. Pervaporation membranes that are highly selective for acetic acid over water. *I&EC Res.* 42: 582–595.

104. Sengupta, A., R. Basu and K. K. Sirkar. 1988. Separation of solutes from aqueous solutions by contained liquid membranes. *AIChE J. 34*(10): 1698–1708.

105. Basu, R. and K. K. Sirkar. 1991. Hollow fiber contained liquid membrane separation of citric acid. *AIChE J.* 37: 383–393.

106. Guha, A. K., C. H. Yun, R. Basu et al. 1994. Heavy metal removal and recovery by contained liquid membrane permeator. *AIChE J.* 40(7): 1223–1237.

107. Basu, R. and K. K. Sirkar. 1992. Pharmaceutical product recovery using hollow fiber contained liquid membrane: A case study. *J. Membr. Sci.* 75(1 & 2): 131–150.

108. Papadopoulos, T. and K. K. Sirkar. 1993. Separation of a 2-propanol/n-heptane mixture by liquid membrane perstraction. *I&EC Res.* 32: 663–673.

109. Feng, X. and J. Ivory. 2002. Hollow fiber and spiral wound contactors for fluid/particle contact and interaction. *Chem. Eng. Comm.* 189: 247–267.

110. Lively, R. P., R. R. Chance and W. J. Koros. 2010. Enabling low-cost CO_2 capture via heat integration. *I&EC Res.* 49(16): 7550–7562.

111. Fan, Y., Y. Labreche, R. P. Lively et al. 2014. Dynamic CO_2 adsorption performance of internally cooled silica-supported poly(ethylenimine) hollow fiber sorbents. *AIChE J.* 60(11): 3878–3887.

112. Brandt, S., R. K. Goffe, S. B. Kessler et al. 1988. Membrane-based affinity technology for commercial purifications. *Biotechnology* 6: 779–782.

113. Roper, D. K. and E. N. Lightfoot. 1995. Separation of biomolecules using adsorptive membranes. *J. Chromatography A.* 702: 3–26.

114. Thommes, J. and M. R. Kula. 1995. Membrane chromatography – an integrative concept in the downstream processing of proteins. *Biotechnol. Prog.* 11: 357–367.

115. Hestekin, J. A., L. G. Bachas and D. B. Bhattacharyya. 2001. Poly (amino acid) functionalized cellulosic membranes: Metal sorption mechanisms and results. *Ind. Eng. Chem. Res.* 40: 2668–2678.

116. Zhang, Y., J. R. Vallin, J. K. Sahoo et al. 2018. High-affinity detection and capture of heavy metal contaminants using

block polymer composite membranes. *ACS Cent. Sci.*4: 1697–1707.

117. Kieffer, R., C. Charcosset, F. Puel et al. 2008, Numerical simulation of mass transfer in a liquid–liquid membrane contactor for laminar flow conditions. *Comp. Chem. Eng.* 32: 1333–1341.

118. Kieffer, R., D. Mangin, F. Puel et al.2009. Precipitation of barium sulphate in a hollow fiber membrane contactor: Part I Investigation of particulate fouling. *Chem. Eng. Sci.* 64(8): 1759–1767.

119. Zarkadas, D. M. 2004. *Crystallization Studies in Hollow Fiber Devices*, PhD Thesis, New Jersey Institute of Technology, New Jersey.

120. Zarkadas, D. M. and K. K. Sirkar. 2006. Anti-solvent crystallization in porous hollow fiber devices. *Chem. Eng. Sci.* 61: 5030–5048.

121. Chen, D., D. Singh, K. K. Sirkar et al. 2015. Continuous synthesis of polymer-coated drug particles by a porous hollow fiber membrane-based antisolvent crystallization. *Langmuir.* 31: 432–441.

122. Chen, D., D. Singh, K. K. Sirkar et al. 2015. Porous hollow fiber membrane-based continuous technique of polymer coating on submicron and nanoparticles via anti-solvent crystallization. *I&E Chem. Res.* 54(19): 5237–5245.

3 Membrane Contactor Device Designs – History and Advancements

A. Sengupta and G. P. Taylor

CONTENTS

3.1 INTRODUCTION

Membrane contactors are special membrane 'modules' that are made typically from hollow fiber hydrophobic membranes. The membrane structure can be homogeneous microporous and non-selective or can be asymmetric microporous with a gas-permeable and gas-selective skin. Membrane contactors have found increasing utility as continuous-contact mass transfer devices to bring two immiscible phases in intimate contact without dispersing one phase into another. The two phases in contact could be a liquid and a gas or two immiscible liquids. The way the membrane functions is distinctly different from that of other membrane processes such as reverse osmosis, ultrafiltration, nanofiltration, microfiltration, electrodialysis, and gas–gas separation. No bulk or convective flow of any fluid takes place across the membrane, and the mass transfer between the phases occurs by diffusion, driven by a concentration differential of each species between the two phases. As such, membrane contactors can act as a device for the stripping of a gas species from a liquid, for absorption (scrubbing) of a gas species into a liquid, or for extracting a component from one liquid to another liquid.

Membrane contactor technology has progressed from early academic curiosity and research, mostly beginning in the 1980s, to a point where the technology has found multiple uses in various industries such as microelectronics, food and beverage, steam and power generation, pharmaceutical and biotechnology, imaging and coatings, oil and gas, and lab and analytical. Applications of membrane contactors have been reviewed in multiple handbooks on membrane technology [1–3]. Membrane contacting as a separation technology possesses certain unique benefits that make it cost-effective for some commercial applications, and it has been accepted by customers as a supplement or replacement for existing processes that may or may not be based on membranes. In some situations, the membrane contactor has emerged as an enabling technology that allows new and creative solutions for previously un-met needs.

By most accounts, the earliest use of a membrane contactor was as an extracorporeal membrane blood oxygenator (ECMO) or artificial lung device, as reviewed by Frischer et al. [4]. Typically, customers do not recognize the term "membrane contactor" but know the process by the functional utility of the device. Examples of these utilities are Membrane Oxygenator, Membrane Deaerator (MDA), Membrane Degasifier (MDG), Gas Transfer Membrane (GTM), Membrane Distillation (MD), Membrane Reactor (MR), Membrane Gas Absorber (MGA), Membrane Solvent Extractor (MSE), etc. The adoption of membrane contactors to large scale industrial processes has only advanced due to the commercial availability of scaled up large contactor sizes with a high flow capacity and due to the scalability of the design.

3.2 GOAL OF THIS CHAPTER

The primary intent of this chapter is to cover membrane contactor physical designs and flow configuration aspects over the years. First, the technology and the principles of operation are discussed briefly, with some remarks on the mass transfer process in membrane contactors. This is followed by a description of various types of membrane contactor designs that have been developed over the years. It should be clearly understood that most types of membrane contactor design and manufacturing information have been considered proprietary and confidential over the years and were initially under intellectual property (IP) protection. Some of the IP protections have expired over time, but even today there is substantial information that contains trade secrets and is not disclosed publicly. In this chapter we have necessarily limited ourselves to information that is available in the public literature.

3.3 DESCRIPTION OF MEMBRANE CONTACTOR

Membrane contactors can be made from flat sheet membranes, and there are some commercial applications of this type of device. However, most common commercial membrane contactors are made from small-diameter, microporous hollow fiber (or capillary) membranes with fine pores that span the hollow fiber wall from the fiber inside surface to the fiber outside surface. Figure 3.1 illustrates a single hollow fiber microporous membrane separating two phases – a 'shell side' and a 'lumen side', meaning outside and inside of the hollow fiber, respectively. The hollow fiber membrane is typically made of hydrophobic materials such as polypropylene, polyethylene, PTFE, PFA, PVDF, PMP, etc.

A membrane contactor would incorporate thousands or hundreds of thousands of hollow fibers like that shown in Figure 3.1. Figure 3.2 represents the general schematic of a membrane contactor of fibers in a cylindrical shell (tube-in-shell) configuration, with inlet and outlet ports identified for the two fluids running through the contactor. A drawing of a large commercially manufactured membrane contactor is shown in Figure 3.3. Advancement of the membrane contactor technology would not happen without the development of such large devices and processes to manufacture them.

The membrane in the membrane contactor acts as a passive barrier and as a means of bringing two immiscible fluid phases (such a gas and a liquid, or an aqueous liquid and an organic liquid, etc.) in contact with each other without dispersion. The phase interface is immobilized at the membrane pore surface as illustrated in Figure 3.1, with

the pore volume occupied by one of the two fluid phases that are in contact. Since it enables the phases to come into direct contact, the membrane contactor functions as a continuous-contact mass transfer device such as a packed column, spray column, or bubble column. However, there is no need to physically disperse one phase into the other, or to separate the phases after separation is completed. Several conventional chemical engineering separation processes that are based on mass exchange between phases (e.g., gas absorption, gas stripping, liquid–liquid extraction, etc.) can therefore be carried out in membrane contactors.

There are variations of membrane contactor processes where the two fluid phases do not directly contact each other. In these cases, two separate phases flow on the two sides of the membrane, but the membrane pores are occupied by a stationary phase immiscible with the two flowing phases. The flowing phases could be liquids or gases, whereas the phase in the pores could be a gas or a liquid, and these are often called 'gas' or 'gas gap' membranes or 'liquid' membranes, respectively. In terms of contactor design, there is no fundamental difference between these scenarios and the 2-phase contact mentioned at the beginning of this paragraph. For example, the contactor shown in Figure 3.3 can be used as a gas–liquid contactor or a liquid–gas–liquid contactor.

3.4 PRINCIPLE OF OPERATION

The principle of membrane contactor operation is based on the natural phenomenon of capillary exclusion. When one side of a hydrophobic microporous membrane is brought in contact with water or an aqueous liquid, the membrane is not spontaneously 'wetted' by the liquid, i.e., the liquid is prevented from entering the pores due to surface tension effect. If a dry microporous hydrophobic hollow fiber membrane with air-filled pores were surrounded by water, there would not be any natural intrusion of water into the pores until the water pressure exceeds a certain critical 'breakthrough' pressure. However, the fluids on either side of a membrane can flow at independent flow rates, while still maintaining continuous contact all along the fiber length. The magnitude of the critical breakthrough pressure 'differential' (between two phases) depends on liquid surface tensions, contact angle, and effective pore diameter [5], and can be quite high.

As an example, considering a water–air–polypropylene system, one can calculate that for a dry membrane with a pore size of 0.03 micron (30 nanometers) the critical entry pressure of water is more than 300 psi (> 20 bar). This unique property of the hydrophobic microporous membrane is utilized effectively by the membrane contactor, so that it can operate within a large operating pressure window. Membrane contactor design and engineering take advantage of the high-pressure operation, so that mass transfer can occur in-line under pressure between two independently flowing fluids without dispersion, which is something that competing technologies cannot do. In situations where pore wetting is expected, such as with liquids of low

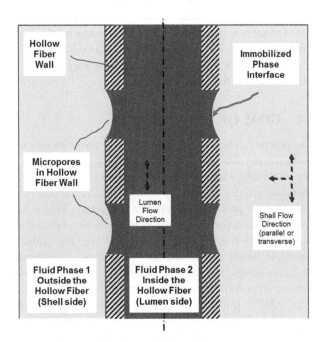

FIGURE 3.1 Single microporous hollow fiber with immobilized phase interface.

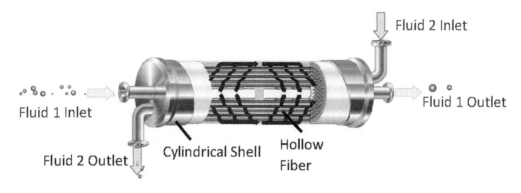

FIGURE 3.2 Membrane contactor of tubular shape: tube-in-shell configuration.

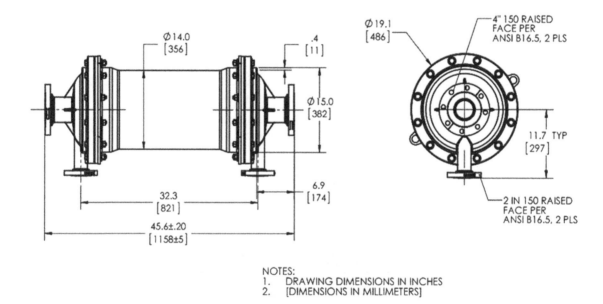

NOTES:
1. DRAWING DIMENSIONS IN INCHES
2. [DIMENSIONS IN MILLIMETERS]

FIGURE 3.3 Drawing of a large commercially manufactured membrane contactor.

surface tension, membrane contactors can be made from dense gas-permeable hollow fiber membranes to accomplish the same benefits.

3.5 BENEFITS OF MEMBRANE CONTACTOR TECHNOLOGY

The special features and resulting benefits that have been driving the increasing adoption of membrane contactor technology in various markets are shown in Table 3.1.

3.6 MEMBRANE CONTACTOR DESIGNS HISTORICAL PERSPECTIVE

Although the membrane is the heart of the membrane contactor principle of operation, the internal design of the contactor 'device' or 'module' is critically important for high mass transfer effectiveness. The internal design dictates how the two phases flow inside the contactor and how the hydrodynamics in each phase is managed. Mass transfer is dependent on the internal hydrodynamics. As the devices become larger to serve bigger commercial-scale process

capacities, the dependence on internal flow management becomes increasingly influential. The device design is also important in developing processes for the large-scale manufacturing of the contactors. In the following section we review various design options investigated over the years.

Designs of membrane contactors with hollow fiber membranes fall into one of two general categories: (a) the primary fluid being treated flows through the inside (lumen) of the hollow fibers, and (b) the primary fluid being treated flows on the outside (shell) of the hollow fibers. Another consideration is the flow direction of the fluid in each phase with respect to the axis of the membrane and with respect to each other. Earliest membrane contactors were of cylindrical shape, with the primary fluid flowing on the lumen side from one end of the fiber to the other and the other fluid flowing on the shell side in parallel direction. This design is therefore called the 'parallel flow' design and is illustrated schematically in Figure 3.4. It is also possible to flow the primary fluid on the shell side in parallel flow configuration, but mass transfer is significantly less efficient and unpredictable.

TABLE 3.1

Features and Benefits of Membrane Contactor Technology

Features	Benefits
Very high active contact area for mass transfer	• Small footprint, profile, and weight of devices and systems • Fits into existing spaces • Ideal for mobile and containerized systems
High efficiency of separation process	• Superior separation performance not always achievable by other competing technologies
No need to disperse or coalesce phases before or after contact	• Extra process step eliminated • More efficient utilization of equipment volume
Flow rates and, to some extent, pressures of contacting phases can be controlled independently	• In-line operation without need to de-pressurize or re-pressurize • Contact area constant irrespective of phase flow rates • More flexible and tolerant to changes in process operating parameters
Modular in nature	• Easier to add system capacity incrementally • Can often be retrofitted into existing systems • Easier scale-up
No chemicals used	• Increased process and operator safety
Rapid process dynamics	• Fast start-up, equilibration, and shut-down
Mass transfer does not depend on gravity	• Contactor can be mounted vertically or horizontally • Contactor will work in microgravity • Able to contact two fluid phases of same densities

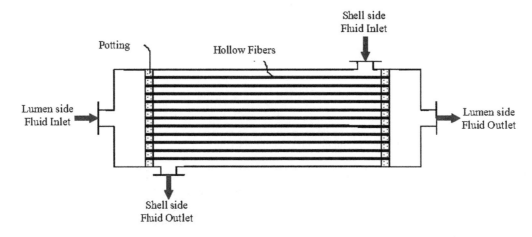

FIGURE 3.4   Membrane contactor – 'parallel flow' design schematic.

The contactors of parallel flow design are relatively easy to manufacture, and fibers are generally encapsulated on ends using centrifugal or static potting methods. Figure 3.5 shows the cutaway view of a typical commercial parallel-flow contactor.

The major drawback of the parallel flow design is that the hollow fibers may not be uniformly spaced. This means that the flow distribution of fluid on the shell side is generally uneven and unpredictable. The shell side often contains 'dead' zones (no flow), leading to flow 'channeling'. This can reduce mass transfer efficiency in the contactor, sometimes severely, and the overall performance of the contactor can vary from unit to unit as well, particularly as the contactor diameter increases. Mass transfer performance can also depend on the orientation of contactor, e.g., vertical or

horizontal. Modeling of mass transfer for the shell side fluid in parallel flow configuration is quite difficult [6]. Because of these issues, most commercial parallel flow contactors are restricted to small size, and scale-up is often not possible.

A significant improvement over this parallel flow design is the 'transverse flow design' where the primary fluid flows on the outside of the hollow fiber membrane at a transverse direction to the fiber axis, and the other fluid flows on the lumen side of the hollow fibers. In another study conducted by Wang and Cussler [7], it was demonstrated that transverse flow on the shell side significantly improves the mass transfer coefficient compared to the parallel flow design. The transverse flow design can be incorporated in contactors that are either rectangular in shape or cylindrical in shape and many

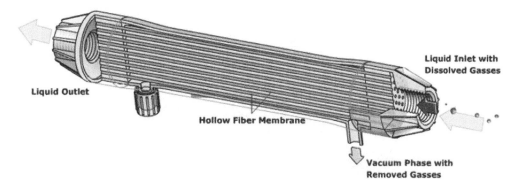

FIGURE 3.5   Cutaway view of a commercial parallel flow contactor for liquid degassing.

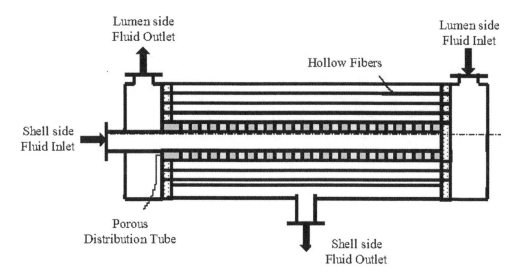

FIGURE 3.6   Possible transverse flow schematic in a membrane contactor of cylindrical shape.

variants are possible. The schematic design of a simple transverse flow contactor of cylindrical shape is shown in Figure 3.6, with the shell side fluid introduced via a core distribution tube to flow radially outward and exiting from the side of the contactor housing. Several commercial contactors in the market today follow such a design. Akasu et al. [8] and Meulen [9] introduced transverse flow membrane contactor designs of rectangular shape (Figure 3.7). Several types of transverse flow contactors with rectangular frame were discussed in [10]. A specific application of a rectangular transverse flow contactor was analyzed by Feron and Jansen [11].

Even though transverse flow on the shell side was proven to provide better mass transfer compared to parallel flow, in practical terms of manufacturing membrane contactors, it was still difficult to ensure that the transverse flow on the shell side was completely uniform along the flow plane, be it a curved plane or a flat plane, and uniform along contactor length. Product developers solved these problems by introducing the concept of a transverse flow contactor of cylindrical shape with flow-directing and flow-reversing internal baffles [12–14]. They also established a method of fabricating such contactors by spirally winding a hollow fiber fabric around a distribution tube and applying multiple glue lines simultaneously. Figure 3.8 schematically represents the design of such a contactor. The cutaway view of a

commercial membrane contactor with this design is shown in Figure 3.9. This is currently the most used design of commercial membrane contactors in the market.

The hollow fiber fabric is a key component of most of these designs. The hollow fibers are knitted or weaved into a planar fabric array (Figure 3.10), which allows for the uniform spacing of fibers and makes the handling of delicate hollow fibers much easier. Uniform fiber-to-fiber spacing in the fabric array substantially improves hydrodynamic efficiency on the shell side and leads to creative new designs to improve the capability of contactors.

The above configurations represented major progress in the design and fabrication method of membrane contactors. The winding of the fabric array around the distribution tube and the simultaneous application of multiple glue lines means that the fabric and fibers remain in place, even as the wound bundle diameter increases in size. The internal baffle design (single or multiple baffles) improved mass transfer efficiency significantly by increasing the shear velocity of fluid across the banks of the hollow fibers. All of these improvements together substantially enhanced the mechanical integrity of the contactor and made the scaling up to large contactor diameter possible without decreasing hydrodynamic efficiency. In addition to design improvement, the contribution of the method of manufacturing

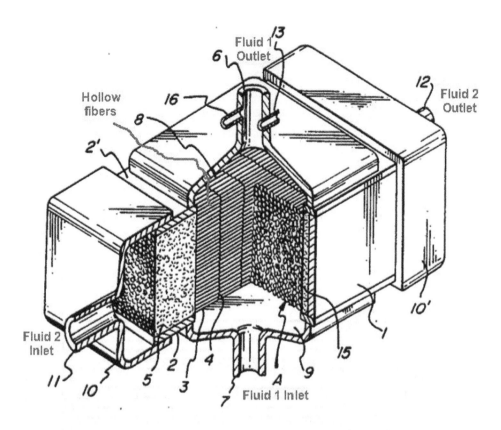

Akasu et al., US Patent 4,911,846 (1990)

FIGURE 3.7   Transverse flow design contactor of rectangular shape.

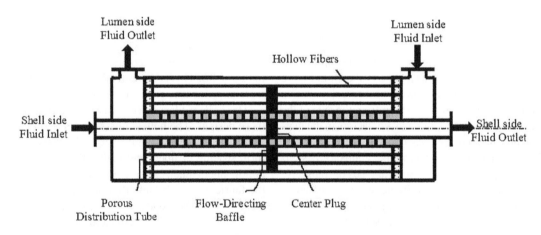

FIGURE 3.8   Transverse flow cylindrical design with flow-directing and flow-reversing baffle.

contactors cannot be overstated. These methods included special components and practical techniques for making large-scale devices at a relatively low cost. Without these, the commercialization of products and the adoption of this technology could not occur.

Yamamura et al. patented a unique baffle design on the gas side of a spiral-wound flat sheet gas permeable membrane module and the method of making the same [15]. This patent does not apply directly to hollow fiber membrane contactor design. However, it is conceivable to use

a variation of this concept to develop unique contactor designs and it is worth mentioning here.

3.7   ADVANCES IN MEMBRANE CONTACTOR DESIGNS AND METHODS OF MANUFACTURE

Many new membrane contactor designs have been proposed and commercialized over the years beyond what was discussed in section 6 above. The drivers for these advances are

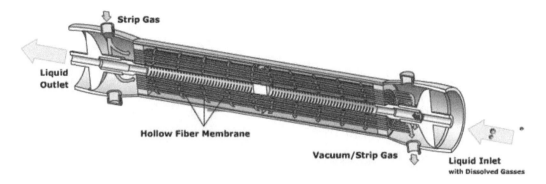

FIGURE 3.9 Cutaway view of a commercial transverse flow tubular contactor for degassing.

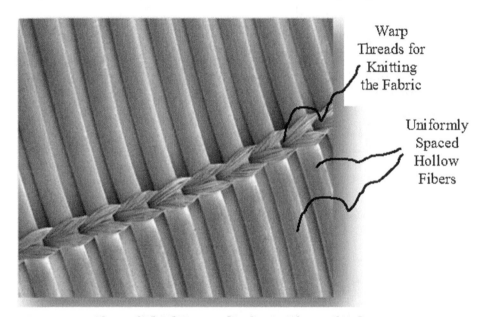

Transverse Flow of Fluid Perpendicular to Plane of Fabric

FIGURE 3.10 Hollow fiber knitted fabric array.

numerous. The design goals include but are not limited to: (1) optimizing the size and shape of contactors for specific uses, (2) creating new designs to expand the functionality of contactors, (3) incorporating new components in contactors to expand their operating capability ranges, (4) changing the manufacturing process to make production more efficient and to reduce the cost of making the contactors, (5) improving contactor design to make the systems built from contactors more compact and efficient, and (6) modifying the contactor components to meet regulatory requirements. Some of these new designs are discussed below in chronological order. As stated earlier, much of the contactor manufacturing processes is proprietary and considered trade secrets. This chapter necessarily restricts itself to publicly available patents and other literature. It is expected that innovative new product designs are being invented and developed at this time that will only become known publicly in future.

Peterson et al. [16] introduced a shell-less contactor design that eliminates the need for individual housing or vessels by directly immersing cartridges in a body of water, which can be a large tank. Two designs are shown in Figure 3.11, with 'E' and 'I' designating the fluid phases external or internal of the hollow fibers, respectively. Several scenarios can be conceived where this would be beneficial since it is not necessary to have a separate flow loop outside of the tank for mass transfer. Instead, the return line for the external phase is emptied within the tank. Phases E and I can flow independently but still contact with each other via the hollow fibers.

Sengupta et al. [17] proposed a device, and the method of making such a device, that can combine two membrane functionalities in a single configuration. In one such embodiment (Figure 3.12), for example, a liquid containing both suspended solids and dissolved gases or bubbles can be treated sequentially using two different types of hollow fiber membranes, first to remove gases or bubbles and second to filter the liquid to produce a clear liquid free of bubbles or gases.

A new 3-port contactor design and method of making such a contactor was introduced by researchers in [18, 19].

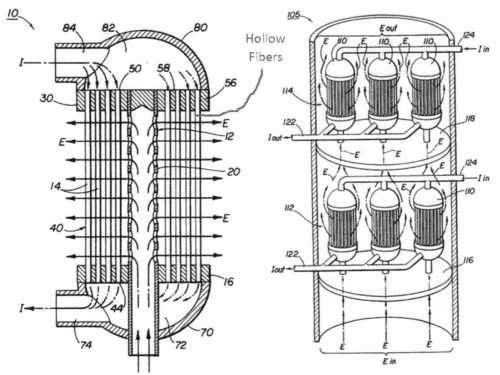

From Peterson et al. US Patent 6,149,817 (2000)

FIGURE 3.11   Contactor design for cartridges without individual housings.

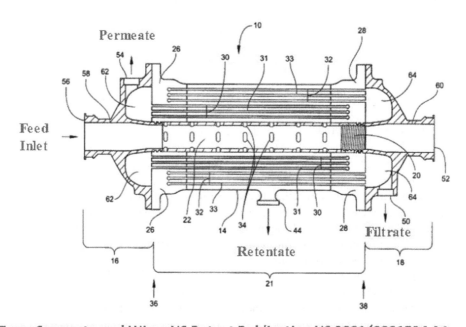

From Sengupta and Wiese US Patent Publication US 2006/0081524 A1

FIGURE 3.12   Bifunctional device such as a filter and a permeator.

This design led to a very compact device and was particularly applicable to smaller contactor sizes without requiring any internal seals. This design was adopted because (a) the design allowed for simpler construction of the device (Figure 3.13), (b) it reduced or minimized possible device failure points during construction, and (c) it minimized pressure drop in the shell side phase. Several commercial products are made today by multiple manufacturers using this design.

A significantly new and innovative contactor design and construction was invented by Taylor et al. [20] where the contactor was made from disc-shaped hollow fiber fabric

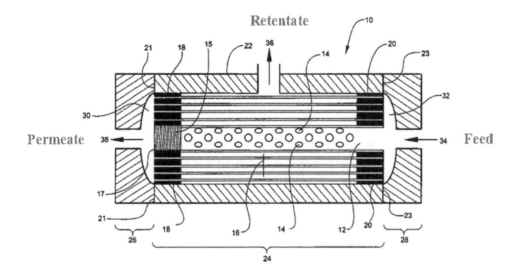

From Sengupta et al. US patent 7,638,049 B2 (2009)

FIGURE 3.13 Three-port contactor design for simpler method of construction.

elements (Figure 3.14). This design provided a large contact area between phases and high mass transfer efficiency in a compact device.

Another novel design of thin rectangular shape, rather than cylindrical shape, was introduced by Taylor et al. [21] as illustrated in Figure 3.15. The advantages of this construction were that the design could be miniaturized, and the thinness allowed such a device to be accommodated in very tight spaces. This expanded the utility of membrane contactors in new application spaces that were not previously accessible. This design has been translated to commercial products. Figure 3.16 shows a photograph of

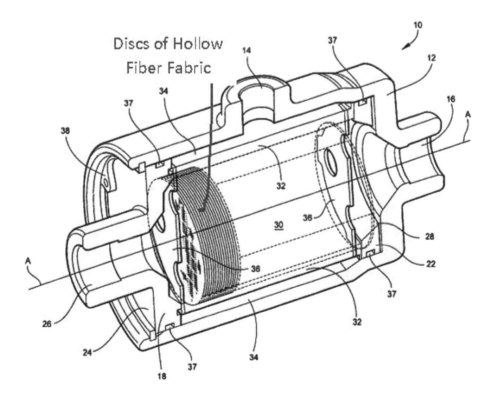

From Taylor et al. US Patent 7,628,916 B2 (2009)

FIGURE 3.14 Contactor design with disc shaped elements.

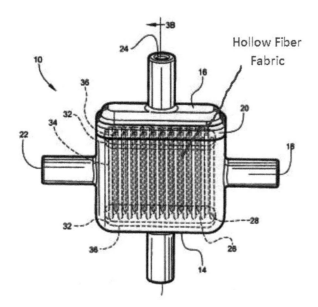

Taylor and Price US Patent 7,641,795 B2 (2010)

FIGURE 3.15 Rectangular and flat shaped contactor design.

such a commercial product used in a degassing application, as well as a cutaway view of the contactor.

In some special applications of membrane contactors, such as described by Majumdar [22], it was necessary to incorporate two sets of membranes in a single contactor. However, fabricating such a contactor was very difficult. Taylor et al. [23] invented a method of fabrication of such a device from two separate hollow fiber fabric rolls. Such a device has up to six fluid ports and three different flow channels through which three fluid streams can flow independently (Figure 3.17). While there is currently no known commercial contactor of this design, it is mentioned here to illustrate the many creative possibilities of contactor design.

A 'wafer' shaped contactor design was introduced by Taylor et al. [24]. This was the first such design of its type.

The contactor could be inserted within a straight piping section in between two standard bolted flange connections (Figure 3.18) and therefore would occupy much less space and have much better mechanical support relative to a contactor built as a separate device. The contactor can be fabricated from hollow fiber fabric disks spaced in parallel with an annular potting. The patent describes several possible uses for such a contactor.

Most commercial membrane contactors are rated for relatively low pressure (<10 bar). A major leap in contactor design was introduced by Taylor and Sengupta in two patents [25, 26] that took advantage of the high-pressure capabilities of commercially available cylindrical Reverse Osmosis (RO) vessels. The design let a cartridge made from hollow fiber fabric array be installed inside a commonly available RO vessel and sealed in such a way that the configuration resembled a dual-walled pipe. The pressure tolerance of the assembled embodiment was the same as that of a well-validated code-rated RO vessel (Figure 3.19). The novelty of this design led to multiple benefits. By co-opting the high-pressure RO vessel, the pressure rating of the contactor could be raised substantially to levels that were hitherto unattainable on a commercial scale. The design allowed multiple cartridges to be installed in a single housing and effectively increased the number of baffles inside the contactor, thus enhancing mass transfer rates. Further, the RO vessels could be connected laterally, resulting in a highly compact skid-built system with a lower footprint, height, and weight relative to competing technologies. The three individual sketches in Figure 3.19, clockwise from top left, schematically represent the three key novelties.

Taylor et al. developed a 'Flat Panel' contactor resembling a HVAC air filter (Figure 3.20) of cuboid shape [27]. The contactor was made from layers of hollow fiber rectangular fabric potted on two ends. The main feature of the design is a high length-to-thickness ratio with two large open faces and two solid sides.

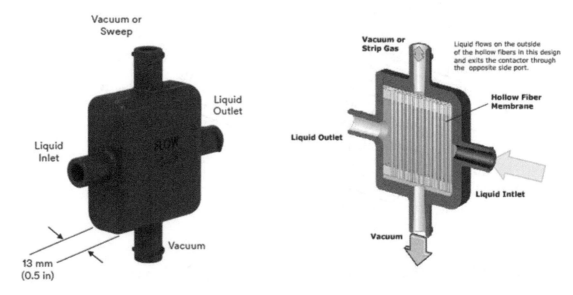

FIGURE 3.16 Transverse flow rectangular contactor for degassing and a cutaway view.

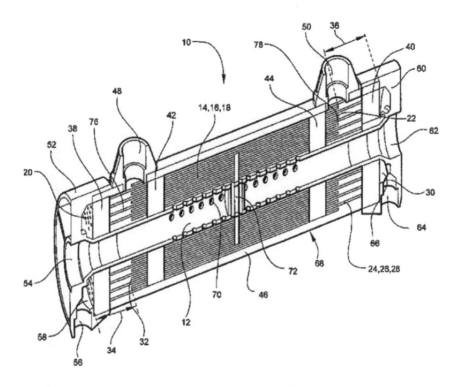

From Taylor et al. US Patent 7,803,274 B2 (2010)

FIGURE 3.17   Design of contactor from dual fabric layers.

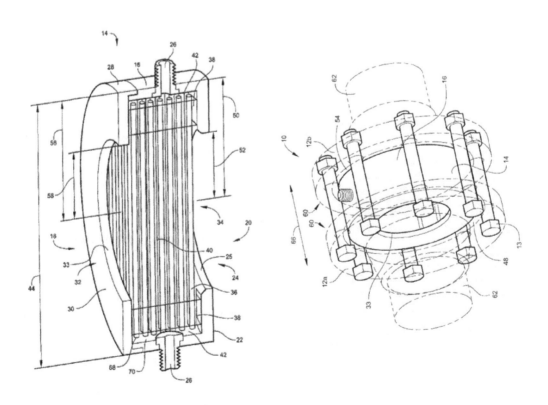

From Taylor and Sengupta US Patent 8,158,001 B2 (2012)

FIGURE 3.18   Wafer-shaped contactor design for inline use.

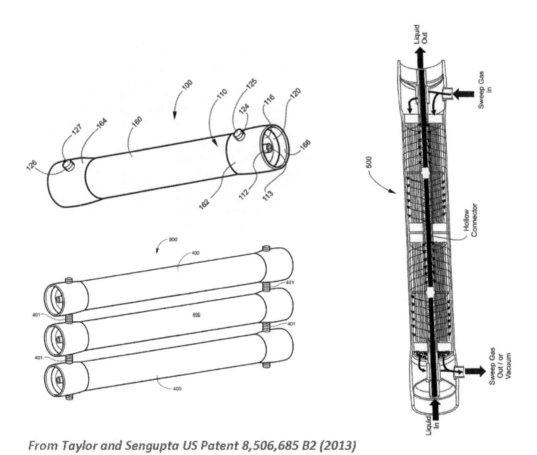

From Taylor and Sengupta US Patent 8,506,685 B2 (2013)

FIGURE 3.19   Contactor design for high pressure tolerance.

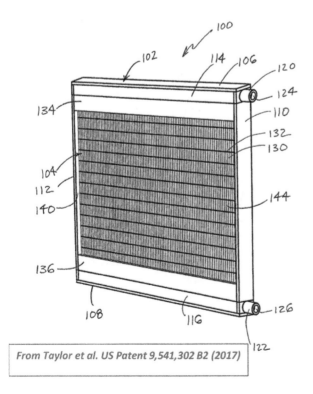

From Taylor et al. US Patent 9,541,302 B2 (2017)

FIGURE 3.20   Flat panel contactor design.

One suitable use for this design is gas–liquid contacting where there is a need to process a gas stream at a high flow rate. The gas phase can be run across the large open faces of the contactor. A large gas flow area and small flow length lead to a low gas side pressure drop, a key requirement in such large air-handling processes. The ends of the potted fibers open into manifolds on each side. The liquid phase can be introduced into one manifold, and liquid flows through the inside of hollow fibers (lumen side) while the gas phase flows outside the fibers in the transverse direction.

3.8 CONCLUDING REMARKS

In the preceding sections, we have attempted to provide a historical perspective of how membrane contactor design and manufacturing has evolved over the years. Membrane contactors will be utilized in increasing frequency in various industries and markets in the coming years. We fully expect that new breakthroughs will be achieved in the future with respect to creating novel designs, introducing new material components, and inventing new manufacturing processes to meet the needs for specific use applications, and to make membrane contactors more economically competitive relative to traditional technologies. We have covered publicly accessible information about membrane contactors and hope that the information provided here will be of utility for scientists and engineers at the cutting edge of this technology.

REFERENCES

* The authors' references to the various patents mentioned in this article do not constitute a grant of a license to practice any of these technologies, nor do they imply the authors' acknowledgment of the validity of any of the referenced patents.

1. K. K. Sirkar, 'Other new membrane processes', in *Membrane Handbook*, Eds. W.S.W. Ho and K.K. Sirkar, Van Nostrand Reinhold, New York (1992), pp. 885–912.

2. B. W. Reed, M. J. Semmens, and E. L. Cussler, 'Membrane contactors', in *Membrane Separation Technology – Principles and Application*, Eds. R. D. Noble and S. A. Stern, Elsevier, Amsterdam (1995), pp. 467–498.

3. A. Sengupta, and R. A. Pittman, 'Application of Membrane Contactors as Mass Transfer Devices', in *Handbook of Membrane Separations*, Eds. A. K. Pabby, S. S. H. Rizvi, and A. M. Sastre. Boca Raton, FL: CRC Press Taylor & Francis Group (2009), pp. 7–24.

4. J. S. Frischer, and C. J. H. Stolar, 'Extracorporeal membrane oxygenation', in *Ashcraft's Pediatric Surgery* (Fifth Edition), Eds. G. Holcomb and J. Patrick Murphy. Philadelphia, PA: Saunders Elsevier (2010).

5. B. S. Kim, and P. Harriott, Critical entry pressure for liquids in hydrophobic membranes, *J Colloid Interface Sci* (1987), 115 (1), 1–8.

6. M. J. Costello, A. G. Fane, P. A. Hogan, and R. W. Schofield. The effect of shell side hydrodynamics on the performance of axial flow hollow fiber modules. *J Membr Sci* (1993), 80, 1–11.

7. K. L. Wang, and E. L. Cussler. Baffled membrane modules made with hollow fiber fabric. *J Membr Sci* (1993), 85, 265–278.

8. H. Akasu, R. Mimura, T. Migaki, T. Yamauchi, and M. Kusachi, "Fluid treating apparatus and method of using it", US Patent 4,911,846, (1990).

9. B. P. T. Meulen, "Transfer device for the transfer of matter and/or heat from one medium flow to another medium flow", US Patent 5,230,796 (1993).

10. G. T. Vladisavljevic, and M. V. Mitrovic, Pressure drops and hydraulic resistances in a three-phase hollow fiber membrane contactor with frame elements, *Chem Eng Process* (2001), 40, 3–11.

11. P. H. M. Feron, and A. E. Jansen, CO2 separation with polyolefin membrane contactors and dedicated absorption liquids: performances and prospects, *Sep Purif Technol* 2002, 27, 231–242.

12. R. Prasad, C. J. Runkle, and H. F. Shuey, "Method of making spiral-wound hollow fiber membrane fabric cartridges and modules having flow-directing baffles", US Patent 5,264,171 (1993).

13. R. Prasad, C. J. Runkle, and H. F. Shuey, "Spiral-wound hollow fiber membrane fabric cartridges and modules having flow directing baffles", US Patent 5,352,361 (1994).

14. X. Huang, J. C. Delozier, R. Prasad, C. J. Runkle, and H. F. Shuey, "Hollow fiber membrane fabric - containing cartridges and modules having solvent-resistant thermoplastic tube sheets, and methods for making the same", US Patent 5,284,584 (1994).

15. H. Yamamura, H. Ikada, Y. Toyoda, K. Nishimura, and K. Imai, "Spiral wound gas permeable membrane module and apparatus and method for using the same", US Patent 5,154,832 (1992).

16. P.A. Peterson, C. J. Runkle, A. Sengupta, and F. E. Wiesler, "Shell-less hollow fiber membrane fluid contactor", US Patent 6,149,817 (2000).

17. A. Sengupta, and F. Wiese, "Membrane contactor and method of making the same", US Patent Application Publication US 2006/0081524 A1 (2006).

18. A. Sengupta, L. I. Holstein, and E. W. Bouldin, "Three-Port high performance mini hollow fiber membrane contactor", US Patent D538,749 S (2007).

19. A. Sengupta, L. I. Holstein, and E. W. Bouldin, "Three-Port high performance mini hollow fiber membrane contactor", US Patent 7,638,049 B2 (2009).

20. G. P. Taylor, R. H. Carroll, T. R. Vido, and T. D. Price, "Hollow fiber module", US Patent 7,628,916 B2 (2009).

21. G. P. Taylor, and T. D. Price, "Membrane contactor", US Patent 7,641,795 B2 (2010).

22. S. Majumdar, K. K. Sirkar, and A. Sengupta, 'Hollow-fiber contained liquid membrane', in *Membrane Handbook*, Ed. W.S.W. Ho and K.K. Sirkar, Van Nostrand Reinhold, New York (1992), pp. 764–808.

23. G. P. Taylor, A. Sengupta, T. D. Price, L. I. Holstein, and T. R. Vido, "Contained liquid membrane contactor and a method of manufacturing the same", US Patent 7,803,274 B2 (2010).

24. G. P. Taylor, and A. Sengupta, "Wafer-Shaped Hollow Fiber Module for In-Line Use in a Piping System", US Patent 8,158,001 B2 (2012).

25. G. P. Taylor, and A. Sengupta, "High Pressure Liquid Degassing Membrane Contactors and Methods of Manufacturing and Use", US Patent 8,506,685 B2 (2013).

26. G. P. Taylor, and A. Sengupta, "High Pressure Liquid Degassing Membrane Contactors and Methods of Manufacturing and Use", US Patent 8,778,055 B2 (2014).

27. G. P. Taylor, T. D. Price, A. Sengupta, P. A. Peterson, and C. G. Wensley, "Flat panel contactors and methods", US Patent 9,541,302 B2 (2017).

Part II

Chapters on Gas–Liquid Contacting

4 Advances in Hollow Fiber Contactor-Based Technology in Gas Absorption and Stripping

P. S. Goh, R. Naim, A. F. Ismail, B. C. Ng, and M. S. Abdullah

CONTENTS

4.1   INTRODUCTION

Gas absorption and stripping are important industry processes for gas mixture separation, impurities removal, and valuable chemical recovery. With the increasing negative effects imparted from the release of greenhouse gases, one of the most widely applied absorption techniques is related to carbon capture from coal-fired power plants. Among the available options, chemical and physical absorption have been well established and commonly used to separate and capture gases from the low-pressure flue gas streams of large scale power plants as well as in the oil, natural gas, and cement industries [1]. However, these techniques have also suffered from several inherent disadvantages that are associated with the requirement of a large footprint, high capital costs, and inefficiencies rooted in the various operational problems. These disadvantages have become some of the critical issues to be concerned with and addressed.

Membrane contactor technology is an attractive alternative to overcome the challenges posed by these conventional technologies owing to the high chemical selectivity and modularity, and compactness of membrane technology–based units. Membrane contactor technology is a hybrid technology that integrates the unique features of membrane separation and the chemical absorption process [2]. Due to the principle of membrane contactor, the system can be feasibly operated with almost no pressurization at a very low level of energy consumption [3]. The absorption through a membrane contactor can offer a high driving force for transport at interface even at trace concentration. The membrane modules with high packing density allow for the maximization of the gas/liquid contact area for efficient absorption. As such, the equipment size can be drastically reduced. The modular concept of membrane technology also allows easy scale-up [4]. In a typical gas–liquid contactor, the porous hydrophobic membrane acts as a non-selective interface between the gas and the liquid absorbent in which the gases are facilitated to pass through the membrane pore and be absorbed into the solution. The liquid absorbents are normally characterized by high surface tension, and low vapor pressure, and possess high CO_2 sorption capacity. Aqueous alkanolamines such as monoethanolamine (MEA) and diethanolamine (DEA) have been commercially used for CO_2 absorption.

Due to the chemical reaction between CO_2 and the absorbent solution, the partial pressure at the permeate side is always close to zero. The transport of CO_2 from the feed side to the absorbent side can be well sustained until the solvent is fully saturated. Hence, compared to the conventionally used membrane gas separation, this principle has tackled the issue related to the pressure ratio problem as the trans-membrane pressure difference required in membrane contactors is much lower. The system design and module fabrication of membrane contactors are also relatively simpler. In terms of gas diffusion mechanisms, it is known that the diffusivity in the gas phase is much higher than in the solid phase, hence a higher flux per unit area can be favorably achieved by membrane contactors. In the case of membrane development, some dense gas separation membranes have been attempted for gas–liquid membrane contactors[5]. However, the gas separation performance is not up to expectation and is not suitable for industrial scale. This is mainly due to the fact that gas–liquid contact happens at the mouths of the membrane pores, hence the creation of mesoporous membrane structure is important to ensure high mass transfer.

A number of membrane configurations such as flat sheet, spiral wound, and hollow fibers have been made possible for membrane contactors. Among all, hollow fiber configurations have received much attention. Hollow fiber membranes offer large surface area/volume to enhance the absorption capacity [6]. The choice of material for membrane contactor application is of primary concern in the system development as membranes play a critical role in maintaining the performances of the process. In general, the membrane materials for gas absorption should possess high porosity to maximize mass transfer and should have low wetting tendency with excellent chemical and thermal resistance. A critical issue facing the development of hollow fiber membranes for gas absorption and stripping through the membrane contactor process is the wetting of membranes which could lead to a substantial increase of mass transfer resistance [7].

The liquid entry pressure (LEP) is a parameter to evaluate the tendency of the solvent to cause membrane wetting. At a trans-membrane pressure greater than the LEP, the absorbent solution can penetrate in the membrane pores and form a stagnant layer. This layer deteriorates the molecular diffusion of the dissolved gases. Various approaches have been established to address this issue by improving the hydrophilicity of the membranes or by developing membranes that are highly resistant to solvent penetration. Generally, two main concepts can be adopted to bestow the membrane with high hydrophobicity and excellent wetting resistance, namely membrane surface morphological alteration and the introduction of hydrophobic groups through various modifications. The former involves the tailoring of dope formulation and membrane fabrication parameters to increase the porosity and roughness of the membranes, while the latter involves the post-modification of the membrane surface to render super hydrophobicity.

In this chapter, the advances in hollow fiber membrane contactors for gas absorption and stripping are presented. The types and fabrications of hollow fiber membranes are first discussed. In the sections that follow, the modifications of hollow fiber membranes for membrane contactor application are reviewed. Next, the performance of hollow fiber membrane contactor systems are evaluated and discussed thoroughly. Finally, the future perspective is highlighted and the conclusion is drawn.

4.2 HOLLOW FIBER MEMBRANE CONTACTORS

4.2.1 MEMBRANE MATERIALS

The hydrophobic membrane used in gas stripping and absorption plays a role in preventing the mixing of the gas stream and liquid absorbent stream. A wide range of hydrophobic polymeric and ceramic membranes has been used for this purpose. Some examples of polymeric membrane materials are polypropylene (PP), polyvinylidene difluoride (PVDF), and polytetrafluoroethylene (PTFE). The membranes made from these polymers can be fabricated through facile membrane fabrication techniques such as phase inversion. They have been widely used for membrane contactors owing to their excellent chemical and thermal resistances. These membranes are also found to be stable in a wide range of corrosive chemicals and organic compounds such as acid and alkali. The comparison made among PP, PVDF and PTFE suggested that the polymeric materials with higher hydrophobicity could achieve higher absorption rates [8]. Among these, microporous polymer materials, PVDF showed the best physical absorption rate. However, when tested for chemical adsorption using MEA aqueous solutions, it was found that PTFE showed the highest efficiency, followed by PVDF and PP. PTFE also has the greatest long-term stability among these hydrophobic materials in which the PTFE hollow fiber membranes used for gas absorption could maintain their stable performance in operation up to 60 hours. Fashandi et al. reported the fabrication of porous poly(vinyl chloride) (PVC) hollow fiber membranes for CO_2 absorption for the first time [9]. Owing to the compelling advantages of PVC material such as good processability, high chemical resistance and low cost that are comparable to commercially available PP and PVDF, PVDF hollow fiber membranes can be feasibly used in gas–liquid membrane contactors for carbon capture.

One major limitation of contactor systems based on polymeric membranes is the tendency of polymeric material to be wetted by the aqueous amine solution which consequently results in the change of membrane structure and morphology. The common absorbents used in membrane contactors for gas absorption are organic solvents such as N- methyl-2-pyrrolidone (NMP) and diethylene glycol (DEG), which have low surface tensions. The hydrophobic polymeric surfaces are characterized by poor chemical resistance and hence are susceptible towards liquids with low surface tension and organic contamination. The vulnerability has prompted the seeping of the solution into the membrane pores and gradually led to membrane wetting. The unfavorable changes have affected the integrity of the system for long-term operation. The porous structure and surface roughness can be greatly altered and hence lead to a reduction in the gas absorption capacity [10]. The most straightforward strategy to tackle this issue is to prevent the penetration of the solvent into the polymeric membrane. Currently, efforts have been focused on minimizing the impacts of wetting by fabricating composite hollow fiber membranes and self-standing polymeric membranes that are highly permeable towards the targeted gas but impermeable towards the liquid absorbent [11]. The thin film composite is designed to have a dense layer that is highly permeable to the targeted gas permeation supported by a typically used microporous substrate [12]. The dense layer with high gas affinity is used to maintain the mass transfer coefficient of the membranes. It has been reported that, when thin film composite hollow fiber membranes were used, membrane wetting was not observed upon contact with MEA aqueous solutions, and the diffusion rate of CO_2 remained high as the membrane pores were free from solvent because the selective layer promoted high CO_2 flux but only permitted very low solvent flux [13]. Chabanon et al. performed a long term study to compare the performance of commercially used membrane contactor fibers, i.e., PP and PTFE with composite fibers fabricated by coating dense polymethylpentene (PMP) and Teflon-AF on microporous PP that act as the support [13]. The desired composite structure has combined high mass transfer performances and high specific interfacial areas for gas–liquid absorption processes. It was found that the composite hollow fiber membranes with dense skin structure demonstrate stable performance even after the operation of 1,000 h for CO_2 absorption using a 30% MEA aqueous solution. The PMP and Teflon-AF thin films have provided a protection to prevent membrane wetting and have also effectively protected the porous PP support from the solvent.

Membrane degradation is another major issue of concern in gas–liquid absorption which is normally based on chemical absorption. In many cases which involve the absorption of acidic gases such as CO_2 or SO_2, the alkanolamine absorbent stream becomes more corrosive after the absorption process due to the formation of acidic products from the reaction between CO_2 and the absorbents [14]. These products tend to affect the membrane chemical structure and gradually lead to membrane degradation. When compared to polymeric materials, ceramic membranes are advantageous due to their chemical and thermal resistance. The overall structure and surface morphology can be well maintained upon contact with the liquid solution. Hence, ceramic membranes have also been widely studied for gas–liquid membrane contactors for the separation of various gases. The outstanding thermal and chemical stability makes ceramic membranes suitable to be used for post-combustion CO_2 capture. As the porosity and pore sizes of membranes can directly influence the mass transfer efficiency, the current development of ceramic membranes is focused on the fabrication or modification of ceramic hollow fiber membranes with high porosity. Another critical issue in using ceramic materials for membrane contactors is the high hydrophilicity of the ceramic membranes due to the presence of surface hydroxyl (–OH) groups. The abundant hydroxyl groups have accelerated the wetting of membrane pores. Using surface modification to introduce hydrophobic entities is a subject of interest in the development of high performance membrane contactors based on ceramic membranes. Abdulhameed et al. prepared hydrophobic ceramic hollow fiber membranes by adding alumina to pure kaolin and followed by FAS grafting [15]. The sintering reaction between kaolin and alumina produced mullite that contributed to the desired characteristics in terms of high thermal and mechanical strength as well as good stability under harsh chemical contact. The successful FAS surface grafting was indicated by the increase in the water contact angle to 140°.

4.2.2 Module Design

A low weight and compact gas–liquid hollow fiber membrane contactor is capable of removing various gas from gas streams based on the variations in gas flow and concentrations of the components. The characteristics and performance of hollow fiber membrane contactors are governed by many factors such as the operating conditions of the gas–liquid system and the properties of the membranes. One of the earliest module designs of polymeric hollow fiber membrane contactors is the Liqui-Cell Extra-Flow module, which has been commercialized by CELGARD LLC [16]. In this module, the hollow fibers typically have an inner diameter of 240 μm and a wall thickness of approximately 30 μm. The microporous polypropylene hollow fibers are woven into a fabric which is then wrapped around a central tube feeder which is responsible for shell side fluid supply. More uniform fiber spacing is provided by the woven fabric thus leading to higher mass transfer coefficients. The baffle positioned at the center of the module could facilitate velocity that is normal to the fibers and minimize shell side bypassing. Figure 4.1 shows the hollow fiber membrane modules with both streams flowing parallel to each other where the fluid can flow concurrently (same direction) or counter currently (opposite direction). Such parallel flow maximizes the interfacial area of the hollow fiber membranes hence significantly reducing the volume of the unit operation. The hollow fiber with parallel flow can be easily manufactured. Furthermore, the fluid dynamics and mass transfer have been well studied and estimated using various computational and experimental approaches [17].

A study conducted by de Montigny et al. pointed out that the module in counter-current mode can achieve a higher mass transfer rate that is up to 20% higher compared to that of the co-current mode [18]. They also observed that the mass transfer could be further improved by flowing the liquid phase in the lumen side of the membrane contactor. However, the pumping cost of the liquid phase will be a major concern especially for the hollow fibers with a small diameter that is in the range of a few hundred microns. It is worth mentioning that the presence of the membrane is responsible for the inevitably increased resistance to the mass transfer process. This resistance has negative impacts on the selectivity and overall removal efficiency. The resistance can be reduced by controlling the thickness of the surface porosity of the membranes. In general, thin hollow fiber membranes with a high surface porosity are favorable to deal with this issue. Additionally, the features of hollow fiber membrane modules in terms of the regularity of hollow fibers, packing density, and the relative flow directions also play critical roles in dictating the efficiency of the adsorption process. Study has shown that in typical hollow fiber membrane module packing, the non-uniform membrane distribution has unfavorably contributed to fluid channeling and bypassing, which in turn resulted in the severe pressure drop and reduction in mass transfer [6].

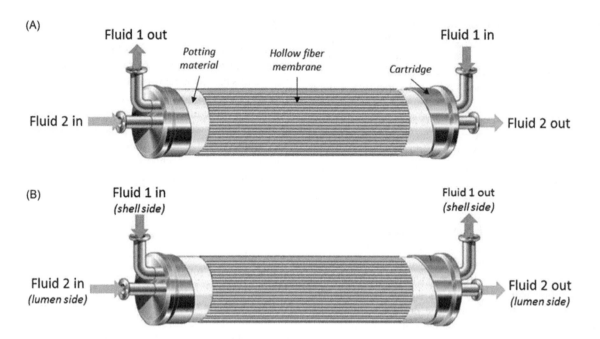

FIGURE 4.1   Hollow fiber membrane contactor modules with (a) counter-current flow and (b) concurrent flow. Copyright (2017) with permission from Elsevier [19].

In a membrane contactor, the absorption and desorption rates are mainly controlled by the mass transfer in gas and liquid phases. The determination of the mass transfer coefficient via diffusion of the gas and liquid phase is crucial to estimate the efficiency of the contactor itself. The overall mass transfer coefficient (K_{ov}) can be estimated by considering the gas, membrane, and liquid mass transfer coefficient, and Henry's law constant. Each of the mass transfer coefficients is a function of the system geometry, fluid properties, and flow velocity. It can be expressed in a series model represented as at [20]:

$$\frac{1}{K_{ov}d_i} = \frac{1}{E k_l d_i} + \frac{1}{H k_m d_{ln}} + \frac{1}{H k_g d_o} \quad (4.1)$$

where K_{ov} is the overall mass transfer coefficient, and k_l, k_m, and k_g are the liquid, membrane, and gas side mass transfer coefficient, respectively. H is described by Henry's constant and E is the enhancement factor, which indicated the influence of a chemical reaction on the mass transfer rate. d_o, d_i, and d_{ln} are the outer, inner, and log mean diameter of the fiber. In hollow fiber membranes, Leveque's correlation is a common equation used to describe the tube side mass transfer. As the liquid absorbent flow inside the hollow fiber is assumed laminar, the mass transfer coefficient in the tube can be estimated from Leveque's correlation [21]:

$$Sh = \frac{k_l d_i}{D_L} = 1.62 \left(\frac{d_i}{L} Re \cdot Sc \right)^{0.33} \quad (4.2)$$

where L is effective length of membrane (m) and D_L is the diffusion coefficient of CO_2 in the liquid phase (m²/s). When a porous membrane with non-wetted mode is assumed, k_m can be described using Fick's law [22]:

$$k_m = \frac{D_{g,eff}\varepsilon}{\delta\tau} \quad (4.3)$$

where $D_{g,\ eff}$ is effective diffusion coefficient of gas in the gas-filled membrane pores and δ, ε, and τ are thickness, porosity, and tortuosity of the membrane, respectively. The gas side mass transfer can be estimated by using the Cussler-Yang equation, which is valid for fluid flows at low Reynolds number [21]:

$$Sh = \frac{k_g d_h}{D_{i,g}} = 1.25 \left(Re \frac{d_h}{L} \right)^{0.93} \left(Sc \right)^{0.33} \quad (4.4)$$

where d_h is hydraulic diameter (m) and L is active length of the hollow fiber membrane (m).

4.3 MODIFICATION OF HOLLOW FIBER MEMBRANES FOR GAS ABSORPTION AND STRIPPING

4.3.1 INCORPORATION WITH ADDITIVES AND NANOMATERIALS

The addition of non-solvent additives in the polymer dope is an easy and efficient method to alter the thermodynamic and kinetic properties during the membrane formation through phase inversion process. Hence it is considered to be a feasible approach to modify the membrane properties in terms of the structure and morphology.

In general, the incorporation of additives and nanomaterials is aimed at achieving two purposes, i.e., (i) to improve the porosity, pore size, and morphology in order to enhance the mass transfer properties at the gas–membrane–liquid interphase; (ii) to improve the surface hydrophobicity of membranes in order to reduce the tendency of membrane wetting. The first purpose can be achieved through the addition of non-solvent additives such as lithium chloride (LiCl), polyerthylene glycol (PEG), phosphoric acid, and the combination of these additives into the polymer dope prior to the hollow fiber spinning. The polymer-solvent-non-solvent system can be altered by the changes in dope viscosity upon the addition of additives with different concentration. The change in viscosity affects the thermodynamic and kinetic behavior of the polymer-solvent-non-solvent additive ternary system [23]. Typically, the increasing concentration of polymer dope can delay the mutual diffusion between solvent and non-solvent. As a result, more sponge-like but less finger-like structures are formed in the porous zone. Thus, the mean pore size of the membrane outer surface is reduced which in turn increases the liquid entry pressure [24].

An interesting strategy to modify the hydrophobicity of a membrane surface involves the use of a relatively new class of surface modifier known as the surface modifying macromolecule (SMM) [25]. SMMs are hydrophobic macromolecules possessing amphipathic structure. The backbone of the SMM is polyurea or polyurethane polymer which contributes to the hydrophilicity of the structure. Both termini of SMMs are end-capped with the hydrophobic fluorine-based polymer chains with low polarity. When the SMM is physically mixed with the polymer dope and cast or spun into a membrane, the low surface energy characteristic of the molecules promotes their migration to the membrane/air interface. With the decrease of the interfacial energy, the surface properties of the resultant membranes can be altered. Bakeri et al. improved the hydrophobicity of polyetherimide hollow fiber membranes by introducing SMMs into the spinning dope and applied for CO_2 absorption in a contactor system. The characterizations showed that the mean pore size and effective surface porosity of the SMM modified membranes had been greatly increased [25].

The progress in nanoscience and material has opened opportunities for the development of a new class of innovative nano-enabled membranes. The direct incorporation of inorganic nanomaterials in the polymeric dope solution represents a simple and diverse method to modify the properties of gas absorption membranes. With the presence of nanomaterials such as clay and silica nanoparticles, the surface and cross-sectional morphological properties of the so-called mixed matrix membranes can be favorably altered in terms of their pore structures, porosity, and hydrophobicity [4]. Nanomaterials with hydrophobic features are of interest in the modification of the surface morphology of polymeric membranes. The first attempt at applying hydrophobic nanomaterials to membrane contactors was made by Razaei et al. [26]. The authors incorporated hydrophobic montmorillonite (MMT) into PVDF polymer matrix for CO_2 absorption. The incorporation of MMT nano clay has disrupted the thermodynamic stability of polymer dope in which the exchange rate of solvent/non-solvent during phase inversion has been sped up. As a result, a more porous skin layer was formed which was advantageous in

reducing mass transfer resistance in the membrane contactor process. The enhancement in surface hydrophobicity of the nano-enabled membranes with MMT loading up to 5 wt% has also increased the critical entry point of water up to 11 bar.

4.3.2 SURFACE MODIFICATION

Attaining membrane surface modifications through techniques such as sol-gel coating, deposition, plasma modification, and polymer grafting is a straight forward approach to introducing new functionalities on the membrane surface to endow wetting resistance and long-term stability to the modified membranes.

The coating of ceramic hollow fiber membrane surfaces with silica aerogel has been extensively reported [27–30]. Mesoporous and hydrophobic silica aerogels can be directly synthesized from various silica-based precursors. For instance, methyltrimethoxysilane (MTMS) precursors have been used to form polymethylsilsesquioxane (PMSQ) aerogels. Silica aerogel is an attractive material for membrane contactor application as it has large specific surface areas, high porosities, and a low density. It has been reported that silica aerogel coating with a pore size in the range of 3–4 nm can effectively prevent membrane wetting by aqueous amine adsorbents hence maintaining the CO_2 absorption flux. It is also known that the pore sizes of silica aerogels can be enlarged by the alkyl-chain bridges of the silane agent. Lin et al. prepared silica aerogel membrane using bis(trimethoxysilyl)hexane (BTMSH), an alkyl-linked bis-silanes, to increase the pores of the aerogel membranes so that the CO_2 gas can easily pass through and reduce the mass transfer resistance for CO_2 absorption [28]. The grafting of fluoroalkylsilane (FAS) with hydrophobic -CF_3 functional groups on the membrane surface is one of the most applicable techniques to enhance the surface hydrophobicity. This modification has been feasibly performed for ceramic membranes as the –OH groups present on the ceramic materials such as alumina, zirconia, and titania can readily react with the ethoxy groups found in the organosilane. In a recent study conducted by Lin et al. [27], triepoxy linker was added to the aerogel to improve the mechanical strength of the membrane for CO_2 absorption in the membrane contactor system shown in Figure 4.2(a). The newly developed FAS-modified tri-epoxy cross-linked aerogel depicted in Figure 4.2(b) was rendered with enhanced surface hydrophobic upon FAS modification.

Surface modification can be done through chemical reaction where the polymeric membranes are etched in a mixture of various inorganic acids and oxidizing agents as well as the mixture of both. Some commonly used reagents are hydrogen peroxide (H_2O_2) and sulfuric acid (H_2SO_4) [31]. During the chemical oxidation process, the reaction conditions such as treatment time and temperature as well as the concentration of reagent play a critical role in determining the successful surface modification of the polymeric membranes. The porous structure and porosity of the membrane can be tailored via the oxidising modification.

(a)   **(b)**

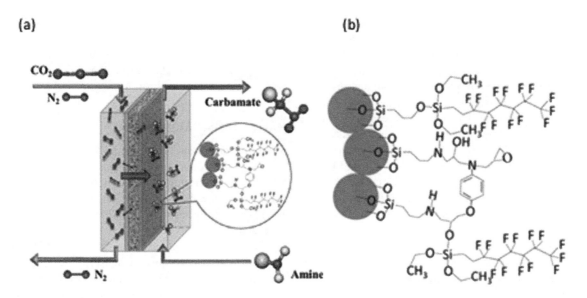

FIGURE 4.2   (a) Schematic diagram of membrane contactor system for gas adsorption, (b) the chemical structure of FAS-modified triepoxy cross-linked silica aerogel. Copyright (2018) with permission from Elsevier [27].

Through the careful control of the reaction condition, the acid modification does not affect the internal macrostructure of the membrane thus maintaining the membrane's integrity and stability. Furthermore, the oxidation process also introduces various types of surface functional group, which allow the secondary modification such as grafting to be performed on the acid-modified membrane surface.

4.4   HOLLOW FIBER MEMBRANE FOR GAS ABSORPTION AND STRIPPING APPLICATIONS

4.4.1   POLYMERIC HOLLOW FIBER MEMBRANES

PVDF hollow fiber membranes with various amounts of SMMs have been fabricated via the dry/wet spinning process [32, 33]. The desired CO_2 absorption properties, i.e., larger pore size and porosity, higher gas permeance, and wetting resistance have been achieved through the incorporation of SMM. As a result, the nitrogen gas permeance increased for all the SMM modified membranes regardless of the concentrations. With the increasing concentration of SMM from 2 to 6 wt%, the maximum CO_2 flux has been increased from 1.7 mol/m^2 s to 5.4 mol/m^{2}s at 300 ml/min of absorbent flow rate. At optimum loading of 6 wt%, the improvement was nearly 650% compared to that with SMM modification. The lowest critical water entry pressure was also observed due to its largest pore size. The effects of non-solvent additives on the morphological and structural change of PVDF hollow fiber membranes have been investigated by Pang et al [34]. Phosphoric acid (PA) in the concentration range of 2–8 wt% were added to the polymer dope prior to the dry-jet wet-spinning. It was found that the presence of PA at a high concentration resulted in the formation of a more sponge-like but less finger-like structure.

The mean pore size near the membrane outer surface also significantly decreased with the increasing concentration of PA additive. With these changes, the wetting resistance has been improved and resulted in low mass transfer resistance. When the membrane incorporated with 8 wt% of PA was applied for the removal of CO_2 in a membrane contacting system using distilled water as the liquid absorbent, the flux of 1.31 mmol/m^{2}s was obtained at a gas inlet flowrate of 130 ml/min. However, a long term stability test of up to 160 hours of operation showed that the CO_2 absorption flux gradually decreased from 1.3 mmol/m^2 to around 1.08 mmol/m^2 before a steady state was achieved due to the occurrence of partial wetting of the membrane pores.

Jin et al. fabricated PVDF hollow fiber membrane for biogas purification using a liquid–gas membrane contactor by coating a layer of superhydrophobic of SiO_2 nanoparticles and polydimethylsiloxane (PDMS) through a spray deposition method [35]. The distribution and homogeneity of the SiO_2 layer was affected by the surface properties of the PVDF hollow fiber in which the micro- and nano-sized porous structure of PVDF membranes have facilitated a uniform coating of SiO_2 layer with an optimized coating solution concentration of 4.44 ppm. The layer of SiO_2 deposition has resulted in the enhancement of surface hydrophobicity with a water contact angle of approximately 155°. When tested using 0.1M MEA aqueous solution as an absorbent, the CO_2 absorption flux of the neat PVDF was decreased by approximately 28% after 80 min of operation but that of the SiO_2 modified PVDF membrane was well maintained at 0.4 mmol/m^2 s even after 300 mins of operations. The long-term stability test using both water and MEA aqueous solution as the absorbent also revealed that the SiO_2 coated PVDF hollow fiber membrane contactor outperformed the neat PVDF membranes. However, it should be pointed out that the membrane mass transfer resistance of the SiO_2

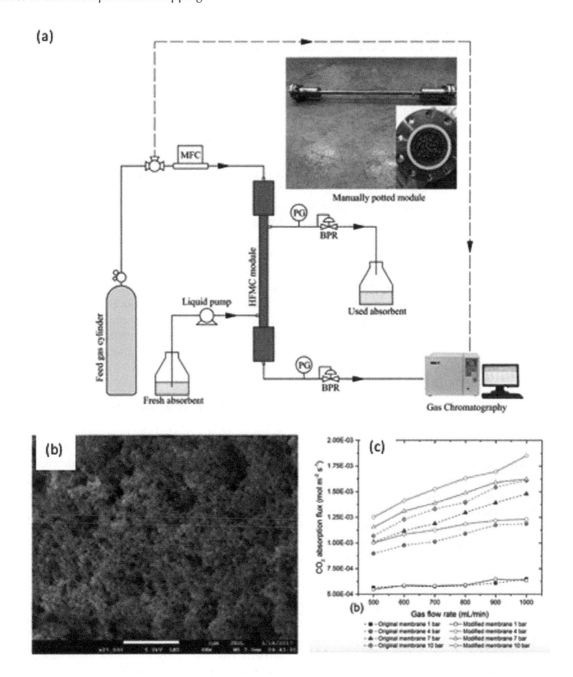

FIGURE 4.3 (a) Experimental set up of the hollow fiber contactor membrane for biogas separation, (b) surface morphology of PTFE membrane deposited with 1.5 wt% of silica nanoparticles, (c) the effect of gas flow rate on the CO_2 absorption flux of membranes at different pressures. Copyright (2018) with permission from Elsevier [36].

coated membrane was higher than that of the neat PVDF membranes, indicating that a hydrophobic layer has unfavorably hindered the mass transfer. The same modification was also adopted by Li et al. in their recent studies to fabricate superhydrophobic hollow fiber membranes for biogas separation [36]. The PTFE membranes deposited with silica nanoparticles were tested using the system shown in Figure 4.3(a) with pressure up to 10 bar using K_2CO_3 as absorbent to enhance the absorption efficiency. A mathematical model was also developed to investigate the absorption performance of silica modified membranes under non-wetted conditions. At silica content of 1.5 wt%, the deposited layer

completely covered the membrane surface pores as shown in Figure 4.3(b). The formation of nano-structured silica on the membrane surface also increased the surface roughness from 86.1 nm to 210.0 nm. This surface topology was responsible for the enhanced hydrophobicity and membrane wetting resistance. As shown in Figure 4.3(c), by fixing the solvent flow rate, the CO_2 absorption flux increased when the gas flow rate was increased from 500 to 1000 ml/min. As a result of higher mass transfer driving force and reduced mass transfer resistance, the silica modified membranes achieved a CO_2 flux of 1.85 mmol/m^2 s with the gas flow rate at 1000 mL /min under 10 bar pressurization. The modeling data has further validated that the superhydrophobic membrane was able to resist membrane wetting even at elevated pressure up to 10 bar.

Razaei et al. performed a study to investigate the effects of hydrophobic montmorillonite (MMT) of different loading on the membrane structure and CO_2 gas absorption performance [26]. Regardless of the loading of MMT, it was observed that the finger-like structure of the PVDF/MMT membranes had been increased. The lengthening of the finger-like pores was favorable to reducing the mass transfer resistance and increasing the gas transport. At the optimum MMT loading of 5 wt%, the surface hydrophobicity and porosity of the mixed matrix membranes have been drastically improved. The increase in both membrane porosity and pore size was ascribed to the presence of clay, which has changed the thermodynamic stability and sped up the solidification of the polymer. As a result, at a flow rate of 200 ml/min, the highest CO_2 flux of 1.59× mmol/m^2s was recorded for PVDF membranes incorporated with 5wt% MMT. This represented an improvement of approximately 21% compared to that of the neat PVDF membrane. Sethunga et al. modified a microporous PVDF hollow fiber membrane for biogas recovery [37]. The PVDF membrane surface was first activated using NaOH prior to the reaction with a perfluoropolyether known as Fluorolink S10. During the activation process, the fluorine atoms in the PVDF polymeric backbone were removed and substituted by hydroxide ions which later act as the active sites for the grafting of Fluorolink S10. Fluorolink S10 possessed low surface energies and binding energies which could lead to greater repellence towards water by forming a smooth hydrophobic surface. Compared to the neat PVDF, the contact angle of the Fluorolink S10 modified PVDF hollow fiber membrane was increased from 84.5° to 111.7°. The dehydrofluorination by NaOH increased the mean pore size of the membrane. However, the bulk porosity was reduced from 84% to 79% because the Fluorolink S10 molecules with high-molecular-weight covered parts of the voids formed on the membrane surface. When tested for biogas recovery, it was found that the membrane exhibited a mass transfer coefficient of 1.53 × 10^{-4} m/s, which was better than the commercial PP hollow fiber membranes. The improvement was attributed to the hydrophobicity of the Fluorolink S10 modified PVDF membrane which had reduced the likelihood of membrane wetting. Furthermore, in the long term stability testing, the methane flux was very stable throughout the 10 days of testing, which suggested a high durability of the modified PVDF hollow fiber membranes for biogas recovery from an aerobic membrane bioreactor effluent.

The effect of heat treatment on PEI membranes for CO_2 absorption has been studied by Ahmadi et al. [38]. The facile heat treatment was conducted by exposing the hollow fiber membranes in a hot air atmosphere with a temperature range of 80–160°C and a duration range of 5–30 mins. Despite the simplicity of the method, the heat treatment was found to be a promising way to alter the overall porosity of the membrane, hence improving the water CEP and

gas absorption behavior of the treated membranes. In particular, at a treatment temperature of 80 °C and a duration of five minutes, the optimum membrane porous structure could be obtained in which the membranes turned out to be denser with the smallest mean pore sizes among the heat-treated PEI membranes. The scaling up of poly(ether ether ketone) (PEEK) hollow fiber membrane modules has been attempted by Li et al. for post-combustion carbon capture [39]. The membrane area was increased by a factor of 90 with the scaling up of the membrane module diameter from 2 inch to 4 inch and the module length from 16 inch to 58 inch. The field test for CO_2 removal from flue gas steam containing 9.6 vol% of CO_2 using activated methyldiethanolamine (aMDEA) has shown a removal efficiency of 93% with a mass transfer coefficient of 1.2 s^{-1}. The mass transfer coefficient was further increased to 1.5 s^{-1} when the membrane module diameter was scaled up to 8 inch. An intrinsic CO_2 permeance up to 2670 GPU using single gas measurement was also observed. It was found that the CO_2 flux and efficiency could be further enhanced by increasing the solvent flow velocity and temperature, as well as the feed pressure.

4.4.2 CERAMIC HOLLOW FIBER MEMBRANES

Magnone et al. reported the preparation of FAS modified alumina hollow fiber membranes for CO_2 absorption for the first time. Using MEA solutions as the adsorbent solvent, a CO_2 absorption flux up to 5.4 mmol/m^2 s was exhibited by the FAS modified ceramic membrane. The absorption flux was higher compared to the conventionally used polymeric membrane for CO_2 absorption owing to the superhydrophobicity of the membranes upon FAS modification [40]. Lin et al. successfully coated silica aerogels on the surface of macroporous alumina hollow fiber membranes using a sol-gel method in order to shrink the membrane pores for CO_2 absorption [30, 41]. The hydrophobicity of silica aerogel tubular membranes was greatly improved through the subsequent FAS surface modifications. The presence of silica aerogels has effectively prevented membrane wetting by the aqueous amine solution. The authors have found that the CO_2 absorption flux of the silica aerogel coated membrane has been improved by at least 300% compared to the uncoated counterpart. The FAS-modified silica aerogel hollow fiber membranes were tested for three continuous consecutive cycles of operation of CO_2 absorption and were found to reach a stable absorption value of approximately 0.6 mmol/m^2. Such observation has proven the durability of the coated surface to facilitate stable and efficient gas absorption. The effects of FAS coating time on CO_2 absorption performance have been investigated by Lee and Park [42]. The porous alumina hollow fiber membranes were first fabricated via phase inversion spinning and sintering and followed by the grafting of FAS with different coating times. It was observed that the hydrophibicity increased with grafting time but remained almost invariable when the coating time was lengthened to 100 hours. The authors reported that two hours was the optimum coating time to achieve sufficient wetting resistance and CO_2 absorption flux. Despite the thin and less dense layer of FAS coating, the layers may add resistance to the gas passage and some membrane pores may be hindered from allowing gas permeation. As a result, the decrease of CO_2 absorption flux was observed with increasing FAS coating time. In this study, the maximum CO_2 absorption of 6mmol/m^2 s at a liquid flow rate of 50 ml/min was achieved using ceramic membrane that was grafted by FAS for two hours. The long term stability test suggested that the FAS modified porous alumina hollow fiber membranes are stable for long term membrane contactor application.

Abdulhameed et al. fabricated super hydrophobicFAS-grafted mullite ceramic hollow fiber membranes for CO_2 capture through a gas–liquid contacting process [43]. The mullite hollow fiber membranes were fabricated by extruding the suspension of kaolin and alumina through phase inversion, followed by sintering at 1450°C. As shown in Figure 4.4(a), the asymmetric ceramic hollow fiber membrane is made up of finger-like structures at the inner surface and sponge-like structures at the outer surface of the fibers. While most of the ceramic membranes do not show the formation of finger-like structures, the finger-like pores obtained with the addition of alumina can favorably promote a higher gas permeation rate. Additionally, the migration of small particles towards the outer surface also allows the desirable pore size distribution. Upon the FAS grafting, the contact angle of the ceramic hollow fiber was increased to 142° and the wettability resistance has been greatly enhanced as shown in Figure 4.4(b). Owing to the desired surface characteristics and porosity, a CO_2 absorption flux as high as 0.18 mol/m^{2}s was achieved at the liquid flow rate of 100 ml/min.

Table 4.1 summarizes the performances of some polymeric and hollow fiber membrane contactors.

4.5 FUTURE PERSPECTIVE AND CONCLUSION

The removal and recovery of various gases from a wide range of sources using a membrane contactor is considered to be an attractive approach to reducing negative environmental effects and adds value to many industrial processes. For decades, membrane contactor technology has served as a promising alternative absorption technology for the mitigation of acid gases produced from the combustion of fossil fuels.

Membrane contactor has been considered as a technically sound option. Nevertheless, long term performance of the membrane contactor operation is an important criterion to be considered to make this technology attractive for industrial processes. In order to attain long term membrane stability, the exploration of new membrane materials and the tailoring of the membrane structure are highly desired to optimize the pore size structure, and hydrophobicity as well as the chemical and thermal stability. Surface modifications of hollow fiber membranes have been widely adopted to modify the membrane surface, particularly to create superhydrophobic surfaces to minimize solvent penetration and membrane wetting. The establishment of surface modification methods such as coating and spray deposition are attractive to achieve this purpose. However, some limitations should be carefully looked into to optimize the performance of the modified membranes. It is known that the presence of an additional layer could unfavorably increase the mass transfer resistance. So it is important to optimize the thickness of the coated or deposited layer in order to maximize the impacts. The durability of the layer is also of great concern where its integrity should be well maintained throughout the absorption process to ensure its long term stability, especially under harsh operating condition that involve acidic gases. Similar limitations should also be considered in the fabrication of thin film composite membranes. Despite the effectiveness in rendering desired hydrophobicity and improved mechanical strength, some modification techniques such as template synthesis and layer-by-layer methods require tedious and multiple steps to obtain the satisfactory results. More attention should also be focused on the simplification of these approaches so that they can be more practically applied to large scale modification and commercial application. Nanomaterials, particularly those with hydrophobic features, are attractive

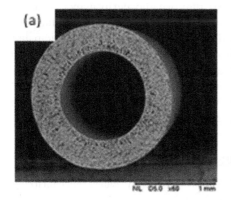
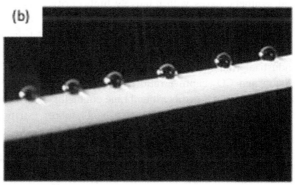

FIGURE 4.4 (a) Cross-sectional scanning electron image of kaolin-alumina ceramic membrane and (b) the water droplet position on kaolin-alumina ceramic membrane. Copyright (2017) with permission from Elsevier [43].

TABLE 4.1

Performances of Some Polymeric and Hollow Fiber Membrane Contactors

Type of Membrane	Feed Stream/Flow Rate	Absorbent/Flow Rate	Absorption Flux	Ref
PVDF with SMM	CO_2/ 100 ml·min^{-1}	Distilled water/ 300 ml· min^{-1}	5.4 mmol·m^{-2}·s^{-1}	[44]
PVDF with SMM	CO_2/ 100 ml·min^{-1}	Distilled water/ 0.03 ml·s^{-1}	0.7 mmol·m^{-2}·s^{-1}	[45]
Electrospun PS coated with PDMS	CO_2+N_2/ 200 ml·min^{-1}	AMP-PZ/ 100 ml·min^{-1}	1.85 mmol·m^{-2}·s^{-1}	[46]
PEI with MMT	CO_2	Distilled water/ 1.4 ml·s^{-1}	1.59 mmol·m^{-2}·s^{-1}	[47]
PVDF with ZSM5	CO_2	Distilled water/1.2 ml·s^{-1}	3.4 mmol·m^{-2}·s^{-1}	[48]
PVDF with graphene	CO_2/ 1.5 L·min^{-1}	Distilled water/ 0.012 ml·s^{-1}	3.0 mmol·m^{-2}·s^{-1}	[4]
PVDF+PFTS	CO_2/CH_4	MEA/ 0.25 ml·s^{-1}	12.7 mmol·m^{-2}·s^{-1}	[49]
Alumina coated with FAS	CO_2+N_2/ 20 ml·min^{-1}	MEA/ 50 ml·min^{-1}	6.0 mmol·m^{-2}·s^{-1}	[42]
Alumina coated with FAS	CO_2+N_2 / 200 ml·min^{-1}	AMP-PZ/ 100 ml·min^{-1}	1.5 mmol·m^{-2}·s^{-1}	[50]
Alumina coated with silica aerogel	CO_2+N_2/ 200 ml·min^{-1}	AMP-PZ/ 100 ml·min^{-1}	1.3 mmol·m^{-2}·s^{-1}	[28]
PEI	CO_2/ 0.12 m·s^{-1}	Distilled water/ 0.3 ml·s^{-1}	2.65 mmol·m^{-2}·s^{-1}	[51]

PS: polysulfone; PEI: polyetherimide; ZSM5: Zeolite Socony Mobil #5; PFTS: perfluorooctyl trichlorosilane; AMP-PZ: 2-amino-2-methyl-1-propanol; PZ: piperazine

additives for membranes to improve the anti-wetting properties. Despite their advantages, a typical preparation of nano-enabled membranes involves the direct incorporation of nanomaterials in the polymer dope via physical mixing or the deposition of nanomaterials onto the surface of membranes through grafting and coating. The leaching of these nanomaterials upon the interaction with the absorbent solvent will be a major concern for the practical usage of nano-enabled membranes for gas–liquid membrane contactors. Currently, the long term stability of nano-enabled membranes for membrane contactor application has not been performed. Hence, more work in this direction is needed to accelerate the adoption of this innovation.

Despite the advantages offered by gas–liquid membrane contactors over conventional gas absorption technology, more work is required to address the technical hurdles that hinder the large scale commercial applications. Hollow fiber membrane contactors have been well investigated based on experimental and theoretical approaches to evaluate their performances and feasibility in lab and industrial scales. Most of the bench-scale studies reported in the open literature mainly focused on the mild operating conditions, use of synthetic gases with relatively high purity, and short-term operation where degradation of absorbent was hardly observed. In pilot-scale studies, the condition of the membrane contactor operation was not applicable to the actual industrial framework due to the relatively low MEA conversion and CO_2 capture ratio [52]. In order to intensify the potential of hollow fiber membrane contactors by considering both economic and technical aspects, more large-scale field studies under relevant industrial conditions are desired. Furthermore, studies and investigations should also be conducted to look into the possibilities of combining several technologies in order to enhance the efficiency of the adsorption and stripping processes.

The competitiveness of membrane contactors in the industrial scale must be considered with the overall efficiency and investment cost. From an economical point of view, a thorough and representative economic analysis of the gas–liquid membrane contactor system must also be performed to justify the potential of membrane contactors to replace the existing technologies. For instance, an economic study of sulfur dioxide gas recovery from gas stream using a ceramic hollow fiber membrane contactor showed that the investment and operating costs are highly dependent on the environmental restrictions and the inlet concentration of sulfur dioxide [53]. One issue with conventional gas–liquid membrane contactors is the limitation on large scale commercial applications due to the challenges in maintenance and production. Some recent studies have looked into the possibilities of scaling up membrane contactors by assembling and arranging the small units of contactors in the desired arrangement [54]. The contactor is properly placed to provide a uniform flow field. For large scale application, the small contactor units can be feasibly plugged into a pre-fabricated frame. Such innovation serves as a straight forward strategy for a maintainable and scalable membrane contactor.

In conclusion, membrane contactor is a proven technology to improve the performance and add value to the existing unit operations, particularly gas absorption and stripping processes. It is anticipated that with the advancements made in the design and fabrication of robust and wetting-resistant hollow fiber membranes, membrane contactors will serve as a technological enabler to drive gain in the field of gas absorption and stripping processes.

REFERENCES

1. C. Dinca, N. Slavu, A. Badea, Benchmarking of the pre/post-combustion chemical absorption for the CO_2 capture, *J. Energy Inst.* 91 (2018) 445–456. doi:10.1016/j.joei.2017.01.008

2. J. Yang, X. Yu, J. Yan, S.-T. Tu, E. Dahlquist, Effects of SO_2 on CO_2 capture using a hollow fiber membrane contactor, *Appl. Energy.* 112 (2013) 755–764. doi:10.1016/j.apenergy.2012.11.052

3. I. Sreedhar, T. Nahar, A. Venugopal, B. Srinivas, Carbon capture by absorption – Path covered and ahead, *Renewable Sustainable Energy Rev.* 76 (2017) 1080–1107. doi:10.1016/j.rser.2017.03.109

4. X. Wu, B. Zhao, L. Wang, Z. Zhang, H. Zhang, X. Zhao, X. Guo, Hydrophobic PVDF/graphene hybrid membrane for CO_2 absorption in membrane contactor, *J. Membr. Sci.* 520 (2016) 120–129. doi:10.1016/j.memsci.2016.07.025

5. D.C. Nymeijer, T. Visser, R. Assen, M. Wessling, Composite hollow fiber gas-liquid membrane contactors for olefin/paraffin separation, *Sep. Purif. Technol.* 37 (2004) 209–220. doi:10.1016/j.seppur.2003.08.002

6. A. Mansourizadeh, A.F. Ismail, Hollow fiber gas-liquid membrane contactors for acid gas capture: A review, *J. Hazard. Mater.* 171 (2009) 38–53. doi:10.1016/j.jhazmat.2009.06.026

7. A.K. Pabby, A.M. Sastre, State-of-the-art review on hollow fibre contactor technology and membrane-based extraction processes, *J. Membr. Sci.* 430 (2013) 263–303. doi:10.1016/j.memsci.2012.11.060

8. S. Khaisri, D. deMontigny, P. Tontiwachwuthikul, R. Jiraratananon, Comparing membrane resistance and absorption performance of three different membranes in a gas absorption membrane contactor, *Sep. Purif. Technol.* 65 (2009) 290–297. doi:10.1016/j.seppur.2008.10.035

9. H. Fashandi, A. Ghodsi, R. Saghafi, M. Zarrebini, CO_2 absorption using gas-liquid membrane contactors made of highly porous poly(vinyl chloride) hollow fiber membranes, *Int. J. Greenhouse Gas Control.* 52 (2016) 13–23. doi:10.1016/j.ijggc.2016.06.010

10. A. Huang, L.H. Chen, C.H. Chen, H.Y. Tsai, K.L. Tung, Carbon dioxide capture using an omniphobic membrane for a gas-liquid contacting process, *J. Membr. Sci.* 556 (2018) 227–237. doi:10.1016/j.memsci.2018.03.089

11. S. Bazhenov, A. Malakhov, D. Bakhtin, V. Khotimskiy, G. Bondarenko, V. Volkov, M. Ramdin, T.J.H. Vlugt, A. Volkov, CO_2 stripping from ionic liquid at elevated pressures in gas-liquid membrane contactor, *Int. J. Greenhouse Gas Control.* 71 (2018) 293–302. doi:10.1016/j.ijggc.2018.03.001

12. L. Ansaloni, A. Hartono, M. Awais, H.K. Knuutila, L. Deng, CO_2 capture using highly viscous amine blends in non-porous membrane contactors, *Chem. Eng. J.* 359 (2019) 1581–1591. doi:10.1016/j.cej.2018.11.014

13. E. Chabanon, D. Roizard, E. Favre, Membrane contactors for post-combustion carbon dioxide capture: A comparative study of wetting resistance on long time scales, *Ind. Eng. Chem. Res.* 50 (2011) 8237–8244. doi:10.1021/ie200704h

14. N. Kladkaew, R. Idem, P. Tontiwachwuthikul, C. Saiwan, Studies on corrosion and corrosion inhibitors for amine based solvents for CO_2 absorption from power plant flue gases containing CO_2, O_2 and SO_2, *Energy Procedia.* 4 (2011) 1761–1768. doi:10.1016/j.egypro.2011.02.051

15. M.A. Abdulhameed, M.H.D. Othman, H.N.A. Al Joda, A.F. Ismail, T. Matsuura, Z. Harun, M.A. Rahman, M.H. Puteh, J. Jaafar, Fabrication and characterization of affordable hydrophobic ceramic hollow fibre membrane for contacting processes, *J. Adv. Ceram.* 6 (2017) 330–340. doi:10.1007/s40145-017-0245-1

16. A. Gabelman, S.T. Hwang, Hollow fiber membrane contactors, *J. Membr. Sci.* 159 (1999) 61–106. doi:10.1016/S0376-7388(99)00040-X

17. K. Dalane, H.F. Svendsen, M. Hillestad, L. Deng, Membrane contactor for subsea natural gas dehydration: Model development and sensitivity study, *J. Membr. Sci.* 556 (2018) 263–276. doi:10.1016/j.memsci.2018.03.033

18. D. deMontigny, P. Tontiwachwuthikul, A. Chakma, Using polypropylene and polytetrafluoroethylene membranes in a membrane contactor for CO_2 absorption, *J. Membr. Sci.* 277 (2006) 99–107. doi:10.1016/j.memsci.2005.10.024

19. E. Favre, H.F. Svendsen, Membranes contactor for intensified post-combustion carbon dioxide capture by gas–liquid absorption process. In *J. Membr. Sci.* 15 (2012), pp. 1–7.

20. H. Kreulen, C.A. Smolders, G.F. Versteeg, Microporous hollow fibre membrane modules as gas-liquid contactors Part 2. Mass transfer with chemical reaction, *J. Membr. Sci.* 78 (1993) 217–238.

21. M. Yang, E.L. Cussler, Designing hollow-fiber contactors, *AIChE J.* 32 (1986) 1910–1916. doi:10.1002/aic.690321117

22. K.K.S.R. Prasad, Dispersion-free solvent extraction with microporous hollow-fiber modules, *AIChE J.* 34 (1988) 177–188. doi:10.1002/aic.690340202

23. H. Pang, H. Gong, M. Du, Q. Shen, Z. Chen, Effect of non-solvent additive concentration on CO_2 absorption performance of polyvinylidenefluoride hollow fiber membrane contactor, *Sep. Purif. Technol.* 191 (2018) 38–47. doi:10.1016/j.seppur.2017.09.012

24. R. Naim, A.F. Ismail, Effect of polymer concentration on the structure and performance of PEI hollow fiber membrane contactor for CO_2 stripping, *J. Hazard. Mater.* 250–251 (2013) 354–361. doi:10.1016/j.jhazmat.2013.01.083

25. G. Bakeri, T. Matsuura, A.F. Ismail, D. Rana, A novel surface modified polyetherimide hollow fiber membrane for gas–liquid contacting processes, *Sep. Purif. Technol.* 89 (2012) 160–170. doi:10.1016/j.seppur.2012.01.022

26. M. Rezaei, A.F. Ismail, S.A. Hashemifard, T. Matsuura, Preparation and characterization of PVDF-montmorillonite mixed matrix hollow fiber membrane for gas-liquid contacting process, *Chem. Eng. Res. Des.* 92 (2014) 2449–2460. doi:10.1016/j.cherd.2014.02.019

27. Y.-F. Lin, Y.-J. Lin, C.-C. Lee, K.-Y.A. Lin, T.-W. Chung, K.-L. Tung, Synthesis of mechanically robust epoxy cross-linked silica aerogel membranes for CO_2 capture, *J. Taiwan Inst. Chem. Eng.* 87 (2018) 117–122. doi:10.1016/j.jtice.2018.03.019

28. Y. Lin, J. Kuo, Mesoporous bis(trimethoxysilyl)hexane (BTMSH)/tetraethyl orthosilicate (TEOS)-based hybrid silica aerogel membranes for CO_2 capture, *Chem. Eng. J.* 300 (2016) 29–35. doi:10.1016/j.cej.2016.04.119

29. Y.-F. Lin, C.-C. Ko, C.-H. Chen, K.-L. Tung, K.-S. Chang, T.-W. Chung, Sol-gel preparation of polymethylsilsesquioxane aerogel membranes for CO_2 absorption fluxes in membrane contactors, *Appl. Energy.* 129 (2014) 25–31. doi:10.1016/j.apenergy.2014.05.001

30. Y.F. Lin, C.H. Chen, K.L. Tung, T.Y. Wei, S.Y. Lu, K.S. Chang, Mesoporous fluorocarbon-modified silica aerogel membranes enabling long-term continuous CO_2 capture with large absorption flux enhancements, *ChemSusChem.* 6 (2013) 437–442. doi:10.1002/cssc.201200837

31. A.A. Ovcharova, V.P. Vasilevsky, I.L. Borisov, V.V. Usosky, V.V. Volkov, Porous hollow fiber membranes with varying hydrophobic–hydrophilic surface properties for gas–liquid membrane contactors, *Pet. Chem.* 56 (2016) 1066–1073. doi:10.1134/S0965544116110128

32. M. Rahbari-Sisakht, A.F. Ismail, D. Rana, T. Matsuura, D. Emadzadeh, Effect of SMM concentration on morphology and performance of surface modified PVDF hollow

fiber membrane contactor for CO_2 absorption, *Sep. Purif. Technol.* 116 (2013) 67–72. doi:10.1016/j.seppur.2013.05.008

33. M. Rahbari-Sisakht, A.F. Ismail, D. Rana, T. Matsuura, A novel surface modified polyvinylidene fluoride hollow fiber membrane contactor for CO_2 absorption, *J. Membr. Sci.* 415–416 (2012) 221–228. doi:10.1016/j.memsci.2012.05.002

34. H. Pang, H. Gong, M. Du, Q. Shen, Z. Chen, Effect of non-solvent additive concentration on CO_2 absorption performance of polyvinylidenefluoride hollow fiber membrane contactor, *Sep. Purif. Technol.* 191 (2018) 38–47. doi:10.1016/j.seppur.2017.09.012

35. P. Jin, C. Huang, J. Li, Y. Shen, L. Wang, Surface modification of poly(vinylidene fluoride) hollow fibre membranes for biogas purification in a gas–liquid membrane contactor system, *R. Soc. Open Sci.* 4 (2017): 171321. doi:10.1098/rsos.171321

36. Y. Li, L. Wang, X. Hu, P. Jin, X. Song, Surface modification to produce superhydrophobic hollow fiber membrane contactor to avoid membrane wetting for biogas purification under pressurized conditions, *Sep. Purif. Technol.* 194 (2018) 222–230. doi:10.1016/j.seppur.2017.11.041

37. G.S.M.D.P. Sethunga, W. Rongwong, R. Wang, T.H. Bae, Optimization of hydrophobic modification parameters of microporous polyvinylidene fluoride hollow-fiber membrane for biogas recovery from anaerobic membrane bioreactor effluent, *J. Membr. Sci.* 548 (2018) 510–518. doi:10.1016/j.memsci.2017.11.059

38. H. Ahmadi, S.A. Hashemifard, A.F. Ismail, A research on CO_2 removal via hollow fiber membrane contactor: The effect of heat treatment, *Chem. Eng. Res. Des.* 120 (2017) 218–230. doi:10.1016/j.cherd.2017.02.013

39. S. Li, T.J. Pyrzynski, N.B. Klinghoffer, T. Tamale, Y. Zhong, J.L. Aderhold, S. James Zhou, H.S. Meyer, Y. Ding, B. Bikson, Scale-up of PEEK hollow fiber membrane contactor for post-combustion CO_2 capture, *J. Membr. Sci.* 527 (2017) 92–101. doi:10.1016/j.memsci.2017.01.014

40. E. Magnone, H.J. Lee, J.W. Che, J.H. Park, High-performance of modified Al_2O_3 hollow fiber membranes for CO_2 absorption at room temperature, *J. Ind. Eng. Chem.* 42 (2016) 19–22. doi:10.1016/j.jiec.2016.07.022

41. Y.F. Lin, J.M. Chang, Q. Ye, K.L. Tung, Hydrophobic fluorocarbon-modified silica aerogel tubular membranes with excellent CO_2 recovery ability in membrane contactors, *Appl. Energy.* 154 (2015) 21–25. doi:10.1016/j.apenergy.2015.04.109

42. H.J. Lee, J.H. Park, Effect of hydrophobic modification on carbon dioxide absorption using porous alumina (Al_2O_3) hollow fiber membrane contactor, *J. Membr. Sci.* 518 (2016) 79–87. doi:10.1016/j.memsci.2016.06.038

43. M.A. Abdulhameed, M.H.D. Othman, A.F. Ismail, T. Matsuura, Z. Harun, M.A. Rahman, M.H. Puteh, J. Jaafar, M. Rezaei, S.K. Hubadillah, Carbon dioxide capture using a superhydrophobic ceramic hollow fibre membrane for gas-liquid contacting process, *J. Clean. Prod.* 140 (2017) 1731–1738. doi:10.1016/j.jclepro.2016.07.015

44. M. Rahbari-Sisakht, A.F. Ismail, D. Rana, T. Matsuura, D. Emadzadeh, Effect of SMM concentration on morphology and performance of surface modified PVDF hollow fiber membrane contactor for CO_2 absorption, *Sep. Purif. Technol.* 116 (2013) 67–72. doi:10.1016/j.seppur.2013.05.008

45. A. Mansourizadeh, Z. Aslmahdavi, A.F. Ismail, T. Matsuura, Blend polyvinylidene fluoride/surface modifying macromolecule hollow fiber membrane contactors for CO_2 absorption, *Int. J. Greenhouse Gas Control.* 26 (2014) 83–92. doi:10.1016/j.ijggc.2014.04.027

46. Y.F. Lin, W.W. Wang, C.Y. Chang, Environmentally sustainable, fluorine-free and waterproof breathable PDMS/PS nanofibrous membranes for carbon dioxide capture, *J. Mater. Chem. A.* 6 (2018) 9489–9497. doi:10.1039/c8ta00275d

47. M. Rezaei DashtArzhandi, A.F. Ismail, T. Matsuura, B.C. Ng, M.S. Abdullah, Fabrication and characterization of porous polyetherimide/montmorillonite hollow fiber mixed matrix membranes for CO_2 absorption via membrane contactor, *Chem. Eng. J.* 269 (2015) 51–59. doi:10.1016/j.cej.2015.01.095

48. M. Rezaei-Dashtarzhandi, A.F. Ismail, P.S. Goh, I. Wan Azelee, M. Abbasgholipourghadim, G. Ur Rehman, T. Matsuura, Zeolite ZSM5-filled PVDF hollow fiber mixed matrix membranes for efficient carbon dioxide removal via membrane contactor, *Ind. Eng. Chem. Res.* 55 (2016) 12632–12643. doi:10.1021/acs.iecr.6b03117

49. Y. Xu, Y. Lin, M. Lee, C. Malde, R. Wang, Development of low mass-transfer-resistance fluorinated TiO_2-SiO_2/PVDF composite hollow fiber membrane used for biogas upgrading in gas-liquid membrane contactor, *J. Membr. Sci.* 552 (2018) 253–264. doi:10.1016/j.memsci.2018.02.016

50. Y.F. Lin, Q. Ye, S.H. Hsu, T.W. Chung, Reusable fluorocarbon-modified electrospun PDMS/PVDF nanofibrous membranes with excellent CO_2 absorption performance, *Chem. Eng. J.* 284 (2016) 888–895. doi:10.1016/j.cej.2015.09.063

51. Y. Zhang, R. Wang, Novel method for incorporating hydrophobic silica nanoparticles on polyetherimide hollow fiber membranes for CO_2 absorption in a gas–liquid membrane contactor, *J. Membr. Sci.* 452 (2014) 379–389. doi:10.1016/j.memsci.2013.10.011

52. D. Albarracin Zaidiza, B. Belaissaoui, S. Rode, E. Favre, Intensification potential of hollow fiber membrane contactors for CO_2 chemical absorption and stripping using monoethanolamine solutions, *Sep. Purif. Technol.* 188 (2017) 38–51. doi:10.1016/j.seppur.2017.06.074

53. P. Luis, A. Garea, A. Irabien, Environmental and economic evaluation of SO_2 recovery in a ceramic hollow fiber membrane contactor, *Chem. Eng. Process. Process Intensif.* 52 (2012) 151–154. doi:10.1016/j.cep.2011.10.006

54. K. He, S. Chen, C.C. Huang, L.Z. Zhang, Fluid flow and mass transfer in an industrial-scale hollow fiber membrane contactor scaled up with small elements, *Int. J. Heat Mass Transfer.* 127 (2018) 289–301. doi:10.1016/j.ijheatmasstransfer.2018.08.039

5 Membrane Contactor Applications in Various Industries

A. Sengupta and P. A. Peterson

CONTENTS

5.1 INTRODUCTION

In a membrane contactor (MC), a gas phase and a liquid phase, or two immiscible liquid phases, come into intimate contact with each other without a need to disperse one phase into the other. Functionally, an MC is equivalent to equipment such as a gas absorption column, gas stripping column, or liquid–liquid extraction column. The use and application of MC technology has been steadily growing over the years because of certain unique characteristics and desirable attributes. In some industries such as microelectronics, the use of MC in some specific process steps has become a standard practice. In other industries such as Steam and Power, Beverage, Pharmaceutical and Biotech, Deionization, Inks and Coating, Oil and Gas, Lab and Analytical, and Environmental Abatement, MC has found varying and increasing levels of adoption. An earlier publication [1] described the wide range of MC applications known at that time. The primary goal of this present chapter is to provide more details and up-to-date examples of commercial MC installations.

Drivers for adoption of MC are not the same in all industries. The following Table 5.1 lists, in no specific order, some features of MC technology and associated benefits of using MC. Customers generally choose MC to supplant or replace existing equipment because of one or more of these benefits. Often, MC has emerged as an enabling technology that filled previously unmet needs.

It should be noted that in many applications the MC device is not even called a contactor but is referred to or better known by other names depending on the specific functionality for which customers use it. Some of these names include Blood Oxygenator (earliest use of MC), Gas Transfer Membrane (GTM), Membrane Degasifier (MDG), Membrane Deaerator (MDA), Membrane-assisted Solvent Extractor (MSE), Membrane Humidifier, etc.

5.2 DESCRIPTION AND PRINCIPLE OF OPERATION OF MC

The principle of operation of MC and the fundamentals of using microporous membranes for various separations in MC have been discussed many times in literature [2–14] so they are not covered in any more details here. The "membranes" used in MC are distinctly different from those in other membrane processes such as reverse osmosis (RO), ultrafiltration (UF), nanofiltration (NF), microfiltration (MF), electro deionization (EDI), dialysis, or membrane gas separation. The contactors shown as an example in Figure 5.1 are generally of tube-in-shell configuration with inlet/outlet ports for the "shell" side and "tube" or "lumen" side. Membranes in MC are mostly non-selective and microporous and are typically made of hydrophobic polymeric materials such as polypropylene (PP), polyethylene (PE), Polytetrafluoroethylene (PTFE), Perfluoroalkoxy Polymer (PFA), Polyvinylidene fluoride (PVDF), Polymethylpentene (PMP), etc. Although technically MC can be made from flat sheet membranes, almost all commercial MC are made from small-diameter microporous hollow fiber (or capillary) membranes with fine pores in the wall extending from fiber interior surface to fiber exterior surface.

TABLE 5.1

Features and Associated Benefits of Membrane Contactor

Features	Benefits
Very high active contact area between phases	Small footprint, profile, and weight of devices and systems; perfect for process intensification, fit into existing spaces, ideal for mobile and containerized systems
High efficiency of separation process	Superior separation performance not always achievable by other competing technologies
No need to disperse or coalesce phases	Extra process step eliminated; more efficient utilization of equipment volume
Flow rates and, to some extent, pressures of contacting phases can be controlled independently	In-line operation without need to de-pressurize or re-pressurize, contact area constant irrespective of phase flow rates, more flexible and tolerant to changes in process operating parameters
Modular in nature	Easier to add system capacity incrementally; can often be retrofitted into existing systems; Easier scale-up
No chemicals used	Increased process and operator safety
Rapid process dynamics	Fast start-up, equilibration, and shut-down
Mass transfer does not depend on gravity	Contactor can be mounted vertically or horizontally, will also work in microgravity, Able to process two fluid phases of same densities

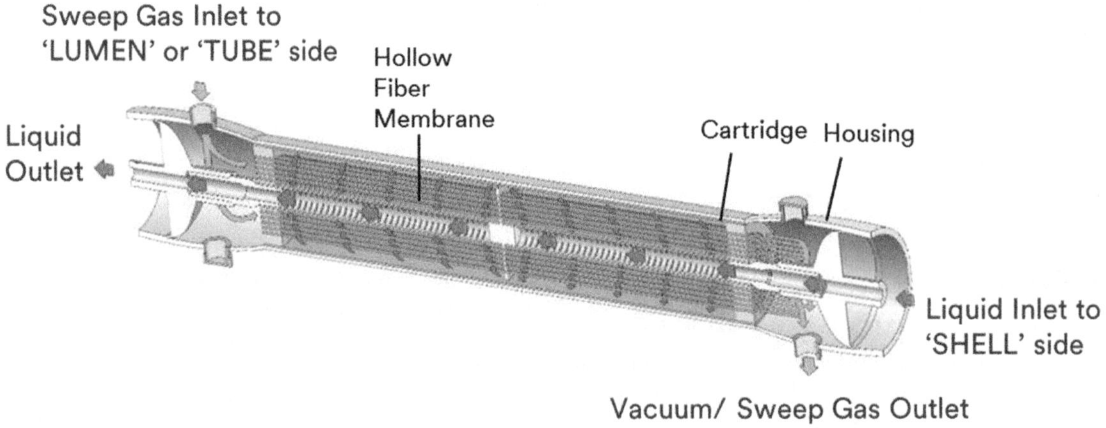

FIGURE 5.1 Tube-in-shell configuration of membrane contactor with shell side and tube (lumen) side.

The membrane in MC acts as a passive barrier and its role is to bring two immiscible fluid phases (such as a gas and a liquid, or an aqueous liquid and an organic liquid, etc.) in contact with each other without dispersion. The phase interface is immobilized at the pore mouth, with the pores filled by one of the two fluid phases that are in contact, as illustrated in Figure 5.2. Since it enables the phases to come in direct contact, the MC functions as a continuous-contact mass transfer device. However, there is no need to physically disperse one phase into the other, or to separate the phases after separation is completed.

The location of the immobilized phase interface and relationship between pressures in the two phases are based on surface tension effect and capillary forces. The critical parameters involved are (a) hydrophobicity or hydrophilicity of the polymeric hollow fiber, (b) surface tension or interfacial tension between two phases, (c) contact angle between the polymer and fluid phases, and (d) effective pore diameter. More information on the interplay between these parameters and the effect of microporous membrane properties on mass transfer between phases is available in references [1–14].

5.3 CURRENT APPLICATIONS OF MC IN DIFFERENT INDUSTRIES

A matrix denoting the main applications of MC at the present time in different industries is shown in Table 5.2. The leftmost column lists the specific use applications of MC and the top row lists various industries. Emerging applications of MC in existing or new industry/markets will be presented in a later section.

Applications of MC in commercial industrial processes are in various stages of development. Early success has come mainly in water deionization, deoxygenation, and debubbling applications. MC systems of a wide range of flow capacities are currently in operation in various parts of the world. Systems with large capacities were possible only after MC devices of a sufficiently large size could be produced commercially on a routine and consistent basis. As a reference, the largest commercially available contactor currently in the market has an active contact area of about 373 m². A picture with the approximate dimensions of this device is shown in Figure 5.3 as a reference. The

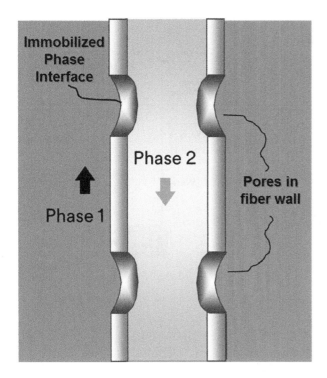

FIGURE 5.2 Microporous hollow fiber with immobilized phase interface.

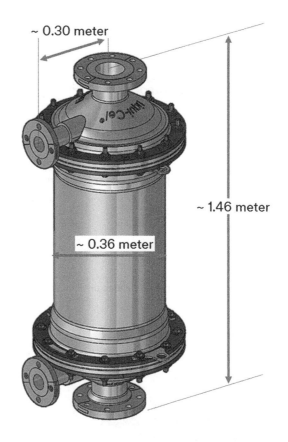

FIGURE 5.3 Example of largest membrane contactor currently in commercial production.

commercial availability of such products has greatly facilitated the large-scale acceptance of this technology.

5.4 MEMBRANE CONTACTOR USE FOR WATER DEIONIZATION (DI)

Water deionization is extensively used as a primary water purification step in many industries, so it is discussed first. The goal of deionization is to remove inorganic ions from water. Two common routes for making DI water are via an ion-exchange (IE) process or via a reverse osmosis (RO) process. Many water sources around the world contain high levels of alkalinity which converts to dissolved CO_2 during purification. Dissolved CO_2 in water is a heavy load on DI processes. Industry has found MC to be very effective in removing CO_2 from water and thereby reducing the cost

of the DI process and minimizing the use and storage of chemicals. In the context of the DI process, MC is often known as Gas Transfer Membrane (GTM) or Membrane Degasifier (MDG). Figure 5.4 shows the various steps in making DI water and identifies the locations within DI trains where MC can replace or eliminate other process steps. For example, MC can eliminate a caustic injection step in a single-pass or two-pass RO system. MC is frequently used just prior to Electro De-Ionization (EDI) to achieve <5 ppm dissolved CO_2 (normally recommended by EDI manufacturers) to make EDI more efficient and to eliminate the need for a supplemental polishing mixed bed

TABLE 5.2
Industry/Major Applications Matrix for Membrane Contactors

Application	De-ionization	Microelectronics	Steam/Power	Beverage	Pharma/Bio-Tech	Ink & Coating	Oil & Gas	Lab/Analytical
O_2 Removal		X	X	X	X		X	
CO_2 Removal	X	X	X	X	X			
Carbonation				X				
Nitrogenation				X				
Bubble Removal		X				X		X
Alcohol Adjustment				X				
Non Condensable Gases (NCG) Removal					X			

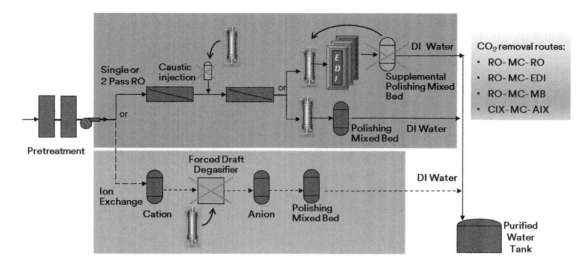

FIGURE 5.4 Typical water de-ionization process steps.

Design Basis:
- Flow rate = 500 m³/h
- Temperature = 15°C

Membrane Contactor System Configuration:
- 11 Parallel Trains of 3 Contactors in Series (33 Total) (Each Contactor with 215 m2 active area)
- 98% CO_2 Reduction
- CO_2 maximum outlet 1.5 ppm

FIGURE 5.5 Membrane contactor system for CO2 removal before mixed bed in water deionization system.

(MB). When positioned in between cation exchanger (CIX) and anion exchanger (AIX), MC can also eliminate the forced-draft degasifier (FDG) in the IX system. MC also greatly reduces the regeneration frequency needed in AIX and MB. It is common for MC customers to get two to five times more DI water between AIX/MB regeneration cycles relative to no-MC systems. This reduces the costs for regeneration, downtime, and resin shipment. In replacing FDG, MC can save space, allow process fluctuations, eliminate de-pressurization/re-pressurization steps, and prevent re-contamination of purified water.

A photograph of a typical commercially installed membrane contactor CO_2 removal system is shown in Figure 5.5.

5.5 MEMBRANE CONTACTOR USE IN MICROELECTRONICS (ME) INDUSTRY

MC is deployed in multiple locations in ME plants. Its main uses are:

A. Degassing of Ultrapure Water (UPW) in make-up, primary, and polishing loops, and at points-of-use (POU)

B. UPW degassing for wafer cleaning and rinsing

C. Degassing of Process Chemistries, such as for etching, plating, etc.

D. UPW resistivity/conductivity control

E. "Functional" water production

F. Semiconductor wastewater treatment, such as removal/recovery of ammonia

UPW Production Process Steps: A generic flow diagram of UPW process where MC is used is presented in Figure 5.6. Some times, both CO_2 and O_2 need to be removed in make-up water system [15].

Specific uses of MC in UPW plant are in removing O_2 and CO_2, and in removing and adding N_2. Table 5.3 summarizes the uses and benefits.

One photograph of a commercially installed membrane contactor system in UPW polishing loop is shown in Figure 5.7.

UPW for Cleaning and Rinsing: An example of the need to remove dissolved O_2 is to minimize the formation and growth rate of silicon oxide and thereby prevent defects on silicon wafers following etching. Minimizing oxide thickness is a critical requirement in wafer processing and

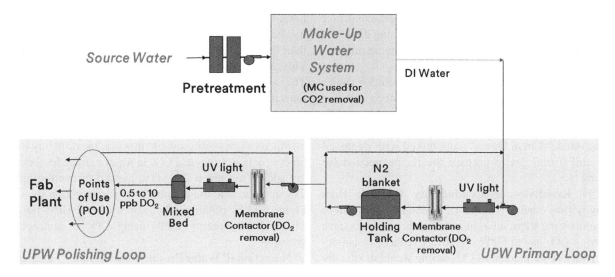

FIGURE 5.6   Generic Ultrapure Water (UPW) production steps.

TABLE 5.3
Process Locations and Benefits of Use for Membrane Contactors in UPW Production

Use	Location of Process	Benefit
Dissolved CO_2 Removal	• Make-up Water	• CO_2 impacts water quality and reduces efficiency of ion exchange equipment and lowers chance of achieving target resistivity specs
Dissolved O_2 Removal	• Make-up Water • Primary Loop • High Purity Polishing Loop • Point of Use	• Deoxygenation protects product from defects, thereby improving production yields • Deoxygenation reduces risk of bacterial growth
Dissolved N_2 Control	High Purity Loop	• Reduced bubble formation

Design Basis:

• **Located In Polishing System**
• **Water Flow = 1600 gpm (360 m³/h)**
• **Temperature = 75°F (24°C)**
• **Inlet O_2: 8.9 ppb**

Membrane Contactor System Configuration:

• **8 Parallel Trains of 3 Contactors in Series (24 Total)**
   (Each Contactor with 131 m2 active area)
• **FRP Housings with PVDF inner surface**
• **Outlet Achieved = < 1 ppb Dissolved O_2**

FIGURE 5.7   Central UPW deoxygenation system – polishing loop.

is the primary driver for use of MC in the semiconductor industry [16–22]. Figure 5.8 shows how reducing dissolved O_2 in UPW impedes oxide growth and allows more processing time [20].

Degassing of Process Chemistries: Figure 5.9 shows a process flow diagram for MC degassing steps in wafer processing [22]. In this specific process cleaning chemicals such as hydrofluoric acid and hydrochloric acid are degassed ("Chem Degas") and mixed with degassed UPW and ozone gas to prepare the recipe needed for cleaning.

UPW Resistivity or Conductivity Control: High resistivity (low conductivity) of UPW can be a problem in some processing steps. It has been found that dosing a trace amount of CO_2 gas to UPW effectively reduces resistivity [23]. A very small amount of CO_2 dosing would drastically

lower the resistivity, as shown in Figure 5.10 where resistivity decreases from 18.2 Meg-Ohm (outside of scale) to less than 0.2 Meg-Ohm by adding only about 15 ppm of CO_2. This chart is based on previously unpublished original data. The symbols in the figure represent different UPW flow rates through MC, demonstrating that change in resistivity is primarily dictated by dissolved CO_2 concentration, and less by UPW flow rate.

Since CO_2 is highly soluble in water, the difficulty in the process is how to control CO_2 in water to trace levels without adding too much CO_2. One method of controlling the CO_2 addition is to use a "split-flow" configuration shown in Figure 5.11 by running only a slip stream of UPW through MC and carbonating UPW using CO_2 at atmospheric pressure.

'Functional' Water Production: The term "Functional" water is sometimes used to designate UPW from which all dissolved gases are first removed to insignificant levels, and then specific gas species are dosed to the degassed UPW in controlled quantities to impart specific functionality to the water which is subsequently used in often proprietary wafer processing steps. The species to be dosed may be various, such as H_2, N_2, NH_3, etc. MC can be used to make such functional water in-line to highly controlled dosage by controlling specific operating parameters such as gas pressure without forming gas bubbles [24]. Figure 5.12 shows some previously unpublished original data from a laboratory trial where hydrogen is added to water using a commercial membrane contactor.

Semiconductor Plant Wastewater Treatment: A unique application of membrane contactor in the microelectronics industry is to remove and recover dissolved ammonia (NH3) from semiconductor plant wastewater. This is an emerging application of MC and as such will be discussed later in this chapter.

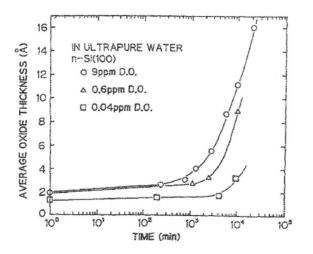

FIGURE 5.8   Effect of dissolved oxygen in UPW on oxide formation, as function of immersion time.

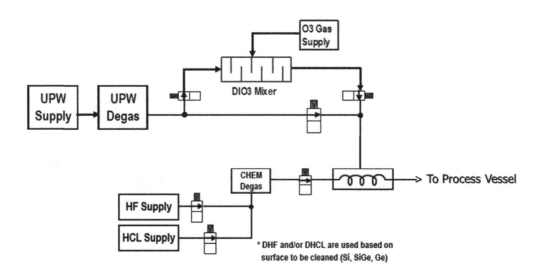

FIGURE 5.9   Schematic diagram for degassing prior to a wafer cleaning step.

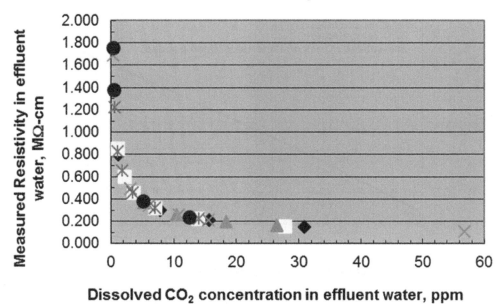

FIGURE 5.10   Reduction of UPW resistivity by CO2 dosing in membrane contactor.

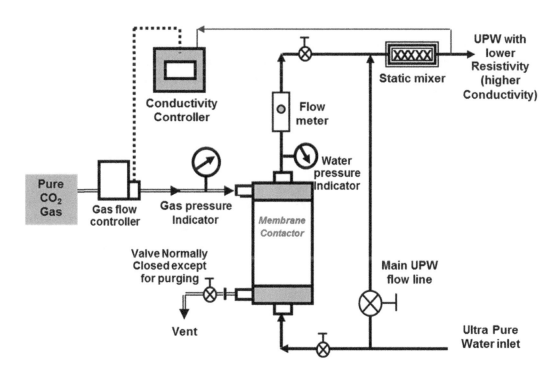

FIGURE 5.11   UPW resistivity reduction using membrane contactor in split-flow configuration.

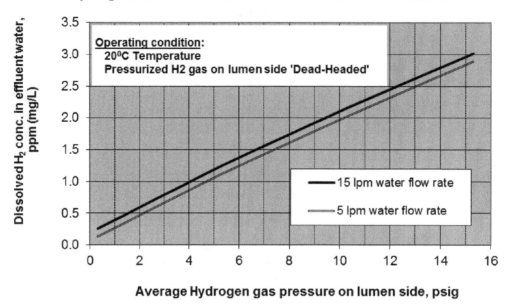

FIGURE 5.12 Illustration of membrane contactor performance in hydrogenation of UPW.

5.6 MEMBRANE CONTACTOR USE IN STEAM AND POWER GENERATION

MC has found increasing use in the steam and power industry, primarily for degassing of boiler feed water (BFW) for steam generation [15, 25]. Steam is used extensively as a utility in many industries, for generating electric power or used directly in refineries, in chemical plants, in sterilization processes, and in district heating, etc. Dissolved oxygen and CO_2 are unwanted in boiler feed water. Lowering the dissolved oxygen level reduces chemical scavenger costs, reduces the need for boiler blow down, and prevents corrosion. Figure 5.13 shows a schematic flow diagram in a combined heat and power plant. As shown in this figure, degassing steps can be implemented in more than one location in such plants.

The specific benefits of using MC are summarized in Table 5.4.

Figures 5.14 and 5.15 show photographs of two installations of MC in the steam/power industry for different purposes: Figure 5.14 is a photograph of a mobile deoxygenation system to make steam in a refinery, and Figure 5.15 is a photograph of a system used for district heating.

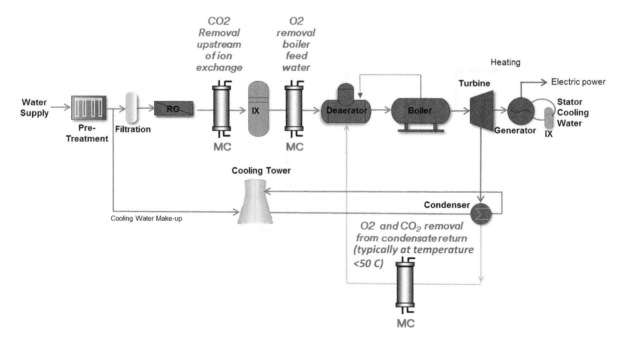

FIGURE 5.13 Possible membrane contactor use locations in combined heat and power plant.

TABLE 5.4

Benefits of Use for Membrane Contactors in Steam and Power Generation

Application	Location of Process	Benefit
Dissolved CO_2 Removal	• Boiler and Heat Removal Recovery Steam Generator (HRSG): make-up Boiler feed water line • Refill water to district heating network	• Reduce risk of corrosion of boiler, piping and downstream components • Decrease chemical use (oxygen scavengers) • Lower water and energy use (reduce boiler blowdown frequency) • Reduce O_2 load on deaerator - lower vent rate, improves OPEX
Dissolved CO_2 Removal	• Boiler make-up before ion exchange • Mobile water treatment systems	• Reduce resin regeneration frequency • Decrease chemical use (i.e. caustic) • Improve EDI water quality • Enabling CO_2 removal in mobile water treatment systems due to compact contactor size • Reduce risk of boiler tube failure caused by formation of carbonic acid

System Data:
- **System use: Condensate Return in Refinery Plant**
- **Design Water Flow Rate = 200 m³/hr**
- **Design Water Temperature = 45°C**
- **22 Parallel Contactors in Code-Rated Vessels (each contactor with 260 m2 active area)**
- **Design maximum Dissolved O2 outlet = 2 ppb**

FIGURE 5.14 Mobile deoxygenation system in refinery.

System data:
- **System Use: O2 removal from Water for District Heating**
- **Design Water Flow Rate = 30 m³/h**
- **Four Membrane Contactors (each contactor with 220 m2 active area)**
- **Design Inlet water O_2 = Air-saturation**
- **Design Outlet water O_2 = 5 ppb**

FIGURE 5.15 Water deoxygenation system for district heating.

5.7 MEMBRANE CONTACTOR APPLICATIONS IN BEVERAGE INDUSTRY

MC processes are used in multitude of areas in the beverage industry. A few publications covering different uses are listed in [26–30].

A. Brewing processes
B. Soft drink production
C. Wine processing
D. Drinking water production
E. Specialty beverages such as coffee

Specific applications of MC in different beverage processes, and associated benefits, are summarized in Table 5.5.

A photograph of a typical installed MC system for the soft drink industry is shown in Figure 5.16, along with the system design data.

MC systems have been installed in several breweries for the deoxygenation of pushing water. High oxygen levels in beer cause issues of taste and shelf life. There are additional uses of deoxygenated water in beer production such as in the pre-coating of filters, pre- and post-filtration, and product chase through pasteurizer. Such an installed degas system and system design parameters are shown in Figure 5.17.

TABLE 5.5
Applications and Benefits of MC in Beverage Industry

Application	Process	Purpose & Benefit
Deoxygenation	• Blending water used with concentrates (syrups, juices, high gravity beer) • Source water degassing • Pushing water or scrubbing water (brewing) » Seal water (brewing) • Red wine production/bottling • Filling	• Extend product shelf life • Consistent product quality • May minimize taste degradation • May help prevent can corrosion (yield loss) • Reduce risk of O_2 pick-up during separation step (centrifuge) • Help minimize aging effect in red wine production • Reduce foaming in filling
Carbonation / Nitrogenation	• Dissolved gas infusion into product (beer, coffee, white wines, and other beverages) • Blending/product water before mixing • Dissolved nitrogen for taste, mouth feel, and head foam stability	• In-line process vs. batch – may reduce processing time • Process simplification • Consistent product quality • Precise concentration levels • Change or improve flavor profile
Decarbonation	• Reduce CO_2 levels in over carbonated product (beer) • CO_2 scrubbing from liquids	• Reduce product waste or wait time • Recovery of CO_2 gas to reduce cost and lower CO_2 footprint

System data:
Design water flow rate = 400 hL/h
Design water temperature = 14° C
Design effluent O2 = <20 ppb

FIGURE 5.16 A typical membrane contactor system in beverage industry.

System data:
  Design Water Flow Rate: 1 - 180 m3/h
  Design water temperature: 0 – 45°C
  Design water pressure: 0 – 8 bar
  Effluent O2 spec.: < 10 – 50 ppb
  System Hot Water Sanitizable to 85 C

FIGURE 5.17 Membrane contactor system for deoxygenating water in beer production.

Excess CO_2 in beer is often undesirable. A photograph of an MC system designed to partially remove CO_2 from beer is shown in Figure 5.18.

The process block diagram in Figure 5.19 illustrates the process steps in a typical hard cider canning operation. As shown in the figure, the hard cider could pick up aerial oxygen in several steps. In one case, a customer encountered DO levels in excess of 1,200 ppb. Such levels of DO increase the probability of corrosion in cans and of can voiding.

A membrane contactor deoxygenation system, shown in Figure 5.20, allows the operator to achieve multiple benefits, such as (a) consistent <200 ppb DO in cans, (b) low pressure loss, (c) low impact on flavor and taste profile, and (d) compact and mobile skid-mounted equipment. The process has been described in [30].

In another illustration for the beverage industry, the MC system shown in Figure 5.21 has been used to nitrogenate beer. It is a rapid in-line process to add N2 gas to beer containing some CO_2 gas. The skid-mounted system is compact and mobile, is simple to operate, and uses low N2 gas pressure.

There are several current applications of membrane contactors for dissolved gas control in wine, as listed below. Specific benefits of MC include reduced processing time, precise control of dissolved gases, efficient gas transfer with reduced gas usage, and gentler impact on wine character.

- O_2 removal from wine to extend shelf life time (< 500ppb DO, sometimes lower)
- CO_2 adjustment for organoleptic taste control, such as
  - Decarbonation in red wines < 0.5 g/L
  - Carbonation in white wines 0.8–10 g/L

System Information:
- Design Water Flow: up to 5 m^3/h
- Design CO_2 spec.: from ~2 v/v to 1 v/v
- Two Membrane Contactors in SS housing (20 m^2 contact area in each)

FIGURE 5.18 Membrane contactors system for controlled removal of CO2 from beer.

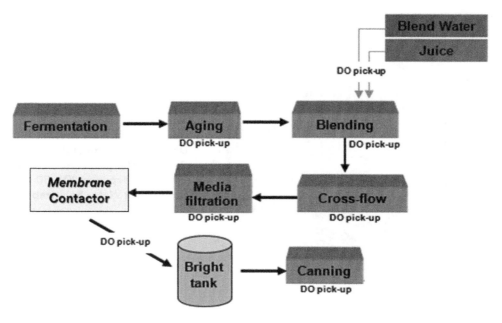

FIGURE 5.19   Membrane contactor use in hard cider canning process.

System Design:
Design Flow Rate: 4.5 m3/hr (20 gpm)
Process Liquid: Clarified Hard Cider (no CO_2)
Design Inlet O2: up to 8 ppm
Design Outlet O2: 200 ppb

FIGURE 5.20   Hard cider deoxygenation system using membrane contactor.

- Carbonation for semi-sparkling wines (max 2.5 bar) and champagne wines (> 3 bar)
- Alcohol "adjustment" (reduction of 2–14% EtOH) in wine

Figure 5.22 shows the photo of a skid-mounted MC system for controlling dissolved gas levels in wine. In addition to removing or adding gases, MC systems can also be configured to simultaneously add one component such as CO_2 while removing a second component such as O_2. These features offer significant system flexibility.

Two unique uses of MCs in the food and beverage industry merit separate mentions [31–34]. One is the ability to reduce alcohol levels in wine or beer without any significant impact on wine flavor or taste. The other is the ability to remove water vapor from and concentrate fruit or vegetable juices at low temperature without damage to the juice. In one such process, described in [32] and illustrated in

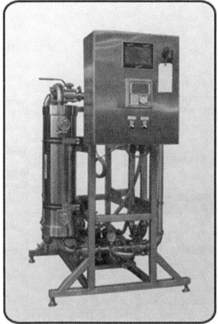

System Design
- **316L SS Contactor**
- **Design liquid flow rate 1-10 m³/h**
- **Design Temperature: 1.6° C**
- **CO₂ inlet concentration: 1.5 v/v**
- **Inlet N₂ pressure: ~3bar (45 psi)**
- **50 ppm effluent dissolved N₂**

FIGURE 5.21 Membrane contactor system to nitrogenate beer.

System Design Data:
Flow rate = 5-20 m³/h
Temperature = 0-40 (50)°C

Membrane Contactors in SS Housing w/ sanitary connections
Effluent CO₂ range: <0.5 to 10 g/L
Effluent O₂ <300 ppb

Capable of Hot Water Sanitization to 85°C
Gas pressure up to 6 bar

FIGURE 5.22 Membrane contactor system for controlled adjustment of gas levels in wine.

Figure 5.23, wine from a storage tank is first taken through an RO unit which allows alcohol and water to permeate through the RO membrane while keeping the flavor components in a wine concentrate. The RO permeate, which is an alcohol-water solution, is next taken through a membrane contactor where a 'strip" water flows on the other side of the membrane in the MC and partially removes the alcohol before the alcohol-lean stream is mixed back with the wine concentrate before returning to the wine tank. The wine is typically run in a batch recirculation process. Precise control and adjustment of the alcohol level can be achieved in the final low-alcohol wine, and it has been reported that an optimized flavor or taste profile can be achieved in wine.

Membrane contactors have also been used in limited cases for drinking water treatment. Specific reported cases include (a) the reduction of excess gases or bubbles from well waters to avoid problems with pumping and distribution, (b) CO₂ removal or adjustment for corrosion control in piping, (c) re-mineralizing and pH adjustment in drinking water to improve the taste. These applications are not yet as well known as the ones described in the previous sections and therefore are not discussed further.

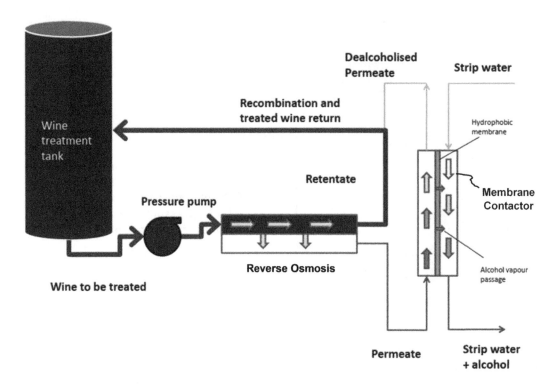

FIGURE 5.23 Flow schematic for wine alcohol adjustment using membrane contactor.

5.8 MEMBRANE CONTACTOR APPLICATIONS IN PHARMACEUTICAL AND BIOTECHNOLOGY INDUSTRY

The pharmaceutical and biotechnology industry needs high-quality water with specific requirements. At present, MC is mostly used in the treatment steps to make purified water, which is subsequently treated through sterilization, endotoxin removal, and final filtration to make water for injection (WFI), or water used in medical procedures, or to prepare parenteral drug formulations.

The flow diagram in Figure 5.24 is the generic depiction of a central purified water production process where MC could be used.

In addition to it use in purified water production, MC is also used in other location points in the water loop. Specific uses and benefits of MC for various locations are summarized in Table 5.6.

A unique application of membrane contactors in the biopharmaceutical industry is to reduce all non-condensable gases (NCG) such as O_2, N_2, CO_2, etc. from purified water before the water is fed to steam sterilizers. NCG species,

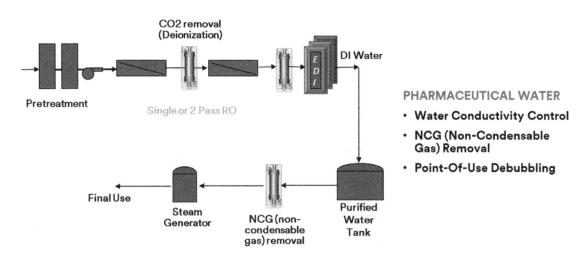

FIGURE 5.24 A generic pure water production process in biopharmaceutical industry.

TABLE 5.6

Uses and Benefits of Membrane Contactors in Biopharmaceutical Industry

Application	Location of process	Benefit
Dissolved CO$_2$ Removal	Commonly pre-EDI	• Chemical free • Continues operation • SS housing for hot water sanitization (to 85°C)
Dissolved O$_2$ Removal	Benchtop UPW Systems	• Easy to expand, due to modular design • Reduction in operation cost
NCG Removal	Pure Steam Generator	• Easy operation • Compact system design
De-bubbling	Clinical Analyzers	• Source of degassed purified water • No interference in analysis from bubbles • Cleanliness, compactness
De-bubbling	Quality control	• No interference in QC process from bubbles • Cleanliness, no contamination

System design:
Flow rate = 0-5 m³/h
Temperature = 10-40°C
Membrane Contactor in SS Housing
NCG < 3.5% v/v per EN 285 standard
FDA approved materials
85°C Hot Water Sanitization

FIGURE 5.25 Non-condensable gases removal system using membrane contactor.

unless removed or reduced from feed water, reduce heat transfer efficiency in steam systems and can also lower the temperature of the generated steam which is used for sterilization. The removal of NCG is, therefore, of particular interest in the pharmaceutical industry. Figure 5.25 shows a photograph of an NCG removal system that incorporates membrane contactors.

All of the pharmaceutical applications described above involved gas–liquid contacts. Much fundamental research has also been done on liquid–liquid extraction in membrane contactors. The principle of the process is described in [12], and [35–42] describe many novel uses.

5.9 MEMBRANE CONTACTOR APPLICATIONS IN OIL AND GAS INDUSTRY

MC is not widely used yet in the Oil and Gas industry, especially where the application is land based. The most likely places where MC success is anticipated are on offshore platforms because of very favorable attributes of space and weight savings compared to conventional technologies. Potential offshore applications of MC include deoxygenation of water (a) to be injected underground for oil field flooding, (b) to be used for crude oil desalting, (c) to supply boiler feed water, etc.

As discussed before, MC enables the design and construction of compact, modular systems, which occupy less space and can be installed in areas where larger deaeration equipment, such as vacuum towers and forced draft towers, is not feasible. Because of the compact size of MC systems, the system weight can be reduced by up to 50% compared to conventional towers. The modular construction enables the design and construction of MC deoxygenation systems that can be quickly adapted for incremental degas capacity requirements after initial installation, bringing an advantage when applied to oil and gas offshore platforms, which are built to operate for decades. A comparison between an MC system and a vacuum tower has been discussed in [43].

Technologies deployed on offshore platforms must meet special criteria and certification, and MC technology is no exception. For example, offshore water degassing units must meet much higher design pressure criteria compared to onshore applications. To meet such challenges, new MC designs have been developed where membrane cartridges can be installed inside code-certified pressure vessels suitable for offshore use. A photograph of such a pilot system built around high-pressure membrane contactors is shown in Figure 5.26.

The performance of seawater deoxygenation using MC has been verified in lab scale, and has been demonstrated in pilot scale, achieving an oxygen concentration of less than 10 ppb at 10 m3/h of low salinity water. An engineering evaluation had shown that for large oxygen removal units (more than 795 m3/day) it is possible for MC to achieve cost savings of approximately 20% [43]. Pilot testing showed that the operation of membrane deaeration technology was simple, and the reproducibility of the process has been demonstrated without the use of chemical scavengers. The technology readiness level of the MC technology is discussed in [44].

5.10 APPLICATIONS OF MEMBRANE CONTACTORS IN INK AND COATING INDUSTRY

Most of the MC uses in the previous sections concentrated on treating water-based liquids. A new and enabling application of MC is for the degassing and debubbling of printing inks, various coatings, solvents, and oils. The primary requirements of the membrane contactors here is that the membrane must be gas-permeable but liquid-impermeable. A sketch of a typical hollow fiber membrane used for this purpose is shown in Figure 5.27. The hollow fiber has a thin "skin" on the outer surface which prevents the penetration of the process liquids into the membrane pore. The process liquid to be degassed flows past one side of the membrane whereas a vacuum is applied to the other side of membrane. Dissolved gases or bubbles in the process liquid are extracted because of the reduced partial pressure on the vacuum side.

Of all such MC applications, the most developed one to date has been ink debubbling or degassing in large commercial inkjet printers. The degassed inks are typically applied to substrates such as tiles, packaging, labels, textiles, printable electronics, coating, etc. Typically, debubbling and degassing are needed to (1) prevent ink starvation and misfires in printers, (2) minimize substrate surface defects, (3) increase printing speed for higher productivity, and (4) reduce downtime and maintenance. In addition, ink degassing is also required to minimize foaming during the filling and packaging of bulk ink into ink cassettes prior to shipment. Some basic information on using membrane contactors for the process of ink debubbling and degassing was presented in [45]. Figure 5.28 shows different locations where MC units are installed for the degassing or debubbling of ink.

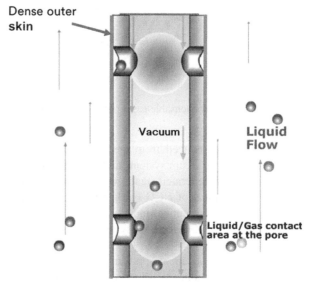

FIGURE 5.26 Pilot system for offshore water degassing using membrane contactor.

FIGURE 5.27 Hollow fiber membrane for degassing inks, coating, etc., in membrane contactor.

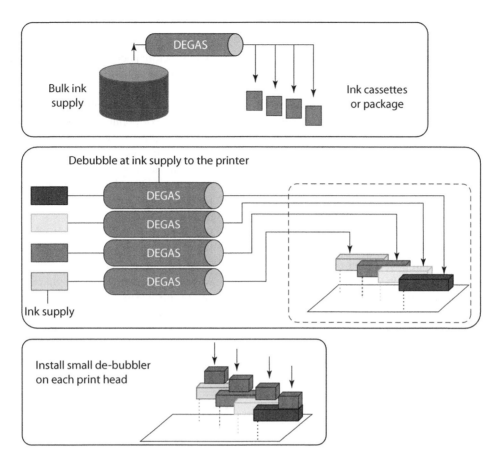

FIGURE 5.28 Locations of membrane contactor use in inkjet printing systems for degassing/debubbling inks.

5.11 EMERGING APPLICATIONS OF MEMBRANE CONTACTORS

There appears to be significant and wide interest in using MCs for a variety of new applications not covered in the previous sections. Most activities are at academic or industrial R&D levels and are at proof-of-concept phase or at bench scale testing phase. Certain applications have been adopted at larger scale but have not yet been replicated by multiple customers. Many of these investigations are for fundamental research and may not necessarily be worth developing to commercial scale. However, there are many others that deserve closer attention because they could conceivably be of great benefit provided that innovative solutions become available in the near future regarding optimum membrane materials and properties, application-specific membrane contactor designs, and low-cost MC manufacturing processes. Some selected applications that we believe to be worth mentioning are listed below. The list is not sorted in any order. Where possible, references to publicly available papers or patents have been made.

- Abatement of pollutants such as NH_3 or H_2S in wastewater using MC in a process called TMCS (Trans-Membrane Chemical Sorption)

  Ammonia removal and recovery using membrane contactor has been found to be of significant interest for some customers under favorable process operating conditions. Process principles and various applications have been published in [46–49]. The removal of H_2S from sour water using a very similar process is presented in [50].

- Enriching fuel value of biogas by removing CO_2 in MC

  Biogas as a renewable source of energy has become a topic of interest in last few years. Biogas contains methane with significant CO_2, and removing the CO_2 improves the fuel value of the methane. Since biogas is available at moderate pressure and temperature, membrane contactors could be a good candidate to scrub out CO_2 from CH4, as discussed in [51–52].

- Debubbling of high-viscosity coatings and resins

  The presence of bubbles can cause much difficulty in the processing and packaging of highly viscous liquids. These could be various coatings and resins for plastic and paper substrates, solvents, food ingredients, soaps, and cosmetics, etc. In-line debubbling is a highly desirable step prior to packaging or bottling because it prevents the re-introduction of bubbles in the liquid. The concept of membrane contactor technology lends itself well to in-line vacuum debubbling; however, it requires the special design of MC devices to compensate for poor gas diffusion and high pressure drop at high viscosity.

- Gas humidification and dehumidification using MC

  MC technology is very suitable for humidifying or dehumidifying air streams along with enthalpy control. The ability to independently set gas and liquid pressures, flow rates, and temperatures in MC allows for some unique capabilities. Research on this application of MC has been presented in [53–58].

- Post-combustion CO_2 capture from flue gas from fossil fuel burning

  The absorption of gas species such as CO_2 in aqueous solutions has been investigated many times over the years [59–64]. Most of the current interest appears to be in the removal of CO_2, a greenhouse gas, from post-combustion flue gas. The process is currently being tested in pilot scale.

- Removal of Volatile Organic Compounds (VOC) and disinfection byproducts such as Tri-Halo Methane (THM) from water

  Several investigations have been conducted over the years on removing volatile compounds from water using MC [65–70]. VOC and THM are often present in source water used for different industries, from semiconductor UPW to drinking water, and removing these compounds is necessary to prevent complications in downstream processing or for health reasons.

- Oxygen removal from ground water prior to re-injection to aquifers

  The need for oxygen removal in the offshore oil and gas industry has already been discussed in Section 8. In separate applications in water-constrained locations, when water is withdrawn from underground aquifers for use above ground for any purpose, from environmental remediation to power generation, mandates are being put in place to re-inject the water back underground after use in the same condition as it was withdrawn. The removal of dissolved oxygen and other volatiles from water is often a part of the requirement. One such example has been described in [71]. Other similar uses cannot be cited here because of proprietary content.

- Oxygen and gas bubbles removal from saline solutions prior to infusion and surgery

  Medical industry customers are beginning to look at MC for in-line removal of gas bubbles from saline solution. The presence of gas bubbles is highly detrimental to surgical or infusion procedures, and the ability of MC to operate "in-line" and "under-pressure" without reintroducing bubbles could be of major benefit. The nature of confidentiality prevents further disclosure, but one example is provided in [72].

- Oxygen and gas bubbles removal from "biosimilar" products such as blood substitutes to improve shelf storage

  An interesting new use of MCs is to remove dissolved oxygen from biological or biosimilar liquids before packaging to increase the shelf storage lifetime of these liquids. No specifics can be disclosed for confidentiality reasons.

- Removal of priority pollutants from wastewater

  Many toxic pollutants in wastewater, such as phenol and naphthene, cannot be effectively removed using distillation. Many academic research articles are available in open literature on using membrane contactors in liquid–liquid extraction configuration for these separations. Some of the older investigations are reported in [73–77].

- Recovery of high-value metal ions in MC using liquid–liquid extraction

  Metals ions often need to be removed from wastewater because they are (a) valuable, (b) rare, and (c) toxic. Since they are non-volatile they cannot be recovered via distillation, so a liquid–liquid extraction process must be employed, primarily using ionic liquids. MC is a potential substitute for mixer-settler technology, which is the most common for extraction. References [78–85] present research work on a range of interesting uses of MC over time.

- CO_2, excess gas, and energy removal from close breathing space or capsules

  The last topic of interest to present is the use of MC to treat close breathing space for life-sustaining purpose [86–88]. Though this is not expected ever to be a financially successful business, it is mentioned here as another illustration of why MC is used in unique applications due to its low mass and compact nature.

5.12 CONCLUDING REMARKS

Membrane contactor technology is now several decades old. It is one of the newest membrane technologies in the market but has a unique set of interesting and useful attributes which is allowing it to serve a diverse range of industries and markets and to create a distinct application base for itself. In many instances, membrane contactors are presenting customers with new opportunities that did not exist before. The current chapter has attempted to provide a historic perspective and to inform readers with the most up-to-date public knowledge about the amazing array of applications developed over the years. It is our hope that this document will continue to generate interest in this technology, and will lead to many creative and innovative ideas that will strengthen existing uses and expand the use of this technology in new applications.

5.13 ACKNOWLEDGMENTS

The authors gratefully acknowledge the teachings and insights of many colleagues over the years, especially those of J. Schneider and M. Ulbricht.

REFERENCES

* The authors' references to the various patents mentioned in this article do not constitute a grant of a license to practise any of these technologies, nor do they imply the authors' acknowledgment of the validity of any of the referenced patents.

1. A. Sengupta and R. A. Pittman, "Application of membrane contactors as mass transfer devices", in A. K. Pabby, S. S. H. Rizvi, and A. M. Sastre, eds, *Handbook of Membrane Separations*, CRC Press Taylor & Francis Group, Boca Raton, FL (2009), 7–24.

2. K. K. Sirkar, "Other new membrane processes", in W. S. W. Ho and K. K. Sirkar, eds, *Membrane Handbook*, Van Nostrand Reinhold, New York, NY (1992).

3. B. W. Reed, M. J. Semmens, and E. L. Cussler, "Membrane contactors", in R. D. Noble and S. A. Stern, eds, *Membrane Separation Technology – Principles and Application*, Elsevier, Amsterdam, The Netherlands (1995), 467–498.

4. M. Shindo, T. Yamamoto and K. Kanada, Gas transfer process with hollow fiber membrane, US Patent 4,268,279 (1981).

5. I. Masano, S. Furusaki and T. Miyauchi, Separation of volatile materials by gas membrane, *Ind Eng Chem Process Des Dev* (1982), 21, 421–426.

6. Z. Qi and E. L. Cussler, Microporous hollow fibers for gas absorption – I. Mass transfer in the liquid, *J Membr Sci* (1985), 23, 321–332.

7. Z. Qi and E. L. Cussler, Microporous hollow fibers for gas absorption – II. Mass transfer across the membrane, *J Membr Sci* (1985), 23, 333–345.

8. B. S. Kim and P. Harriot, Critical entry pressure for liquids in hydrophobic membranes, *J Colloid Interface Sci* (1987), 115 (1), 1–8.

9. U. Baurmeister and M. Pelger, Device for separating gas bubbles from fluids, US Patent 4,828,587 (1989).

10. E. L. Cussler, R. W. Callahan and P. R. Alexander, Liquid/liquid extractions with microporous membranes, US Patent 4,966,707 (1990).

11. K. K. Sirkar, Immobilized-interface solute transfer process, US Patent 4,997,569 (1991).

12. R. Prasad and K. K. Sirkar, "Membrane-based solvent extraction", in W. S. W. Ho and K. K. Sirkar, eds, *Membrane Handbook*, Van Nostrand Reinhold, New York, NY (1992), 727–741.

13. F. E. Wiesler and R. A. Sodaro, Degasification of water using novel membrane technology, *Ultrapure Water* (1996), 35, 53.

14. T. Aanazawa, M. Miyashita, K. Murata, H. Akasu and S. Mimura, Hollow fiber membrane type phase contact apparatus, Japanese Patent 2,725,311 (1997).

15. S. H. Macklin, W. E. Haas and W. S. Miller, Carbon dioxide and dissolved oxygen removal from makeup water by gas transfer membranes, In *International Water Conference*, Pittsburgh, PA, Oct 30 (1995).

16. Y. Yagi, T. Imaoka, Y. Kasama, T. Ohmi, Advanced ultrapure water systems with low dissolved oxygen for native oxide free wafer processing. *IEEE Trans Semicond Manuf* (1992), 5 (2), 121–127.

17. M. S. L. Tai, I. Chua, K. Li, W. J. Ng and W. K. Teo, Removal of dissolved oxygen in ultrapure water production using microporous membrane modules, *J Membr Sci* (1994), 87, 99–105.

18. J. Bujedo and P. A. Peterson, Removing dissolved oxygen from ultrapure water, *Semiconductor International*, October (1997).

19. A. Sengupta, P. A. Peterson, B. D. Miller, J. Schneider and C. W. Fulk, Large-scale application of membrane contactors for gas transfer from or to ultrapure water, *Sep Purif Tech* (1998), 14, 189–200.

20. M. Morita, T. Ohmi, E. Hasegawa, and A. Teramoto, Native oxide growth on silicon surface in ultrapure water and hydrogen peroxide, *Jpn J Appl Phys* (1990), 29 (Part 2), 12.

21. F. E. Wiesler and C. T. Kao, Meeting water quality specifications for 300 mm processing, Semiconductor China, Beijing, China, Mar 28–29 (2001).

22. A. Sengupta, S. Willis, K. Elsherif and R. Pagliaro, Method for preparing parts-per-trillion level dissolved O_2 in UPW using membrane degassifiers to enable superior wafer surface preparation, *Presented at Ultra-Pure Micro*, Austin, TX, May (2018).

23. K. Sakai, H. Kato and T. Kanbe, Method and device for controlled Resistivity of ultrapure water, Japanese Patent 11,057,707 (1999).

24. M. Sell, M. Bischoff, A. Mann, R. D. Behling, K. V. Peinemann and K. Kneifel, Method of introducing hydrogen into aqueous liquids without forming bubbles, US Patent 5,523,003 (1996).

25. P. D'Angelo, Oxygen removal – Theory and potential use of deoxygenation membranes in the utility industry, *Ultrapure Water* (1995), 13, 60–63.

26. J. Mackey and J. Mojonnier, CO_2 injection using membrane technology, Presented at *Eighth Annual Conference on the Operation of Technologically Advanced Beverage Plants and Warehouses*, Location Not Cited, Mar 21–*23* (1995).

27. F. Breitschopf, W. Brau, S. Dittrich and R. Koukol, A new kind of water deaeration based on hollow fiber membranes, *Brauwelt Int*, (1997) 5, 431.

28. C. B. Gill and I. D. Menneer, Advances in gas control technology in the Brewery, *Brewer* (1997), 83 (987), 77–84.

29. Technical Feature Article, Non-dispersive diffusion for nitrogenation, *Brauwelt Int* (2000), 18 (2), 129–130.

30. A. Kapadia and B. E. Calvi, Membrane contactors manage dissolved oxygen in canning process, *Water Technology Journal*, July 26 (2017).

31. P. A. Hogan, R. P. Canning, P. A. Peterson, R. A. Johnson and A. S. Michaels, A new option: Osmotic distillation, *Chem Eng Prog*, July (1998).

32. Wollan, D. (2010) Membrane and other techniques for the management of wine composition. In: *Managing Wine Quality*, Ed. A.G. Reynolds (Woodhead Publishing, Cambridge, UK) pp. 133–163.

33. P. Russo, L. Liguori, D. Albanese, A. Crescitelli and M. Di Matteo, Investigation of Osmotic Distillation technique for beer dealcoholization, *Chem Eng Trans* (2013), 32, 1735–1740.

34. A. Cassano, E. Drioli, G. Galaverna, R. Marchelli, G. Di Silvestro and P. Cagnasso, Clarification and concentration of citrus and carrot juices by integrated membrane processes, *J Food Eng* (2003), 57, 153–163.

35. A. A. Wald, J. L. Lopez and S. L. Matson, Membrane-mediated antibiotic extraction using liquid ion exchangers, Presented at *North American Membrane Society (NAMS) Conference*, Syracuse, NY (1989).

36. R. Prasad and K. K. Sirkar, Hollow fiber solvent extraction of pharmaceutical products – A case study, *J Membr Sci* (1989), 47, 235–259.

37. L. Dahuron and E. L. Cussler, Protein extractions with hollow fibers, *AIChE J* (1988), 34 (1), 130–136.

38. H. B. Ding, P. W. Carr and E. L. Cussler, Racemic leucine separation by hollow-fiber extraction, *AIChE J* (1992), 38 (10), 1493–1498.

39. G. Vatai and M. N. Tekic, Membrane-based ethanol extraction with hollow fiber module, *Sep Sci Technol* (1991), 26 (7), 1005–1011.

40. C. J. Wang, R. K. Bajpai and E. L. Iannotti, Non-dispersive extraction for recovering lactic acid, *Appl Biochem Biotechnol* (1991), 28–29, 589–603.

41. R. Kertesz, S. Schlosser and M. Simo, Membrane bases solvent extraction and stripping of phenylalanine in HF contactors, In *Euromembrane 2004*, Hamburg, Germany, Sep 28–Oct 1 (2004).

42. A. Baudot, J. Floury and H. E. Smorenburg, Liquid-liquid extraction of aroma compounds with hollow fiber contactor, *AIChE J* (2001), 47 (8), 1780–1793.

43. S. van Pelt and H. Churman, Offshore membrane deaeration as a replacement for vacuum tower deaeration – A comparative study, Presented at *SPE Annual Technical Conference and Exhibition*, Houston, TX, Sep (2015).

44. Anonymous, Technology qualification of 3M™ LIQUI-CEL™ membrane contactors for deoxygenation of low salinity water for polymer flooding, water, Online. Available at: https://www.wateronline.com/doc/technology-qualificationof-liquicel-membrane-contactors-for-deoxygenation-of-low-salinity-water-for-polymerflooding-0001, Mar 26 (2018).

45. A. Sengupta, Ink debubbling and degassing using membrane degassers: Design and process considerations, Presented at *The Ink Jet Conference*, Chicago, IL, April (2018).

46. M. Stasiak, J. Munoz and A. Sengupta, Ammonia removal from industrial wastewater using membrane contactor, In Paper Presented at *International Congress on Membranes and Membrane Processes (ICOM)*, Seoul, Korea, Aug 21–26 (2005).

47. M. Stasiak, M. Ulbricht, J. Schneider, J. Munoz, A. Sengupta, B. A. Kitteringham and F. E. Wiesler, Using 'Trans Membrane Chemi Sorption' (TMCS) for ammonia removal from industrial waste waters, *New Fab Technol J*, July (2011).

48. M. Ulbricht, J. Schneider, M. Stasiak, and A. Sengupta, Ammonia recovery from industrial wastewater by TransMembraneChemiSorption, *ChemieIngenieur Tech* (2013), 85 (8), 1259–1262.

49. M. Ulbricht, G. Laknerb, J. Laknerc and K. Belafi-Bako, Trans-membrane chemisorption of ammonia from sealing water in Hungarian powder metallurgy furnace, *Desalin Water Treat* (2017), 75, 253–259.

50. J. Minier-Matar, A. Janson, A. Hussain, and S. Adham, Application of membrane contactors to remove hydrogen sulfide from sour water, *J Membr Sci* (2017) 541, 378–385.

51. S. Vogler, A. Braasch, G. Buse, S. Hempel, J. Schneider, and M. Ulbricht, Biogas conditioning using hollow fiber membrane contactors, *ChemieIngenieur Tech* (2013), 85 (8), 1254–1258.

52. A. Park, Y. M. Kim, J. F. Kim, P. S. Lee, Y. H. Cho, H. S. Park, S. E. Nam, and Y. I. Park, Biogas upgrading using membrane contactor process: Pressure-cascaded stripping configuration, *Sep Purif Technol* (2017), 183, 358–365.

53. U. Bonne, D. W. Deetz, J. H. Lai, D. J. Odde and J. D. Zook, Membrane dehumidification, US Patent 4,915,838 (1990).

54. T. Meulen and P. Berend, Transfer device for the transfer of matter and/or heat from one medium flow to another medium flow, US Patent 5,230,796 (1993).

55. B. Wagner, W. Jehle, J. Steinwandel, T. Staneff T and J. Herczog, Continuous membrane supported cabin air dehumidification process. Presented at *Euromembrane 97*, University of Twente, Enschede, The Netherlands, June 26 (1997).

56. K. Nowak, K. Kneifel, R. Waldemann, J. Wind, W. Albrecht, R. Hilke, R. Just, K. V. Peinemann, Hollow fiber membrane contactor for air humidity control. Presented at *Euromembrane 2004*, *Hamburg, Germany*, Sep 28–Oct 1 (2004).

57. D. A. Thompson, Enthalpy pump, US Patent US 6,684,649 B1 (2004).

58. T. Takagi, Humidifying membrane module, European Patent EP 2 258 464 B1 (2013).

59. D. O. Cooney and C. C. Jackson, Gas absorption in a hollow fiber device, *Chem Eng Commun* (1989), 79, 153–163.

60. P. H. M. Feron, A. E. Jansen and R. Klaassen, Membrane technology in carbon dioxide removal, TNO Environmental and Energy Research, Publication number 92–075 (1992).

61. P. H. M. Feron and A. E. Jansen, CO_2 separation with polyolefin membrane contactors and dedicated absorption liquids: Performances and prospects, *Sep Purif Technol* (2002), 27, 231–242.

62. Y. Dindore, D. W. F. Brilman, P. H. M. Feron, G. F. Versteeg, CO_2 absorption at elevated pressures using a hollow fiber membrane contactor, *J Membr Sci* (2004), 235, 99–109.

63. K. A. Hoff, O. Juliassen, O. Falk-Pedersen, H. F. Svendsen, Modeling and experimental study of carbon dioxide absorption in aqueous alkanolamine solutions using a membrane contactor, *Ind Eng Chem Res* (2004), 43 (16), 4908–4921.

64. J. P. Ciferno, T. E. Fout, A. P. Jones and J. T. Murphy, Capturing carbon from existing coal-fired power plants, Chem Eng Prog (2009), April, 33.

65. M. J. Semmens, R. Qin and A. Zander, Using a microporous hollow fiber membrane to separate VOCs from water, *J Am Water Works Assoc* (1989), 81, 162–167.

66. G. C. Sarti, C. Gostoli and S. Bandini, Extraction of organic components from aqueous streams by vacuum membrane distillation, *J Membr Sci* (1993), 80, 21–33.

67. R. Klaassen, A. E. Jansen, F. Oesterholt and B. A. Bult, *Pertraction of Hydrocarbons from Wastewater Streams*, Final Report 94-039, *TNO Environmental and Energy Research*, Delft, The Netherlands, Mar (1994).

68. D. Schwarz, TOC reduction using membrane technology, *Ultrapure Water* (2000), 18, 60–63.

69. A. Dey, G. Thomas, K. A. Kekre and T. Guihe, Removing THMs by membrane contactors in a UPW plant – A case study, *Water Cond Purif* (2002), February 120–123.

70. B. Raczko, Update on innovative use of gas permeable hydrophobic membranes for volatile contaminant and THM control, Presented at *AWWA Annual Conference & Exposition, Dallas, TX, June* (2012).

71. C. E. Koch, Groundwater remediation system pilot and operating data, In Paper 36 Presented at the *79th International Water Conference, Scottsdale, AZ, Nov.* (2018).

72. C. F. E. Kirsch and B. W. Reed, Device for removal of gas bubbles and dissolved gasses in liquid, US Patent 6,503,225 (2003).

73. T. Porebski, S. Tomzik, W. Ratajczak, A. Wieteska, M. Zebrowski and M. Karabin, Industrial application of hollow fiber modules in the process of phenol extraction from the hydrocarbon fraction, In *Proceedings International Conference PERMEA 2003*, Tatranske Slovakia, Sept 7–11 (2003).

74. T. J. Chen and J. R. Sweet, Selective separation of naphthenes from paraffins by membrane extraction, US Patent 5,107,056 (1992).

75. C. J. Tompkins, A. S. Michaels and S. W. Peretti, Removal of p-nitrophenol from aqueous solution by membrane-supported solvent extraction, *J Membr Sci* (1992), 75, 277–292.

76. A. K. Zander, J. S. Chen and M. J. Semmens, Removal of hexachlorocyclohexane isomers from water by membrane extraction into oil, *Water Res* (1992), 26 (2), 129–137.

77. C. H. Yun, R. Prasad and K. K. Sirkar, Membrane solvent extraction of priority organic pollutants from aqueous waste streams, *Ind Eng Chem Res* (1992), 31, 1709–1717.

78. B. M. Kim, Membrane-based solvent extraction for selective removal and recovery of metals, *J Membr Sci* (1984), 21, 5–19.

79. M. Matsumoto, H. Shimauchi, K. Kondo and F. Nakashio, Kinetics of copper extraction with Kelex-100 using a hollow fiber membrane extractor, *Solvent Extr Ion Exch* (1987), 5 (2), 301–323.

80. P. R. Alexander and R. W. Callahan, Liquid-liquid extraction and stripping of gold with microporous hollow fibers, *J Membr Sci* (1987), 35, 57–71.

81. J. T. F. Keurentjes, T. G. J. Bosklopper, L. J. van Dorp and K. van'tRief, The removal of metals from edible oil by a membrane extraction procedure, *J Am Oil Chem Soc* (1990), 67 (1), 28–32.

82. M. Goto, F. Kubota, T. Miyata and F. Nakashio, Separation of Yttrium in a hollow fiber membrane, *J Membr Sci* (1992), 74, 215–221.

83. C. H. Yun, R. Prasad, A. K. Guha and K. K. Sirkar, Hollow fiber solvent extraction removal of toxic heavy metals from aqueous waste streams, *Ind Eng Chem Res* (1993), 32 (6), 1186–1195.

84. D. Kim, L. E. Powell, L. H. Delmau, E. S. Peterson, J. Herchenroeder and R. R. Bhave, Selective extraction of rare earth elements from permanent magnet scraps with membrane solvent extraction, *Environ Sci Technol* (2015), 49, 9452–9459.

85. X.-L. Lu, "Membrane Contactor", In *Membrane-Based Separations in Metallurgy*, Chap 13, (Eds.) Lan Ying Jiang and Na Li, 335–356, Amsterdam, Netherlands: Elsevier Inc. (2017).

86. G. Noyes, A mobile liquid venting membrane separator for carbon dioxide, humidity, and waste heat removal from spacesuits and manned spacecraft. In *23rd International Conference on Environmental Systems*, Colorado Springs, CO, July 12–15 (1993).

87. T. O. Leimkuehler, C. Spelbring, D. R. Reeves and J. M. Holt, Development of the next generation gas trap for the space station internal thermal control system, Presented at SAE International Conference, Paper 2003-01-2566 (2003).

88. K. Li, M. S. L. Tai and W. K. Teo, Design of a CO_2 scrubber for self-contained breathing systems using a microporous membrane, *J Membr Sci* (1994), 86, 119–125.

Part III

Chapters on Liquid–Liquid Contacting

6 Computational Modeling of Mass Transfer in Hollow Fiber Membrane-Based Separation Processes

B. Swain, A. M. Sastre, and Anil K. Pabby

CONTENTS

6.1 INTRODUCTION

A number of separation processes are being carried out by using hollow fiber membrane contactors (HFMCs) in the place of conventional equipment. HFMCs are used widely in different applications and are even suitable for use at an industrial scale because of their high and constant specific interfacial area and their geometrical configuration for easy fabrication [1, 2]. In HFMCs, the membrane mainly acts as a physical permeable barrier between the two fluids. In this case, the membrane will not have any effect on partition coefficients unlike most membrane operations. HFMCs are independent of the fluid phases in contact and are also independent from corrosion due to being made up of plastic material. Because they have no moving parts, high modularity and compatibility, easy control, and straight forward scale-up, everyone's attention is drawn towards HFMCs. Overall, HFMCs bring modernization to the extraction processes with a step towards process intensification. Process intensification is discussed briefly in **Chapter 8** of this book. HFMCs provide considerable extraction at micro as well as macro level of solute concentrations [3–7]. Assessing the technical maturity of HF contactors, it is required a comprehensive understanding of the separation process and identification of the influential

parameters responsible for the design, scale-up, and optimization. Mathematical modeling is a tool used to understand the process in-depth and also to identify the key influential parameters.

6.2 DIFFERENT SCALE OF MASS TRANSFER MODELING

Modeling quantifies the mass transfer under dynamic equilibrium condition with each parametric effect including contactor module geometry. Design and operating process parameters can be optimized for the effective output and durability. One should have an in-depth understanding of modeling aspects for a better command of it. Modeling can be broadly categorized at different scales for different applications as below: [8]

- At nanoscale: molecular modeling, to better control the surface states of catalysts and activator, to increase selectivity, and to facilitate asymmetrical syntheses (chiral technologies), or to explain the relationships of structure / activity at the molecular scale in order to control the crystallization, coating, and agglomeration kinetics, etc.

- At microscale: mathematical modeling based on the computational chemistry, which is very useful to gain a better knowledge of complex media such as non-Newtonian liquids, melted salts, supercritical fluids, multiphase dispersions, and suspensions, and more generally all systems whose properties are controlled by rheology and interfacial phenomena such as emulsions, colloids, gels, froths, foams, hydro-soluble polymers, and particulate media such as powders, aerosols, and charged and viscous liquids. This scale of modeling also covers fractal structures of porous media and their influence on mass and heat transfer, and on chemical and biological reactions.

- At meso and macroscales: the design of new operating modes for existing equipment such as reversed flow, cyclic processes, unsteady operations, extreme conditions, i.e., high temperature, high pressure technologies, and supercritical media can be modeled at this scale. The Computational Fluid Dynamics (CFD)-based model is a more suitable approach for the design of new equipment or unit operations especially by seeking to render the process step multifunctional with higher yields in coupling the chemical reaction with separation or heat transfer, which provides a considerable economic benefit. More generally CFD is of assistance when it concerns the design of new equipment based on new principles of coupling or uncoupling elementary operations (transfer, reaction, separation). This approach also reduces the headache of rigorous experimentation.

- At the scale of production units and multiproduct plants: dynamic simulation and computer tools are more and more useful to analyse the operating conditions of the equipment of each of the production units and to predict both the material rows and states and residence times within individual pieces of equipment in order to simulate the whole production in terms of time and energy costs. This allows for an interactive walk to predict in a few seconds the new performances (product quality and final cost) obtained by any change due to a blocking step or a bottleneck in the supply chain. Many different scenarios may be tested within a short time, thus allowing the rapid identification of an optimal solution. For instance, the simulation of an entire production year takes less than 10 minutes on a computer. It is clear that such computer simulations enable the design of individual steps, show the structure of the whole process at the mega scale, and place the individual process in the overall context of production. Current breakthroughs in information collecting and processing are the results of previous modeling, simulation, transcription, translation, and interpretation at different scales.

6.3 MODELING OF HOLLOW FIBER MEMBRANE CONTACTOR-BASED EXTRACTION

In HFMCs, the mass transfer of species between two sides of a membrane is a landmark. The present chapter covers two configurations of hollow fiber contactor, i.e., non-dispersive solvent extraction (NDSX) and hollow fiber supported liquid membrane technique (HFSLM). The hydrophilic or hydrophobic nature of the porous membrane represents a relationship between the aqueous phase and the organic phase. One of the main advantages of HFMCs is that the membrane will develop scattering and distribution without any contact, so that the emulsion is not formed during the separation. Moreover, the rate of both phases can be chosen independently. As a result, problems such as undeveloped flow and foaming will not occur in these contactors. The ability to use a selective organic extractant as the separating agent of the system has increased the use of hollow fiber membranes. This process is called non-dispersive solvent extraction [1, 2, 6, 9–14]. In this process, a porous membrane provides contact between the feed and organic extractant. On one side of the membrane, the feed or aqueous solution flows, on the other side, an organic extractant flows co-currently or counter-currently. NDSX is an equilibrium-based mode of operation, where solute can be extracted selectively and can be extracted in micro as well as macro concentration. Scrubbing of the loaded solvent, if required, is possible in this technique only to enhance the selectivity. Generally in NDSX modes of operations, the feed flows on the lumen side and the solvent on the shell side. Hydrophobic membranes are made of polypropylene (PP), polytetrafluoroethylene (PTFE), and polyvinyllidenefluoride (PVDF). So the non-wetting aqueous solution is not able to diffuse into the membrane pores as long as the pressure is kept below the critical pressure, the solvent will fill all membrane pores. During the operation of NDSX mode, a pressure difference of 0.2 bars between the aqueous and organic phase is always ensured to avoid entrainment of the organic into the aqueous phase. In this case, it can be noted that the driving force for this process is the solute concentration gradient between the feed and the extracting solution. This process consists of four different stages for transferring solutes throughout the membrane and could be clarified by following the direction of the mass transfer of solute from the feed flowing in the hollow fiber lumens into the organic phase flowing in the shell side. The first stage is transfer of solute through the feed boundary layer; the second stage is extraction into the organic at the feed-membrane interface; the third stage is diffusion through the porous membrane fiber; and finally the fourth stage is the transfer of solute through the boundary layer into the bulk organic phase flowing in the shell side. Figure 6.1 shows a view of the general principles of mass transfer in the NDSX mode based on film model [4].

In HFSLM technique, two sides of the membrane are in contact with the feed phase and the strip phase whereas

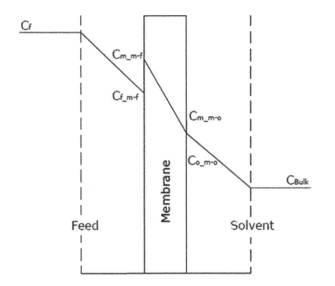

FIGURE 6.1 Mass transfer in NDSX mode based on film model.

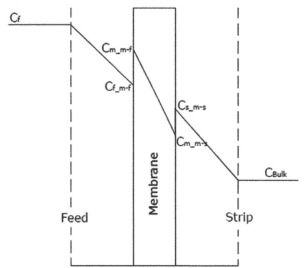

FIGURE 6.2 Liquid–liquid mass transfer in HFSLM mode based on film model.

the membrane phase is impregnated within the pores of the lumens and acts as a barrier between the two phases. The carrier present in the membrane phase binds the component from the feed phase at the mouth of the micro-pores and recovers it at the strip. In the case of selective transports, the affinity between the desired component and the carrier has to be higher than for the other compounds contained in the feed solution. The performance of the process is strongly affected by the properties of the membrane phase, the carrier concentration, the selectivity, the fluid dynamic of phases, and the membrane characteristics. This technique has the advantage of membrane stability over emulsion liquid membrane when operated at constant pressure on both the sides of the membrane. This technique is a non-equilibrium based extraction process, where uphill transportation of the solute is possible because the concentration gradient prevails across the membrane phase during the entire duration of the operation. Figure 6.2 shows a view of the general principles of mass transfer in the HFSLM mode based on film model.

The modeling of mass transfer in hollow fiber membrane contactors has been investigated by many researchers. Swain et al. [15] described different approaches of mass transfer modeling established for HFMCs. Two approaches of mathematical modeling have been broadly practiced for mass transfer in separation processes using hollow fiber contactors. The first approach is based on resistance-in-series model, through which the solute transport mechanism can be explained by considering four actions:

(i) transport species flow through the feed boundary layer; (ii) crossing of the aqueous-solvent interface within the membrane; (iii) transport species diffusion into the pores filled with the solvent phase; (iv) transport species pass through the solvent boundary layer into the organic phase flowing in the shell side in the case of NDSX mode or into the stripping phase in the case of HFSLM mode. The mass transfer resistances in the feed phase boundary

layer, the porous membrane, and the solvent phase or the strip phase are considered in series [16, 17]. Consideration of plug flow on the lumen side and the shell side further simplifies the model.

Table 6.1 contains literature on NDSX and HFSLM techniques based on a resistance-in-series model. This approach is not able to provide the flow field and concentration distribution with their spatial and temporal variations.

The second approach is based on fundamental equations for conservation of mass. These are two-dimensional equations in which axial as well as radial variations of the flow field and concentration are considered. This model also takes into account the variation of fluid properties in both of the directions. Very few works are reported based on this approach because of the complexities involved in it.

With the development of hardware and the availability of computational commercial software, it is possible today to develop models for processes of complex geometries with the involvement of multiphysics. CFD-based modeling solves simultaneously flow, mass, and heat, etc., to obtain spatial and temporal variations of variables in the computational domain. This model gives a better understanding of the process and identifies the key variables responsible for design, optimization, and scale-up. This model is able to quantify the variables within the computational domain as well as at interfaces [15]. CFD-based modeling approach can also be broadly classified with assumptions as equilibrium at interfaces and rate-based approach considering chemical reaction at the interfaces. Details of CFD modeling based on equilibrium at interface in hollow fiber membrane contactors operated on the NDSX mode and HFSLM mode of operations are discussed in depth. This chapter also describes the rate based modeling approach considering chemical reaction at the interface. Some of the literature based on CFD models using hollow fiber contactors is listed in Table 6.2.

TABLE 6.1

Some Literature Based on Resistance-in-Series Model

Authors	Mode of operation	Applications
Ansari et al., 2016 [11]	NDSX	U(VI) and Th(IV) by TBP and dialkyl amides
Boributh et al. 2011 [18]	NDSX	CO_2 by distilled water
Kandwal et al., 2011 [19]	HFSLM	Cs (I) extracted by Calix-[4]-bis (2,3-naptho)-crown-6
Boributh et al. 2012 [20]	NDSX	CO_2 by MEA (Monoethanolamine)
Kandwal et al., 2012 [21]	HFSLM	Sr(II) by DTBCH18C6
Suren et al., 2012 [22]	HFSLM	Pb(II) by D2EHPA (di-2-ethylhexyl phosphoric acid)
Sunsandee et al., 2012 [23]	HFSLM	S (HCl /H_2SO_4/ HNO_3)-amlodipine by Chiral selector O,O'-Dibenzyl-(2s,3s)-tartaric acid((+)DBTA)
Ambre et al., 2013 [24]	NDSX	Nd(III) extracted by DNPPA & TOPO
Jagdale et al., 2013 [25]	HFSLM	Nd(III) extracted by TODGA
Sunsandee et al., 2013 [26]	HFSLM	Levocetirizine by O,O'-Dibenzyl-(2R,3R)-tartaric acid((-)DBTA)
Ferreira et al., 2014 [27]	NDSX	Thiols˙ extracted by [C_2.mim][CF_3SO_3]
Wannachod et al., 2014 [28]	HFSLM	Nd(III) extracted by HEHEPA
Wannachod et al., 2015 [29]	HFSLM	Nd(III) extracted by PC88A
Chaturabul et al., 2015 [30]	HFSLM	Hg(III) extracted by Aliquate 336
Wang et al., 2017 [31]	NDSX	Tube side mass transfer at low Graetz range
Yadav et al., 2019 [32]	HFSLM	Terbium by EHEHPA
Aligwe et al., 2019 [33]	NDSX	NH_3 by H_2SO_4

6.4 CFD MODELING BASED ON ASSUMPTION OF EQUILIBRIUM

6.4.1 NDSX / HFSLM TECHNIQUE IN ONCE THROUGH MODE OF OPERATIONS

Two-dimensional CFD-based models are used to investigate and quantify the solute transport using hollow fiber membrane contactors based on conservation equations. The computational domain is comprised of three parts: lumen, membrane, and shell. Figure 6.3 shows the schematic view of the computational domains for CFD studies of NDSX or HFSLM modes of operation. In the NDSX mode, the solvent flows in the shell side and will be replaced by the strip solution for the HFSLM.

Velocity profiles in the computational domain are required to study the effect of flow rates on mass transfer.

TABLE 6.2

Some Literature Reported on CFD-Based Modeling Using Hollow Fiber Contactors

Reference	Unit Operation	Applications
Ghasem et al. (2019) [34]	Gas absorption	CO_2 absorption by CNT-water based nanofluids
Swain et al., (2019) [35]	Solvent extraction	Uranium by TBP with *n*-dodecane
Hajilary et al. (2018) [36]	Gas absorption	CO_2 absorption by water-based nanofluids
Haghshenas et al. (2016) [37]	Solvent extraction	Aroma compound by water, miglyol, sunflower oil, and n-hexane
Razavi et al. (2016) [38]	Gas absorption	CO_2 absorption by ammonium ionic liquid
Ghadiri et al. (2015) [39]	Solvent extraction	Rubidium by using Dicyclohexano-18-Crown-6
Ghadiri et al. (2013) [40]	Solvent extraction	Cesium by using dicyclohexano 18 crown 6 with chloroform
Miramini et al. (2013) [41]	Solvent extraction	Acetone by using super critical fluid; propane.
Shirazian et al. (2012) [42]	Solvent extraction	Ammonia from water by using sulphuric acid
Rezakazemi et al. (2012) [43]	Solvent extraction	Dissolved ammonia by using sulphuric acid.
Marjani et al. (2011) [44]	Solvent extraction	Copper by using di(2-ethylhexyl) phosphoric acid (D2EHPA)
Shirazian et al. (2011) [45]	Solvent extraction	Ethanol or acetone by a dense solvent; propane
Shirazian et al. (2009) [46]	Gas absorption	CO_2 absorbed by pure water

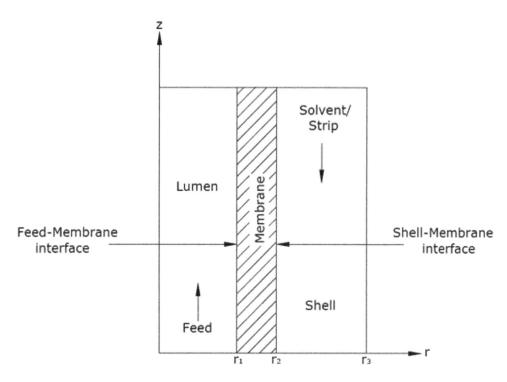

FIGURE 6.3 Schematic view of computational domain; NDSX / HFSLM technique.

This can be obtained by solving Navier–Stokes equations under steady state condition.

$$\nabla.\eta\left(\nabla v+\left(\nabla v\right)^{\mathrm{T}}\right)-\rho\left(v.\nabla\right)v-\nabla P+\mathbf{F}=0 \quad (6.1)$$

$$\nabla.v = 0 \quad (6.2)$$

Where, ρ, v, η, P and $\mathbf{F}$ represent density of the fluid, velocity vector, dynamic viscosity, pressure, and volumetric body force respectively. Navier–Stokes equations are not solved in the domain representing membrane because the liquid in the pores of membrane can be assumed to be stagnant. The steady state species transport equation is given by **Equation 6.3**.

$$\nabla.\left(Cv\right)+\nabla.\mathbf{J}=0 \quad (6.3)$$

Where, C and $\mathbf{J}$ are the concentration and diffusive flux of solute, respectively. The first and second terms in **Equation 6.3** represent the convective transport and diffusive transport of solute, respectively. Diffusive flux can be expressed by **Equation 6.4**.

$$\mathbf{J} = -D\nabla C \quad (6.4)$$

The total molar flux, which is comprised of convective and diffusive fluxes, can be expressed as given by **Equations 6.5**.

$$\Phi = -D\nabla C + Cv \quad (6.5)$$

where, D and Φ denote the diffusion coefficient and total flux, respectively. Convective and diffusive fluxes are considered in the lumen as well as in the shell whereas only diffusive flux is considered in the membrane.

Thus the total flux on the membrane phase is equal to diffusive flux as written by **Equation 6.6**.

$$\Phi_{\mathrm{m}} = -D_{\mathrm{m}}\nabla C \quad (6.6)$$

The boundary conditions required to solve the Navier–Stokes equations and mass transfer equation mentioned above are listed in Table 6.3. The model inputs required for CFD modeling are mentioned in Table 6.4.

6.4.2 NDSX/HFSLM Mode in Recycle Mode of Operations

Mass transfer through a hollow fiber contactor operated on recycle mode can be predicted by using the CFD model. Figure 6.4 shows the schematic representation of the computational domain used for NDSX or HFSLM configuration. In this case, mass is balanced for the feed as well as solvent/strip reservoirs along with the equations of the model by using **Equations 6.7–6.9**. These equations basically relate the inlet and outlet concentrations of the lumen side and the shell side. Implementation of these equations along with the equations of the CFD model basically changes the concentration of inlet boundary conditions with the progress of operation.

$$V_{\mathrm{f}}\frac{dC_{\mathrm{f}}}{dt} = Q_{\mathrm{f}}\left(C_{\mathrm{fout}}-C_{\mathrm{ft}}\right) \quad (6.7)$$

$$V_{\mathrm{o}}\frac{dC_{\mathrm{o}}}{dt} = Q_{\mathrm{o}}\left(C_{\mathrm{oout}}-C_{\mathrm{ot}}\right) \quad (6.8)$$

TABLE 6.3
Boundary and Initial Conditions for Once through Mode of Operation

Position	Tube (Momentum)	Tube (Mass transfer)	Membrane (Mass transfer)	Shell (Momentum)	Shell (Mass transfer)
$z = 0$	$\left.\begin{array}{l} v_{fz} = v_{fz0} \\ v_{fr} = 0 \end{array}\right\}$ (NDSX/ HFSLM)	$C_f = C_{f0}$ (NDSX/ HFSLM)	$\dfrac{\partial C_m}{\partial z} = 0$ (NDSX/ HFSLM) (Insulation)	$P = 0$ (NDSX/HFSLM) (Outlet Pressure)	$\Phi_o = C_o v_o$ (NDSX) $\Phi_s = C_s v_s$ (HFSLM)
$z = L$	$P = 0$ (NDSX / HFSLM) (Outlet Pressure)	$\Phi_f = C_f v_f$ (NDSX/ HFSLM)	$\dfrac{\partial C_m}{\partial z} = 0$ (NDSX/ HFSLM) (Insulation)	$\left.\begin{array}{l} v_{oz} = v_{oz0} \\ v_{or} = 0 \end{array}\right\}$ (NDSX) $\left.\begin{array}{l} v_{sz} = v_{sz0} \\ v_{sr} = 0 \end{array}\right\}$ (HFSLM)	$C_o = 0$ (NDSX) $C_s = 0$ (HFSLM)
$r = 0$	$\left.\begin{array}{l} \dfrac{\partial v_{fz}}{\partial r} = 0 \\ \text{(Symmetric)} \\ v_{fr} = 0 \end{array}\right\}$ (NDSX/HFSLM)	$\dfrac{\partial C_f}{\partial r} = 0$ (NDSX/ HFSLM)	-	-	-
$r = r_1$	$v_f = 0$ (NDSX/HFSLM) (No slip)	$\left.\begin{array}{l} C_f = C_m/K_d \\ \Phi_f = -\Phi_m \end{array}\right\}$ (NDSX / HFSLM)	$\left.\begin{array}{l} C_m = C_f.K_d \\ \Phi_f = -\Phi_m \end{array}\right\}$ (NDSX/HFSLM)	-	-
$r = r_2$	-	-	$\left.\begin{array}{l} C_m = C_o \\ \Phi_m = -\Phi_o \end{array}\right\}$ (NDSX) $\left.\begin{array}{l} C_m = C_s.K_s \\ \Phi_m = -\Phi_s \end{array}\right\}$ (HFSLM)	$\begin{array}{l} v_o = 0 \text{ (NDSX)} \\ v_s = 0 \text{ (HFSLM)} \\ \text{(No slip)} \end{array}$	$\left.\begin{array}{l} C_o = C_m \\ \Phi_m = -\Phi_o \end{array}\right\}$ (NDSX) $\left.\begin{array}{l} C_s = C_m/K_s \\ \Phi_m = -\Phi_s \end{array}\right\}$ (HFSLM)
$r = r_3$	-	-	-	$\left.\begin{array}{l} \dfrac{\partial v_{oz}}{\partial r} = 0 \\ \text{(Symmetric)} \\ v_{or} = 0 \end{array}\right\}$ (NDSX) $\left.\begin{array}{l} \dfrac{\partial v_{sz}}{\partial r} = 0 \\ \text{(Symmetric)} \\ v_{sr} = 0 \end{array}\right\}$ (HFSLM)	$\dfrac{\partial C_o}{\partial r} = 0$ (NDSX) $\dfrac{\partial C_s}{\partial r} = 0$ (HFSLM)

$$V_s \frac{dC_s}{dt} = Q_s \left(C_{sout} - C_{st} \right) \qquad (6.9)$$

Where V is the volume of the reservoir, Q is the flow rate, and C is the concentration of the solute.

Velocity profiles in the computational domain can be obtained by solving Navier–Stokes equations:

$$\rho \frac{\partial v}{\partial t} - \nabla . \eta \left(\nabla v + \left(\nabla v \right)^{T} \right) + \rho \left(v . \nabla \right) v + \nabla P - \mathbf{F} = 0 \qquad (6.10)$$

$$\frac{\partial \rho}{\partial t} + \nabla . \left(\rho v \right) = 0 \qquad (6.11)$$

TABLE 6.4
CFD Model Inputs of NDSX/HFSLM Configuration on Once through Mode of Operation

Model Parameters	Unit	Configuration
Lumen inner radius (r_1)	m	(NDSX/HFSLM)
Lumen outer radius (r_2)	m	(NDSX/HFSLM)
Shell radius (r_3)	m	(NDSX/HFSLM)
Lumen length (L)	m	(NDSX/HFSLM)
Porosity of lumens (ε)	%	(NDSX/HFSLM)
Tortuosity of lumen (τ)	-	(NDSX/HFSLM)
Diffusion coefficient of feed fluid (D_f)	m²/s	(NDSX/HFSLM)
Diffusion coefficient of solvent (D_o)	m²/s	(NDSX/HFSLM)
Diffusion coefficient of membrane phase $\left(D_m = D_o \left(\dfrac{\varepsilon}{\tau} \right) \right)$	m²/s	(NDSX/HFSLM)
Diffusion coefficient of strip (D_s)	m²/s	(NDSX/HFSLM)
Density of feed (ρ_f)	kg/m³	(NDSX/HFSLM)
Density of Solvent (ρ_o)	kg/m³	(NDSX/HFSLM)
Viscosity of feed (η_f)	Pa.s	(NDSX/HFSLM)
Viscosity of solvent (η_f)	Pa.s	(NDSX/HFSLM)
Density of strip (ρ_s)	kg/m³	(HFSLM)
Viscosity of strip (η_f)	Pa.s	(HFSLM)
Distribution coefficient of extraction (K_d)	-	(NDSX/HFSLM)
Distribution coefficient of extraction (K_s)	-	(HFSLM)
Inlet concentration of feed (C_{f0})	mol/m³	(NDSX/HFSLM)
Inlet concentration of solvent (C_{o0})	mol/m³	(NDSX)
Inlet concentration of strip (C_{s0})	mol/m³	(HFSLM)

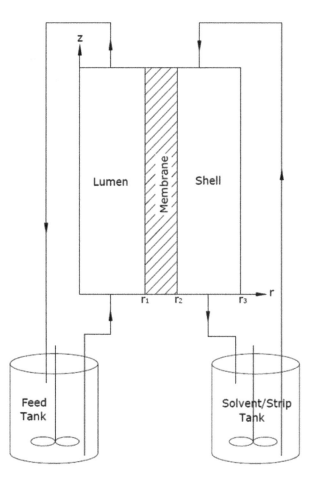

FIGURE 6.4 Schematic representation of computational domain, NDSX / HFSLM technique.

where, ρ, v, η, P, and **F** represent density of the fluid, velocity vector, dynamic viscosity, pressure, and volumetric body force respectively. The membrane present within the pores of the lumens is stagnant, so the Navier–Stokes equations are not solved in this domain. The solute transport can be written as:

$$\frac{\partial C}{\partial t} + \nabla.(Cv) + \nabla.\mathbf{J} = 0 \qquad (6.12)$$

Where, C, and **J** are the concentration of the solute, and the diffusive flux of solute, respectively. The terms in the left hand side in **Equation 6.12** represent accumulation term, convective, and diffusive transport of the solute, respectively. Diffusive flux can be expressed as:

$$\mathbf{J} = -D\nabla C \qquad (6.13)$$

By using **Equation 6.13**, **Equation 6.12** can be rewritten as:

$$\frac{\partial C}{\partial t} + \nabla.(Cv) - \nabla.(D\nabla C) = 0 \qquad (6.14)$$

Membrane phase domain considers diffusive flux whereas lumen side and shell side consider both convective as well as diffusive flux. So, the species transport equation on the membrane phase can be modified as:

$$\frac{\partial C}{\partial t} - \nabla.(D_m \nabla C) = 0 \qquad (6.15)$$

Navier–Stokes equations and the mass transfer equation written above can be solved by using the boundary conditions listed in Table 6.5. CFD model inputs are given in Table 6.6.

As the feed and solvent inlet concentration varies with time, it is required to solve the mass balance equations of feed and solvent/strip reservoirs along the equations of the CFD model. Rezekazemi et al. [43] studied removal of ammonia from an industrial waste water stream on recycle mode of operation by using COMSOL along with MATLAB. MATLAB is used to change the inlet feed boundary concentration with increase in time of operation. Figure 6.5 shows the algorithm of the numerical simulation. The distribution of concentration and velocity in the computational domain can be obtained from the developed model.

TABLE 6.5

Boundary and Initial Conditions for Recycle Mode of Operation

Position	Tube (Momentum)	Tube (Mass transfer)	Membrane (Mass transfer)	Shell (Momentum)	Shell (Mass transfer)
$z = 0$	$\left.\begin{array}{l} v_{fz} = v_{fz0} \\ v_{fr} = 0 \end{array}\right\}$(NDSX/ HFSLM)	$C_f = C_{ft}$ (NDSX/HFSLM)	$\dfrac{\partial C_m}{\partial z} = 0$ (NDSX/ HFSLM) (Insulation)	$P = 0$ (NDSX/HFSLM) (Outlet Pressure)	$\Phi_o = C_{oout} v_o$ (NDSX) $\Phi_s = C_{sout} v_s$ (HFSLM)
$z = L$	$P = 0$ (NDSX / HFSLM) (Outlet Pressure)	$\Phi_f = C_{fout} v_f$ (NDSX/HFSLM)	$\dfrac{\partial C_m}{\partial z} = 0$ (NDSX/ HFSLM) (Insulation)	$\left.\begin{array}{l} v_{oz} = v_{oz0} \\ v_{or} = 0 \end{array}\right\}$(NDSX) $\left.\begin{array}{l} v_{sz} = v_{sz0} \\ v_{sr} = 0 \end{array}\right\}$(HFSLM)	$C_o = C_{ot}$ (NDSX) $C_s = C_{st}$ (HFSLM)
$r = 0$	$\left.\begin{array}{l} \dfrac{\partial v_{fz}}{\partial r} = 0 \\ \text{(Symmetric)} \\ v_{fr} = 0 \end{array}\right\}$ (NDSX/HFSLM)	$\dfrac{\partial C_f}{\partial r} = 0$ (NDSX/HFSLM)	-	-	-
$r = r_1$	$v_f = 0$ (NDSX/HFSLM) (No slip)	$\left.\begin{array}{l} C_f = C_m / K_d \\ \Phi_f = -\Phi_m \end{array}\right\}$ (NDSX / HFSLM)	$\left.\begin{array}{l} C_m = C_f \cdot K_d \\ \Phi_f = -\Phi_m \end{array}\right\}$ (NDSX/HFSLM)	-	-
$r = r_2$	-	-	$\left.\begin{array}{l} C_m = C_o \\ \Phi_m = -\Phi_o \end{array}\right\}$(NDSX) $\left.\begin{array}{l} C_m = C_s \cdot K_s \\ \Phi_m = -\Phi_s \end{array}\right\}$ (HFSLM)	$v_o = 0$ (NDSX) $v_s = 0$ (HFSLM) (No slip)	$\left.\begin{array}{l} C_o = C_m \\ \Phi_m = -\Phi_o \end{array}\right\}$ (NDSX) $\left.\begin{array}{l} C_s = C_m / K_s \\ \Phi_m = -\Phi_s \end{array}\right\}$ (HFSLM)
$r = r_3$	-	-	-	$\left.\begin{array}{l} \dfrac{\partial v_{oz}}{\partial r} = 0 \\ \text{(Symmetric)} \\ v_{or} = 0 \end{array}\right\}$(NDSX) $\left.\begin{array}{l} \dfrac{\partial v_{sz}}{\partial r} = 0 \\ \text{(Symmetric)} \\ v_{sr} = 0 \end{array}\right\}$(HFSLM)	$\dfrac{\partial C_o}{\partial r} = 0$ (NDSX) $\dfrac{\partial C_s}{\partial r} = 0$ (HFSLM)

Time	Feed tank concentration		Solvent/Strip tank concentration	
$t = 0$	$C_{ft} = C_{f0}$ (NDSX/ HFSLM)		$C_{ot} = C_{o0}$ (NDSX) $C_{st} = C_{s0}$ (HFSLM)	

TABLE 6.6

CFD Model Inputs for NDSX/HFSLM Configuration on Recycle Mode of Operations

Model Parameters	Unit	Configuration
Lumen inner radius (r_1)	m	(NDSX/HFSLM)
Lumen outer radius (r_2)	m	(NDSX/HFSLM)
Shell radius (r_3)	m	(NDSX/HFSLM)
Lumen length (L)	m	(NDSX/HFSLM)
Porosity of lumens (ε)	%	(NDSX/HFSLM)
Tortuosity of lumen (τ)	-	(NDSX/HFSLM)
Diffusion coefficient of feed fluid (D_f)	m²/s	(NDSX/HFSLM)
Diffusion coefficient of solvent (D_o)	m²/s	(NDSX/HFSLM)
Diffusion coefficient of membrane phase $\left(D_m = D_o\left(\dfrac{\varepsilon}{\tau}\right)\right)$	m²/s	(NDSX/HFSLM)
Diffusion coefficient of strip (D_s)	m²/s	(NDSX/HFSLM)
Density of feed (ρ_f)	kg/m³	(NDSX/HFSLM)
Density of solvent (ρ_o)	kg/m³	(NDSX/HFSLM)
Viscosity of feed (η_f)	Pa.s	(NDSX/HFSLM)
Viscosity of solvent (η_f)	Pa.s	(NDSX/HFSLM)
Density of strip (ρ_s)	kg/m³	(HFSLM)
Viscosity of strip (η_f)	Pa.s	(HFSLM)
Distribution coefficient of extraction (K_d)	-	(NDSX/HFSLM)
Distribution coefficient of extraction (K_s)	-	(HFSLM)
Inlet concentration of feed (C_{f0})	mol/m³	(NDSX/HFSLM)
Inlet concentration of solvent (C_{o0})	mol/m³	(NDSX)
Inlet concentration of strip (C_{s0})	mol/m³	(HFSLM)
Volume of feed (V_{f1})	m³	(NDSX/HFSLM)
Volume of solvent (V_{o1})	m³	(NDSX)
Volume of feed (V_{s1})	m³	(HFSLM)

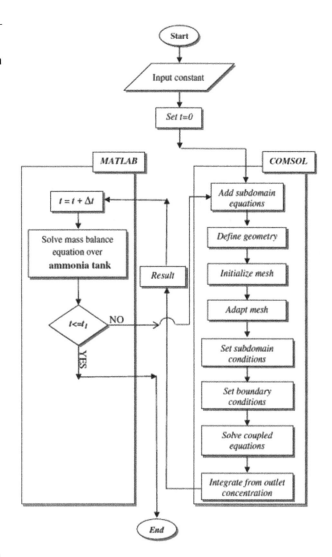

FIGURE 6.5 Algorithm for CFD simulation of NDSX recycle mode of operations [Reprinted from Desalination 285, M. Rezakazemi, S. Shirazian, A. S. Nezameddin, Simulation of ammonia removal from industrial waste water streams by means of a hollow fiber membrane contactor, 383–392, Copyright (2012) with permission from Elsevier [43]].

6.5 CFD MODELING ON RATE-BASED APPROACH

A rate-based CFD modeling approach considers chemical reaction at interfaces rather than assumption of equilibrium as considered in the previous section. The chemical reaction at feed-membrane interface takes place when the extractant reacts with the solute of the feed solution given by **Equation 6.16**. In the case of the HFSLM technique, the complex species also reacts with the stripping solution at the membrane-strip interface

$$aA_{Aq} + bB_O \xrightarrow{k} cC_O \qquad (6.16)$$

where A is solute ion, B is extractant, and C is complex species of solute ion, and a, b, c are stoichiometric coefficients of A, B, and C, respectively.

Reaction rate (R_A) can be expressed as:

$$-R_A = k\, C_A^n \qquad (6.17)$$

where, k is the reaction rate constant and n is the order of reaction.

The facilitated transport mechanism, which covers solute species transportation, complexion, and diffusion, and recovery of the solute, normally follows these steps: (1) the solute species in the feed solution are transported to the feed-membrane interface and subsequently reacted to form complex, (2) the complex species diffuse to the opposite interface of the membrane due to concentration gradient, (3) for HFSLM techniques, the complex species react with the stripping solution at the membrane-strip interface and transported into the stripping solution, but in the NDSX technique, the complex species transfer into the solvent phase.

Rate-based CFD modeling can be attempted by incorporating the reaction term in the species transport equation (**Equation 6.3 and Equation 6.12**). Results can be obtained by solving modified equations with appropriate boundary conditions. Panchanaron et al. [47] reported a rate-based model by considering chemical reaction for recovery of copper in the HFSLM technique.

6.6 NUMERICAL SIMULATIONS

To obtain the numerical simulations for any of the techniques discussed in the previous section, it is required to solve the model equations with their appropriate boundary conditions. Researchers who developed CFD models to consider a single fiber for their studies assumed that the performance of all fibers is identical. Governing equations along with boundary conditions are solved by using COMSOL Multiphysics, which is a finite element method. It is required to conduct a grid independent test before simulations to avoid any uncertainty or computational error due to grid density. Figure 6.6 shows a typical mesh element size by Ghadiri et al. [39] for prediction of rubidium (Rb) extraction.

6.7 RESULTS OBTAINED FROM CFD MODELS

6.7.1 Velocity Field in the Extractor

Shell side velocity contours by Marjani et al. [44] are shown in Figure 6.7, and Figure 6.8 shows the profile of the organic phase in the shell side at different axial locations. Velocity field is the solution of Navier–Stokes equations. The profiles obtained from the CFD model are almost parabolic in nature. Velocity profiles of the feed side also can be obtained from the model.

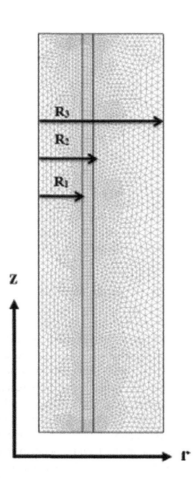

FIGURE 6.6 Typical mesh element used for Rb simulation [Reprinted from J. Water Process. Eng. 6, M. Ghadiri, M. Asadollahzadeh, A. Hemmati, CFD simulation for separation of ion from waste water in a membrane contactor, 144–150, Copyright (2015) with permission from Elsevier [39]].

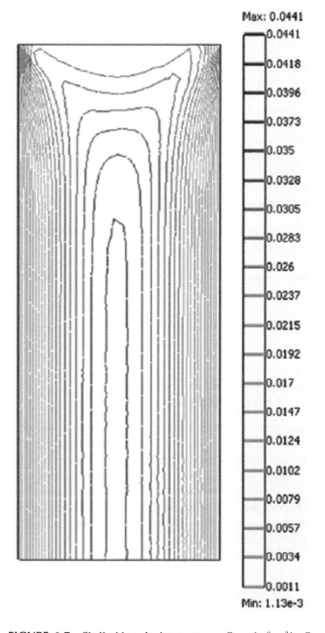

FIGURE 6.7 Shell side velocity contours, Q_f = 1e^{-8} m^3/s, Q_o = 1e^{-7} m^3/s [Reprinted from Desalination 281, A. Marjani, S. Shirazian, Simulation of heavy metal extraction in membrane contactors using computational fluid dynamics, 422–428, Copyright (2011) with permission from Elsevier [44]].

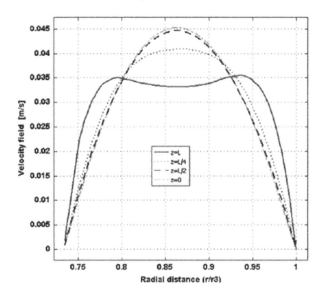

FIGURE 6.8 Velocity profiles in the shell side at different axial locations, $Q_f = 1e^{-8}$ m³/s, $Q_o = 1e^{-7}$ m³/s [Reprinted from Desalination 281, A. Marjani, S. Shirazian, Simulation of heavy metal extraction in membrane contactors using computational fluid dynamics, 422–428, Copyright (2011) with permission from Elsevier [44]].

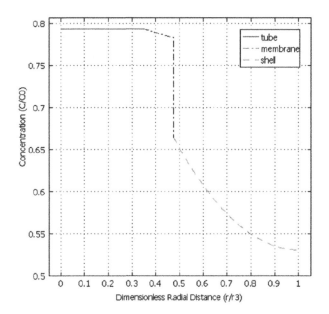

FIGURE 6.9 Radial concentration profile of CO_2 at middle of the contactor; r_1=0.1mm, r_2=0.15mm, r_3=0.3mm, L=20cm, m=0.84, $D_{CO2\text{-}tube}$=1.85e^{-5}, $D_{CO2\text{-}membrane}$=3.7e^{-6}, $D_{CO2\text{-}shell}$=1.9e^{-9}, Q_f & Q_o = 400 ml/min; each, T=298 K, p=1atm [Reprinted from Simul. Model. Pract. Th. 17, S. Shirazian, A. Modhadassi, S. Moradi, Numerical simulation of mass transfer in gas–liquid hollow fiber membrane contactors for laminar flow conditions, 708–718, Copyright (2009) with permission from Elsevier [46]].

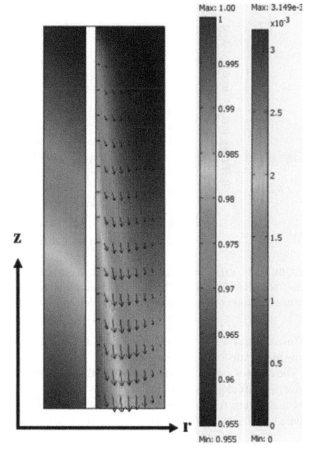

FIGURE 6.10 Concentration distribution of Rb; $Q_f = 1e^{-7}$ m³/s & Q_o=1e^{-8} m³/s [Reprinted from J. Water Process. Eng. 6, M. Ghadiri, M. Asadollahzadeh, A. Hemmati, CFD simulation for separation of ion from waste water in a membrane contactor, 144–150, Copyright (2015) with permission from Elsevier [39]].

6.7.2 Concentration and Mass Flux Profile in the Contactor

The radial concentration profile of CO_2 in the contactor; tube shell and membrane phase, as simulated by Shirazian et al. [46], is shown in the Figure 6.9. It can be observed that the change in concentration in the tube and on the membrane phase is low and is significant in the shell side. This can be attributed to the higher resistance to gas transport in the liquid phase compared to the gas phase, which is because of the higher diffusion coefficient in the membrane and tube side as compared to shell side.

Figure 6.10 shows the concentration distribution of Rb in the tube and shell side studied by Ghadiri et al. [39]. It is evident from the figure that the inlet concentration is at its highest value and axial concentration decreases due to

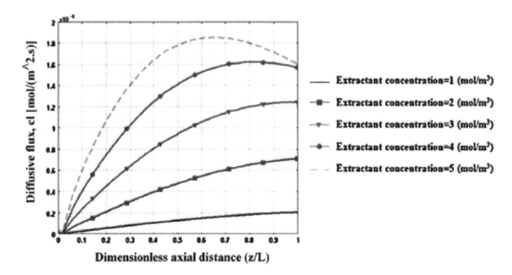

FIGURE 6.11 Effect of extract concentration on diffusive flux of cesium, $Q_f = 1e^{-7}$ m³/s & $Q_o=1e^{-8}$ m³/s, $C_f = 1$ mol/m³ [Reprinted from Chem. Eng. Process. 69, M. Ghadiri, S. Shirazian, Computational simulation of mass transfer in extraction of alkali metals by means of nano-porous membrane extraction, 57–62, Copyright (2013) with permission from Elsevier [40]].

diffusive transport of the solute. Arrows in the shell side show mass transfer flux in the shell side.

The effect of carrier concentration on the diffusive flux and convective flux along the axis of the module is shown in Figure 6.11 and Figure 6.12, respectively. It can be observed from Figure 6.11 that the diffusive flux of cesium increases with an increase in carrier concentration. This can be attributed to the increase in distribution coefficient enhancing mass transfer of cesium. As more and more cesium is getting extracted into the solvent phase with an increase in carrier concentration, it is obvious that the convective flux in the feed side will decrease. The same trend can be witnessed in Figure 6.12.

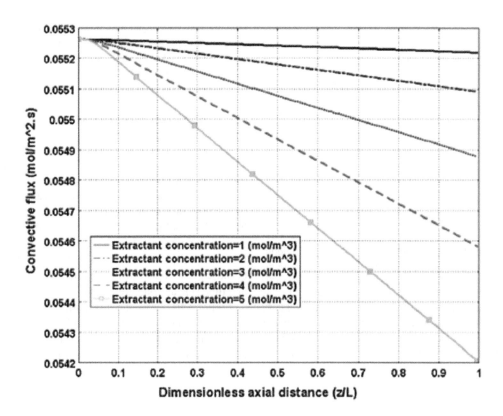

FIGURE 6.12 Effect carrier concentration on convective flux of cesium in the feed side, $Q_f = 1e^{-7}$ m³/s & $Q_o=1e^{-8}$ m³/s, $C_f = 1$ mol/m³ [Reprinted from Chem. Eng. Process. 69, M. Ghadiri, S. Shirazian, Computational simulation of mass transfer in extraction of alkali metals by means of nano-porous membrane extraction, 57–62, Copyright (2013) with permission from Elsevier [40]].

6.8 CONCLUSIONS AND FUTURE PERSPECTIVE

The rapid development in hardware and commercial software gives a new horizon to CFD-based modeling. These models are able to handle two-dimensional and three-dimensional complex sophisticated geometry with the multiphysics involved in them. The separation process of once through as well as recycle mode of operations can be studied by using the CFD model. The distribution of velocity, pressure, concentration, and mass flux in the computational domain can be obtained by solving Navier–Stokes equations coupled with mass transfer equation. Variables can be quantified easily within the computational domains as well as boundaries. Post processing of the results allows for interpretations, which gives an in-depth understanding of the processes and identify the responsible variables. Overall, this modeling approach gives better insights into separation processes and helps in determining the optimum design, operation, and scale-up with a wide range of applications and also reduces rigorous experimentation.

NOMENCLATURES

C: Concentration of solute [mol /m^3]
D: Diffusion coefficient [m^2/s]
F: Volumetric body Force [N/m^3]
H$^+$: Acidity, M
J: Diffusive flux [mol./m^2.s]
k: Reaction rate constant [m^3/mol.s]
K$_d$: Distribution coefficient of extraction
K$_s$: Distribution coefficient of stripping
L: Length of module [m]
P: Pressure [Pa]
Q: Flow rate [lph]
r: Radial direction of module
r$_1$: Inner radius of lumen [m]
r$_2$: Outer radius of lumen
r$_3$: Inner radius of shell [m]
v: Velocity vector [m/s]
z: Axial direction of module
t: Time [s]

GREEK LETTERS

Φ: Molar flux [mol /m^2.s]
ρ: Density [kg/m^3]
ε: Porosity of lumens [%]
η: Viscosity [Pa.s]
τ: Tortuosity of lumens

SUBSCRIPTS AND SUPERSCRIPTS

Aq: Aqueous
0: Initial / Inlet
f: Feed
l: Lumen
m: Membrane

m-f: Membrane-feed interface
m-o: Membrane-organic interface
m-s: Membrane-strip interface
o: Organic
out: Outlet
s: strip
t: time [s]
T: Transpose of vector

ABBREVIATIONS

CFD: Computational fluid dynamics
D2EHPA: Di-2-ethylhexyl phosphoric acid
DTBCH18C6: 4, 4' (5')-di-tert-butyl-dicyclohexano-18 -crown-6
HEHEPA: 2-ethylhexyl-2-ethylhexyl phosphoric acid
HFMCs: Hollow fiber membrane contactors
HFSLM: Hollow fiber supported liquid membrane
MDEA: Methyldiethanolamine
MEA: Monoethanolamine
NDSX: Non-dispersive solvent extraction
PC88A: 2-Ethylhexylphosphonic acid mono-2-ethylhexyl ester
TODGA: N, N, N', N'-Tetraoctyl diglycolamide

REFERENCES

1. Pabby, A. K., Swain, B., Sastre, A. M., 2017. Recent advances in smart integrated membrane assisted liquid extraction technology, *Chem. Eng. Process.* 120: 27–56.
2. Pabby, A. K., Rizvi, S. H. S., Sastre, A. M. (Eds.), 2015. *Handbook of Membrane Separations, Chemical, Pharmaceutical, Food and Biotechnological Applications*, 2nd edition, CRC Press, Boca Raton, FL.
3. Gupta, S. K., Rathore, N. S., Sonawane, J. V., Pabby, A. K., Janardan, P., Changrani, R. D., Dey, P. K., 2007. Dispersion-free solvent extraction of U(VI) in macro amount from nitric acid solutions using hollow fiber contactor, *J. Membr. Sci.* 300: 131–136.
4. Estay, H., Bocquet, S., Romero, J., Sanchez, J., Rios, G. M., Valenzuela, F., 2007. Modeling and simulation of mass transfer in near-critical extraction using a hollow fiber membrane contactor, *Chem. Eng. Sci.* 62: 5794–5808.
5. Sarrade, S., Guizard, C., Rios, G. M., 2002. Membrane technology and supercritical fluids: Chemical engineering for coupled processes, *Desalination* 144: 137–142.
6. Sarrade, S., Rios, G. M., Carlés, M., 1996. Nano-filtration membrane behaviour in a supercritical medium, *J. Membr. Sci.* 114: 81–91.
7. Afrane, G., Chimowitz, E. H., 1996. Experimental investigation of a new supercritical fluid-inorganic membrane separation process, *J. Membr. Sci.* 116: 293–299.
8. Charpentier, J., 2002. The triplet "molecular processes–product–process" engineering: The future of chemical engineering, *Chem. Eng. Sci.* 57: 4667–4690.
9. Conradie, E. W., Westhuizen, D. J. V. D., Nel, J. T., Krieg, H. M., 2018. The hafnium selective extraction from a zirconium (hafnium) heptafluoride ammonium solution using organophosphorous-based extractants, *Solvent Extr. Ion Exch.* 36: 658–673.
10. Alguacil, F. J., López, F. A., Garcia-Diaz, I., Rodríguez, O., 2016. Cadmium (II) transfer using (TiOAC) ionic liquid as

carrier in a smart liquid membrane technology, *Chem. Eng. Process.* 99: 192–196.

11. Ansari, S. A., Kumari, N., Raut, D. R., Kandwal, P., Mohapatra, P. K., 2016. Comparative dispersion-free solvent extraction of Uranium(VI) and Thorium(IV) by TBP and di-alkyl amides using a hollow fiber contactor, *Sep. Purif. Technol.* 159: 161–168.

12. Rout, P. C., Sarangi, K., 2015. Comparison of hollow fiber membrane and solvent extraction techniques for extraction of cerium and preparation of ceria by stripping precipitation, *J. Chem. Technol. Biotechnol.* 90: 1270–1280.

13. Fouad, E. A., Bart, H. J., 2007. Separation of zinc by a non-dispersive solvent extraction process in a hollow fiber contactor, *Solvent Extr. Ion Exch.* 25: 857–877.

14. Bothun, G. D., Knutson, B. L., Strobel, H. J., Nokes, S. E., Brignole, E. A., Díaz, S., 2003. Compressed solvents for the extraction of fermentation products within a fiber membrane contactor, *J. Supercrit. Fluids* 25: 119–134.

15. Swain, B., Singh, K. K., Pabby, A. K., 2018. Mass transfer modeling in hollow fiber liquid membrane separation process, in: Sridhar, S., Kumar, S. R., (Eds.), *Membrane Technology Sustainable Solutions in Water, Health, Energy and Environmental Sectors*, CRC Press, Boca Raton, FL, Chapter 12, pp. 253–275.

16. Bocquet, S., Torres, A., Sanchez, J., Rios, G. M., 2005. Modeling the mass transfer in solvent-extraction processes with hollow-fiber membranes, *AIChE J.* 51: 1067–1079.

17. Bringas, E., Roman, M. F. S., Irabien, J. A., Ortiz, I., 2009. An overview of the mathematical modeling of liquid membrane separation processes in hollow fiber contactors, *J. Chem. Technol. Biotechnol.* 84: 1583–1614.

18. Boributh, S., Assabumrungrat, S., Laosiripojana, N., Jiraratananon, R., 2011. A modeling study on the effects of membrane characteristics and operating parameters on physical absorption of CO_2 by hollow fiber membrane contactor, *J. Membr. Sci.*, 380: 21–33.

19. Kandwal, P., Dixit, S., Mukhopadhyay, S., Mohapatra, P. K., 2011. Mass transport modeling of CS(I) through hollow fiber supported liquid membrane containing calyx-[4]-bis (2,3-naptho) crown-6 as the mobile carrier, *J. Chem. Eng.* 174: 110–116.

20. Boributh, S., Rongwong, W., Assabumrungrat, S., Laosiripojana, N., 2012. Mathematical modeling and cascade design of hollow fiber membrane contactor for CO_2 absorption by monoethanolamine, *J. Member. Sci.* 401–402: 175–189.

21. Kandwal, P., Ansari, S. A., Mohapatra, P. K., 2012. A highly efficient supported liquid membrane system for near quantitative recovery of radio-strontium from acidic feeds. Part II: Scale up mass transfer modeling in hollow fiber configuration, *J. Member. Sci.* 405–406: 85–91.

22. Suren, S., Wongsawa, T., Pancharoen, U., Lothongkum, A. W., 2012. Uphill transport and mathematical model of Pb(II) from dilute synthetic lead-containing solutions across hollow fiber supported liquid membrane, *J. Chem. Eng.* 191: 503–511.

23. Sunsandee, N., Leepipatpiboon, N., Ramakul, P., Pancharoen, U., 2012. The selective separation of (S) – Amlodipine via a hollow fiber supported liquid membrane: Modeling and experiment verification, *J. Chem. Eng.* 180: 299–308.

24. Ambre, D. N., Ansari, S. A., Anitha, M., Kandwal, P., Singh, D. K., Singh, H., Mohapatra, P. K., 2013. Non-dispersive solvent extraction of neodymium using hollow fiber contactor: Mass transfer and modeling studies, *J. Membr. Sci.* 446: 106–112.

25. Jagdale, Y. D., Vernekar, P. V., Patwardhan, A. W., Patwardhan, A. V., Ansari, S. A., Mohapatra, P. K., Manchanda, V. K., 2013. Mathematical model for the extraction of metal ions using hollow fiber supported liquid membrane operated in a recycle mode, *J. Sep. Sci. Technol.* 48: 1–14.

26. Sunsandee, N., Leepipatpiboon, N., Ramakul, P., 2013. Selective enantio separation of levocetrizine via a hollow fiber supported liquid membrane and mass transfer prediction, *Korean J. Chem. Eng.* 30 (6): 1312–1320.

27. Ferreira, A. R., Neves, L. A., Ribeiro, J. C., Lopes, F. M., Coutinho, J. A. P., Coelhoso, I. M., Crespo, J. G., 2014. Removal of thiols from model jet-fuel streams assisted by ionic liquid membrane extraction, *Chem. Eng. J.* 256: 144–154.

28. Wannachod, T., Leepipatpiboon, N., Pancharoen, U., Nootong, K., 2014. Separation and mass transport of Nd (III) from mixed rare earths via hollow fiber supported liquid membrane: Experiment and modeling, *Chem. Eng. J.* 248: 158–167.

29. Wannachod, T., Leepipatpiboon, N., Pancharoen, U. Phatanasri, S., 2015. Mass transfer and selective separation of neodymium ions via a hollow fiber supported liquid membrane using PC88Aas extractant, *J. Ind. Eng. Chem.* 21: 535–541.

30. Chaturabul, S., Wannachod, T., Leepipatpiboon, N., Pancharoen, U., Kheawhom, S., 2015. Mass transfer of simultaneous extraction and stripping of mercury (II) from petroleum produced water via HFSLM, *J. Ind. Eng. Chem.* 21: 1020–1028.

31. Wang, C. Y., Mercer, E., Kamranvand, F., Williams, L., Kolios, A., Parker, A., Tyrrel, S., Cartmell, E., Mcadam, E. J., 2017. Tube side mass transfer for hollow fiber membrane contactors operated in the low Graetz range, *J. Membr. Sci.* 523: 235–246.

32. Yadav, K. K., Singh, D. K., Kain, V., 2019. Separation of terbium from aqueous phase employing hollow fiber supported liquid membrane with EHEHPA as carrier, *J. Sep. Sci. Technol.* 54 (9): 1521–1532.

33. Aligwe, P. A., Sirkar, K. K., Canlas, C. J., 2019. Hollow fiber gas membrane-based removal and recovery of ammonia from water in three different scales and types of modules, *J. Sep. Sci. Technol.* 224: 580–590.

34. Ghasem, N., 2019. Modeling and simulation of CO_2 absorption enhancement in hollow fiber membrane contactor using CNT-water based nanofluids, *J. Membr. Sci. Res.* 5: 295–302.

35. Swain, B., Singh, K. K., Pabby, A. K., 2019. Numerical simulation of uranium extraction from nitric acid medium using hollow fiber contactor, *Solvent Extr. Ion Exch.* 37 (7): 526–544.

36. Hajilary, N., Rezakazemi, M., 2018. CFD modeling of CO_2 capture by water-based nanofluids using hollow fiber membrane contactor, *Int. J. Greenhouse Gas Control* 77: 88–95.

37. Haghshenas, H., Sadeghi, M. T., Ghadiri, M., 2016. Mathematical modeling of aroma compound recovery from natural source using hollow fiber membrane contactors with small packing fraction, *Chem. Eng. Process.* 102: 194–201.

38. Razavi, S. M. R., Rezakazemi, M., Albadarin, A. B., Shirazian, S., 2016. Simulation of CO_2 by solution of ammonium ionic liquid in hollow fiber contactors, *Chem. Eng. Process.* 108: 27–34.

39. Ghadiri, M., Asadollahzadeh, M., Hemmati, A., 2015. CFD simulation for separation of ion from waste water in a membrane contactor, *J. Water Process. Eng.* 6: 144–150.

40. Ghadiri, M., Shirazian, S., 2013. Computational simulation of mass transfer in extraction of alkali metals by means of nanoporous membrane extraction, *Chem. Eng. Process.* 69: 57–62.

41. Miramini, S. A., Razavi, S. M. R., Ghadiri, M., Mahdavi, S. Z., Moradi, S., 2013. CFD simulation of acetone separation from an aqueous solution using supercritical fluid in a hollow fiber membrane contactor, *Chem. Eng. Process.* 72: 130–136.

42. Shirazian, S., Pishnamazi, M., Rezakazemi, M., Nouri, A., Jafai, M., Noroozi, S., Marjani, A., 2012. Implementation of finite element method for simulation of mass transfer in membrane contactors, *Chem. Eng. Technol.* 35: 1077–1084.

43. Rezakazemi, M., Shirazian, S., Nezameddin, A. S., 2012. Simulation of ammonia removal from industrial waste water streams by means of a hollow fiber membrane contactor, *Desalination* 285: 383–392.

44. Marjani, A., Shirazian, S., 2011. Simulation of heavy metal extraction in membrane contactors using computational fluid dynamics, *Desalination* 281: 422–428.

45. Shirazian, S., Marjani, A., Fadai, F., 2011. Supercritical extraction of organic solutes from aqueous solutions by means of membrane contactors: CFD simulation, *Desalination* 277: 135–140.

46. Shirazian, S., Modhadassi, A., Moradi, S., 2009. Numerical simulation of mass transfer in gas-liquid hollow fiber membrane contactors for laminar flow conditions, *Simul. Model. Pract. Theory* 17: 708–718.

47. Pancharoen, U., Wongsawa, T., Lothongkum, A. W., 2011. A reaction flux for extraction of Cu (II) with LIX84I in HFSLM, *J. Sep. Sci. Technol.* 46: 2183–2190.

7 Treatment of Metal-Bearing Aqueous Solutions *via* Hollow Fibers Processing

F. J. Alguacil

CONTENTS

7.1 INTRODUCTION

Minimizing the impact of pollutants discharged into the environment either in the form of gases or metal-bearing liquid effluents is of concern industrially. However, in the case of metal-bearing effluents, a number of solutions, in the form of traditional and advanced separation technologies, can be used to remediate these unwelcome situations.

Besides the traditional, advanced separation technologies include:

i) liquid–liquid extraction,
ii) ion-exchange, adsorption with smart adsorbents,
iii) membranes and liquid membranes,

and among liquid membranes, the various operational approaches include:

i) bulk liquid membranes,
ii) emulsion liquid membranes,
iii) supported liquid membranes.

In practise, this last option seems to be the most popular for the treatment of metal-bearing hazardous liquid contaminants, or even waste solutions containing valuable metals. Thus, in liquid membranes, two main operational options can be used:

i) supported liquid membranes, and
ii) non-dispersive extraction with strip dispersion,

both in batch (flat supports) or continuous (hollow fiber supports) devices.

A hollow fiber supported liquid membrane (HFSLM) is a viable technique for metal extraction when present in the aqueous solution at very low concentrations, because it has a larger mass transfer per unit surface area, high selectivity, low energy consumption, and necessitates a lower amount of extractant to be involved in the metal recovery separation process.

This chapter deals with findings encountered in the period of 2014 to 2019, on the use of hollow fibers for the treatment of a series of metal-bearing aqueous solutions.

7.2 CHEMISTRY IN RELATION TO THE PROCESSING OF METAL-BEARING AQUEOUS SOLUTIONS *VIA* HOLLOW FIBER OPERATIONS

To be effective for the removal of metals from aqueous wastes, effluents, etc., the fibers must be impregnated by organic extractants, which are responsible for:

i) the enhancement of the transport of a given metal or metals,
ii) the separation in selective form of the targeted metal from the unwanted accompanying species/ metals.

Thus, together with the kinetic aspects of the transport processes, the chemistry involved in all these transport processes plays a key role in the success of a given system. Below, a brief explanation of how these extractants extract or transport metals is presented; this explanation is considered on the basis of the type of chemical equilibrium involved within the respective family of extractants or carriers.

7.2.1 Acidic and Acid-Chelating Extractants

The acidic type of carriers transport metals *via* a cation exchange mechanism which is an effective process. For this process, the main condition is that the metal must be in the aqueous solution as a cation. The cation exchange mechanism under equilibrium involved in the exchange can be represented by the general form as given below:

$$M_{aq}^{n+} + nHR_{org} \Leftrightarrow MR_{n_{org}} + nH_{ac}^{+} \qquad (7.1)$$

As seen in equation (7.1), M represents the metal in the aqueous solution and HR the extractant. Also, irrespective of the chemical structure of an extractant, the extractant always has a proton, which is interchangeable with the metal cation. It is evident that for metals with valence n^{+} one needs n extractant molecules to maintain the neutral charge of the extracted-transported complex. It is also evident from the above equation that during this reaction protons are released to the aqueous solution; this situation needs to be taken into account because the pH of the aqueous solution changes along the process, and some type of neutralization of this solution is necessary.

A special case of acidic extractants belongs to those acidic extractants which form chelating compounds with metals. Chelation of a metal is due to the presence of an atom in the organic molecule, which acts as an electron pair donor to the metal, and thus, a chelate is formed. Aside from this peculiarity, the equilibrium involved in the transport is the same as mentioned above. A typical case of acidic-chelating extractants are oximes that, in the case of Cu^{2+}, form a very stable 2:1 oxime:copper complex (Figure 7.1).

With acidic or acidic-chelating extractants, a stripping reaction, that is a shift of Equation (7.1) to the left, is completed with acidic aqueous solutions. This reaction, which regenerates the extractant and strips the metal in aqueous solution (product), is also known as back extraction.

Organic derivatives of phosphoric acid, phosphonic acids, and phosphinic acids, etc., constitute acidic extractants. On the other side, for extracting copper(II), oximes and β-diketones (acidic-chelating extractants) are often used to extract this element from aqueous solutions.

7.2.2 Basic Extractants

In this category of compounds, either amines or quaternary ammonium salts are included. On the other hand, quaternary ammonium salts are also included into ionic liquids category. As is clear from the literature, amines can also be

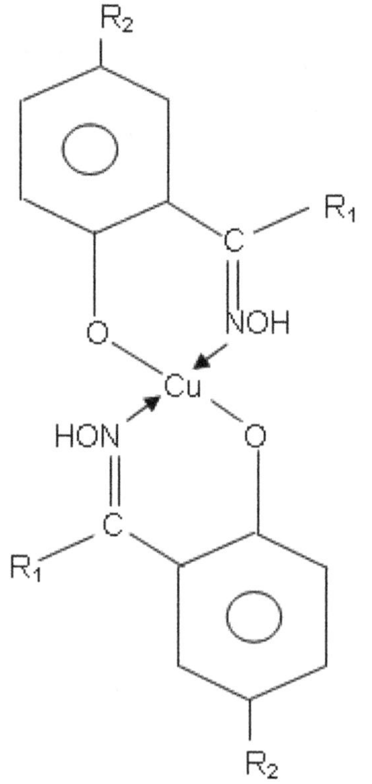

FIGURE 7.1 2:1 (oxime:copper) complex. The arrows represent the donation of an electro pairs from the nitrogen atoms to the copper atom. See two six-member-rings formed around copper(II) ion which give the maximum stability to the chelate compound, R1 = CH3 (ketoximes), or H (aldoximes). R2 = alkylchains

considered as ionic liquid extractants or carriers because of their chemical form, the amine salt, in which they extract metals, which fully resemble and match the stoichiometry of quaternary ammonium salts. This is because these types of reagents extract metals when they form anionic complexes. Thus, the amines are of three types: primary, secondary, and tertiary. All three extract metals by an anion exchange reaction which is given below:

$$R_3NH^+X_{org}^- + MX_{n_{oprg}}^- \Leftrightarrow R_3NH^+MX_{n_{org}}^- + X_{aq}^- \quad (7.2)$$

As seen in above reaction, when amines are pre-equilibrated with mineral acid, the reaction takes place when contacted with an aqueous solution containing a mineral acid:

$$R_3N_{org} + H_{aq}^+ + X_{aq}^- \Leftrightarrow R_3NH^+X_{org}^- \qquad (7.3)$$

Under these experimental conditions, it is possible to form an amine salt, which is the active compound in the anionic exchange reaction. It is clear that these amine salts, in reality, are some type of ammonium salts, which behave like ionic liquids. In the above reactions (Equations (7.2) and (7.3)), the organic compound is a tertiary amine; but with primary and secondary amines the same type of equilibria occurs. Also in the above reactions, the anion X^- can be the different, i.e., chloride in the organic and cyanide in the aqueous, nitrate in the organic and bisulfate in the aqueous, etc. With this type of organic compound, the stripping reaction is carried out with a variety of reagents like acidic solutions, complexants, etc.

7.2.3 SOLVATION EXTRACTANTS

The chemistry associated with this group of reagents involves the solvation of neutral metal compounds by organic compounds which contain at least one electron-donor atom. Due to this solvation the solubility of the inorganic compounds in the organic solution increases (this solvation behavior was first observed by Peligot around 1840, and it was the basis for the development of modern solvent extraction practise in the United States in around 1942).

Solvation reagents are typically divided into two main groups: compounds which have oxygen or sulfur bonded to phosphorous atoms and those in which oxygen atoms are bonded to carbon atoms. Under this classification, the first group included phosphates, phosphonates, phosphinates, and phosphine oxides whereas phosphine sulfide belongs to the sulfur bonded to phosphorous group. The second main group of solvation reagents includes ethers, esters, ketones, alcohols, and amides.

Considering these series of reagents, it is not possible to give a general extraction reaction like the acidic or basic reagents. Further, water molecules associated with the metallic compounds play a key role in the solvation mechanism. Phosphorous containing compounds compete with water molecules and replace them in the coordination sphere associated with the metals, whereas in the case of oxygen-containing extractants, water molecules are part of the solvated compound and probably form bridges between the metal compound and the organic compound.

The stripping of loaded metals from the organic phase can be performed by a variety of reagents or metal-complexing agents, specific for each system.

7.2.4 IONIC LIQUIDS (ILs)

These extractants emerged today as a fourth class of extractants for metals. By definition, ionic liquids are compounds which contain ions solely and are liquids at temperatures below 100°C. As mentioned earlier, they contain ions, cations, and the complementary anions where both cation and anion can be of inorganic and organic nature. These ILs are considered as *green solvents* due to their properties related to negligible vapor pressure, high thermal stability, and relatively high viscosity, etc. However, in the opinion of the author of this chapter, this high viscosity may be the major drawback for the practical use of some of these ILs, as it is necessary to dilute them with organic solvents to facilitate phase separation (solvent extraction practice) or metal transport (membranes operation). The chemistry involved with these types of

reagents includes extraction-transport by ion pair or by anion exchange (analogous to basic extractants, the amine salts associated with the extraction of metals are ionic liquids in nature). Metal stripping can be accomplished by a variety of reagents specific to a given system.

For a better performance of the above systems, the extractants are normally diluted in organic solvents or diluents. Dilution is necessary, among other things, to:

i) meet the requirement for a selective range of extractant concentrations for any specific case,
ii) decrease the viscosity of the organic phase, and thus increase the transport of a given metal,
iii) reduce the initial extractant inventory,
iv) improve the dispersion and coalescence properties of the organic phase; this is important when phase separation is necessary, such as in liquid membrane operations with strip dispersion.

The diluents used in practise are kerosene-type diluents, and aliphatic and aromatic diluents. There is no specific rule, besides experimentation, to choose a particular type of diluent, though with acidic and acidic-chelating extractants, kerosene and aliphatic diluents are often used, whereas in the case of solvation reagents, aromatic diluents may be the choice.

In practise, an unexpected situation occurs during phase separation several times, and this is when a third phase, or second organic phase, appears, either in the extraction or stripping stages, in the given system (Figure 7.2). In such a case, the operation is simply not practical, and it is necessary to change the diluent or at best to use a third organic component which is known as a modifier. This is added to the organic phase and leads to the disappearance of the third phase. Normally, the modifiers used are alcohols and even TBP (tributylphosphate, an extractant in its own right), though there is no general rule as to what modifier must be used, or as to the correct amount of modifier to be added to the organic phase to avoid the third phase. This has to be determined experimentally.

FIGURE 7.2 Third phase or second organic phase formation (left) after organic-strip phases separation in the pseudo emulsion phase reservoir tank against a normal two phases system (right)

7.3 HF-BASED APPLICATIONS

7.3.1 SYSTEM USING ACIDIC AND CHELATING EXTRACTANTS

The transport of Ag^+ from acidic pharmaceutical wastewater *via* HFSLM (hollow fiber supported liquid membrane) was investigated [1]. The wastewater contained 30 mg/L of Ag^+ and 120 mg/L of Fe^{3+}, and was subjected to HFSLM operation as a feed solution. LIX 84-I (2-hydroxy-5-nonylacetophenone oxime) dissolved in kerosene was used to impregnate the fibers, whereas a $Na_2S_2O_3$ solution was the receiving phase. The influence of Fe^{3+} on Ag^+ transport was investigated, firstly using solutions with the high Fe^{3+} concentration and later, with a solution in which iron was precipitated with a phosphoric acid solution. The best transport of Ag^+ was achieved by using 0.1 M LIX 84-I and 0.5 M $Na_2S_2O_3$·$5H_2O$ solutions, being the pH of the feed and receiving solutions 3.5 and 2, respectively. The flow rates of the feed and receiving solutions were 0.2 L/min. A residual 0.6 mg/L Ag^+ remained in the solution; however, it was below the mandatory discharge limit. No effect of Fe^{3+} concentration in the aqueous solution on Ag^+ transport was observed. The key parameters were defined to confirm the performance of the system. The controlling transport regime of Ag^+ across HFSLM was attributed to the diffusion flux and reaction flux models.

This same LIX 84-I extractant, also dissolved in kerosene, was used in a comparative study between solvent extraction and hollow fiber membrane technologies in the extraction of V(V) from aqueous chloride solutions [2]. Vanadium extraction in the presence of various other metals using hollow fiber supported liquid membrane (HFSLM) operations has been investigated. The influence of various experimental variables such as pH of the feed solution (1–2.5), flow rate (105–410 mL/min), LIX 84-I concentration (0.08–0.96 M), vanadium concentration (10.3–98.4 mol/m^3), and receiving phase concentration (0.01–0.1 M NaOH) on vanadium transport has been evaluated. Increasing the feed solution pH up to 1.75, the extractant concentration up to 0.32 M, and the flow rate up to 290 mL/min, the flux value increased up to a certain level and then decreased. The extracted species was determined to be VO_2R·HR. Vanadium can be extracted selectively and separated completely from a multi-elemental bearing solution by using this membrane technology, and the separation factor values were found to be in the order: $\beta_{V/Cu} < \beta_{V/Fe} < \beta_{V/Zn} < \beta_{V/Co} < \beta_{V/Mn} < \beta_{V/Ni}$. From the receiving phase, high purity (99%) V_2O_5 was produced by precipitation with sulfuric acid at pH value around 2.

The transport of zinc from sulfate solutions using pseudo-emulsion-based hollow fiber strip dispersion technique was investigated [3]. The reagent Ionquest 801 (phosphoric acid (2-ethylhexyl)-mono(2-ethylhexyl)ester)) was used as an extractant in these experiments. The influence of several parameters like the flow rates (135–320 mL/min and 100–450 mL/min for the feed and pseudo-emulsion phases, respectively), the initial concentration of Zn(II) (0.1, 0.5,

TABLE 7.1

Overall Mass Transfer Coefficients (K_p) in the Permeation of Zn(II) Using Various Carriers and in the Presence of Ca(II) in the Aqueous Feed Solution [3]

[a]Carrier	$K_p \times 10^6$, m/s	[b]$K_p \times 10^6$, m/s
Cyanex 272	1.6	0.95
DEHPA	4.5	1.0
Ionquest 801	4.3	2.5

[a] Organic phase: 10% v/v extractant and 5% v/v TBP in Shellsol D70
[b] In the presence of 0.5 g/L Ca(II)

and 1 g/L), the extractant concentration (4–20% v/v), the pH (2.1–4), the volume ratio of feed to strippant, and the presence of Ca(II) in the feed phase was studied. A few tests with the extractants Cyanex 272 and DEHPA were also carried out for comparison. Some of the results are summarized in Table 7.1. The decrease in the K_p of zinc(II) in the presence of calcium(II) can be explained by the calcium co-transport, being the difference more pronounced in the DEHPA > Ionquest 801 > Cyanex 272. Moreover, the values of K_p for calcium were given as: 4.7×10^{-10}, 3.9×10^{-8}, and 5.6×10^{-7} m/s for Cyanex 272, Ionquest 801, and DEHPA, respectively.

It was concluded that Cyanex 272 presented the best overall data for zinc/calcium separation.

The transport of yttrium from aqueous nitrate solutions was investigated using dinonyl phenyl phosphoric acid (DNPPA) dissolved in petrofin (an aliphatic fraction of refined kerosene) as a carrier phase of the liquid membrane in a microporous hydrophobic hollow fiber supported liquid membrane (HFSLM) module [4]. Various experimental variables were considered in order to investigate their potential influence on the permeability and flux of Y(III), i.e., feed composition and acidity, carrier concentration, strip acid concentration, and flow rate of feed phase. The strip solution was sulfuric acid. The effect of DNPPA concentration was studied on Y(III) transport, which was found to increase, whereas this percentage was decreased when the metal concentration in the feed was increased. Best, optimized conditions were obtained with a lower acidity in the feed solution and 3 M sulfuric acid concentration in the stripping phase. This is due to the increase in mass transfer driving forces under these experimental conditions. The percentage of Y(III) transport increased with an increase in flow rates up to 100 mL/min. Under optimized conditions of 0.2 M DNPPA, 1 g/L Y(III), 0.05 M HNO_3 feed acidity, 3 M sulfuric acid as strip solution, and 100 mL/min flow rate, the percentage of metal transport exceeded 95% in 5 hours of operation. The overall chemical reaction involved in the process can be described by the following reaction:

$$Y_{aq}^{3+} + 3\left(HR\right)_{2_{org}} \Leftrightarrow Y\left(HR_2\right)_{3_{org}} + 3H_{aq}^+ \qquad (7.4)$$

where HR represents the extractant molecule and $(HR)_2$ its dimmer form, and subscripts aq and org stand for the aqueous and organic phases, respectively. Thus, metal transport from the feed to the organic phase is carried out by shifting the equilibrium to the right (low acidity), and in the receiving solution (high acidity) the equilibrium is shifted to the left, releasing yttrium ions to the strip solution, whereas the extractant is regenerated and remains in the carrier phase.

The separation of Co(II) and Li(I) by non-dispersive solvent extraction using a hollow fiber supported liquid membrane has been investigated [5]. The parameters of the process were optimized to achieve quantitative separation of Co(II) over Li(I) with Cyanex 272 diluted with kerosene. For both the hollow fiber and the flat sheet supported liquid membrane processes, the aqueous feed pH of 6.0 and 0.75 M Cyanex 272 in the membrane phase were the best conditions for transport across the membrane phase, whereas the best stripping results were obtained with 0.1 and 0.025 M sulfuric acid concentrations. In the case of the dispersive solvent extraction process, the quantitative separation of the metals was achieved at an equilibrium pH of 5.5 using 0.1 M Cyanex 272, and a strip solution of 0.01 M sulfuric acid. Under the best experimental conditions, the separation factor $\beta_{Co/Li}$ was found to be 18, 178, and 180 for hollow fiber supported liquid membrane, flat sheet supported liquid membrane, and dispersive solvent extraction, respectively. From different models and the mathematical analyses in both the dispersive solvent extraction and non-dispersive solvent extraction processes, a cation-exchange mechanism, with $(Co(HL_2)_2)$ stoichiometry for the complex formed in the organic phase, is responsible for the Co(II) transport or extraction.

A hollow fiber supported liquid membrane operation was used to investigate the mass transfer of Nd(III) [6]. At pH 4.5, using as the carrier 0.5 M 2-ethylhexyl phosphonic acid mono-2-ethylhexyl ester (PC88A) dissolved in octane and 1M sulfuric acid as the receiving solution, optimum extraction and stripping reached yields of 98% and 95%, respectively. Mass transfer resistance due to chemical reaction proved to be of the highest value, i.e., $Re = 5.9 \times 10^2$ s/cm, indicating that the mass transfer resistance from the extraction was most likely the controlling step of the overall process. The experimental data fitted well with the proposed mass transfer model.

For many years, the solvent extraction of copper(II) with oximes has been attracting the attention of scientists and industry, and logically this interest had been mirrored in the application of liquid membranes for copper using oximes. The next reference [7], described the non-dispersive extraction study of copper(II), with the aldoxime Acorga M5640 dissolved in Shellsol D70 (aliphatic diluent), using a unique hollow fiber module. The organic phase flowed through the shell side, whereas the feed phase flowed by the tube side of the module. As reported in the literature, in this type of operation, the interface was stabilized in the pores by applying a higher pressure to the feed phase (in the order of 2 bar). The effects on the extraction rate of the flow rates

(having little effect in the present experimental conditions, Figure 7.3), the concentration of copper (percentage of copper transport decreased on increasing metal concentration over time), the role of extractant concentration (increasing the kinetics for 0.1 g/L solutions in the 5–10% v/v range and also in the case of 1 g/L Cu up to 10% v/v, however, in the last case a further increment of the extractant concentration to 20% v/v did not affect the transport kinetics), the pH of feed solution (increasing the metal transport from pH 1 to 3), and the presence of sodium sulfate in the same feed phase (and increase in the ionic strength reduced the transfer of copper from the feed to the organic solution) were investigated.

The extraction process was found to be governed by diffusion in the aqueous boundary layer and also by chemical reaction. The kinetic data obtained were used to simulate the extraction of copper with the pseudo-emulsion-based hollow fiber with strip dispersion technique (sulfuric acid as a receiving phase), using the same variables as in the single-extraction operation. The agreement between the results thus calculated and the experimental data was found to be good enough, particularly for feed solutions with a copper content below 1 g/L.

The pseudo-emulsion-based hollow-fiber strip dispersion (PEHFSD) technique was investigated as an alternative to conventional solvent extraction for the simultaneous separation and concentration of a cobalt(II)–nickel(II) mixture using Cyanex 272, 5% TBP, and kerosene as the organic phase, sulfuric acid solution was used as receiving phase [8]. The experimental work was carried out by continuous recirculation of the feed and pseudo-emulsion phases through the hollow-fiber module. The separation factor, $\beta_{Co/Ni}$, increased rapidly with time, being the maximum value, 128 at feed pH of 6.5, obtained after two hours of operation,

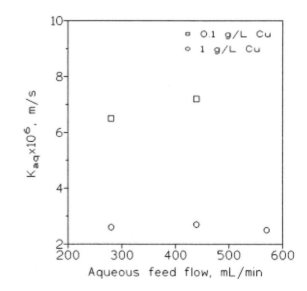

FIGURE 7.3 Influence of the aqueous feed flow rate on the overall mass transfer coefficient (Kaq). Aqueous feed solution: 0.1 or 1 g/L Cu(II) at pH 1.4. Organic phase: 10% v/v Acorga M5640 in Shellsol D70. Organic feed flow: 320 mL/min

and using a 0.2 M Cyanex 272 concentration in the organic phase. The process was dominated by the mass transfer resistance from the extraction reaction. A higher separation factor and extraction rates were achieved using PEHFSD in comparison to conventional solvent extraction.

Meerholz et al. [9] aimed at designing and constructing an automated membrane–based solvent extraction system for use in Zr(IV) and Hf(IV) extraction. The objective was to attain independent automated control of flow rate and pressure, while improving the accuracy and repeatability of extraction results. Flow rate (100 mL/min aqueous phase and 850 mL/min organic phase) and pressure (maximum 200 kPa) were controlled using the pressure PID control algorithms and optimized using the Cohen–Coon tuning method. After optimization, a case study for the extraction of Zr(IV) and Hf(IV), using Cyanex 301 (bis(2,4,4-trimethylpentyl)dithiophosphinic acid) as a carrier dissolved in cyclohexane, was conducted. It was shown that the automated system was able to accurately control the flow rate and pressure. This improvement of accuracy led to highly reproducible extraction results with the standard deviations varying by less than 1.2%. From the above, it can be concluded that the automated system was successfully implemented with independent control of the flow rate and pressure. The extraction of Hf(IV) was 90%, whereas that of Zr(IV) reached 57%. Unfortunately, no stripping investigations were carried out; thus, the study has to be further investigated, as this seems incomplete.

A new approach to the use of hollow fiber modules is reported when the feed and the organic phases are mixed, and the acceptor phase flows separately across the system. This approach is called hollow fiber renewal liquid membrane (HFRLM), and this methodology was used in the extraction and stripping of Th(IV) [10]. The organic solution of bis(2,4,4-trimethylpentyl) phosphininc acid (Cyanex 272) diluted in kerosene was used as the carrier phase which could transport Th^{4+} from the feed nitrate solution to the receiving solution containing sulfuric acid. In the experimentation, the mixture of the feed and organic phases, at a volume ratio of A/O= 20, flowed through the tube side while the receiving phase was pumped through the shell side of the module. The results indicated that the stability of the HFRLM process during 10 h of continuous operation was satisfactory. Furthermore, the results showed that the increase in the flow of the feed and organic phases mixture in the tube side, the concentration of Cyanex 272 in the liquid membrane phase, and the pH of the feed led to the increase in the mass transfer flux; but, the increase in the sulfuric acid concentration of the receiving phase from 0.2 to 1 M, and the initial concentration of thorium(IV) were not relevant in the process. The overall mass transfer coefficient was calculated in the range of 5.7×10^{-8} to 2.4×10^{-7} m/s.

A novel extraction gel membrane (EGM) was prepared via gelation technique by forming a polydimethylsiloxane-di(2-ethylhexyl)phosphoric acid (PDMS-D2EHPA) extraction gel layer on the outer surface of PVDF ultrafiltration hollow fibers [11]. This special design aims at achieving both efficient mass transfer within the thin extraction gel layer, and improvement in operating stability by preventing the carrier (D2EHPA) loss from EGM. The physical and chemical properties of the prepared EGM were studied *via* a series of characterization techniques, and the fixed site jumping carrier mechanism for EGM was confirmed. The influences of the EGM preparation parameters on its performance in nickel separation were investigated systematically. The optimal extraction efficiency and stability of EGM were achieved by the EGM prepared from the ratio of 10:8.0:0.2 for PDMS/TEOS/DBTL, 24 wt% PDMS concentration, 40 min-coating, and a low level (15–25%) of relative humidity. Using EGM prepared from these conditions, long-term operating stability was tested, and the flux results were compared to those of the conventional supported liquid membrane (SLM) process. It was shown that the flux attenuation of the EGM was only 34.11% within 120 h, while that of the conventional SLM was 100% within 45 h. Simultaneously, nickel initial flux using EGM was as high as 2436.30 mg m^{-2} h^{-1}, which was 7.1 times that for the conventional SLM. The results of this study indicated evident advantages of the novel EGM process over the traditional SLM by showing higher flux and operating stability. The whole experimentation was done using a 0.5 M sulfuric acid solution as a receiving phase.

In the next work [12], the recovery of Y(III), from acidic nitrate media, by PEHFSD system using bis(2-ethylhexyl) phosphoric acid (D2EHPA) as a carrier was investigated, being the diluent of the organic phase kerosene. The effects of several operating parameters, including the initial concentration of Y(III) in the feed phase ($1-4 \times 10^{-3}$ M), the flow rate of the feed, the stirring speed, and the volumetric ratio of feed to strip on Y(III) separation were studied. In this experimental series, fixed conditions were: 0.1 M D2EHPA, 0.5 M nitric acid feed acidity, 3 M nitric acid receiving solution. The Y(III) transport was analyzed on the concentration ratio of Y(III) ions, percentage extraction, percentage stripping, and overall mass transfer coefficient K_p. The PEHFSD system surpasses HFSLM operation considering separation performance and stability. K_p of HFSLM system decreased after the second run, with values of K_p of 8.5×10^{-6} versus 6.7×10^{-6} m/s, for the first and second run respectively, but K_p of PEHFSD system remained constant even at the fifth run, averaging 8.3×10^{-6} m/s. It was concluded that the dispersed droplets in the strip dispersion phase in the PEHFSD system enhanced the separation performance and stability of the membrane module.

Hollow fiber membrane (HFM) operation in non-dispersive solvent extraction (NDSX) mode has been used for the separation of dysprosium from NdFeB magnetic scrap material using EHEHPA (2-ethylhexyl-2-ethylhexylphosphonic acid) dissolved in normal heavy paraffin as a carrier phase [13]. The effect of various hydrodynamic parameters, including aqueous phase acidity (0.1–0.5 M nitric acid), carrier concentration (0.2–0.83 M), dysprosium concentration (0.25–2 g/L) in the feed, phase ratio (O/A:

400 mL/400–1600 mL), and flow rate (50–200 mL/min) were investigated to optimize the conditions for quantitative recovery of dysprosium from the leach liquor obtained by the dissolution of hard disk drive (HDD) in nitric acid media. In conclusion, 0.5 M EHEHPA, 0.3 M HNO$_3$ as aqueous feed phase, 100 mL/min flow rate, and phase ratio of 1:1 were found to be the optimum for both of the cycle test runs. It was claimed by the authors that Dy(III) can be separated from Nd(III) and Pr(III), though this separation decreased at lower nitric acid concentrations in the feed phase (Table 7.2).

Scarce stripping data were given. It was only mentioned that 1.5 M nitric acid solution recovered the dysprosium into the organic solution after 150 min processing of a 0.5 M EHEHPA solution loaded with 3.3 g/L of the rare earth. Taking the advantage of the fast extraction kinetics of dysprosium, two cycle HFM-NDSX approaches were adopted to concentrate this rare earth from 20 to 83% in the first cycle and increase the purity to >97% in the second cycle. The overall process also yielded neodymium rich by-product for its further purification.

Hollow fiber membrane (HFM) operation in supported liquid membrane mode has been used for the separation of terbium from nitrate aqueous solutions by using EHEHPA (PC88A) diluted in heavy paraffin as carrier phase [14]. Different variables affecting the metal transport process were considered: aqueous-phase acidity (0.1–0.75 M nitric acid), carrier concentration (0.25–1 M), and flow rate (100–300 mL/min). Under optimized conditions of 1 M carrier in a heavy paraffin, 0.1 M HNO$_3$ in the aqueous feed phase, flows in feed and strip phases of 100 mL/min, and a strip phase of 3.5 M nitric acid, the recovery of terbium was greater than 95%. The metal (1 g/L) was fully transported from the feed to the membrane phase in around 30 min, whereas 40 to 60 min were needed to recover (98%) terbium in the strip phase. In the presence of other rare earths, the transport order found was Tb>Gd>La. From the experimental results, several diffusional parameters were estimated. The reaction responsible for transport of Tb(III) was defined as:

$$Tb_{aq}^{3+} + 3\left(H_2R_2\right)_{org} \Leftrightarrow Tb\left(HR_2\right)_{3org} + 3H_{aq}^+ \quad (7.5)$$

TABLE 7.2
Percentage Extraction of Dy(III), Nd(III), and Pr(III) as a Function of the Feed Phase Acidity [15]

Rare Earth	0.1 M HNO$_3$	0.3 M HNO$_3$
Dy(III)	>99	>99
Pr(III)	69	8
Nd(III)	79	13

Feed solution: 0.2, 0.4, and 0.4 g/L Dy, Nd, and Pr in nitric acid. Organic phase: 0.5 M EHEHPA. Time 120 min

where (H_2R_2) represented the dimer form of the carrier. According to the equation, Tb(III) transport occurred by shifting the reaction to the right, and the metal was stripped shifting the reaction to the left.

7.3.2 Systems Using Basic Extractants

Not all of the investigations are based on experimental work as it is demonstrated in a theoretical work related to the transport of Ag(CN)$_2^-$ in a hollow fiber extractor using LIX 79 dissolved in n-heptane as a carrier phase [15]. LIX 79 is a guanidine derivative, where active group being N,N-bis(2-ethylhexyl)guanidine, the transport process of the silver-cyanide complex from the feed to the organic phase is related to an anion exchange mechanism. The basis for this simulation is the experimental work described elsewhere [16]. A two-dimensional mathematical model was developed for the transport of Ag(CN)$_2^-$ ions through porous membrane extractors. Simulations were done using computational fluid dynamics of momentum and mass transfer in all subdomains of a hollow-fiber membrane extractor by COMSOL Multiphysics software. The latter uses a finite element method for numerical simulations. Parabolic velocity profile was used for the aqueous feed in the tube side and the solvent flow in the shell side that was characterized by Happel's equation. The distribution of concentration was obtained for the solute in the membrane module. Simulation results indicated that increasing feed flow rate reduced the extraction efficiency of silver from aqueous phase to organic phase from 90% at 1×10^{-6} m^3/s to near 50% at 10×10^{-6} m^3/s. Dimensionless concentration distribution [Ag]$_t$/[Ag]$_0$ of the silver-cyanide complex in the tube side of the membrane extractor in axial and radial direction shows that it moves to the membrane due to the concentration difference, and then it is swept by the moving carrier in the shell.

Trialkylamine (N235) diluted in kerosene/n-heptane and TBP (tributylphosphate) as modifier was used to investigate the removal of Cr(VI) from aqueous solutions both with solvent extraction and hollow fiber strip dispersion technologies [17]. Both the conditions for Cr(VI) extraction and the influence of coexisting contaminants on extraction have been investigated. The results indicated that the separation of Cr(VI) could be carried out with amine N235, and the recovery rate could be increased by adding TBP as the modifier due to its strong bonding force against Cr(VI) species and thus synergistic extraction. Thus, in this investigation the addition of TBP has two benefit effects: (i) it acts as a modifier of the organic phase; and (ii) it acts as a co-extractant for Cr(VI) (comments of the author of the chapter). There is a negative influence of accompanying anions in the aqueous solution; thus, it is described that inorganic phosphorus could inhibit the chemical reaction between tertiary amine and Cr(VI) ions owing to the competition of the adsorption sites, whereas sulfate and nitrate anions could inhibit extraction of Cr(VI) because of competition coordination (Figure 7.4), and probably some negative

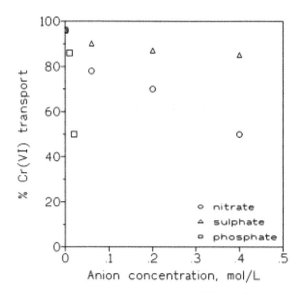

FIGURE 7.4 Influence of the anion concentration in the aqueous feed phase on Cr(VI) transport. Aqueous feed phase: 0.01 g/L Cr(VI) in 0.1 M HCl and the respective anion. Flow rate: 100 mL/min. Pseudo-emulsion phase: 1% v/v N235 in n-heptane and 1 M NaOH. Flow rate: 80 mL/min. Time: 2 h

salting-out effect caused by the presence of these three anions in the aqueous solution (the author of this chapter). Moreover, the separation of Cr(VI) from coexisting metal ions could be achieved with trialkylamine except Fe^{2+}, and this should be due to the reductive properties of this cation against Cr(VI). Based on solvent extraction results, the feasibility of pseudo-emulsion-based hollow fiber strip dispersion used for separation and recovery of Cr(VI) has been investigated, and the transport properties of Cr(VI) and coexistence with interfering ions have been explored.

The experimental data (extraction behavior and interference of accompanying ions) obtained in the solvent extraction step is practically the same as those obtained in the membrane operation.

Probably as a consequence of the previous reference (comment by the author of this chapter), in the next work [18], the transport of Cr(VI) from acidic media *via* a pseudo-emulsion-based hollow fiber strip dispersion (PEHFSD) technique using a carrier phase of tri-octylamine (N235) dissolved in heptane, was investigated. The effects of various hydrodynamic and chemical parameters, including the flow rates, variation in the feed pH, initial Cr(VI) concentration, and N235 concentration in the pseudo-emulsion, on Cr(VI) transport were investigated. The optimum conditions for Cr(VI) separation from hydrochloric acidic media were also estimated. The results indicated that the transport is controlled by the diffusion of Cr(VI) species in the stagnant film between the aqueous solution and the liquid membrane.

Hollow fiber renewal liquid membrane (HFRLM) operation in continuous and recycling modes for the extraction of uranium(VI) from acidic sulfate solution has also been investigated [19]. Tertiary amine Alamine 336 diluted in

kerosene was used as the carrier solution in the liquid membrane phase. In the batch experiments, the effect of sulfuric acid, extractant, and uranium(VI) concentration were studied, and the optimum concentration of the feed and carrier phases were determined as 0.15 and 0.0125 M, respectively. Various parameters affecting the HFRLM performance, including the tube and shell side flow rates, organic/aqueous volume ratio, stripping phase type, and concentration of carrier and stripping phase were studied. The mass transfer flux increases with increasing the tube side flow rates, whereas the shell side flow rate did not have any significant effect on this value. The uranium transfer flux increases with increasing O/A ratio, strip, and Alamine 336 concentration, reaching a maximum value at 1/20 O/A relationship, 0.5 M and 0.0125 M, respectively. A further increase in these parameters results in uranium transfer decrease. The results show that the liquid membrane phase is the rate-controlling step. Among the investigated strip phases, 0.5 M NH_4Cl results in 60.4% uranium(VI) recovery in the recycling mode operation.

The performance of HFRLM for uranium transport was investigated after the addition of different surfactant agents either to the feed or receiving aqueous solutions [20]. The influence of several variables on metal transport was investigated: sulfuric acid concentration (0.01–0.5 M) in the feed solution, Alamine 336 concentration (0.01.0.02 M), and type of stripping agent (NaCl, NH4Cl, and NaHCO3). In the recycling mode of HFRLM, maximum uranium recovery of 63% was obtained at 0.15 M sulfuric acid, 0.0125 M Alamine 336 in kerosene, and 0.25 M of ammonium chloride as feed, membrane, and receiving phases, respectively. The fluxes with HFRLM operation were greater than the flux obtained with HFSLM. The uranium transfer would be improved by adding concentrations of 0.05 mM of SDS, CTAB, and LAE-7 to the feed phase and 10 mM of CTAB and LAE-7 to the receiving phase.

The tertiary amine Alamine 336 dissolved in kerosene and decanol was used as carrier phase to investigate the transport of germanium, in a hollow fiber device, from a solution of pH 2.3, containing 0.06 g/L germanium, Ni, Zn, Cd, Co, and tartaric acid, to a strip solution of HCl 1 M [21]. The flows were of 23 L/h for the pseudo-emulsion phase (organic and strip solutions) circulating in the lumen side, and 28 L/h for the feed phase circulating by the shell side. A concentration factor for germanium of near 4 was reached after approximately 150 min, and no transport of Ni, Zn, Cd, and Co was observed. Using the same experimental conditions, the removal of germanium from the aqueous solution was compared using liquid–liquid extraction, HFSLM, and FSSLM; the results summarized in Table 7.3 showed that the efficiencies were comparable using LLE and HFSLM, and were greater than that of FSSLM operation.

7.3.3 Systems Using Solvating Extractants

Hollow fiber supported liquid membrane (HFSLM) technology was used to investigate the co-transport of neodymium

TABLE 7.3

Removal of Germanium(IV) from the Feed Phase Using Various Separation Technologies [21]

Technology	% Germanium Removal
Liquid–liquid extraction	90
Hollow fiber supported liquid membrane	86
Flat-sheet supported liquid membrane	25

Time: 30 min

and uranium ions from aqueous nitrate media [22]. The carrier solution impregnating the membrane pores consisted of N,N,N',N'-tetraoctyl diglycolamide (TODGA), isodecanol as a modifier, and dodecane as a diluent. The strip solution was distilled water. The results suggest that there is competition between neodymium and uranium ions for complexation with TODGA. The initial rate of extraction of Nd^{3+} was found to be approximately six times that of UO_2^{2+}. Experimental data were explained by a mathematical model for simultaneous transport of two metal ions. A model was derived and found to be in good agreement with the experimental data when the diffusivities of neodymium-TODGA complex (Dnm) and uranium-TODGA complex (Dum) in the membrane pore were 1.1×10^{-11}, and 4×10^{-12} m²/s, respectively. Furthermore, the authors claimed that the present model can be used to predict the extraction behavior in the co-transport of cobalt and nickel, but neither the experimental conditions, nor the carrier phase, were the same (comment of the author of this chapter).

In a very similar investigation to the above with respect to the metals, carrier, and strip phase involved in it, different feed and strip operating modes have been compared for the transport of neodymium (Nd^{3+}) and uranyl (UO_2^{2+}) ions in nitric acid media [23]. The flux values of the uranyl cation for feed in recycling and once through mode with fresh strip were 2.26×10^{-9} kmol/m²s and 2.33×10^{-9} kmol/m²s, respectively. So, these configurations resulted in low separation factors. The flux values of Nd^{3+} for feed and strip, both in recycling and in once through mode, were 2.21×10^{-9} kmol/m²s and 2.14×10^{-9} kmol/m²s, respectively. The operating modes, both feed and strip solutions in recycling and both feed and strip phases in once through operation mode, were found to be the best configurations if the flux of Nd and the separation factors are the targets of the investigation.

A review paper has been published that deals with the removal of arsenic by the use of liquid membranes technologies [24]. Many people are aware that water contamination with toxic arsenic compounds, at both (III) and (V) oxidation states, represents one of the most serious hazards for many years. The natural occurrence of the toxic metal has been revealed recently for a large number of countries; the risk of arsenic poisoning being particularly high in Bangladesh and India, although recently Europe (i.e. Spain) is facing a

similar problem. From years ago, liquid membranes (LMs) surged as a promising alternative to the existing removal processes, i.e., chemical precipitation, adsorption, solvent extraction, etc., showing numerous advantages in terms of energy consumption, efficiency, selectivity, and operational costs. The development of different LM configurations has been fully considered by several research groups, and this includes the removal of As(III) and As(V) from aqueous solutions. For these two, most of these LM systems are based on the use of phosphine oxides as carriers when the arsenic removal is from sulfuric acid media, and both As(III) and As(V) are in the aqueous solution as neutral species, that is, their corresponding acids H_3AsO_3 and H_3AsO_4. When arsenic forms anionic species, that is, from low acidic to alkaline solutions, the above neutral extractants are not suitable for arsenic removal from the solution, and here Aliquat 336 (a quaternary ammonium salt), and probably other ionic liquids of similar characteristics, are suitable for this task. Particularly promising for water treatment is the hollow fiber supported liquid membrane (HFSLM) configuration, which offers high selectivity, easiness in the transport of the targeted metal ions, large surface area, and non-stop flow process. Emulsion liquid membrane (ELM) systems have not been extensively investigated, and this is probably due to their ability, as membrane operations have largely been surpassed by other more advanced liquid membrane configurations. For such ELM configurations, the most relevant step toward process efficiency is the choice of the surfactant type and its concentration, which supports the emulsion stability. Since, ELM investigation are beyond the scope of this chapter, they are not included here.

The transport of Hg^{2+} through hollow fiber supported liquid membrane was investigated using calix[4]arene nitrile and NOPE (2-nitrophenyl octylether) as a diluent [25]. Optimum conditions were reached using pH of 4.5 in the feed solution (flowing by the tube side), 0.004 M of calix[4]arene nitrile, deionized water as stripping solution (flowing by the shell side), and an operating temperature of 40°C. In this operation, the feed and strip solution flowed in counter-current mode. The complex species formed in the organic phase presented the next stoichiometry: [Hg(C₆H₂(NO₂)₃O)₂·L], being L= calix[4]arene, whereas in the stripping stage with water, mercury is released to the strip phase as Hg^{2+} and the extractant is regenerated; in this strip phase, OH^- is formed as a consequence of the chemical reaction. The percentages of extraction and stripping of mercury(II) ions are 99.5 and 97.5%, respectively. The stability of the liquid membrane was investigated and showed to be stable over 24 hours. After the process, the content of Hg^{2+} in the outlet solution was found to be below the legislation limit of 5ppb.

Another investigation reports the selective separation of zinc over iron from spent pickling wastes or effluents using two different membrane-based solvent extraction process configurations, non-dispersive solvent extraction (NDSX) and emulsion pertraction technology (EPT) [26]. The goal of the process is to obtain a highly concentrated zinc solution with a negligible content of iron to allow the outlet solution to enter the Zn electrowinning step to yield zinc cathode as end and saleable product. The effect of the following process variables on the kinetics and selectivity of zinc separation has been evaluated: (i) process configuration NDSX and EPT, (ii) extractant concentration, tributylphosphate (TBP) in the range of 20 to 100% v/v, and (iii) stripping phase/feed phase volume ratio in the range V_s/V_f 0.2–2. The transport of iron, chloride, and free acid has also been monitored to gain insight into the separation fundamentals. EPT configuration overcame NDSX in terms of zinc and iron separation kinetics, although separation selectivity (at 30 min) was higher for NDSX configuration, $\beta_{Zn/Fe}$ of 22, compared to EPT process $\beta_{Zn/Fe}$ of 15. The optimum TBP content in the extractant phase was found to be 50% v/v. A further increase did not improve the Zn recovery kinetics and reduced the Zn/Fe selectivity. The increase in the V_s/V_f ratio improved the process efficacy in terms of kinetics and zinc recovery.

The next reference [27], deals with the analysis of an integrated zinc recovery process by means of electrowinning of the receiving solutions from the treatment of spent pickling baths by hollow fiber strip dispersion methodology (HFSD), which increases the initial Zn/Fe molar ratio. TBP is used again as the extractant, which dissolves in Shellsol D70 (aliphatic diluent) and water conforms the receiving phase. The pickling bath contained: 120 g/L Zn^{2+}, 1.7 g/L Fe_{TOTAL} (0.04 g/L Fe^{3+}), 8.5 g/L Cl^-, and 1.1 M H^+, whereas the pseudo-emulsion phase was formed by 800 mL of the organic solution and 200 mL of water. Under different experimental conditions in the membrane operation (20 and 50% TBP, operational time from 1 to 3 h), various receiving solutions containing different concentrations of zinc and iron ([Zn^{2+}]/[Fe^{2+}] molar relationships from 2.9 to 32.5) in acid media (pH of the solution from 0.60 to 1.67)) are obtained, and are subjected to electrowinning to investigate the efficiency and selectivity of zinc electrowon over iron under different operational conditions.

Interest in the recovery of rare earth elements from different secondary sources is still growing, and the hollow fiber supported liquid membrane operation was selected to investigate the recovery of neodymium, praseodymium, and dysprosium from commercial NdFeB magnets and industrial scrap magnets [28]. The operation runs as a single step with the feed and strip solutions circulating continuously through the system. The effects of several experimental variables on the transport of metals such as flow rate, concentration of the metals in the feed solution, membrane configuration, and feed and strip acidities were investigated. The first experimental results led to consideration of a multi-membrane module configuration, with these metals dissolved in aqueous nitric acid solutions, as the operational mode to obtain a better selectivity for their transport, and no co-transport of other accompanying metals in the feed solution. The use of aqueous hydrochloric acid solution, as leaching agent for the solid material, resulted in the co-transport of these other metals, due to the

formation of their chloroanion complexes. The corresponding oxides of neodymium, praseodymium, and dysprosium were recovered from the strip solution through precipitation, drying, and annealing steps. The above investigation was carried out by using Cyanex 923 (phosphine oxide) and TODGA (amide) as carriers in the liquid membrane phase.

The authors of the previous reference reported, in a new publication [29], the development of a supported liquid membrane system which used polymeric hollow fiber modules to extract rare earth elements (REEs) from neodymium-based magnets with neutral extractants such as tetraoctyldigylcol amide (TODGA) and Cyanex 923 (phosphine oxide) diluted in isoparaffin (Isopar L commercial diluent); tributylphosphate (TBP) also formed part of the carrier phase. Previous solvent extraction investigations indicated that with TODGA, best selectivity was obtained with respect to that reached with the use of Cyanex 923, thus, the amide extractant was further used in the liquid membrane system. The effect of process variables such as REEs concentration, type and concentration of the acid in the feed phase, and membrane area on REEs recovery was investigated. Best results were obtained with nitric acid solutions against hydrochloric acid solutions, thus, typical nitric acid concentrations in the feed and strip phases were of 3 and 0.5 M, respectively. To improve the results first obtained, systems containing series or four or eight membrane modules were tested (Figure 7.5). The modules' (Membrana) characteristics were: 700 polypropylene hollow fibers with 100 cm^2 area, 0.25 mm internal diameter, and wall thickness of 0.03 mm. In the operation, the pressure on the feed phase was maintained at approximately 200 kPa, whereas the strip phase was maintained at atmospheric pressure. Results derived from this investigation indicated that Nd recovery in the strip phase increased using 0.5 M HNO_3 as strip phase and 1 g/L Nd in the feed phase. These results were optimized using a series of eight modules, which further confirmed the high selectivity of the carrier phase for these rare earth elements against non-rare earth elements, i.e., iron and boron present in the feed solution. The extracted REEs were then recovered by precipitation followed by the annealing step to obtain crystalline REE oxide powders in nearly pure form. It is not clear the role of TBP in the process (comments of the author of the chapter).

^{99}Mo production waste having a unique actinide and fission product composition including enriched uranium are not used for the recovery of uranium appearing in these commercial alkaline molybdenum solutions. Thus, nondispersive membrane-based solvent extraction (MBSX), which is how the authors of the work named the hollow fiber supported liquid membrane (HFSLM) operation, is used to resolve this situation [30]. The residue was dissolved in nitric acid and extracted using TBP; 0.1 M acetohydroxamic acid (AHA) solution was added to the feed phase to prevent co-extraction of plutonium(IV).

Complete U extraction with MBSX was achieved at low U(VI) concentrations, i.e., 0.1 g/L, in the feed phase while

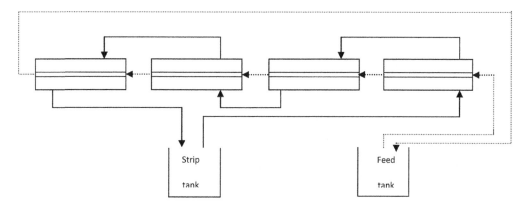

FIGURE 7.5 Scheme of the four modules in series operation for the separation of REE from non-REE. Feed phase: tube side. Strip phase: shell side. Operational conditions: feed phase (2 L): 1 or 5 g/L Nd in 6 M HNO3. Strip phase (200 mL): 0.5 or 3 M HNO3. Organic phase: 30% TODGA and 30%TBP in Isopar L (40%). Time: 24 h

more than one MBSX module was required in the treatment of high U(VI) concentrations, i.e., 98 g/L. Ammonium carbonate was used as an alternative strippant to dilute nitric acid. Almost quantitative uranium stripping (99%) from an organic phase containing 7.5 g/L U(VI) was obtained in a single contacting step with 0.5 M ammonium carbonate solution. Practically none of the characteristic fission products present in the ^{99}Mo production residue was extracted when AHA was added to the feed phase and neither influencing uranium nor fission product extraction. This investigation showed that MBSX methodology was suitable for the selective extraction (with AHA added) of uranium(VI) and stripping (0.5 M ammonium carbonate) of this metal from a simulated ^{99}Mo production.

Due to their characteristic of bonding with metals, calixarenes are always a topic of investigation, and in this investigation bis-octyloxy-calix[4]arene-mono-crown-6 (CMC) dissolved in a mixture of iso-decanol and n-dodecane was used for radio-cesium separation from acidic feed solutions using a hollow fiber supported liquid membrane (HFSLM) operation [31]. Solvent extraction and flat sheet supported liquid membrane (FSSLM) studies were first used to optimize the conditions required for the subsequent HFSLM investigation; all of the above operations used water as stripping reagent. In hollow fiber operations, the feed solution flew in the tube side and the strip phase in the shell side of the fibers, both solutions flew in counter-current mode. The permeation rate of cesium in HFSLM increased steadily with nitric acid concentration from 1 to 6 M. At a moderate nitric acid concentration of 4 M, which is the acid concentration of the high level waste, near quantitative (exceeding 99%) transport of cesium was possible within 2 h of operation for a 300 mL feed volume. Transport was suppressed under simulated high level waste conditions due to the presence of large concentrations of cesium (0.32 g/L), but more than 90% transport was observed in six hours. Stability of the membrane was excellent as results from 12 days of continuous operation showed. The permeability coefficient, mass transfer coefficient, and various diffusional parameters of the diffusing metal/ligand species

were calculated to understand the permeation behavior of cesium. This work confirmed the possible use of CMC in hollow fiber contactor for radioactive waste treatment where the ligand inventory is very low. It was described that the extraction or transport reaction was:

$$Cs_{aq}^+ + NO_{3_{aq}}^- + L_{org} \Leftrightarrow \left(Cs{\cdot}L\right)\left(NO_3^-\right)_{org} \quad (7.6)$$

being L the extractants and the subscript aq and org the species in the aqueous feed and carrier phases, respectively.

The liquid membrane operation of dispersion-free solvent extraction of U(VI) and Th(IV) from nitric acid medium was investigated using microporous hollow fiber membranes impregnated with various monoamide extractants: N,N-di-2-ethylhexyl isobutyramide (D2EHiBA), N,N-dihexyl hexanamide (DHHA), N,N-dihexyl decanamide (DHDA), and N,N-dihexyl octanamide (DHOA), and the results were compared with those obtained with TBP [32]. The investigation was done in a single hollow fiber module, in which the organic solution flew by the shell side, whereas the feed phase by the tube side, and apparently the same configuration was used in the separate strip operation. In all of the cases, the transport of the metals from the feed phase to the organic solution responded to the formation of species with $UO_2(NO_3)_2L$ and $Th(NO_3)_4L$ stoichiometries, being L the extractant. The effects of flow rate and feed metal ion concentration on the extraction rate were investigated. The extraction was principally controlled by the distribution coefficient of the metal ions with the carriers. The studies were performed under different hydrodynamic conditions, and the overall mass transfer coefficients of the transported species were calculated and found to be independent of the flow rates, thus, it seems that the overall extraction rates were governed by the viscosity and density of the organic phase. The separation factor between U(VI) and Th(IV) was excellent with branched chain monoamides, as D2EHiBA, followed by DHOA. The results indicated the selective recovery of uranium in the presence of large concentrations of thorium by monoamides in dispersion-free solvent

TABLE 7.4

Transport and Strip of U(VI) and Th(IV) Using Monoamides and TBP as Extractants [32]

Carrier	Element	% Extraction[a]	% Stripping[b]
DHOA	U(VI)-Th(IV)	100–3.9	51[c]–82[d]
DHHA	U(VI)-Th(IV)	84–8	no data
DHDA	U(VI)-Th(IV)	77–1.3	no data
D2EHiBA	U(VI)- Th(IV)	37–0.2	no data
TBP	U(VI)- Th(IV)	100–54.3	75[c]–67[d]

[a] Time: 90 min. [b]Time: 120 min. [c]Stripping agent: 1M Na$_2$CO$_3$. [d]Stripping agent: distilled water

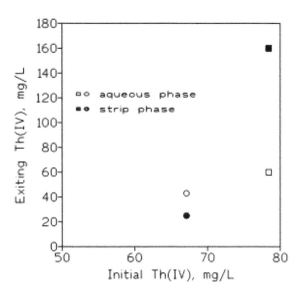

FIGURE 7.6 Influence of the Th(IV) concentration in the inlet aqueous feed phase on the metal outlet concentration in the feed and strip phases. Feed aqueous solution: Th(IV) in 2 M NaNO3. Organic phase: 30% v/v TBP in kerosene. Strip solution: 0.1 M HNO3. Tube side velocity: 0.025 m/s and shell side velocity: 0.003 m/s [Th(IV)]= 67.1 mg/L. Tube side velocity: 0.085 m/s and shell side velocity: 0.0009 m/s [Th(IV)]= 78.4 mg/L). Time: 40 min

extraction method. A not high efficiency of only 50% uranium recovery from the loaded organic phases was found in the stripping stage with sodium carbonate. Next, Table 7.4 showed some of the results obtained with the amides and TBP at 1.1 M in n-dodecane, 10 g/L metal in 3 M nitric acid as a feed phase, and 50 mL/min flow of both phases. The same flow was used in the stripping experiments.

It can be seen that in the case of U(VI) only DHOA extractant gives similar extraction yields to TBP, whereas in the case of Th(IV), TBP presented the best results. In terms of stripping recoveries, discrete uranium results were obtained either with TBP and DHOA.

In a recently published work on the recovery of actinides from aqueous solutions [33], Nd(III) was used to simulate the behavior of Am(III). Thus, a calixarene derivative was used in the transport of Nd (Am(III)) from nitric acid feeds using hollow fiber supported liquid membrane (HFSLM) containing a diglycolamide-functionalized calix[4]arene (C4DGA) as the carrier extractant. The effect of feed acidity and Nd(III) concentration (as it is said before, used to represent Am(III)) in the feed on the permeation of Nd(Am(III)) was investigated. Complete permeation of Nd(Am(III)) from the source to the receiving phase was possible within 30 min at tracer scale. The permeation of the metal ion was unaffected with the feed HNO$_3$ concentration in the range of 2–4 M, while the presence of macro quantities of Nd(Am(III)) in the feed solution suppressed the metal ion transport rates. The permeability coefficient, mass transfer coefficient, and various diffusional parameters of the diffusing metal-ligand complex were calculated to understand Nd(Am(III)) transport behavior. The transport rates of Nd(Am(III)) were predicted by a mathematical model with an excellent match between the experimental and calculated data. The present study gives an opportunity to use rare ligands for radioactive waste treatment due to the very low extractant inventory using HFSLM operation.

In another study [34], Th(IV) is transported through a hollow fiber renewal liquid membrane (HFRLM) operation, TBP and 0.1 M nitric acid are a carrier and an acceptor phase, respectively. The effects of operating mode, flow rates of the tube, metal concentration (Figure 7.6), and the shell sides

and volume ratio of the organic to the aqueous phase on the mass transfer performance of HFRLM process were investigated. Experimental results indicated a satisfactory stability of HFRLM process during 12 h of continuous operation. The mass transfer rate of Th(IV) in the operating mode in which a stirred mixture of feed and organic phases passes through the tube side was higher than the value reached when the stirred mixture of the acceptor and the organic phases flew through it. It was found that the mass transfer rate in the HFRLM operation was about twice its valor than that in the hollow fiber supported liquid membrane operation.

With an acidic waste raffinate of zirconium purification plant contains about 2 g/L of Zr in 2 M nitric acid, the conventional solvent extraction process cannot be applied to recover Zr(IV) from this lean stream due to low distribution coefficient of Zr(IV) with tributylphosphate at low acidity. Thus, an in-house synthesized ligand, mixed alkyl phosphine oxide (MAPO), was found to be suitable for the recovery of the metal from the above stream [35]. Using 0.1 M MAPO at 2 M feed acidity, the distribution coefficient of Zr(IV) is found to be 20. Batch solvent extraction studies revealed that MAPO forms 3:1 complex with Zr. The best operational conditions from batch studies were used to investigate the applicability of dispersion liquid membrane in hollow fiber contactor for Zr recovery from Zr-bearing. Hollow fiber contactor employing dispersion liquid membrane, in once through operational mode, yields 24% transport of Zr from the aqueous feed containing 1.2 g/L Zr using 0.1 M MAPO. Moreover, a mathematical model was developed to predict the transport phenomena of Zr in nitrate media through MAPO imbibed into the pores of the membrane in the contactor.

In another study [36], Roman et al. propose a flowsheet based on the combination of membrane processes for the effective recovery of value-added components contained in mining effluents with high concentrations of hydrochloric acid and metal anionic and cationic chloro-complexes. A representative case study has been selected consisting of a solution of zinc and iron that under the studied conditions was solubilized forming anionic and cationic chloro-complexes. The high complexity of the system required a selective membrane-based solvent extraction step to successfully achieve the separation of cationic iron from a solution containing the acid together with anionic species of zinc followed by a diffusion process through ion conductive membranes for acid recovery. The liquid membrane operation consisted in hollow fiber modules, one for the extraction, and the other for stripping. In the extraction step, the feed solution containing zinc(II), iron(III), and HCl was put into contact with an undiluted TBP; from this, a raffinate containing iron(III) and an organic solution of zinc(II) and HCl was obtained. The organic phase fed the second module in which the stripping step with water occurred; this step left an organic solution which was recycled back to the first HF module and an exiting zinc and HCl solution which fed the electrodialysis operation, which allowed the recovery of the acid and eventually zinc *via* a further electrowinning operation. It is claimed that, although the quantitative results are case dependent, the process can be well extended to any mining leaching effluent coming from the use of HCl as a leaching agent and containing metal chloro-complexes.

Hollow fiber membrane with strip dispersion operation was used to recover and purify uranium from simulated waste streams containing nitric acid, high uranium concentrations, and radionuclide contaminants [37]. The carrier phase was TBP, and ammonium carbonate was used as a strippant for this system. The radionuclide surrogates (Co, Ru, Sb, Cs, and Sr at 1 g/L concentration each) were not extracted, and complete uranium recovery was obtained from 15 g/L U(VI) in 3 M nitric acid solutions using a 1:4 ratio of 30% v/v TBP in kerosene and 0.75 M of the ammonium salt. All of the uranium in the stripping solution was precipitated (yield >99%).

Am(III) and Eu(III) were separated using phenyl sulphonic acid functionalized bis-triazinyl pyridine (SO$_3$-Ph-BTP) in a supported liquid membrane with varying concentrations of N,N,N',N'-tetraoctyl diglycolamide [38]. Further, the separation of both elements was attempted using hollow-fiber experimentation. In the process, Am(III) was selectively concentrated with a volume reduction factor of greater than 1,000, though the experimentation seemed to be strange and not reliable (in the opinion of the authors of the chapter), since a glass reactor containing the feed solution was used, and small pieces (4–5 cm long) of the hollow fiber soaked with a TODGA solution filling the pores, and a nitric acid solution (strip phase) in the tube side were submerged and agitated. Using a more conventional hollow fiber module, Eu(III) was selectively transported from a mixture of Am(III) and Eu(III). However, to achieve this separation, 10 mM of SO$_3$-Ph-BTP were added to the feed phase (100 mL. lumen side), containing the metals in 1 M nitric acid, and in order to complex Am(III), the strip phase (100 mL, shell side) was of 0.01 M nitric acid. Both solutions flowed at 50 mL/min. The organic phase filling the membrane pores was of 0.1 M TODGA in dodecane and 5% v/V iso-decanol.

Pseudo-emulsion-based hollow fiber strip dispersion (PEHFSD) technology was used to recover zinc(II) from chloride aqueous media [39]. As a carrier, pyridine oxime-ether, N-decoxy-1-(pyridin-3-yl)ethaneimine (D-3EI), was used; the carrier was diluted in a mixture of toluene and decanol. As is usual in this type of investigation, several parameters were investigated to asses their influence (or not) on metal transport: flow rates (2120–400 mL/min feed phase, 180–360 mL/min pseudo-emulsion phase), feed (0.3–5 g/L Zn(II), HCl(0–3 M) NaCl (1–2 M), and membrane (0.05–1 M carrier) phases composition, and type of strippant. In the experiments, the volume of the feed phase was of 800 mL, whereas that of the pseudo-emulsion phase was also of 800 mL (400 mL each of the organic and strip solutions). Based on the experimental results, the best zinc(II) permeabilities were found under the next conditions: 200–300 mL/min as flowrates for both feed and pseudo-emulsion phase, decrease of the metal transport as the initial zinc(II) concentration increased from 1 to 5 g/L, increase of the metal transport as the HCl concentration in the feed phase increased from nil to 1–3 M, 0.1 M carrier in toluene/decanol as organic phase, and a 5% w/v Na$_2$SO$_4$ solution as strippant.

The carrier transported zinc *via* a solvation reaction:

$$ZnCl_{2_{aq}} + n\left(D-3EI\right)_{org} \Leftrightarrow ZnCl_2 \cdot \left(D-3EI\right)_{n_{org}} \quad (7.7)$$

However, this carrier can be protonated:

$$H^+_{aq} + D-3EI_{org} \Leftrightarrow H\left(D-3EI\right)^+_{org} \quad (7.8)$$

and reacts with zinc-chloro anionic complexes:

$$ZnCl^{2-n}_{n_{aq}} + \left(n-2\right)H\left(D-3EI\right)^+_{org} \Leftrightarrow ZnCl^{2-n}_n\left(n-2\right)H\left(D-3EI\right)^+_{org} \quad (7.9)$$

with n = 3, 4.

Fe(II) was not transported, and whereas the transport of Fe(III) occurred, the separation Zn(II)/Fe(III) was good.

7.3.4 Systems Using Ionic Liquids

The separation of mercury(II) from petroleum produced water from the Gulf of Thailand was carried out using a hollow fiber supported liquid membrane system (HFSLM) [40]. Optimum parameters for feed treatment were 0.2 M HCl, 4% v/v Aliquat 336 (quaternary ammonium salt) in toluene as carrier phase, and 0.1 M thiourea as a receiving phase or stripping solution. The best percentage obtained for extraction was 99.7% and for stripping 90.1%. The overall separation efficiency was of 94.9% taking into account both extraction and stripping stages. The metal transport from the feed to the carrier solution followed an anion exchange mechanism (being $HgCl_4^{2-}$ as anionic species in feed). The stripping stage was followed by regeneration of the extractant, and mercury was released to the strip solution as $HgCl_2(NH_2CSNH_2)$ compound. The model derived in the work indicated that the overall process needed to consider whether convection, diffusion, and reaction played a key role for an accurate prediction in this unsteady state model.

The extraction of cerium by a hollow fiber supported liquid membrane using trioctyl methylammonium chloride (TOMAC) as a mobile carrier was investigated [41], and the performance was compared with solvent extraction results. In terms of cerium extraction, the efficiency of the hollow fiber supported liquid membrane operation was greater than that of solvent extraction. In the membrane operation, the effect of different parameters such as pH, flow rates, extractant concentration, metal ion concentration, and strip solution concentration on cerium extraction was investigated. These parameters were optimized for maximum cerium flux and were found to be: pH 1.0, 0.1 M TOMAC in kerosene, 290 mL/min feed solution flow rate, 150 mL/min strip solution flow rate, and 0.9 M sulfuric acid used as strip solution. It was determined that the species formed in the organic phase was $(R_3NCH_3^+)_2Ce(SO_4)_3^{2-}$. The selective separation of cerium in the presence of other metal ions using a hollow fiber supported liquid membrane (HFSLM) was also investigated and the separation factor values Ce/metal (in all the cases with values greater than 10000) were found to be in the following order $Ni^{2+}> Co^{2+}> Zn^{2+}> Mn^{2+}> Cu^{2+}> Fe^{3+}$. Finally, highly pure (99.99%) CeO_2 was prepared by stripping precipitation with oxalic acid and thermal decomposition at 900°C for 1 h.

Organic solutions of triisooactylammonium chloride (TiOAC) in Exxsol D100 (aliphatic diluent) was used as a carrier in a transport investigation of Cd(II) from HCl solutions using a hollow fiber contactor [42]. The metal-bearing feed solution was passed through the tube side, and the pseudo-emulsion of the organic solution and NH_4OH (strip solution) was passed through the shell side in counter-current mode. In the pseudo-emulsion-based hollow fiber membrane with strip dispersion (PEHFMSD) technology, the strip solution was dispersed in the organic phase, the latter being a continuous phase. Different hydrodynamic and chemical variables: variation in feed acidity (zero to 1 M HCl), cadmium concentration in the feed (0.01–1 g/L), carrier concentration in the organic phase (2.2×10^{-4}–0.22 M), variation in the feed, strip and organic phase volume ratios, and variation in the feed and pseudo-emulsion flows, were investigated. The ammonium hydroxide was selected as a strippant due to the fact that it facilitated the stripping of the metal due to the formation of cationic complexes with cadmium(II), i.e., $Cd(NH_3)_4^{2+}$. Experimental data indicated that using 0.22 M extractant in 5% v/v iso-decanol and Exxsol D100 organic solution and 0.5 M ammonium hydroxide as a receiving phase, the efficient removal of the metal with an overall mass transfer coefficient of 2.7×10^{-4} cm/s was accomplished.

In the investigation of indium(III) transport using the pseudo-emulsion-based hollow fiber strip dispersion technology [43], the ionic liquid $RNH_3^+HSO_4^-$, generated by the reaction of the primary amine Primene JMT dissolved in Solvesso 100 (aromatic diluent) and sulphuric acid, was used. The reaction is represented by the equilibrium:

$$RNH_{2_{org}} + H_{aq}^+ + HSO_{4_{aq}}^- \rightarrow RNH_3^+HSO_{4_{org}}^- \quad (7.10)$$

where RNH_2 represented the active group of amine, i.e., Primene JMT, and the subscripts aq and org referred to the respective aqueous and organic phases. In the operation, the feed solution containing the metal was flowed through the tube side, and the pseudo-emulsion of ionic liquid and sulphuric acid was passed through the shell side in counter-current mode. Thus, a contactor for extraction and stripping was used. Several hydrodynamic and chemical parameters, such as variation in flows, feed pH (0.6–2), indium concentration in feed (0.01–0.1 g/L), carrier concentration (0.0025–0.25 M), etc., were investigated. The diluent in the organic phase was Solvesso 100 (aromatic diluent), whereas the strip solution was a sulphuric acid solution. Best experimental conditions for the extraction of indium (exceeding 80%) and stripping (65%) were found to be as follows: 0.4 L/min (feed flow rate), pH 2 in the feed solution, and a pseudo emulsion formed by 0.25 M ionic liquid in Solvesso 100 and 2 M sulphuric acid and maintaining O/A = 0.75 and flow rate at 0.1 L/min. The transport of indium(III) was also investigated using various extractants (Figure 7.7). It can be seen that with the aldoxime Acorga M5640, the best results are claimed by the authors.

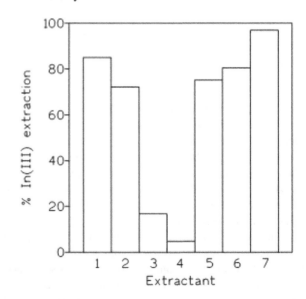

FIGURE 7.7 Indium(III) extraction using different extractants. Aqueous feed phase: 0.01 g/L In(III) at pH 2. Flow rate: 400 mL/min. Pseudo-emulsion phase: 10% v/v extractant (1-primary amine Primene JMT, 2-primary amine Primene 81R, 3-secondary amine Amberlite LA2, 4-tertiary amine Hostarex A324, 5-organic phoshinic acid Cyanex 272, 6-organic phosphoric acid DP-8R, 7-aldoxime Acorga M5640) in Solvesso 100 and 2 M sulphuric acid. Flow rate: 100 mL/min. Time: 3 h

In order to reduce the environmental impact of the use of toxic diluents, the non-toxic diluents: coconut oil, soybean oil, and sunflower oil were investigated and compared with toxic diluents such as benzene, cyclohexane, chlorobenzene, etc., for the separation of platinum(IV) through a hollow fiber supported liquid membrane and using Aliquat 336 as an extractant [44]. Thiourea in HCl solutions was used as a receiving phase. Once the carrier phase impregnated the membrane pores, and the system was washed with water to remove any excess of the carrier phase, the feed (tube side) and the strip solutions (shell side) were pumped to the module. The extraction was carried out *via* an anion exchange mechanism, whereas the stripping occurs by complexation of the platinum(IV)-chloride complex with thiourea and formation of $PtCl_4((NH_2))_2CS)_2$ species in the receiving phase. At 30°C, the permeability obtained with the toxic diluents is better than that of non-toxic diluents; however, an increment of this variable yields better results with the non-toxic diluents (Table 7.5). It is observed that the best results, 91.4% with respect to platinum transport and 81% stripping, are achieved with sunflower oil at 55°C. The extraction process is exothermic and spontaneous. Within this diluent, the overall mass transfer resistance (R) is 146.6×10^4 s/cm, and the transport is controlled by the mass transfer resistance from the liquid membrane, this control step being the same for all the diluents employed in this work. Further, a modified Apelblat model gave an excellent match for predicting the extraction behavior of platinum(IV) at 1.2818% root-mean square deviation (sunflower oil).

A supported liquid membrane system was developed for the separation of Cd from either high salinity or acidity aqueous media [45]. Three extractants were first considered, Aliquat 336, trioctylamine, and trilaurylamine, and having Aliquat 336 give the best performance for cadmium transport. Based on this study, this was selected to continue the investigation for parametric study. The effect of carrier concentration (0–0.5 M), organic solvent (decaline dihexylether, dodecane plus 4% dodecanol and cumene), and feed and receiving solutions on the metal permeability were investigated. A membrane consisting of a Durapore (polyvinylidene difluoride) polymeric

support impregnated with a 0.5 M Aliquat 336 solution in decaline (decahydronaphtalene) was chosen as the more appropriate for cadmium removal from the feed solution. This system allows the effective transport of trace levels of Cd through the formation of $CdCl_4^{2-}$ complex, which is the predominant species responsible for the transport process, in both NaCl and HCl solutions. The above results were scaled up to a hollow fiber configuration, in which the feed phase, flowing by the tube side, contained 2 mg/L Cd(II) in 1 M NaCl solution, and the stripping solution, flowing in the shell side, was water. After 22 hours running time, the transport of Cd(II) was of more than 90% for organic solutions containing 0.05 or 0.5 M Aliquat 336 in decaline; however, the cadmium recovery in the strip solution was low (about 30%) for the 0.5 M Aliquat 336 solution, but near quantitative for the more diluted organic solution. This behavior was not observed in the flat-sheet configuration, which was to be attributable to the higher viscosity of the 0.5 M ammonium salt organic solution that resulted in increased membrane resistance. Further results indicated that this carrier phase, in a hollow fiber supported liquid membrane configuration, allowed for the enrichment and separation of trace levels of Cd from spiked seawater samples, facilitating the analytical determination of this toxic metal.

Different task-specific ionic liquids: 3-[1-(hydroxyimine)undecyl]-1-propylpyridinium bromide and 3-[1-(hydroxyimine)undecyl]-1-propylpyridinium chloride, diluted in toluene and 10% v/v 1-decanol, as the carrier phases in the Zn(II) transport from chloride media through pseudo-emulsion-based membrane strip dispersion (PEHFSD) were investigated [46]. The transport of zinc(II) was analyzed as a function of various experimental variables: flow rates of phases, ionic liquid structure in the organic phase, initial zinc(II) concentration (0.3–5 g/L), and chloride ions in the feed phase. The removal of zinc from the feed solution at around 91%, and the overall mass transfer coefficient of permeation was calculated from the experimental data, the values being found in the range of 2.5×10^{-7} to 8.0×10^{-7} m/s. The recovery of Zn(II) in the stripping solution (5% w/v Na_2SO_4) was around 70 to 80% for most conditions tested. The results of using the task-specific ionic liquids as carrier were also compared to that obtained using the commercial extractant TBP, and experimental data indicated that these ionic liquids perform equally as TBP, however with better results for a much lower extractant concentration in the organic solution.

Aliquat 336 dissolved in kerosene was used as carrier phase to investigate the mass transfer of arsenic(V) through asymmetric polyvinylidene fluoride hollow fiber membrane contactors [47]. The fibers were prepared and characterized prior to their use in a membrane contactor; different variables were investigated: pH (4.4–12.5), temperature (20–50 C), and initial concentration As(V) (0.02–0.1 g/L) in the feed phase. Arsenic(V) transport was influenced by the three variables. No data about stripping was included in the investigation. The transport of arsenic(V) was based in the

TABLE 7.5

Effect of Temperature on Pt(IV) Transport Yields Using Aliquat 336 Dissolved in the Non-toxic Diluents as the Carrier Phase [44]

Temperature, °C	Coconut Oil, %	Soybean Oil, %	Sunflower Oil, %
30	45.3	50.9	65.5
45	53.1	56.4	76.0
55	60.4	80.0	91.4

Feed phase: 0.005 mg/L at pH 5. Carrier phase: 10% v/v Aliquat 336 in the diluents. Receiving phase: 0.8 M thiourea in 1 M HCl. Flow rates: 100 mL/min each feed and strip phases. Time: 30 min

extraction of some of the negatively charged arsenic species, thus, in the pH range 6.98–8.2:

$$(n-1)R_4N^+Cl_{org} + H_{(n-1)}AsO_{4_{aq}}^{(1-n)} \Leftrightarrow$$

$$(R_4N^+)_{(n-1)} H_{(n-1)}AsO_{4_{org}}^{(1-n)} + (n-1)Cl_{aq}^- \qquad (7.11)$$

with n = 2 or 3; however, at pH values around 12, there was no arsenic transport because the predominant As(V) species, AsO_4^{3-}, was not extracted.

An ionic liquid ($R_4N^+L^-$) derived from DEHPA and Aliquat 336 was used to separate Sc(III) from a leachate containing 2.5 g/L Sc(III), 25 g/L Mg(II), and 0.5 M HCl, coming from the treatment of a Mg–Sc alloy [48]. The ionic liquid was synthetized accordingly to:

$$R_4N^+Cl_{org}^- + HL_{org} + NaHCO_{3_{aq}}$$

$$\rightarrow R_4N^+L_{org}^- + NaCl_{aq} + CO_2 + H_2O \qquad (7.12)$$

where HL stands for DEHPA. The organic phase of the extractant in kerosene filled the fiber pores. In the experiments, 250 mL of each of the feed and strip phases were circulated through the tube and shell sides of the fibers, respectively. Maximum Sc(III) flux was obtained under the next experimental conditions: pH 4 in feed solution, carrier concentration: 0.4 M, strip phase: 6 M NaOH, and flow rate of 150 mL/min. The transport was related to the next reaction:

$$Sc_{aq}^{3+} + 3Cl_{aq}^- + 3(R_4N^+L^-)_{org} \Leftrightarrow ScCl_3 \cdot 3R_4N^+Cl_{org}^- \qquad (7.13)$$

Magnesium was not co-transported with Sc(III). From the strip solution, and after adding oxalate to precipitate $Sc_2(C_2O_4)_3$, scandium oxide (Sc_2O_3) was obtained by calcination of the oxalate salt. Comment from the author of the chapter: note the discrepancy between the composition of the leach solution (0.5 M HCl) and the best pH value(4) of the feed phase to obtain best Sc(III) transport.

Flat sheet (FSSLM) and hollow fiber supported ionic liquid membrane (HFSLM) technologies were used to investigate the selective separation of germanium from simulated water leach liquors of CGFA containing Zn(II), Ni(II), Cd(II), and Co(II) [49]. Aliquat 336 extractant was used as the carrier. In the hollow fiber device, the next conditions were used, aqueous feed (2L): 0.1 g/L (each) Ge, Ni, Cd and Co, 1 g/L Zn, and tartaric acid in a [tartaric acid]/[Ge] molar relationship of 2, organic phase (50 mL): Aliquat 336 5% v/v in kerosene and decanol, strip phase (500 mL): HCl 1 M. The carrier phase was dispersed into the strip phase. The system reached the steady state approximately after 30 min of reaction, with all the germanium transferred to the strip phase after 240 min; thus, a germanium concentration factor of 4 was achieved. With the addition of tartaric acid to the feed solution, the species $Ge(OH)_2T^{2-}$ was formed, which reacted with the ionic liquid in the form:

$$Ge(OH)_2 T_{aq}^{2-} + 2R_4N^+Cl_{org}^-$$

$$\Leftrightarrow (R_4N^+)_2 Ge[OH]_2 T_{org}^{2-} + 2Cl_{aq}^- \qquad (7.14)$$

As a consequence of the stripping reaction, Ge(OH)₂HT⁻ species was released to the strip phase, and the ionic liquid was regenerated. The transport rate using HFSLM was much faster than in the FSSLM operation, i.e., 97% of metal transport was achieved after 30 min or 10 h for HFSLM and FSSLM technologies, respectively.

7.3.5 Miscellaneous Carriers

A review paper published by Garcia et al. [50] focused on two wastes generated from different processes, such as acid mine drainage and cyanide tailings. These wastes are considered critical with respect to the separation processes mainly dealing with purification and recovery of both heavy metals and the cyanide present in them. The metal mining of certain ores is associated with the generation of acid mine drainage waters and effluents containing cyanides that can have long-term negative impacts on the environment. The important aspects of these processes are firstly that obtaining profitable components from those effluents would improve the resource efficiency and secondly the improvement in environmental performance of the mine operation. The processes used are adsorption and membrane technology to obtain valuable compounds from acid mine drainage waters and ion-exchange, acidification-volatilization-reneutralization, sulfidization-acidification-recycling-thickening, and membrane technology from cyanide tailings. Moreover, the use of the 12 green engineering principles for improving the environmental performance of the purification techniques is also described by authors. It can be concluded that the energy and chemical consumption and the generation of waste are the main environmental disadvantages of the purification and recovery techniques, whereas these limitations can be surpassed both improving the efficiency of the processes and by the use of renewable energy and materials; the use of hybrid processes in the purification and recovery operations may be also of benefited.

The recovery of zinc from chloride solutions using pseudo-emulsion-based hollow fiber strip dispersion (PEHFSD) technique was investigated [51]. The novel extractant, 1-(3-pyridyl)undecan-1-one oxime (3PC10), and TBP, were used in the processes. With the feed phase flowing in the tube side and the pseudo-emulsion flowing in the shell side, the influence of several parameters, including the initial concentration of Zn(II) and sodium chloride in the aqueous phase and the type of extractant on Zn(II) extraction was studied. In the case of the oxime, toluene was the diluent of the organic phase, and 5% w/v of sodium sulfate solution was the receiving phase, for TBP, Shellsol D70 (aliphatic) was the diluent and water was used as the stripping phase. The transport of zinc(II) from the feed to the organic phase followed a solvation reaction, and formation of $ZnCl_2(HL)_2$ being HL the oxime, or, according to an ion-pair extraction, where the single or diprotonated forms (H_2L^+) and (H_3L^{2+}), respectively, formed complexes with $ZnCl_3^-$ and $ZnCl_4^{2-}$. At high chloride concentrations, TBP extracted zinc(II) by forming the species $H_2ZnCl_4\cdot2TBP$, whereas decrease in acidity lead to the formation of different species in the carrier solution, i.e., $ZnCl_2\cdot2TBP$. The Zn(II) permeation/transport was analyzed on the basis of the overall mass transfer coefficient (Table 7.6).

Based on solvent extraction experiments showing the synergistic effect of D2EHPA and TBP mixtures for the extraction of Li^+ in earlier studies, a new paper was published [52] in which the above system was implanted in hollow fiber supported liquid membrane experiments with low concentrations of Li^+, Na^+, and K^+ in the feed phase, and the system showed higher flux for Li^+. However, at high concentrations of Na^+ and K^+ in the aqueous solution, the flux of Li^+ was noted to be decreased.

These results indicated that the oxime is a potential carrier of zinc from chloride solutions, and results were comparable with those obtained with the organophosphorous derivative, despite the lower concentration of the oxime with respect to the one used with TBP.

In another interesting paper, several extractants were used to investigate their performance in the transport of uranium from monazite leach solution in two steps [53]. Under this experimentation, at first pure single extractant was used and later, the best extractants were used in synergic mixtures to investigate their effect on uranium transport *via* hollow fiber supported liquid membrane operation (HFSLM). Dodecane was the organic diluent used, whereas sulfuric acid was the strippant phase. Several acidic and solvation extractants were, thus, investigated to assess their ability to transport uranium(VI) (Figure 7.8). This first series of experiments indicated that Cyanex 272 (phosphinic acid) and TOPO (phosphine oxide) were the best extractants in each class of reagents. Optimum conditions for synergistic extraction were found to be pH 5 in the feed solution (0.045 g/L U(VI)), 0.1M Cyanex 272 + 0.1M TOPO in dodecane as the carrier phase, and 2M sulfuric acid as the stripping solution at 30°C, and with flows of 100 mL/min for each feed and receiving solution, the

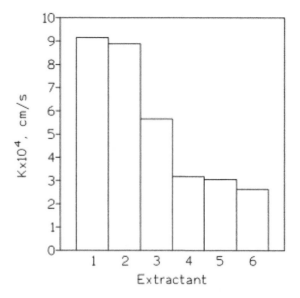

FIGURE 7.8 Influence of the extractant on the overall mass transfer coefficient (K). Aqueous feed phase: uranium(VI) at pH 5. Organic phase: 0.1 M extractant (1 – organic phosphinic acid Cyanex 272, 2 – organic phosphonic acid PC88A, 3 – organic phosphoric acid D2EHPA, 4 – phosphine oxide TOPO, 5 – phosphine oxide Cyanex 923, 6 – phosphoric ester TBP) in dodecane. Strip phase: 2 M sulfuric acid. Flow rates: 100 mL/min for both feed and strip solutions

operation resulting in a overall mass transfer coefficient (K_p) of 1.4×10^{-3} cm/s. After a fourth cycle, the accumulative extraction and stripping yields of uranium reached 95 and 90%, respectively.

The fractional resistance due to extraction reaction proved to be the highest, indicating that mass transfer resistance from the extraction was the controlling step and thus the process is chemical reaction dependent.

In a another review paper [54], novel processes based on SIMALE (Smart Integrated Membrane Assisted Liquid Extraction) have been proposed as effective methods for the selective separation of different chemical species such as metal ions, organic/biologically important compounds, and gas mixtures from different waste streams including nuclear waste. The industrial use of supported liquid membranes (SLMs) based on conventional liquids is limited by their relative instability and short lifetime. Under SIMALE techniques, the stability of the SLMs is enhanced by a modified SLM with pseudo-emulsion-based hollow fiber strip dispersion or non-dispersive solvent extraction techniques. In order to promote operational stability, SIMALE, using ionic liquids, as a liquid membrane phase could overcome these inconveniences due to their negligible vapor pressure and the possibility of minimizing their solubility in the surrounding phases. Finally, this review also discuss a series of applications including scale up, process intensification aspects, current status of the technology, and future directions.

The recovery of Zn(II) from chloride solutions using pseudo-emulsion-based hollow fiber strip dispersion

TABLE 7.6

Overall Mass Transfer Coefficients (K_p) at Various Experimental Conditions [51]

Carrier[a]	HCl, M	NaCl, M	Zn, g/L	K_p, m/s
3PC10	1	1	1	9.4×10^{-7}
	1	1	5	2.8×10^{-7}
	-	2	5	8.0×10^{-8}
TBP	1	1	1	5.7×10^{-7}
	1	1	5	5.6×10^{-7}
	-	2	5	2.8×10^{-7}

[a] Carrier phase: 0.1 M 3PC10 in toluene or 2.9 M TBP in Shellsol D70

(PEHFSD) technique with pyridine extractants was investigated [55]. The following extractants were used in the research: 1-(3-pyridyl)undecan-1-one oxime and its quaternary salts 3-[1-(hydroxyimine)undecyl]-1-propylpyridinium bromide and 3-[1-(hydroxyimine)undecyl]-1-propylpyridinium chloride, whereas toluene and 1-decanol were the diluent and modifier for the organic solution, respectively. Several parameters including hydrodynamic conditions (with little effect on metal transport), extractant, metal and strippant concentrations (best zinc recoveries using 5% sodium sulphate solution), as well as the extractant structure on the zinc(II) transport were investigated, also determining the surface properties of the various aqueous/organic systems and mass transfer resistances. All of the above reagents seemed to be appropriate carriers in the PEHFSD process, and the values of the overall mass transfer coefficient of permeation were found to be in the range of 2.5×10^{-7} to 1.1×10^{-6} m/s.

In this research, 1-(3-pirydyl)undecan-1-one, 1-propyl-3-undecanoylpyridinium bromide, and 1-propyl-3-undecanoylpyridinium chloride, as well as 1-(3-pirydyl)undecan-1-one oxime with its chloride and bromide salt were investigated as new extractants for the recovery of copper from aqueous solutions [56]; again toluene and 1-decanol were the respective diluent and modifier in the organic phase. It was found that the increasing chloride ions concentration, from nil to 4 M, in the aqueous solution increased the copper(II) extraction. The investigation also included the speciation of the complexes transported into the organic phase, for the neutral extractant the chemical equilibrium involved in the process is described as:

$$CuCl_{2_{aq}} + nHL_{org} \Leftrightarrow CuCl_2 \cdot (HL)_{n_{org}} \qquad (7.15)$$

Subscript aq and org represent the aqueous and organic solutions, respectively, and HL the extractant. The quaternary ammonium salts can extract copper–chloride complexes, but with higher affinity towards $CuCl_4^{2-}$ complex, being represented by an anion exchange reaction as:

$$CuCl_{4_{aq}}^{2-} + 2\left(LR^+Br^-\right)_{org} \Leftrightarrow \left(CuCl_4^{2-}\right)\left(LR^+\right)_{2_{org}} + 2Br_{aq}^-$$

$$(7.16)$$

Here, the subscripts have the same significance as in the above equation (7.15), whereas LR^+Br^- represents the molecule of the ionic liquid. At acidic conditions, the stoichiometry of the extracted species evolved to $CuCl_nH_m(LR^+X^-)$, being X the counter anion. Also the selectivity of the quaternary bromide salts towards iron ions was investigated. Moreover, the laboratory-scale experiments indicated that pyridinium bromide derivative was an efficient and selective copper extractant and it could be proposed to copper removal from the solution obtained from the chloride leaching of copper sulfides.

Another published review in 2018 deals with the opportunities of mining from cities as one of the future directions for sustainable resource management [57]. The separation of highly dispersed, low level concentration of valuable metals requires highly efficient extraction technology, where membrane extraction has shown significant advantages owing to its potentially high selectivity and low footprint. This review summarized the recent progress in membrane extraction with respect to the development of membrane materials and the process configurations with special focus on preventing the loss of the organic extractant and/or degradation of the membrane materials.

Also, the development of different membrane materials, for extending the membrane performance, and developments on membrane configuration and process are presented, i.e., ion exchange membranes, membrane contactors, and composite hollow fiber membranes. Hydrophilic/hydrophobic blend polymers and block co-polymers have shown a much more extended lifetime, which is promising for potential development with respect to future direction. This work provides not only the recent development on membrane materials but also the application process/scheme that can help to improve the performance of membrane extraction processing, which might be useful for scientists and industry that are looking for smart alternative solutions for city mining. There is a common thought that the key problem to be solved for a practical (industrial) use of new membrane extraction technologies is a long-term membrane stability associated with these liquid membrane technologies. The newly developed membrane processes are continuously addressing the challenges faced by membranologists.

An interesting study carried out by Kittisupakorn et al. focused on neodymium, which is a key component in rare earth magnets. Being among rare earth elements (REEs), recovery is of paramount importance for such materials [58]. Neodymium magnets (also known as NdFeB) are permanent magnets made of neodymium, iron, and boron alloy. These magnets are used in several products which require low magnet mass or strong magnetic fields such as microphones, in-ear headphones, and computer hard disks. The above is the basis of research aimed to study the transport of Nd^{3+} through the HFSLM, by means of synergistic extraction systems, to increase extraction or transport efficiency. The separation of Nd(III) from nitric solution is carried out by a mixture of 0.5 M of D2EHPA and 0.5 M of TOPO as the carrier phase immobilized in the support, and 1 M of sulfuric acid in the stripping solution. To achieve good separation of Nd(III) throughout the continuous process, a Model Predictive Control (MPC) technique is applied to control the concentration of extracted neodymium at a desired set point by manipulating the flow rate of the feed solution, and its control performance is compared with that of a Proportional Integral Derivative (PID) controller. Simulation study has shown that the optimum extraction of 94.5% can be achieved, and the MPC and PID controllers can control the concentration of neodymium ions at the desired set point. However, the control responses by MPC are better than the PID ones in both nominal and parameter mismatch cases.

7.4 CONCLUDING REMARKS AND FUTURE TRENDS

In the treatment of metal-bearing aqueous solutions, carrier-impregnated hollow fiber modules offer the advantages of both liquid–liquid extraction and membrane separations; this is due to the fact that they combine the effects of mass transfer and chemical reaction with selectivity inherent in the above two separation processes. Operationally, hollow fiber modules are used in various operation modes: (i) as two contactors, one for the extraction and other for the strip steps, (ii) as one contactor for both simultaneous steps (HFSLMs), and (iii) as a single module in the most advanced non-dispersive solvent extraction with strip dispersion (NDSXSD, HFNDSX, PEHFSD) operational mode for the extraction and removal of valuable or hazardous metals from different aqueous solutions and in radioactive waste processing. The three methodologies are characterized as having surface-to-volume ratios in the order of several thousands m^2/m^3. Ionic liquids (ILs) are taking their position as carriers to replace the *more conventional* extractants used in liquid–liquid extraction.

Future use of the carrier-impregnated hollow fiber module methodologies will probably involve the developing of the next topics:

i) improving the fiber characteristics, to allow for the treatment of more chemically aggressive aqueous media,

ii) the use of ionic liquids as "green solvents", to make the technologies more environmentally friendly,

iii) extending the use of the methodologies on real solutions, giving a more realistic view of the performance of the carrier-impregnated hollow fiber modules,

iv) scaling up the methodologies to pilot plant or larger scale.

REFERENCES

1. Wongsawa, T., Sunsandee, N., Pancharoen, U., Lothongkum, A.W. 2014. High-efficiency HFSLM for silver-ion pertraction from pharmaceutical wastewater and mass-transport models. *Chemical Engineering Research and Design* 92:2681–2693. DOI: 10.1016/j.cherd.2014.01.005

2. Rout, P.C., Sarangi, K. 2014. Separation of vanadium using both hollow fiber membrane and solvent extraction technique-A comparative study. *Separation and Purification Technology* 122:270–277. DOI: 10.1016/j.seppur.2013.11.010

3. Agarwal, S., Reis, M.T.A., Ismael, M.R.C., Carvalho, J.M.R. 2014. Zinc extraction with Ionquest 801 using pseudo-emulsion based hollow fibre strip dispersion technique. *Separation and Purification Technology* 127:149–156. DOI: 10.1016/j.seppur.2014.02.039

4. Vijayalakshmi, R., Chaudhury, S., Anitha, M., Singh, D.K., Aggarwal, S.K., Singh, H. 2015. Studies on yttrium permeation through hollow fibre supported liquid membrane from nitrate medium using di-nonyl phenyl phosphoric acid as the carrier phase. *International Journal of Mineral Processing* 135:52–56. DOI: 10.1016/j.minpro.2015.02.003

5. Swain, B., Mishra, C., Jeong, J., Lee, J.-C., Hong, H.S., Pandey, B.D. 2015. Separation of Co(II) and Li(I) with Cyanex 272 using hollow fiber supported liquid membrane: A comparison with flat sheet supported liquid membrane and dispersive solvent extraction process. *Chemical Engineering Journal* 271:61–70. DOI: 10.1016/j.cej.2015.02.040

6. Wannachod, T., Leepipatpiboon, N., Pancharoen, U., Phatanasri, S. 2015. Mass transfer and selective separation of neodymium ions via a hollow fiber supported liquid membrane using PC88A as extractant. *Journal of Industrial and Engineering Chemistry* 21:535–541. DOI: 10.1016/j.jiec.2014.03.016

7. Agarwal, S., Reis, M.T.A., Ismael, M.R.C., Carvalho, J.M.R. 2015. Extraction of Cu(II) with acorga M5640 using hollow fibre liquid membrane. *Chemical Papers* 69:679–689. DOI: 10.1515/chempap-2015-0076

8. Mondal, A., Bhowal, A., Datta, S. 2016. Mass transfer and immobilized organic phase instability studies for cobalt(II)–nickel(II) separation by PEHFSD. *Chemical Engineering Communications* 203:1269–1277. DOI: 123.0.1080/00986445.2016.1174856

9. Meerholz, K., van der Westhuizen, D.J., Krieg, H.M. 2017. Automation of membrane based solvent extraction unit for Zr and Hf separation. *Separation and Purification Technology* 179:204–214. DOI: 10.1016/j.seppur.2017.01.064

10. Allahyari, S.A., Minuchehr, A., Ahmadi, S.J., Charkhi, A. 2017. Thorium pertraction through hollow fiber renewal liquid membrane (HFRLM) using Cyanex 272 as carrier. *Progress in Nuclear Energy* 100:209–220. DOI: 10.1016/j.pnucene.2017.06.012

11. Ren, X., Jia, Y., Lu, X., Shi, T., Ma, S. 2018. Preparation and characterization of PDMS-D2EHPA extraction gel membrane for metal ions extraction and stability enhancement. *Journal of Membrane Science* 559:159–169. DOI: 10.1016/j.memsci.2018.04.033

12. Pirom, T., Arponwichanop, A., Pancharoen, U., Yonezawa, T., Kheawhom, S. 2018. Yttrium (III) recovery with D2EHPA in pseudo-emulsion hollow fiber strip dispersion system. *Scientific Reports* 8:7627. DOI: 10.1038/s41598-018-25771-4

13. Yadav, K.K., Anitha, M., Singh, D.K., Kain, V. 2018. NdFeB magnet recycling: Dysprosium recovery by non-dispersive solvent extraction employing hollow fibre membrane contactor. *Separation and Purification Technology* 194:265–271. DOI: 10.1016/j.seppur.2017.11.025

14. Yadav, K.K., Singh, D.K., Kain, V. 2019. Separation of terbium from aqueous phase employing hollow fibre supported liquid membrane with EHEHPA as carrier. *Separation Science and Technology* 54:1521–1532. DOI: 10.1080/01496395.2018.1541471

15. Daraei, A, Aghasafari, P., Ghadiri, M., Marjani, A. 2014. Modeling and transport analysis of silver extraction in porous membrane extractors by computational methods. *Transactions of the Indian Institute of Metals* 67:223–227. DOI: 10.1007/s12666-013-0339-6

16. Kumar, A., Haddad, R., Alguacil, F.J., Sastre, A.M. 2005. Comparative performance of non-dispersive solvent extraction using a single module and the integrated membrane process with two hollow fiber contactors. *Journal of Membrane Science* 248:1–14. DOI: 10.1016/j.memsci.2004.09.003

17. Li, Y., Cui, C. 2015. Extraction behavior of Cr(VI) and coexistence with interfering ions by trialkylamine. *Journal of Water Reuse and Desalination* 5:494–504. DOI: 10.2166/wrd.2015.110

18. Wang, Y., Li, Y., Zhong, Y., Cui, C. 2017. Chromium(VI) removal by tri-n-octylamine/n-heptane via a pseudo-emulsion-based hollow fiber strip dispersion technique. *Desalination and Water Treatment* 63:103–112. DOI: 10.5004/dwt.2017.20150

19. Zahakifar, F., Charkhi, A., Torab-Mostaedi, M., Davarkhah, R. 2018. Performance evaluation of hollow fiber renewal liquid membrane for extraction of uranium(VI) from acidic sulfate solution. *Radiochimica Acta* 106:181–189. DOI: 10.1515/ract-2017-2821

20. Zahakifar, F., Charkhi, A., Torab-Mostaedi, M., Davarkhah, R., Yadollahi, A. 2018. Effect of surfactants on the performance of hollow fiber renewal liquid membrane (HFRLM): A case study of uranium transfer. *Journal of Radioanalytical and Nuclear Chemistry* 318:973–983. DOI: 10.1007/s10967-018-6082-z

21. Kamran, H., Irannajad, M., Fortuny, A., Sastre, A.M. 2019. Non-dispersive selective extraction of germanium from fly ash leachates using membrane-based processes. *Separation Science and Technology* 54:2879–2894. DOI: 10.1080/01496395.2018.1555170

22. Vernekar, P.V., Jagdale, Y.D., Sharma, A.D., Patwardhan, A.W., Patwardhan, A.V., Ansari, S.A., Mohapatra, P.K. 2014. Simultaneous extraction of neodymium and uranium using hollow fiber supported liquid membrane. *Separation Science and Technology* 49:1509–1520. DOI: 10.1080/01496395.2014.890628

23. Sharma, A.D., Patwardhan, A.W., Ansari, S.A., Mohapatra, P.K. 2015. Comparison of different HFSLM configurations for separation of neodymium and uranium. *Separation Science and Technology* 50:332–342. DOI: 10.1080/01496395.2014.973524

24. Marino, T., Figol, A. 2015. Arsenic removal by liquid membranes. *Membranes* 5:150–167. DOI: 10.3390/membranes5020150

25. Chaturabul, S., Wannachod, T., Mohdee, V., Pancharoen, U., Phatanasri, S. 2015. An investigation of Calix4.arene nitrile for mercury(II) treatment in HFSLM application. *Chemical Engineering and Processing: Process Intensification* 89:35–40. DOI: 10.1016/j.cep.2015.01.003

26. Laso, J., Garcia, V., Bringas, E., Urtiaga, A.M., Ortiz, I. 2015. Selective recovery of zinc over iron from spent pickling wastes by different membrane-solved extraction process. *Industrial and Engineering Chemistry Research* 54:3218–3224. DOI: 10.1021/acs.iecr.5b00099

27. Carrillo-Abad, J., Garcia-Gabaldon, M., Ortiz-Gandara, I., Bringas, E., Urtiaga, A.M., Ortiz, I., Perez-Herranz, V. 2015. Selective recovery of zinc from spent pickling baths by the combination of membrane-based solvent extraction and electrowinning technologies. *Separation and Purification Technology* 151:232–242. DOI: 10.106/j.seppur.2015.07.051

28. Kim, D., Powell, L.E., Delmau, L.H., Peterson, E.S., Herchenroeder, J., Bhave, R.R. 2015. Selective extraction of rare earth elements from permanent magnet scraps with membrane solvent extraction. *Environmental Science & Technology* 49:9452–9459. DOI: 10.1021/acs.est.5b01306

29. Kim, D., Powell, L., Delmau, L.H., Peterson, E.S., Herchenroeder, J., Bhave, R.R. 2016. A supported liquid membrane system for the selective recovery of rare earth elements from neodymium-based permanent magnets. *Separation Science and Technology* 51:1716–1726. DOI: 10.1080/01496395.2016.1171782

30. Fourie, M., Meyer, W.C.M.H., van der Westhuizen, D.J., Krieg, H.M. 2016. Uranium recovery from simulated molybdenum-99 production residue using non-dispersive membrane extraction. *Hydrometallurgy* 164:330–333. DOI: 10.1016/j.hydromet.2016.07.001

31. Jagasia, P., Ansari, S.A., Raut, D.R., Dhami, P.S., Gandhi, P.M., Kumar, A., Mohapatra, P.K. 2016. Hollow fiber supported liquid membrane studies using a process compatible solvent containing calix4.arene-mono-crown-6 for the recovery of radio-cesium from nuclear waste. *Separation and Purification Technology* 170:208–216. DOI: 10.1016/j.seppur.2016.06.036

32. Ansari, S.A., Kumari, N., Raut, D.R., Kandwal, P., Mohapatra, P.K. 2016. Comparative dispersion-free solvent extraction of Uranium(VI) and Thorium(IV) by TBP and dialkyl amides using a hollow fiber contactor. *Separation and Purification Technology* 159:161–168. DOI: 10.1016/j.seppur.2016.01.004

33. Ansari, S.A., Mohapatra, P.K., Kandwal, P, Verboom, W. 2016. Diglycolamide-functionalized Calix4.arene for Am(III) recovery from radioactive wastes: Liquid membrane studies using a hollow fiber contactor. *Industrial and Engineering Chemistry Research* 55:1740–1747. DOI: 10.1021/acs.iecr.5b04148

34. Allahyari, S.A., Minuchehr, A., Ahmadi, S.J., Charkhi, A. 2016. Th(IV) transport from nitrate media through hollow fiber renewal liquid membrane. *Journal of Membrane Science* 520:374–384. DOI: 10.1016/j.memsci.2016.08.009

35. Pandey, G., Paramanik, M., Dixit, S., Mukhopadhyay, S., Singh, R., Ghosh, S.K., Shenoy, K.T. 2017. Extraction of zirconium from simulated acidic nitrate waste using liquid membrane in hollow fiber contactor. *Desalination and Water Treatment* 90:63–69. DOI: 10.5004/dwt.2017.21404

36. Fresnedo San Román, M., Ortiz-Gándara, I., Bringas, E., Ibañez, R., Ortiz. I. 2018. Membrane selective recovery of HCl, zinc and iron from simulated mining effluents. *Desalination* 440:78–87. DOI: 10.1016/j.desal.2018.02.005

37. Fourie, M., van der Westhuizen, D.J., Krieg, H.M. 2018. Uranium recovery and purification from simulated waste streams containing high uranium concentrations with dispersion liquid membranes. *Journal of Radioanalytical and Nuclear Chemistry* 317:355–366. DOI: 10.1007/s10967-018-5860-y

38. Bhattacharyya, A., Ansari, S.A., Prabhu, D.R., Kumar, D., Mohapatra, P.K. 2019. Highly efficient separation of Am^{3+} and Eu^{3+} using an aqueous soluble sulfonated BTP derivative by hollow-fiber supported liquid membrane containing TODGA. *Separation Science and Technology* 54:1512–1520. DOI: 10.1080/01496395.2019.1578803

39. Reis, M.T.A., Ismael, M.R.C., Wojciechowska, A., Wojciechowska, I., Aksamitowski, P., Wieszczycka, K., Carvalho, J.M.R. 2019. Zinc(II)recovery using pyridine oxime-ether – Novel carrier in pseudo-emulsion hollow fiber strip dispersion system. *Separation and Purification Technology* 223:168–177. DOI: 10.1016/j.seppur.2019.04.076

40. Chaturabul, S., Srirachat, W., Wannachod, T., Ramakul, P., Pancharoen, U., Kheawhom, S. 2015. Separation of mercury(II) from petroleum produced water via hollow fiber supported liquid membrane and mass transfer modeling. *Chemical Engineering Journal* 265:34–46. DOI: 10.1016/j.cej.2014.12.034

41. Rout, P.C., Sarangi, K. 2015. Comparison of hollow fiber membrane and solvent extraction techniques for extraction of cerium and preparation of ceria by stripping precipitation. *Journal of Chemical Technology and Biotechnology* 90:1270–1280. DOI: 10.1002/jctb.4426

42. Alguacil, F.J., López, F.A., García-Díaz, I., Rodriguez, O. 2016. Cadmium(II) transfer using (TiOAC) ionic liquid as carrier in a smart liquid membrane technology. *Chemical Engineering and Processing: Process Intensification* 99:192–196. DOI: 10.1016/j.cep.2015.06.007

43. García-Díaz, I., López, F.A., Alguacil, F.J. 2017. Transport of indium(III) using pseudo-emulsion based hollow fiber strip dispersion with ionic liquid $RNH_3^+HSO_4^-$. *Chemical Engineering Research and Design* 126:134–141. DOI: 10.1016/j.cherd.2017.08.012

44. Wongkaew, K., Mohdee, V., Pancharoen, U., Arpornwichanop, A., Lothongkum, A.W. 2017. Separation of platinum(IV) across hollow fiber supported liquid membrane using non-toxic diluents: Mass transfer and thermodynamics. *Journal of Industrial and Engineering Chemistry* 54:278–289. DOI: 10.1016/j.jiec.2017.06.002

45. Pont, N., Salvadó, V., Fontàs, C. 2018. Applicability of a supported liquid membrane in the enrichment and determination of cadmium from complex aqueous samples. *Membranes* 8:21. DOI: 10.3390/membranes8020021

46. Wojciechowska, A., Reis, M.T.A., Wojciechowska, I., Ismael, M.R.C., Gameiro, M.L.F., Wieszczycka, K., Carvalho, J.M. 2018. Application of pseudo-emulsion based hollow fiber strip dispersion with task-specific ionic liquids for recovery of zinc(II) from chloride solutions. *Journal of Molecular Liquids* 254:369–376. DOI: 10.1016/j.molliq.2018.01.135

47. Bey, S., Semghouni, H., Criscuoli, A., Benamor, M., Drioli, E., Figoli, A. 2018. Extraction kinetics of As(V) by aliquat-336 using asymmetric PVDF hollow-fiber membrane contactors. *Membranes* 8:53. DOI: 10.3390/membranes8030053

48. Parhi, P.K., Behera, S.S., Mohapatra, R.K., Sahoo, T.R., Das, D., Misra, P.K. 2019. Separation and recovery of Sc(III) from Mg–Sc alloy scrap solution through hollow fiber supported liquid membrane (HFLM) process supported by Bi-functional ionic liquid as carrier. *Separation Science and Technology* 54:1478–1488. DOI: 10.1080/01496395.2018.1520730

49. Kamran, H., Irannajad, M., Fortuny, A., Sastre, A.M. 2019. Selective separation of Germanium(IV) from simulated industrial leachates containing heavy metals by non-dispersive ionic extraction. *Minerals Engineering* 137:344–353: DOI: 10.1016/j.mineng.2019.04.021

50. García, V., Häyrynen, P., Landaburu-Aguirre, J., Pirilä, M., Keiski, R.L., Urtiaga, A. 2014. Purification techniques for the recovery of valuable compounds from acid mine drainage and cyanide tailings: Application of green engineering principles. *Journal of Chemical Technology and Biotechnology* 89:803–813. DOI: 10.1002/jctb.4328

51. Wieszczycka, K., Regel-Rosocka, M., Staszak, K., Wojciechowska, A., Reis, M.T.A., Ismael, M.R.C., Gameiro, M.L.F., Carvalho, J.M.R. 2015. Recovery of zinc(II) from chloride solutions using pseudo-emulsion based hollow fiber strip dispersion (PEHFSD) with 1-(3-pyridyl) undecan-1-one oxime or tributylphosphate. *Separation and Purification Technology* 154:204–210. DOI: 10.1016/j.seppur.2015.09.017

52. Sharma, A.D., Patil, N.D., Patwardhan, A.W., Moorthy, R.K., Ghosh, P.K. 2016. Synergistic interplay between D2EHPA and TBP towards the extraction of lithium using hollow fiber supported liquid membrane. *Separation Science and Technology* 51:2242–2254. DOI: 10.1080/01496395.2016.1202280

53. Wannachod, T., Wongsawa, T., Ramakul, P., Pancharoen, U., Kheawhom, S. 2016. The synergistic extraction of uranium ions from monazite leach solution via HFSLM and its mass transfer. *Journal of Industrial and Engineering Chemistry* 33:246–254. DOI: 10.1016/j.jiec.2015.10.006

54. Pabby, A.K., Swain, B., Sastre, A.M. 2017. Recent advances in smart integrated membrane assisted liquid extraction technology. *Chemical Engineering and Processing: Process Intensification* 120:27–56. DOI: 10.1016/j.cep.2017.06.006

55. Staszak, K., Wojciechowska, A., Reis, M.T.A., Wojciechowska, I., Wieszczycka, K., Ismael, M.R.C., Carvalho, J.M.R. 2017. Recovery of zinc(II) from chloride solutions using pseudo-emulsion based hollow fiber strip dispersion with pyridineketoxime extractants. *Separation and Purification Technology* 177:152–160. DOI: 10.1016/j.seppur.2016.12.046

56. Wojciechowska, A., Wieszczycka, K., Wojciechowska, I. 2017. Efficient recovery of copper from aqueous solutions with pyridine extractants (oxime, ketone) and their quaternary pyridinium salts. *Separation and Purification Technology* 185:103–111. DOI: 10.1016/j.seppur.2017.05.020

57. Song, J., Huang, T., Qiu, H., Niu, X., Li, X.-M., Xie, Y., He, T. 2018. A critical review on membrane extraction with improved stability: Potential application for recycling metals from city mine. *Desalination* 440:18–38. DOI: 10.1016/j.desal.2018.01.007

58. Kittisupakorn, P., Konaem, W., Suwatthikul, A. 2018. Improvement of synergistic extraction of neodymium ions via robust model predictive control. *Computer Aided Chemical Engineering* 43:967–972. DOI: 10.1016/B978-0-444-64235-6.50170-4

8 Process Intensification in Integrated Use of Hollow Fiber Contactor-Based Processes

Case Studies of Sparkling Water Production and Membrane Contactor-Based Strip Dispersion Process

Anil K. Pabby, S. R. Wickramasinghe, Kamalesh K. Sirkar, and A. M. Sastre

CONTENTS

8.1 INTRODUCTION

Process intensification (PI) may be achieved by new modes of production, which are based on scientific principles. New operating modes have been studied in the laboratory and/or at pilot stages: hollow fiber contactor (HFC)-based membrane separations, reverse osmosis and nanofiltration, reversed flow for reaction-regeneration, unsteady operations, cyclic processes, extreme conditions, pultrusion, low-frequency vibrations to improve gas–liquid contacting in bubble columns, high temperature, and high pressure technologies, and supercritical media are now seriously considered for practical applications [1]. In this chapter, our focus will be mainly on HFC-based membrane separations with respect to their applications in metal extraction/separation through a detailed description of case studies implementing process intensification. Before going into an in-depth discussion of process intensification and its components, the basics of process intensification are discussed below.

8.1.1 WHAT IS PROCESS INTENSIFICATION (PI)?

One of the woodcuts in the famous sixteenth century book by Georgius Agrícola [2] described the process of retrieving gold from gold ore. The similarities between some of the devices used in the process and the basic equipment of today's chemical process industries (CPI) is striking. Indeed, Agricola's drawing indicates that process intensification, no matter how we define it, does not seem to have had much impact in the field of mixing technology over more than four centuries. But, what actually is process intensification?

As shown in Figure 8.1, the field generally can be divided into two areas [3]:

- *process-intensifying equipment*, such as novel reactors, and intensive mixing, heat-transfer, and mass-transfer devices; and
- *process-intensifying method*, such as new or hybrid separations, integration of reaction and separation, heat exchange, or phase transition (in so-called multifunctional reactors), techniques using alternative energy sources (light, ultrasound, etc.), and new process-control methods (like intentional unsteady state operation).

Obviously, there can be some overlap. New methods may require novel types of equipment to be developed and *vice*

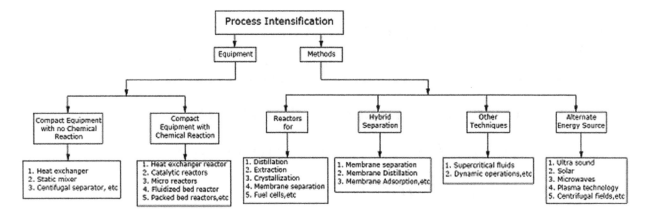

FIGURE 8.1 Process intensification and its components.

versa, while novel apparatuses already developed can make use of new, unconventional processing methods.

Membrane technologies address the requirements of so-called process intensification because they allow improvements in manufacturing and processing, substantially decreasing the equipment-size/production-capacity ratio, energy consumption, and/or waste production resulting in cheaper, sustainable technical solution. Drioili (2001, 2005) [4, 5] document the state-of-the-art and include progress and perspectives on integrated membrane operations for sustainable industrial growth. The first studies on membrane reactors used membranes for distributing the feed of one of the reactants to a packed bed of catalyst. They were used in order to improve selectivity in partial oxidation reactions.

In the PI framework by Stankiewicz and Moulijn [3, 6], any development in process equipment or methods that contribute to dramatic improvements in manufacturing and processing qualifies as process intensification. It must ultimately result in a cheaper and more sustainable technology. There are two specific types of developments. The first type of development leads to dramatic improvements in processing by eliminating or bypassing limitations of conventional chemical engineering equipment.

A second type of development consists of a combination of two or more processes or functions in one device or process resulting in PI. It is expected that such developments will lead to smaller/compact devices for the same production goal. A demonstration of the first type of development leading to dramatic improvements in processing by eliminating or bypassing limitations of conventional equipment is provided by the processes of membrane absorption, membrane extraction, and membrane stripping. The devices for such processes are broadly identified as membrane contactors. In conventional devices based on phase dispersion, there are limits of phase flow rate ratios beyond which there would be flooding. By having immobilized gas–liquid interface [7,8] or liquid–liquid interface [9] at the pore mouths of porous hydrophobic membranes, the limitation on the flow rate ratio of the two phases is removed as long as appropriate pressure conditions are maintained and breakthrough

pressure difference between the phases is avoided. The other important factor with hollow fiber membrane is membrane performance which is also a valuable component of process intensification. Membrane performance can be enhanced by suitable modification of the surface. This can effectively prevent the loss of liquid membrane phase without offering additional resistance to mass transfer. In this direction, Sun et al. [10] treated polypropylene hollow fiber membrane (PP-HFM) using heptadecafluoro-1,1,2,2-tetrad ecyltrimethoxysilane (FAS) and SiO_2 to increase the surface hydrophobicity. The modification scheme for PP-HFM is shown in Figure 8.2. The modification conditions were investigated and the grafted PP-HFM was used as the support of HF-SLM to removal phenols from coal gasification wastewater (CGW) After being grafted, the influence of hydrophobic modification on the stability performance of HF-SLM was studied in a pilot-scale experiment. The results were claimed to be promising by the authors.

There are a number of other basic advantages in such membrane contactors: a much enhanced surface area per unit equipment volume leading to highly compact devices when hollow fiber membranes are used; modular equipment for easy scale up or scale down; non-dispersive operation eliminating foaming and weeping in gas–liquid systems; prevention of emulsification in liquid–liquid systems; elimination of the need for coalescence in liquid–liquid contacting processes; no need for density difference in liquid–liquid systems.

The earliest equipment of the second type of development is provided by membrane distillation (MD) [11–13] and membrane crystallization (MC) [14]. The process of membrane distillation, especially direct contact membrane distillation (DCMD), combines evaporation of a volatile species on one surface of a porous non-wetted hydrophobic membrane with its condensation in the distillate stream on the other surface of the membrane; the primary application has been removal of water from hot brine into a cold distillate stream.

In air gap membrane distillation (AGMD), the two functions take place in two distinct parts of one separation device: evaporation takes place on one surface of the porous

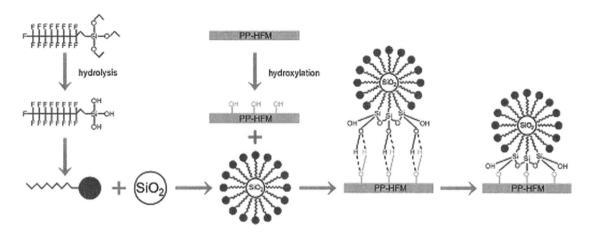

FIGURE 8.2 The modification scheme for PP-HFM [Reprinted from *Chem. Eng. Res. Design*, 126, H. Sun, J.Yao, H. Cong, Q. Li, D. Li, B. Liu, Enhancing the stability of supported liquid membrane in phenols removal process by hydrophobic modification, 209–216, Copyright (2017) with permission from Elsevier [11]].

hydrophobic membrane as the water vapor emerging from the other side of the non-wetted membrane encounters a contiguous cold surface and condenses. In membrane crystallization, the process of membrane distillation is utilized to concentrate the solution initiating crystallization.

An even earlier design of the second type of development is found in a supported liquid membrane (SLM) or an immobilized liquid membrane (ILM), or later in a contained liquid membrane (CLM). In an SLM, the pores of a porous/microporous membrane are filled with a liquid which is held by capillary forces. Feed liquid solution, meant to undergo solvent extraction, contacts one surface of the pore liquid on one side of the membrane where solvent extraction takes place. The feed liquid and the pore liquid (solvent) must be immiscible. On the other side of the membrane, back extraction of the extracted solute takes place from the pore solvent into a back extraction solvent with which the pore liquid must be immiscible.

The phrase ILM is used more often when a gas mixture is separated through the liquid membrane immobilized in the pores of the support membrane. Contained liquid membrane achieves separation in a similar fashion except that the liquid membrane is now placed between two separate membranes or two separate sets of porous hollow fibers.

8.2 DEFINITION OF NEW METRICS

In particular, new metrics for comparing membrane performance with that of traditional operations are tentatively proposed by Criscuoli and Drioli [15]. The comparison is performed in terms of the following: productivity/size ratio; productivity/weight ratio; flexibility; modularity. Referring to the flexibility of the plant, two different aspects are considered: the ability to handle variations that might occur during the life of the plant (such as variation of the pressure, the temperature, or the feed composition) and the ability to be applied in different cycles of production.

The first metric proposed compares the productivity (*P*)/ size ratio (PS) of membrane units with that of conventional operations (**Equation 8.1**). Future plants should be characterized by high productivities and low sizes; therefore, when the PS metric is higher than 1, membrane operations should be preferred, whereas, for PS values lower than 1, traditional units should be chosen.

$$PS\left(Productivity/size\ ratio\right) = \frac{P/size\left(membranes\right)}{P/size\left(traditional\right)} \quad (8.1)$$

Another important parameter in installing new plants is the weight of the operating units involved, especially for installations off shore or for plants that are built in remote areas. To take into account this aspect in making the plant design, the productivity/weight ratio (PW) is defined (see **Equation 8.2**) that compares the productivity/weight ratio of membrane units with that of conventional systems. Plants with high productivities and low weight are preferred; then, a PW higher than 1 is in favor of membranes, whereas a PW lower than 1 means that traditional system are performing better.

$$PW\left(Productivity/weight\ ratio\right) = \frac{P/weight\left(membranes\right)}{P/weight\left(traditional\right)}$$

(8.2)

The flexibility metric (see **Equation 8.3**) compares membranes and traditional devices in terms of variations for which the existing plants are able to work, without requiring any type of adjustment. As for the previous metrics, membranes are preferred when a value higher than 1 is obtained. The modularity takes into account the changes of the plant size related to variations of the plant productivity. This is expressed as flexibility1 and defined below:

$$\text{Flexibility1} = \frac{\text{variation handled}(\text{membrane})}{\text{variation handled}(\text{traditional})} \qquad (8.3)$$

Equation 8.4 expresses the flexibility2 as the ratio between the number of processes that membrane units are able to perform and that related to traditional systems. Again, a flexibility2 value higher than 1 is in favor of membranes.

$$\text{Flexibility2} = \frac{\text{N Process performed}(\text{membrane})}{\text{N Process performed}(\text{traditional})} \qquad (8.4)$$

A typical property of membrane operations is its modularity. Modularity takes into account the changes of the plant size due to variations of the plant productivity. To compare the modularity of membranes with that of traditional units, modularity metric is defined (see **Equation 8.6**). Given a variation of the plant productivity [e.g., from productivity1 (calculated for membranes) to productivity2 (calculated for traditional system)] (see **Equation 8.5**), this metric compares the variations of the area (for membranes) with those of the volume (for conventional systems). The membrane system has a higher modularity if the modularity metric is lower than 1; modularity values higher than 1 are in favor of the traditional system.

$$\text{M I (Modularity index)} = \frac{\text{Productivity2}}{\text{Productivity1}} \qquad (8.5)$$

$$\text{M(Modularity)} = \frac{\left[\dfrac{Area2}{Area1}(\text{membranes}) - \text{MI}\right]}{\dfrac{Vol.2}{Vol.1}(\text{traditional}) - \text{MI}} \qquad (8.6)$$

8.3 CASE STUDIES ON PROCESS INTENSIFICATION DEALING HF-BASED APPLICATIONS

8.3.1 Case Study 1: Sparkling Water Production

As a case study, the new metrics are applied to sparkling water production. Although the new metrics have been defined for membrane operations, the proposed approach can be more generally applied to evaluate the "degree of intensification" achievable with any other process of interest.

The sparkling water production is usually carried out in packed columns where the water to be carbonated is deoxygenated by using a stripping stream consisting of carbon dioxide. The deoxygenated water, slightly carbonated, is then carbonated to the desired value by injecting carbon dioxide under pressure in the pipeline. Figure 8.3 shows a scheme for the traditional process.

The same operation can be carried out by using membrane contactors. In this case, hydrophobic membranes are used. Water flows on one side of the membrane while carbon dioxide flows on the other side. Due to a difference in partial pressures, carbon dioxide is transferred from the gas side to the liquid and, simultaneously, the oxygen is removed from the water to the gas side. With this system, the deoxygenated water at the exit of the membrane module already contains the desired value of dissolved carbon dioxide, and there is no need for further CO_2 injection (see Figure 8.4). The proposed new metrics have been calculated by considering the same feed stream characteristics (Table 8.1) and the same targets (Table 8.2) for the two systems. Concerning the membrane system, hollow fiber modules of 130 m^2 in series have been considered, each one with a size of 0.08 m^3 and a weight of

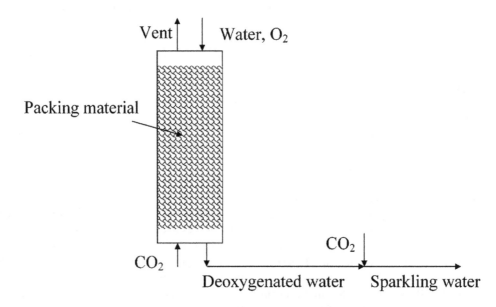

FIGURE 8.3 Scheme for sparkling water production by traditional systems [Reprinted from *Ind. Eng. Chem. Res.* 46, A. Criscuoli, and E. Drioli, New Metrics for Evaluating the Performance of Membrane Operations in the Logic of Process Intensification, 2268–2271, Copyright (2007) with permission from American Chemical Society [15]].

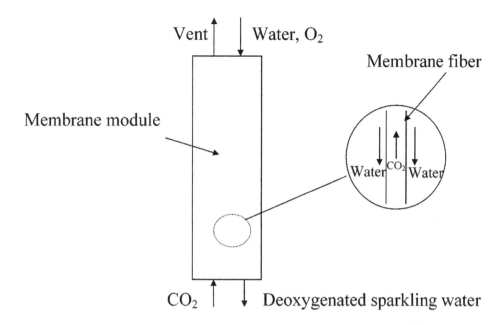

FIGURE 8.4 Scheme for sparkling water production by membrane systems [Reprinted from *Ind. Eng. Chem. Res.* 46, A. Criscuoli, and E. Drioli, New Metrics for Evaluating the Performance of Membrane Operations in the Logic of Process Intensification, 2268–2271, Copyright (2007) with permission from American Chemical Society [15]].

TABLE 8.1
Feed Stream Characteristics

S.No.

1.	Water flow rates-productivity (m³/h) 30
2.	Pressure (bar) 1
3.	O_2 content (ppm) saturation value 9.24
4.	Temperature (°C) 20

Source: Reprinted from *Ind. Eng. Chem. Res.* 46, A. Criscuoli, and E. Drioli, New Metrics for Evaluating the Performance of Membrane Operations in the Logic of Process Intensification, 2268–2271, Copyright (2007) with permission from American Chemical Society [15]]

TABLE 8.2
Target Values for the Exit Stream

S.No.

1.	Water flow rates-productivity (m³/h) 30
2.	CO_2 content (g/L) 4.3
3.	O_2 content (ppm) <0.05
4.	Temperature (°C) 20

Source: Reprinted from *Ind. Eng. Chem. Res.* 46, A. Criscuoli, and E. Drioli, New Metrics for Evaluating the Performance of Membrane Operations in the Logic of Process Intensification, 2268–2271, Copyright (2007) with permission from American Chemical Society [15]]

100 kg. The internal diameter of the fibers was about 200 μm and the pore size around 0.03 μm. Data on the traditional system have been obtained from industry [15].

In Tables 8.3 and Table 8.4 the carbon dioxide flow rate and pressure, the size, and the weight required, respectively, for the membrane and the traditional systems, are reported. By using these values the PS and PW metrics have been calculated (**see** Table 8.5). Both metrics are higher than 1, meaning that membranes are preferable, providing the same productivity as traditional devices with lower size and weight.

For calculating the flexibility, defined as the ability to handle the changes in operating conditions of the plant, three possible variations for the feed stream have been taken into account (see Table 8.6): water temperature, water flow rate (productivity), and oxygen content in the water. In particular, for the water flow rate, a variation of 50% of the flow rate used for the units design (30 m³/h) has been considered. For all of the variations analysed, membrane units

guarantee the final target, whereas the traditional system is not able to handle an increase of the water flow rate with respect to the design value. This leads to a flexibility value higher than 1 (specifically, flexibility1 estimated to be 1.39).

Concerning the different applications potentially covered by the systems, the traditional system could also be applied to liquid–liquid operations, whereas the membrane units (membrane contactors) can find application for liquid–liquid and gas-vacuum operations, as well as membrane/osmotic distillation. Membrane units are, therefore, more flexible than the traditional column with a flexibility2 value of 3. For calculating the modularity of the two systems, a productivity of 3 m³/h has been considered. Table 8.7 reports the areas and the volumes required for treating the design specification as well a production rate that is 10 times lower. By using the results from **Equations 8.5 and 8.6**, the modularity metric obtained is lower than 1, meaning that membranes are more modular than the traditional

TABLE 8.3

Data of the Traditional System for the Sparkling Water Production

S.No.

1.	CO_2 flow rate (Kg/h) 130
2.	CO_2 pressure (bar) 7
3.	Size (m^3) 1.55
4.	Weight (kg) 700

Source: Reprinted from *Ind. Eng. Chem. Res.* 46, A. Criscuoli, and E. Drioli, New Metrics for Evaluating the Performance of Membrane Operations in the Logic of Process Intensification, 2268–2271, Copyright (2007) with permission from American Chemical Society [15]]

TABLE 8.4

Data of the Membrane System for the Sparkling Water Production

S.No.

1.	CO_2 flow rate (Kg/h) 110
2.	CO_2 pressure (bar) 1–2.5
3.	Size (m^3) 0.24
4.	Weight (kg) 300

Source: Reprinted from *Ind. Eng. Chem. Res.* 46, A. Criscuoli, and E. Drioli, New Metrics for Evaluating the Performance of Membrane Operations in the Logic of Process Intensification, 2268–2271, Copyright (2007) with permission from American Chemical Society [15]]

TABLE 8.5

PS and PW Metrics

S.No.

1.	PS 6.46
2.	PW 2.33

Source: Reprinted from *Ind. Eng. Chem. Res.* 46, A. Criscuoli, and E. Drioli, New Metrics for Evaluating the Performance of Membrane Operations in the Logic of Process Intensification, 2268–2271, Copyright (2007) with permission from American Chemical Society [15]]

unit. The performance of the two systems has also been compared by applying existing metrics. In particular, with the available data, the mass intensity has been calculated in both cases. The mass intensity is defined as the ratio of the total mass used in the process to the mass of the product:

$$mass\,intensity = \frac{total\,mass\,(kg)}{mass\,of\,product\,(kg)} \quad (8.7)$$

TABLE 8.6

Variations in the Operating Conditions Considered

S.No.

1.	Water flow rates-productivity (m^3/h) 15–45
2.	Water temperature (°C) 5–20
3.	O_2 content at 20 °C (ppm) 1–9.24

Source: Reprinted from *Ind. Eng. Chem. Res.* 46, A. Criscuoli, and E. Drioli, New Metrics for Evaluating the Performance of Membrane Operations in the Logic of Process Intensification, 2268–2271, Copyright (2007) with permission from American Chemical Society [15]

TABLE 8.7

Variations in the Plant Size Due to a Variation of the Productivity

Selected flow rate	Volume at traditional system (m^3)	Area of membrane system (m^2)
For treating 30 m^3/h	1.55	390
For treating 3 m^3/h	0.098	38

Source: Reprinted from *Ind. Eng. Chem. Res.* 46, A. Criscuoli, and E. Drioli, New Metrics for Evaluating the Performance of Membrane Operations in the Logic of Process Intensification, 2268–2271, Copyright (2007) with permission from American Chemical Society [15]

A process with a low mass intensity value is naturally preferred. To make the comparison, the mass intensity ratio has been obtained, which is the ratio of the mass intensity of membranes to the mass intensity of the traditional system:

$$mass\,intensity\,ratio = \frac{mass\,intensity\,(membrane)}{mass\,intensity\,(traditional)} \quad (8.8)$$

For the selected case study, the mass of product is that of the water treated and the mass used is that of the stripping gas. By using the CO_2 and the water values for the two systems, a mass intensity ratio of 0.85 is obtained.

8.3.2 Case Study 2: Intensified Treatment of Spent Pickling Solutions from Hot-Dip Galvanizing Processes

8.3.2.1 Evaluation of the Degree of Intensification of the EPT Process versus a Conventional Solvent Extraction Process

In this example, two indices are applied to compare the performance of membrane-based solvent extraction processes and traditional solvent extraction units following the metrics defined by Criscuoli and Drioli [15]. The first proposed

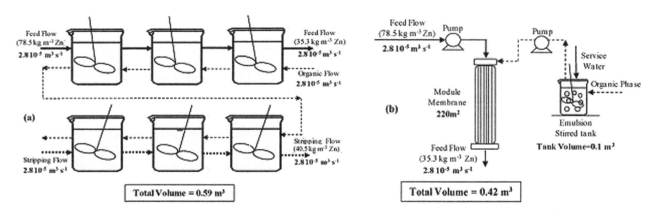

FIGURE 8.5 Flow diagrams of conventional (a) and membrane-based solvent extraction (b) processes [Reprinted from *Process Intensif. Chem. Eng. Process.* 67, E. Bringas, M.F. San Román, A.M. Urtiaga, I. Ortiz, Integrated use of liquid membranes and membrane contactors: enhancing the efficiency of L-L reactive separations, 120–129, Copyright (2013) with permission from Elsevier [16]].

index, PS, compares the ratio between the productivity and the equipment size for both the membrane units and the conventional operations. As expressed in **Equation 8.1** and **Equation 8.6**, PS and modularity were calculated by these authors respectively. To compare the modularity of membrane contactors versus traditional units, the modularity index (*M*) is defined according to **Equation 8.5**. Given a specific variation of the plant productivity (e.g., from productivity1 to productivity2), this metric compares the variations of the membrane area [(for membrane contactors – productivity1)] with those of the volume [(for conventional mixer settlers in the case of liquid extraction operations – productivity2)].

In this developed process, if M < 1, membrane contactors should be preferred, while when M > 1, the traditional operation would be chosen. These metrics have been applied to evaluating the degree of intensification of the EPT process versus conventional zinc recovery technology [16]. The recovery of zinc from spent pickling solutions (SPS), with reference to the best available techniques (BAT) recommended by the European Union (EU) for the metal processing industry [17], can be carried out using conventional solvent extraction in a cascade of extraction and stripping stirred tanks. Stocks et al. [18] discussed in detail dealing with the material and energy flows in the process, which contained three extraction stages for completing zinc extraction and three more stages for zinc stripping and solvent recovery.

In this work, the total volume of the traditional system was calculated following the methodology proposed by Torab-Mostaedi et al. [19], by defining the same values of productivity ($2.8 \times 10^{-7} m^3 s^{-1}$) and final zinc concentration in the stripping phase ($40.5 kg m^{-3}$) as those obtained in the EPT process at bench scale. The total volume of the mixing tanks was calculated to be $2.7 \times 10^{-2} m^3$, while the volume of the membrane contactor (Liqui-Cel® Extra-Flow 2.5×8, Membrane) and emulsion tank was only $2.3 \times 10^{-3} m^3$.

The resulting value of the productivity/size ratio was PS = 11.7, noticeably higher than 1 and thus meaning that the EPT technology is preferable versus the conventional

extraction in mixer settlers due to the lower size of equipment required to achieve the same productivity. For calculating the modularity of the two systems, an increase of productivity up to $2.8 \times 10^{-5} m^3 s^{-1}$ (productivity2) was considered. A model of the EPT process was used to obtain the new membrane area that fulfills the proposed productivity2, while maintaining the same target for the attainable concentration of zinc in the stripping phase, with the result of Size2 (refer to scaled up system), (membrane area) = 220 m^2. Likewise as for the PS parameter, the volume for the traditional operation was obtained following the same procedure reported above with the result of Size2, (volume of mixing tanks) = 0.59 m^3. By substituting the values of membrane area (EPT process) and volume of mixing tanks (solvent extraction process) into **Equation 8.5**, the modularity index obtained was M = 0.73, a value lower than 1 which means that the EPT process is more modular than the traditional process based on stirred tanks facilitating the process scale up. Figure 8.5 shows the flow diagrams, including size and productivities values, of both the conventional and membrane-based solvent extraction processes.

8.4 LIMITATIONS AND BARRIERS IN ADOPTING PROCESS INTENSIFICATION AND CHALLENGES

Despite all of the potential advantages and a number of successful commercial applications of process intensification principles, there are still several important barriers hindering deeper changes in the chemical process industry including membrane-based industry. For example, the R&D efforts in chemical companies are primarily focused on new products (chemistry) and much less on the new manufacturing methods (chemical engineering). Chemical manufacturers are not interested in developing the novel types of equipment or processing techniques. It is simply not their key business. On the other hand, equipment manufacturers and engineering companies are not sufficiently active in the field of process intensification. Many novel apparatuses and processing methods are not yet proven on the industrial scale. It is well known that plant managers must minimize risk and are hesitant to introduce new technologies that have not been proved. Further discussion on the limitations of PI can be found in Reference [20]. The good news is that many of the above-mentioned obstacles can be removed or at least minimized, and universities, along with non-academic research centers, can play an important role in this process.

As far as membrane-based contacting processes are concerned, they have demonstrated considerable process intensification. Sometimes the limitation is the integration of two processes, which is anticipated when comparing the performance of two separate processes. This has been achieved by one or more of the following: equipment volume reductions sometimes by an order of magnitude for the same production level; achieving multistep processes in one device; non-dispersive processing in phase contacting-based processes; enhancement of operational flexibility by avoiding flooding, loading, reduction in energy required for processing.

8.5 CONCLUSION AND FUTURE PERSPECTIVES

New metrics to be used for comparing membrane process to traditional unit operations in the logic of the process intensification have been proposed. These metrics take into account the size, the weight, the flexibility, and the modularity of the plants and can be easily obtained, once data on system performance are available. They do not replace the existing indicators, referring to other aspects of the production plants. Therefore, it is important to point out that the final evaluation of the "sustainability" of processes must be always carried out by considering the new metrics together with the environmental, economic, and societal indicators. The new metrics would represent, in fact, "something more to think about" in performing the overall analysis of processes. In the first case study, as an example calculation, the new metrics have been obtained for the sparkling water production. It is important to underline that the proposed approach can be used not only for membrane operations, but also for comparing, in terms of process intensification, any other process of interest to traditional operations. In the second case study, this work proposes a computer-based simulation methodology to analyse the benefits of employing membrane facilitated transport separation processes based on the combination of reactive liquid membranes and membrane contactors. From the initial hypothesis and the previous knowledge both on the experimental behavior and mathematical modeling of the EPT process, the design and scale up of liquid membrane processes when they are incorporated into the normal operation of existing industrial processes has been firstly addressed, followed by the evaluation of the degree of intensification provided by the use of liquid membranes

FIGURE 8.6 Vision of a modern petrochemical plant located in Switzerland [Reprinted from *Chem. Eng. Journal.* 134, J.-C. Charpentier, In the frame of globalization and sustainability, process intensification, a path to the future of chemical and process engineering (molecules into money), 84–92, Copyright (2007) with permission from Elsevier [23]].

as compared to conventional solvent extraction processes. The first objective was tackled by illustrating the performance of the scaled up EPT process applied to the regeneration of chromium-based passivation baths. On the other hand, the comparison, through suitable intensification metrics, of conventional solvent extraction and EPT both applied to the recovery of valuable components from the spent pickling solutions generated in the hot-dip galvanizing process has been analyzed. These promising results confirm the effectiveness of EPT when it is incorporated into the operation of existing surface treatment processes and by extension into other industrial processes.

As future perspectives, for process intensification, many new technologies are being developed in the chemical industries motivated by improved chemistry, enhanced safety, improved processing energy, and environmental benefits, low inventories, capital cost reduction, enhanced corporate image, novel or enhanced products, and value to customers [21–22]. In Reference [3] a vision of how a future plant employing process intensification is proposed and compared to a conventional plant. In an important review published by Charpentier [23] improvements in the design of modern petrochemical plants which are actually accomplished today in certain desired cases are described (see Figure 8.6 showing a TAMOIL plant located in Switzerland, including a Fluid Catalytic Cracking unit for the valorization of heavy crude oil charges).

Several important limits must be overcome before process intensification is widely adapted, such as the maturity and economic competitiveness of the new technologies compared to the conventional technologies. The conservatism of plant owners using batch processors means they will not easily accept continuous processing solutions.

REFERENCES

1. J.-C. Charpentier, Process intensification by miniaturization, *Chem. Eng. Technol.* 28 (2005) 255–258.

2. G. Agricola, *De Re Metallica Libri Xll.* Basel, Switzerland: Froben & Episopius (1556).

3. A. I. Stankiewicz, J. A. Moulijn, Process intensification: Transforming chemical engineering, *Chem. Eng. Prog.* 96 (2000) 22–34.

4. E. Drioli, M. Romano, Progress and new perspectives on integrated membrane operations for sustainable industrial growth, *Ind. Eng. Chem. Res.* 40 (2001) 1277–1300.

5. E. Drioli, E. Curcio, G. Di Profio, State of the art and recent progresses in membrane contactors, *Chem. Eng. Res. Des.* 83 (2005) 223–233.

6. K. K. Sirkar, A. G. Fane, R. Wang, S. R. Wickramasinghe, Process intensification with selected membrane processes, *Chem. Eng. Process.* 87 (2015) 16–25.

7. K. Esato, B. Eiseman, Experimental evaluation of Gore-Tex membrane oxygenator, *J. Thorac. Cardiovasc. Surg.* 69 (5) (1975) 690–697.

8. T. Tsuji, K. Suma, K. Tanishita, H. Fukazawa, M. Kanno, H. Hasegawa, A. Takahashi, Development and clinical evaluation of hollow fiber membrane oxygenator, *Trans. Am. Soc. Artif. Intern. Organs* 27 (1981) 280–284.

9. A. Kiani, R. R. Bhave, K. K. Sirkar, Solvent extraction with immobilized interfaces in a microporous hydrophobic membrane, *J. Membr. Sci.* 20 (1984) 125–145.

10. D. W. Gore, Gore-Tex membrane distillation, Proc. 10th Annual Convention of Water Supply Improvement Association, July 25–29, Honolulu, HI (1982).

11. H. Sun, J. Yao, H. Cong, Q. Li, D. Li, B. Liu, Enhancing the stability of supported liquid membrane in phenols removal process by hydrophobic modification, *Chem. Eng. Res. Des.* 126 (2017) 209–216.

12. K. Schneider, T. S. V. Gassel, Membrane destillation, *Chem. Ing. Tech.* 56 (1984) 514–521.

13. R. W. Schofield, A. G. Fane, C. J. D. Fell, Heat and mass transfer in membrane distillation, *J. Membr. Sci.* 33 (1987) 299–313.

14. Y. Wu, E. Drioli, The behavior of membrane distillation of concentrated aqueous solution, *Water Treat.* 4 (1989) 399–415.

15. A. Criscuoli, E. Drioli, New metrics for evaluating the performance of membrane operations in the logic of process intensification, *Ind. Eng. Chem. Res.* 46 (2007) 2268–2271.

16. E. Bringas, M. F. San Román, A. M. Urtiaga, I. Ortiz, Integrated use of liquid membranes and membrane contactors: Enhancing the efficiency of L-L reactive separations, *Chem. Eng. Process. Process Intensif.* 67 (2013) 120–129.

17. European Commission, Reference document on best available techniques for surface treatment of metals and plastics, http://eippcb.jrc.es (accessed 15.11.19).

18. C. Stocks, J. Woodb, S. Guyb, Minimisation and recycling of spent acid wastes from galvanizing plants, *Resour., Conserv. Recycl.* 44 (2005) 153–166.

19. M. Torab-Mostaedi, S. J. Safdari, M. A. Moosavian, M. Ghannadi Maragheh, Mass transfer coefficients in a Hanson mixer-settler extraction column, *Braz. J. Chem. Eng.* 25 (2008) 473–481.

20. J.-C. Charpentier, Process intensification, *Ind. Eng. Chem. Res.* 41 (2002) 1920–1924.

21. C. Tsouris, J. V. Porcelli, Process intensification—Has it time finally come? *Chem. Eng. Prog.* 10 (2003) 50–55.

22. A. I. Stanckiewicz, J. A. Moulijn, *Reengineering the Chemical Process Plant: Process Intensification*, Marcel Dekker, New York (2004).

23. J.-C. Charpentier, In the frame of globalization and sustainability, process intensification, a path to the future of chemical and process engineering (molecules into money), *Chem. Eng. J.* 134 (2007) 84–92.

9 Application of Hollow Fiber Contactor for Polishing of Lean Streams Containing Uranium from Nuclear Industry
Scale up and Demonstration

S. Mukhopadhyay, S. Mishra, K. T. Shenoy, and V. A. Juvekar

CONTENTS

NOMENCLATURE

ε: porosity of hollow fiber
DLM: dispersion liquid membrane
HF: hollow fiber
HFC: hollow fiber contactor
TBP: Tri-n-butyl phosphate

9.1 INTRODUCTION

At present, the recovery of metal ions is a necessity because of stringent environmental regulations. Metals like copper, cadmium, nickel, and chromium, have carcinogenic effects. Likewise, radioactive waste resulting from nuclear facilities threatens the ecosystem. Uranium is the principal source of this type of pollution. Even under strict policies, regulations, and vigilance, industries indiscreetly dispose of several effluent streams containing toxic or otherwise hazardous metal ions into the river streams; such practices are particularly followed in countries which are either less developed or developing. Although one may argue for strict enforcement of pollution abatement laws, the problems are also related to the inadequate technologies available to treat such effluents. Conventional processes like solvent extraction are unable to treat such streams; this is particularly so when these polluting metal ions are present in very low concentrations. Further, in some cases, the process may prove uneconomical and ecologically unfriendly because of a requirement of large amounts of organic solvent. Membrane separations are addressing a few such problems; in particular, processes like schemes adopting facilitated transport [1].

The recovery of valuable metals such as uranium from lean acidic raffinate is a major task in the nuclear industry in view of the limited resources and the strict environmental regulations for the disposal of aqueous effluents. The existing methods used for this purpose viz. solvent extraction (Figure 9.1) and ion exchange have their own limitations of large phase ratio, number of steps, poor selectivity, etc. The solvent extraction process depicted in Figure 9.1 needs separate extraction and stripping steps, multistage contactor for recovery from lean streams, large density difference between the phases (for efficient phase separation), and large solvent inventories. Moreover, there are entrainment

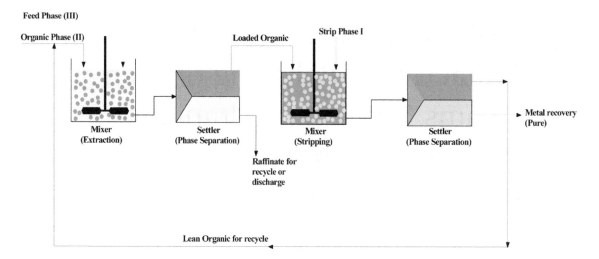

FIGURE 9.1 Conventional solvent extraction process.

losses due to emulsification and loss in settler due to permanent emulsion formation. In the ion exchange process, there are problems such as resin fouling, throughput limitations, and poor selectivity, which limit its applicability. In addition, the driving force for mass transfer in these processes is low due to equilibrium limitations [2].

9.2 LIQUID MEMBRANES

In a liquid membrane, the liquid phase is acting as membrane to permeate a specific material or solute selectively from a source liquid phase placed on one side of it to the similar phase placed on the other side, through diffusive transport. The source phase containing the solute is called the feed phase and the receiving phase is called the strip phase. The membrane phase is chosen such that it is insoluble with the feed and the strip phase, and the solute gets selectively extracted in the membrane phase, diffuses through it, and gets back extracted in the strip phase. Liquid membrane systems can be of two types:

1. W/O/W (organic liquid phase as membrane with aqueous feed and aqueous strip phase).
2. O/W/O (Aqueous phase acting as liquid membrane between organic feed and organic strip phase).

The transport of solute takes place through selective solubility in the membrane phase or by complexation with a suitable carrier agent present in the membrane phase. The process, often called facilitated transport, aids simultaneous extraction and stripping and uphill transport (transport of solute from low to high concentration with the aid of positive concentration gradient of the co/counter transporting agent).

9.2.1 CONFIGURATIONS OF LIQUID MEMBRANE

Major liquid membrane configurations include supported liquid membrane (SLM) as shown in Figure 9.2, emulsion liquid membrane (ELM) as shown in Figure 9.3, and dispersion liquid membrane (DLM) depicted in Figure 9.4. In all of these configurations the solute transport takes place by a cyclic mechanism involving complexation at the source side of the membrane and release at the receiving side [3]. Other configurations include polymer inclusion membrane (PIM), contained liquid membrane [4], and ion exchange membrane [5].

In supported liquid membrane [6, 7], the organic carrier is immobilized by the capillary action in pores of microporous polymeric support. Such a configuration allows for the removal of the enriched stripping solution without disruption of the membrane. The supported liquid membranes

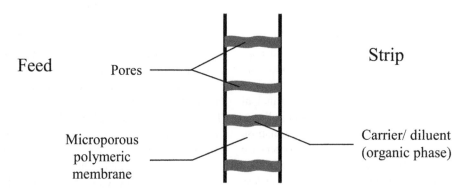

FIGURE 9.2 Supported liquid membrane.

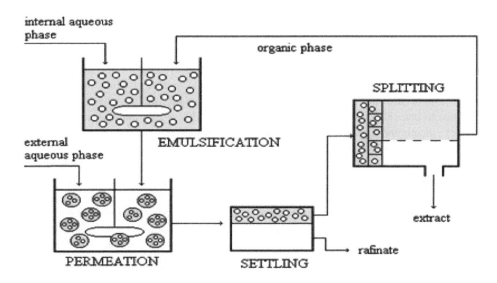

FIGURE 9.3 Emulsion liquid membrane.

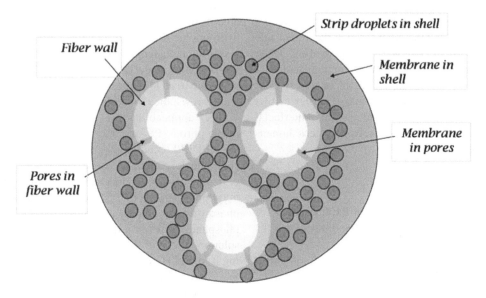

FIGURE 9.4 Dispersion liquid membrane in hollow fiber contactor.

have many inherent advantages like a much reduced organic requirement, low capital and operating costs, low energy requirements, and lower maintenance costs. Despite such obvious advantages, it has a constraint of loss of organic phase from pores of support that limit their effective application [8–11]. This loss is because of the solubility of the organic phase (present in pores) in adjacent aqueous phases and the pressure difference between phases present across the pores.

Emulsion liquid membrane works on a double emulsion principle; the organic carrier phase forms the wall of an emulsion droplet separating the aqueous feed from the aqueous product solution [12]. The emulsion is stabilized by the addition of the thickeners and surfactants. The emulsion liquid membrane process comprises four main operations. First, stripping solution is emulsified in the organic membrane phase. This water/oil emulsion

then enters a large mixer vessel, where it is again emulsified with feed solution to form water/oil/water emulsion. Solute in the feed solution then permeates by facilitated transport through the walls of the emulsion to the stripping solution. The mixture then passes to a settler tank where the oil droplets separate the solute-depleted feed solution. The emulsion concentrate then passes to a de-emulsifier where the organic membrane and stripping solutions are separated. The regenerated organic membrane is recycled. The emulsion liquid membrane delivers a very fast rate of mass transfer operation due to the large specific surface and thin liquid membrane.

Although it is a rapid transport method, the potential of the emulsion liquid membrane process lies mainly in pollution control rather than in the recovery of solutes from aqueous streams [13]. If recovery of solutes is the objective, the emulsion must be broken up, the recovered product solution should be treated further, and the organic phase that constituted the liquid membrane recycled to form new emulsion. Operationally this leads to many steps which make its application difficult for the recovery of solutes.

In another configuration called dispersion liquid membrane (DLM), a dispersion of strip phase in a continuous organic phase is confined to one side, and the aqueous feed phase to the other side of a membrane pore. The organic phase fills the pores of the micro porous polymeric membrane and acts as a liquid membrane. In DLM, continuous replenishment of the pores by the membrane liquid allows elimination of the problem of stability faced by SLM due to loss of the liquid membrane from the pores [14]. Due to short residence times of the dispersion in the contactor, DLM does not require the use of surfactant to stabilize the dispersion. This eases the product recovery process. The dispersion of strip phase in the organic phase is created externally in a mixer device.

The hollow fiber contactor is advanced equipment with numerous polymeric micro porous lumens bundled up inside a polymeric shell [15] applicable to intensification of the solvent extraction process with one phase passing through the shell side and the other through the lumen side contacting in a non-dispersive way. Transport of the solute takes place through the pores of lumen impregnated with the solvent phase by molecular diffusion. In this type of contactor, very high specific area of mass transfer can be achieved without mechanical agitation. This interfacial area, A, is well defined in form of porous polymeric lumens and can be calculated as

$$A = 2\pi R_i l F \varepsilon$$

where, R_i denotes the internal radius of the fiber (cm), l is the length of the fiber (cm), F is the number of fibers, ε is the membrane porosity.

Due to their modular design, enhancement of purity and throughput by numbering up in series and in parallel, respectively, is easy with these contactors. The hollow fiber contactor is widely used for liquid membrane-based separations for its compactness and ease of operation [16]. Both SLM and DLM configurations can be operated in the hollow fiber modules.

Hollow fiber dispersion liquid membrane (HFDLM) process employs dispersion liquid membrane (DLM) in hollow fiber (HF) contactors (Figure 9.4). Here, a dispersion of the strippant in the organic extractant is passed through the shell side of the hollow fiber contactor, while the feed solution is passed through the lumen side [17, 18]. This configuration provides a large interfacial area of stripping because of numerous tiny droplets of strip phase dispersed in the organic membrane phase in the shell side.

9.2.2 ROLE OF WETTABILITY OF THE PORES OF HOLLOW FIBER ON SOLUTE TRANSPORT

Depending on the wettability of the material of the microporous support by the organic and aqueous phases, the pores get impregnated fully or partially with the aqueous or the organic phase. The location of this interface is important in deciding the rate of extraction. Since the diffusion coefficient of the metal cation in the aqueous feed phase is much higher than that of the metal-carrier complex in the organic phase, a higher rate of extraction can be achieved by maintaining the interface as near the shell end of the pore as possible so that diffusion path length in the organic phase is kept as low as possible. For application of an aqueous strip in organic membrane type of dispersion, the wettability of the organic phase to the membrane material is very important. Low wettability results in a shorter transport path in the organic phase and better rates of extraction. On the other hand, low wettability causes a reduction of the breakthrough pressures and therefore requires more stringent control of the differential pressure across the membrane. Adequate wettability of polymeric membrane is essential for its ability to perform carrier-facilitated transport of metal species. Between the two polymeric membrane materials available in the form of hollow fiber, polysulfone (PS) is less hydrophobic than polypropylene (PP), and would provide a shorter path length in the organic phase. Polysulfone, PS, is a better material for working in the nuclear environment due to its toughness, its stability to nuclear radiation, and its resistance to oxidation.

The applicability of hollow fiber contactors with micro-porous PS and PP fibers has been studied using the HFDLM technique for recovery of uranium from lean acidic nuclear waste streams [19]. Water-in-oil dispersion of 1 M $NaHCO_3$ in dodecane containing 30% v/v Tri-n-butyl phosphate (TBP) is used as extractant. The organic phase was prepared by dissolving the 30% v/v Tri-n-butyl phosphate (TBP) in n-dodecane. From uranyl nitrate pure solution, a feed of 0.5 g/L U(VI) was prepared by dissolving it in 1 N HNO_3. A raffinate stream of a uranium refining plant having acidity 1 N was also used as the feed. Raffinate contains mainly U(VI) (around 200 ppm) along with some other ions in trace quantities. An aqueous solution of 1 M sodium bicarbonate was used as the strippant. The dispersion was prepared by adding an equal volume of strippant to the organic phase in drop wise manner, while agitating the organic phase at 3,000 RPM [20]. Five different hollow fiber contactors were used. Two contactors with PS lumens, PS1 and PS2, and the other three contactors, consisting of polypropylene lumens, PP1, PP2, and PP3 were tested. Dimensional details for the contactors are presented in Table 9.1.

A schematic of HFDLM experimental set up operated working in re-circulating mode is shown in Figure 9.5. It consists of a hollow fiber module, a feed and dispersion reservoir, a feed and dispersion pump, and an agitator for dispersing the strippant into the organic phase. The feed is passed through the lumen side and the water-in-oil dispersion (1 M $NaHCO_3$ dispersed in dodecane containing 30 % v/v Tri-n-butyl phosphate) is circulated through the shell. The organic phase partially fills the microspores of

TABLE 9.1

Dimensional details for the contactors

Module	PS1	PS2	PP1	PP2	PP3
Diameter of the shell (cm)	5.0	5.5	5.0	8.5	8.8
Effective length, l (cm)	40	42	16	25	62
Fibre MOC	PS	PS	PP	PP	PP
Fibre OD, $2r_2$ (cm)	0.15	0.15	0.03	0.03	0.03
Fibre ID, $2r_1$ (cm)	0.125	0.10	0.024	0.024	0.024
Fibre log mean diameter, $2r_{lm}$ (cm)	0.14	0.12	0.027	0.027	0.027
Fibre wall thickness, L $(= r_2 - r_1)$ (cm)	0.0125	0.025	0.003	0.003	0.003
Pore diameter (nm)	9	9	50	50	50
Porosity, ε %	70	70	40	40	40
Number of fibres, n_f	250	520	10,000	30,000	31800
Mass transfer area $(2\pi r_{lm} l \varepsilon n_f)$ (m²)	0.302	0.592	0.543	2.54	6.69
Contactor volume (m³)	7.85×10^{-4}	9.98×10^{-4}	3.14×10^{-4}	1.42×10^{-3}	3.77×10^{-3}
Mass transfer area/contactor volume (m²/m³)	385	593	1729	1789	1774.5

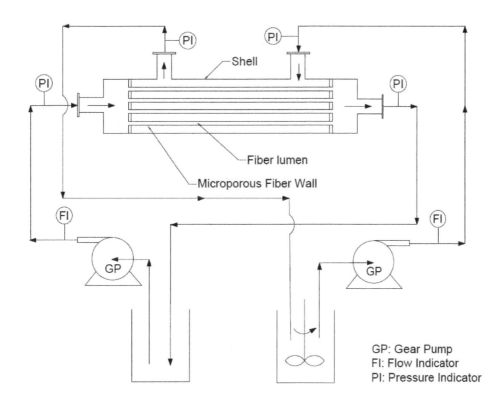

FIGURE 9.5   Schematic of HFDLM process in re-circulating mode.

the hollow fibers through capillary action. Positive pressure differential should be maintained between the lumen-side of the hollow fibers and the shell in order to prevent seepage of the organic phase into the lumen. At the end of the experiment, the agitator is turned off and the dispersion is allowed to separate into two phases viz. the lean organic solution and the loaded strip solution. Uranium content in the aqueous solutions is determined using ICP-OES or colorimetry.

The photograph of the experimental setup is shown in Figure 9.6.

The HFDLM process has been employed for the transport of uranium using feed phase as pure uranyl nitrate solution containing 0.5 g/L U(VI) in 1 N HNO_3, organic phase as 30% v/v TBP in dodecane, and strip phase as 1 M $NaHCO_3$. The flow rate of the feed phase through fiber lumen and the flow rate of dispersion through shell has been maintained at 200 ml/min. Equal volumes of the organic and the strip phases are used for preparing the dispersion. Mass transfer coefficients for aqueous and organic phases can be estimated below through available correlations in the literature [17–19].

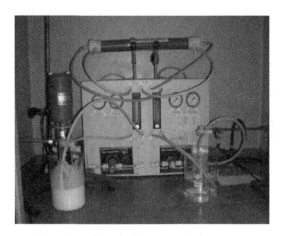

(a) Experimental set up with
polysulfone contactor

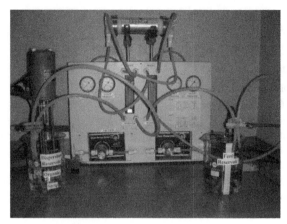

(b) Experimental set up with
polypropylene contactor

FIGURE 9.6 Experimental set up of HFDLM process; (a) PS1 contactor, (b) PP1contactor.

9.2.3 MATHEMATICAL MODEL

In carrier-facilitated transport through HFDLM, metal cation diffuses through the aqueous feed boundary layer and reacts reversibly with the carrier agent at the feed-membrane interface resulting in the formation of metal-carrier complex $UO_2(NO_3)_2.2TBP$. The complex then diffuses through the membrane due to its own concentration gradient and consequently at membrane-strip interface, it releases metal ion in the form of $Na_4(UO_2)(CO_3)_3$. The carrier thus left over diffuses back towards the feed-membrane interface, and the cycle continues. Released cation diffuses through the aqueous strip film.

The extraction of U(VI) using TBP as the extractant from nitric acid medium can be represented by the following equilibrium reaction:

$$UO_2^{2+}(aq) + 2NO_3^-(aq) + 2TBP(org)$$

$$\xleftarrow{\quad K_{eq}\quad} UO_2(NO_3)_2.2TBP(org) \tag{9.1}$$

Concentrations at the interface are related as per **Equation 9.2**, considering fast chemical reaction:

$$K_{eq} = \frac{\left[UO_2(NO_3)_2.2TBP\right]_{org}}{\left[UO_2^{2+}\right]_{aq}\left[NO_3^-\right]_{aq}^2\left[TBP\right]_{org}^2} = \frac{K_{df}}{\left[NO_3^-\right]^2\left[TBP\right]^2} \tag{9.2}$$

Mass balance for a single hollow fiber inside its lumen over differential length Δz and time Δt under no mass accumulation gives:

$$\dot{V}C_{Bf}\big|_z \Delta t = \dot{V}C_{Bf}\big|_{z+\Delta z}\Delta t + J_{org}(2\pi R_{Alm}\Delta z\varepsilon)\Delta t \tag{9.3}$$

Where,

$$R_{Alm} = \frac{(R_o - R_i)}{\ln\left(\frac{R_o}{R_i}\right)}$$

Where, $\dot{V}$ denotes volumetric flow rate, C_{Bf} is UO_2^{2+} concentration in bulk feed solution flowing through fiber lumens, J_{org} signifies UO_2^{2+} flux through liquid membrane present in pores, R_i, R_o and R_{Alm} are the internal radius, outer radius, and log-mean radius of the hollow fiber tube, respectively. ε is the porosity of hollow fiber.

Similarly, mass balance for a single hollow-fiber outside the lumen in shell side over differential length Δz and time Δt under no mass accumulation gives:

$$\dot{V}_s C_{Bs}\big|_z \Delta t + F(J_{aq,s}(2\pi R_o\Delta z\varepsilon)\Delta t) = \dot{V}_s C_{Bs}\big|_{z+\Delta z}\Delta t \tag{9.4}$$

where, C_{Bs} is uranium concentration in bulk strip solution, F stands for the total number of hollow fibers in the contactor, V_s is flow rate of strip solution in shell side, $J_{aq,s}$ is the flux through aqueous strip film outside the wall of fiber.

At a given axial position, there is no mass accumulation of the species and the rate of mass transfer through the aqueous boundary layer must be equal to the rate of mass transfer through the membrane and equal to the overall mass transfer rate in the system. Mass transfer resistances in series of feed film, organic membrane in pores, and strip film are taken into account [17].

$$N\,(mol/s) = \frac{(K_{df}C_{Bf} - K_{ds}C_{Bs})}{\dfrac{K_{df}}{k_{aqf}A_1} + \dfrac{1}{k_{org}A_{A,lm}} + \dfrac{K_{ds}}{k_{aqs}A_2}} \tag{9.5}$$

where, N is the rate of mass transfer in mol/s, k_{aqf} is mass transfer coefficient of feed phase in the tube side (cm/s), and k_{aqs} is the receiver phase mass transfer coefficient in the shell side, k_{org} is the membrane mass transfer coefficient, K_{df} is distribution coefficient at feed-membrane interface and K_{ds} is distribution coefficient at membrane-strip interface, C_{Bs} is concentration of uranium in bulk strip solution, surface area of hollow fiber as, $A_I = 2\pi R_i l\varepsilon$, $A_{A,lm} = 2\pi R_{A,lm} l\varepsilon$ and $A_2 = 2\pi R_o l\varepsilon$, where l is the effective fiber length.

For tube side, from **Equations 9.3 and 9.5,**

$$\frac{dC_{Bf}}{dz} = \left(-\frac{1}{\pi R_i^2 v l}\right)\frac{K_{df}C_{Bf} - K_{ds}C_{Bs}}{\zeta} \qquad (9.6)$$

where, v denotes the fluid flow velocity.

For shell side, from **Equation 9.4 and 9.5,**

$$\frac{dC_{BS}}{dz} = \left(\frac{F}{\dot{V}_s 1}\right)\frac{K_{df}C_{Bf} - K_{ds}C_{Bs}}{\zeta} \qquad (9.7)$$

Where, $\zeta = \dfrac{K_{df}}{k_{aqf}A_1} + \dfrac{1}{k_{org}A_{A,lm}} + \dfrac{K_{ds}}{k_{aqs}A_2}$

ζ is the resistance that appears during the transport process. It constitutes three resistances. One of them is the resistance when the liquid is flowing through the hollow fiber lumen. The second resistance is due to the diffusion of the metal ligand complex across the liquid membrane which is immobilized in the porous wall of the fiber. The third resistance is presented by the strip solution and organic interface at outside of the fiber (shell side). **Equations 9.6 and 9.7** form an initial value problem (IVP) in two variables C_{Bf} and C_{Bs}. This IVP can be solved to get variation of C_{Bf} and C_{Bs} with axial distance, z along the fiber. Initial condition is taken as $(C_{Bfo}, C_{Bso} = 0)$ at $z = 0$. Where, C_{Bfo} and C_{Bso} are initial uranium concentration of feed and strip, respectively. Distribution coefficients, K_{df}, and K_{ds} have been generated through independent liquid–liquid extraction experiments.

9.2.3.1 Estimation of Mass Transfer Coefficient of Feed Phase

The Wilke–Chang correlation can be used for estimation of diffusivity of uranyl ion in the aqueous phase as following [21] for solute molar volume less than 0.5 m³/kmol,

$$D_f = 1.173 \times 10^{-16}\frac{(\varphi M_B)^{1/2}T}{\mu_B V_A^{0.6}} \qquad (9.8)$$

Where, M_B is the molecular weight of solvent B, μ_B is the viscosity of solvent in Pa.s, V_A is the solute molar volume in m³/kmol and φ is an association parameter of the solvent. Substituting, $M_B = 63$ kg/kmol, $\varphi = 1$ for nitric acid medium, $T = 298$ K, $\mu_B = 1 \times 10^{-3}$ Pa.s and $V_A = 9.35 \times 10^{-2}$ m³/kmol [22], D_f is estimated as 1.15×10^{-9} m²/s. The mass transfer coefficient inside the fibers is described by the Leveque

equation (for Graetz number greater than four) considering fast chemical reaction under laminar flow as following [15],

$$Sh = 1.62 \ (4r_I^2 v/ (lD_f))^{1/3} \qquad (9.9)$$

Where, Sh is the Sherwood number $(=2k_{fl}r_I/D_f)$, k_{fl} is mass transfer coefficient of aqueous feed film, r_I is the internal radius of fiber, v is the linear flow velocity of the feed solution through fiber lumen, l is the length of the fibre and D_f is the diffusion coefficient of uranyl ion in the aqueous phase. Substituting D_f $(=1.15 \times 10^{-9}$ m²/s) along with fiber characteristic properties (r_I and l) and feed flow velocity (v) in Leveque equation, k_{fl} has been estimated for PS and PP contactors and listed in Table 9.2.

9.2.3.2 Estimation of Mass Transfer Coefficient of Organic Membrane Phase

The diffusion coefficient of uranyl complex, D_{oeff}, in dodecane containing TBP has been calculated using the Stokes–Einstein equation for porous medium as following [21] for solute molar volume greater than 0.5 m³/kmol,

$$D_{oeff} = D_o(\varepsilon/\tau) = \frac{9.96 \times 10^{-16}\ T}{\mu V_A^{1/3}}(\varepsilon/\tau) \qquad (9.10)$$

Where, D_o is the diffusion coefficient of $UO_2(NO_3)_2.2TBP$ in the bulk organic phase and μ is the viscosity of organic liquid membrane in Pa.s. Substituting $V_A = 0.64$ m³/kmol [22], μ for Dodecane containing 30 % v/v TBP $= 1.75 \times 10^{-3}$ Pa.s and $T = 298$ K, D_o is estimated as 1.968×10^{-10} m²/s. Calculated values of D_f and D_o are in accordance with that found in the literature [23].

The viscosity of organic liquid membrane of varying composition is measured using the Ostwald viscometer and presented in Figure 9.7. Viscosity is used for estimation of the mass transfer coefficient. On an increase in the concentration of ligand, the viscosity of the organic phase increases, which leads to an increase in the mass transfer resistance of the organic phase as per the Stokes–Einstein equation [24].

9.2.3.3 Estimation of Mass Transfer Coefficient of Diffusion Film at the Organic Membrane Interface Shell Side

Average shell side velocity, v_s is estimated using superficial velocity divided by free flow area $(1 - \varphi)$. Shell side equivalent diameter, d_e is estimated through $4 \times$ flow area/total fibre circumference. Using d_e and v_s along with density and viscosity of shell side liquid, Reynolds number has been calculated and shown in Table 9.2. At low Reynolds number (<10), mass transfer coefficients around a stationary cylinder can be obtained through following correlation [25, 26],

$$St\ Sc^{0.67} = 3.42\ Re^{-0.672} \qquad (9.11)$$

Where, $St = k_{ol}/v_s$

TABLE 9.2

List of parameters for contactors

Parameters	PS1	PS2	PP1	PP2	PP3
Fiber inner diameter, $2r_1$ (cm)	0.125	0.10	0.024	0.024	0.024
Fiber outer diameter, $2r_2$ (cm)	0.15	0.15	0.030	0.030	0.030
Fiber wall thickness, L (cm)	0.0125	0.025	0.003	0.003	0.003
Effective length of fiber, l (cm)	40	42	16	25	62
Diameter of the shell (cm)	5	5.5	5	8.5	8.8
Fiber packing fraction, φ	0.225	0.387	0.45	0.45	0.45
Feed velocity, v (cm/s)	1.087	0.82	0.74	1.23	1.16
No of fibers, n_f	250	520	10000	30000	31800
Porosity, ε	0.7		0.4		
Tortuosity, τ	1.5		2		
Diffusion coefficient in aqueous phase, D_f (cm²/s)	1.15×10^{-5}				
Diffusion coefficient in bulk organic phase, D_o (cm²/s)	1.97×10^{-6}				
Aqueous phase mass transfer coefficient, k_{f1} (cm/s)	9.93×10^{-4}	9.58×10^{-4}	2.05×10^{-3}	2.10×10^{-3}	7.05×10^{-3}
Average velocity shell side, v_s (cm/s)	0.219	0.229	0.38	0.40	0.55
Equivalent diameter shell side, d_e (cm)	0.517	0.24	0.036	0.058	0.039
Re_{shell} $(=d_ev_s\rho/\mu)$	8.47	4.07	1.03	1.735	1.65
Sc_{shell} $(=\mu/\rho D_o)$	6794		1163		
Organic phase mass transfer coefficient shell side, k_{o1} (cm/s)	4.88×10^{-4}	8.29×10^{-4}	3.50×10^{-3}	2.59×10^{-3}	3.68×10^{-3}

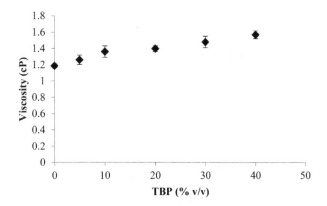

FIGURE 9.7 Viscosity of organic phase for varying concentration of TBP dissolved in dodecane.

Mass transfer coefficient of diffusion film at the organic membrane interface, k_{o1} is thus estimated for PS and PP contactors and presented in Table 9.2.

9.2.4 TRANSPORT OF URANIUM IN ONCE THROUGH MODE OPERATION OF HOLLOW FIBER CONTACTOR

Polysulfone contactors (PS1 and PS2) have been employed for the recovery of uranium from nitrate medium through the HFDLM process using 30% v/v TBP in dodecane.

Using PS1 and PS2 contactors, a percent extraction of 28% and 40% has been achieved, respectively, in a single pass of the feed solution as indicated in Figure 9.8.

Similar studies have been conducted using microporous contactors with polypropylene lumen (PP1, PP2, and PP3) and a percent extraction of 55%, 75%, and 95% has been observed, respectively. The results are shown in Figure 9.9. The details of the contactors are presented in Table 9.2. The hollow fiber contactor is a promising device for application in the recovery of values like uranium from dilute streams using DLM configuration. This separation process has been applied for treatment of lean acidic streams such as raffinate from uranium refining plant [19].

A mathematical model has been developed [19], which can be used for the prediction of the extraction rate of other ions and other types of polymeric membrane. The model is based on continuity equations along the tube and shell side and Fick's first law of diffusion. Assumptions involved in the development of the model are fast reaction, steady state condition, and huge stripping phase interfacial area. The data needed for the design are, the carrier mediated distribution coefficient of the ion in the organic phase, diffusion coefficient of the ion in the aqueous phase, and diffusion coefficient of the ion-carrier complex in the organic phase. The model developed for U-HNO₃-TBP system is a versatile one and is applicable to any solute-solvent system with similar complexation chemistry.

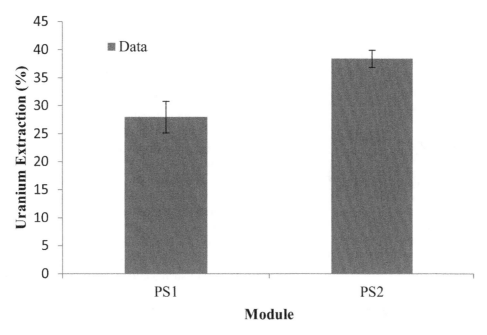

FIGURE 9.8 Extraction of uranium from nitrate medium using polysulfone (PS) contactors in once-through mode; organic: 30 % v/v TBP dissolved in dodecane, aqueous strip: 1 M NaHCO₃, W/O ratio in dispersion is 1:1 v/v; agitation rpm 3000.

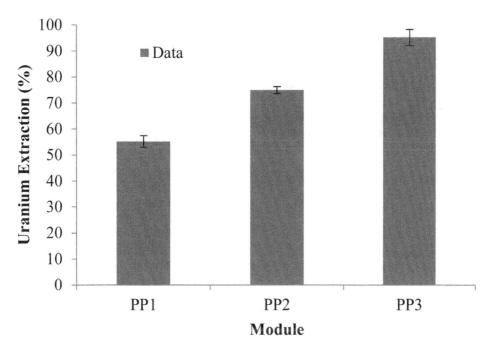

FIGURE 9.9 Extraction of uranium from nitrate medium using polypropylene (PP) contactors in once-through mode; organic: 30 % v/v TBP dissolved in dodecane, aqueous strip: 1 M NaHCO₃, W/O ratio in dispersion is 1:1 v/v; agitation rpm 3000.

9.3 APPLICATION OF MULTIPLE HOLLOW FIBER CONTACTORS FOR HIGHER EFFICIENCY

In order to enhance the purity of streams, hollow fiber contactors are added in series. To understand the system better, firstly, Figure 9.10 shows the set up with single hollow fiber contactor, PP1. The set up consists of a hollow fiber module,

a feed and dispersion reservoir, a feed and dispersion pump, and an agitator for dispersing the strippant into the organic phase. By series multiplication of contactors, lumen length or interfacial area for extraction increases, and thus transport increases. It is observed that in the array of contactors PP1 shown in Figure 9.11, a percent extraction of uranium in raffinate at the end of first, second, and third contactor is 50%, 81%, and 92%, respectively.

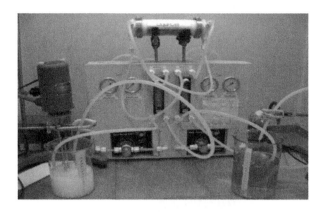

FIGURE 9.10 Single hollow fiber contactor set up with PP1 operating in recirculation mode; Flow rate: 12 LPH; feed: Uranium raffinate; dispersion: aq/org 1:1, aq (1 M NaHCO₃), org (30 % v/v TBP in Dodecane).

FIGURE 9.11 Experimental set up of three HF contactors (PP1) in series operating in once-through mode; flow rate: 12 LPH; feed: Uranium raffinate; dispersion: aq/org 1:1, aq (1 M NaHCO₃), org (30 % v/v TBP in Dodecane).

9.3.1 Scale Up: Application of HFDLM Using Multiple Large Contactors in Series

In order to achieve a higher extent of extraction with a single pass flow, it is important to understand the role of lumen length in the transport of the solute. The mathematical model developed for the same system earlier [19] was used to predict the extent of transport of uranium using HFDLM in smaller contactors PP1 (refer Table 9.2 for details of contactors). To validate the findings, experiments were conducted with uranium raffinate feed having 1,000 ppm of uranium in 1M nitric acid to find the extent of extraction of uranium as a function of the length of the module. The strip and organic phase was pumped into a 5 liter glass vessel and agitated at 4,000 rpm using a high speed agitator. The dispersion was pumped from this vessel to the hollow fiber cluster in continuous counter current mode with the pre-filtered feed solution passing through the lumen side. The aqueous samples were collected from the exit ports and

analysed for uranium content. It was found that with four modules in series more than 95% uranium was recovered in the product stream.

Figure 9.12 shows the experimental results as well as the model predictions for the transport of uranium along the lumen length for the array of four HFC's PP2 combined in series and operated in single pass mode. Mass transfer coefficients evaluated from data generated with PP1 are used here for simulation. For each contactor PP2, effective fiber length is 25 cm and feed flow velocity is 1.23 cm/s. Flux was 9.24e-03 mol/m²/hr. Efficiency of module was calculated as 89%. It is found that placing four PP2 in series will lead to more than 98% transport of uranium from feed to strip in single pass mode as per model prediction. It can be seen that the prediction ties in well with the data. Utilizing the model developed, transport performance of such a contactor of any size can be estimated at its effective fiber length for a given carrier concentration.

Figure 9.12 shows that with the addition of modules in series, the uptake of Uranium from feed solution increases, and with four modules in series, the uptake is almost complete. This prediction has been proven to be true in the bench scale trials. Fine adjustment of flow rate and pressure in shell and tube side ensured no trans-membrane convective flow through the pores and thus avoided contamination of the shell and lumen side phases. The speed of agitation during the formation of the strip-organic solvent dispersion is optimised such that the dispersion is stable throughout the four contactors. This is important for achieving adequate efficiency of separation.

Based on series of the lab scale experimental findings discussed above with application of TBP based dispersion liquid membrane, a bench scale experimental unit has been set up, with four hollow fiber modules PP2 in series. This set up, shown in Figure 9.13, has been used for the demonstration of the recovery of uranium from lean acidic raffinate stream of uranium purification plant (termed as "feed" hereafter) having about 700–1,000 ppm of uranium in 1M nitric acid medium. Stability of flow through the module, stability of dispersion, and efficiency of uranium transport have been observed.

The preliminary bench scale experiment with a 7 litre batch volume of the feed at 60 L/h throughput gave 99.85% extraction of uranium, bringing down the final HF raffinate concentration to 1 ppm of uranium. The dispersion quality is satisfactory even after the fourth module. Matching of the volume of feed phase going in and coming out of the contactors showed that the organic membrane in pores is stable and there is no phase mixing. These results provided the motivation for the bench scale trial with a larger batch volume.

After the success of this preliminary run, the bench scale set up was operated with a 200 litre batch size of filtered UMP raffinate with 60 L/h throughput using dispersion liquid membrane configuration in once through mode for both feed and dispersion. The process conditions were, O/A of dispersion: 1, Organic: 30% TBP in Dodecane, Strippant:

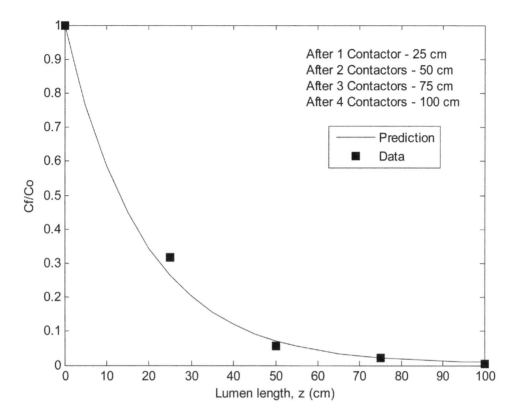

FIGURE 9.12 Transport of uranium with lumen length using four hollow fiber contactor, PP2, in series: feed: Uranium raffinate; dispersion: aq/org 1:1, aq (1 M NaHCO$_3$), org (30 % v/v TBP in dodecane); flow rate: 1000 ml/min; C$_f$ is concentration of Uranium in raffinate after one pass; C$_0$ is initial concentration of Uranium in feed in ppm.

FIGURE 9.13 Bench scale set up: Four PP2 HF modules in series.

1 M sodium bi carbonate solution, Dispersion speed: 4,000 RPM, Flow Rate: 60 L/h for both the feed (lumen side) and the strip in organic dispersion (shell side). The organic phase impregnated in the pores of the lumen acted as the liquid membrane for diffusive transport of uranium and nitric acid from the feed to the strip phase. During operation, transmembrane pressure has been maintained at 1–1.5 psig. Figure 9.14 shows the exit concentration of uranium in the raffinate phase during single pass flow through each of the four modules starting with a feed of 700 ppm of uranium.

The strip phase gets loaded with twice the concentration of uranium in the feed stream with single pass counter current mode of operation. The aqueous raffinate from the hollow fiber array showed a decrease in uranium concentration from 0.7 gpl to 0.03 gpl. The concentration of loaded strip phase is found as 2 gpl. The concentration of uranium in the product phase may be increased with alkali make up and recirculation of the strip phase in semi batch mode. The nitric acid concentration of the raffinate and the strip (product) phase alkalinity is found to be 0.9 M and 0.7 M, respectively. The uranium concentration in the product phase is increased up to 5.5 gpl by a recycle of the product for fresh stripping make up of alkalinity with addition of measured amount of sodium bicarbonate in the dispersion tank. The rise in concentration of the strip product with every round of recycle is shown in Figure 9.15. Figure 9.16 presents the flow-sheet developed for the present process.

The salient observations include:

i) Efficient transport of uranium is achieved from the feed to the strip product phase at 60 L/h throughput using the HFDLM system.

ii) The concentration of uranium in the raffinate can be reduced through use of longer lumen.

iii) With four hollow fiber modules in series, the dispersion is stable and no trans-membrane phase

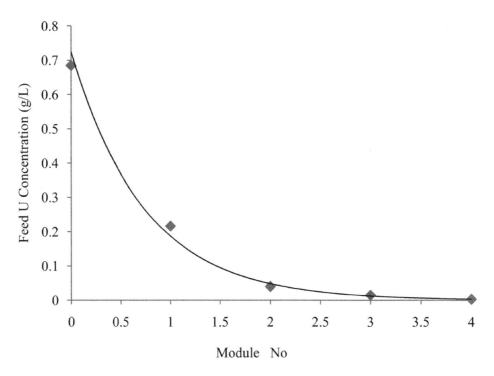

FIGURE 9.14 Concentration of Uranium in raffinate at module exit. feed: Uranium raffinate; dispersion: aq/org 1:1, aq (1 M NaHCO$_3$), org (30 % v/v TBP in Dodecane); flow rate: 60 LPH; four PP2 modules in series.

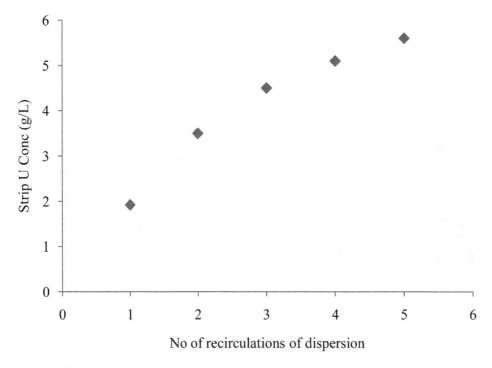

FIGURE 9.15 Concentration of Uranium in product after alkali make up and recycles feed: Uranium raffinate; dispersion: aq/org 1:1, aq (1 M NaHCO$_3$), org (30 % v/v TBP in Dodecane); flow rate: 60 LPH; four PP2 HF modules in series with strip re-circulation.

mixing is observed in general except a small carry over of organic phase in the raffinate due to the pressure fluctuation at the start up.

iv) Phase separation of loaded dispersion is fast and efficient.

v) Nitric acid gets co-transported along with uranium from feed to strip phase, which is under control for the present set up and scale. This is an important aspect to be observed and considered for larger scale operations.

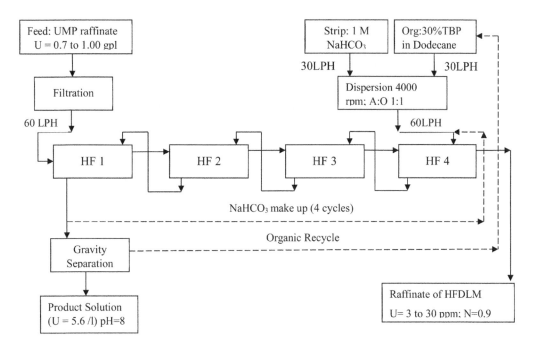

FIGURE 9.16 Flow sheet for recovery of Uranium from Uranium plant raffinate using HFDLM technique.

9.4 CONCLUSION

Hollow fiber contactors are promising pieces of equipment for the application of liquid membrane techniques in continuous countercurrent mode. The location of the organic-aqueous interface in the micro pores is very important for efficient transport of the solute through the membrane phase. A simple mathematical model has been developed for the prediction of the location of the interface in the pores, rate of extraction, etc. The model validated on a small scale has been successfully used for scale up of the process for higher throughput in bigger contactors employing dispersion liquid membranes for the polishing of uranium bearing lean streams.

Multiple modules can be employed in series with countercurrent flow of the organic and aqueous phases based on the required extent of removal of the solute. Maintaining the quality of dispersion is important for avoiding interstage dispersification and pumping. A fine adjustment of the flow rate and cross membrane pressure is needed to avoid reduction in efficiency due to phase mixing across the pores. The co-transport of nitric acid is an important aspect which needs attention during operations at the higher scale.

REFERENCES

1. Kentish, S.E., Stevens, G.W. 2001. Innovations in separations technology for the recycling and re-use of liquid waste streams. *Chem. Eng. J.* 84:149–159.
2. Hayworth, H.C., Ho, W.S., Burns, W.A., Li, N.N. 1983. Extraction of uranium from wet process phosphoric acid by liquid membranes. *Sep. Sci. Technol.* 18:493–521.
3. Yang, X.J., Fane, A.G., Soldenhoff, K. 2003. Comparison of liquid membrane processes for metal separations: Permeability, stability, and selectivity. *Ind. Eng. Chem. Res.* 42:392–403.
4. Sengupta, A., Basu, R., Sirkar, K.K. 1988. Separation of solutes from aqueous solutions by contained liquid membranes. *AIChE J.* 34(10):1698.
5. Xu, T. 2005. Ion exchange membranes: State of their development and perspective. *J. Membr. Sci.* 263:1.
6. Gyves, J., Miguel, E.R.S. 1999. Metal ion separations by supported liquid membranes. *Ind. Eng. Chem. Res.* 38: 2182.
7. Kocherginsky, N.M., Yang, Q., Seelam, L. 2007. Recent advances in supported liquid membrane technology. *Sep. Purif. Technol.* 53:171–177.
8. Danesi, P.R., Reichley-Yinger, L., Rickert, P.G. 1987. Lifetime of supported liquid membranes: The influence of interfacial properties, chemical composition and water transport on the long-term stability of the membranes. *J. Membr. Sci.* 31:117.
9. Hill, C., Dozol, J.-F., Rouquette, H., Eymard, S., Tournois, B. 1996. Study of the stability of some supported liquid membranes. *J. Membr. Sci.* 114:73–80.
10. Zha, F.F., Fane, A.G., Fell, C.J.D. 1995. Effect of surface tension gradients on stability of supported liquid membranes. *J. Membr. Sci.* 107:75–86.
11. Neplenbroek, A.M., Bargeman, D., Smolders, C.A. 1992. Mechanism of supported liquid membrane degradation: Emulsion formation. *J. Membr. Sci.* 67:133.
12. Patnaik, P.R. 1995. Liquid emulsion membranes: Principles, problems and applications. *Biotechnol. Adv.* 13(2):175.
13. Largman, T., Sifniades, S. 1978. Recovery of copper(II) from aqueous solutions by means of supported liquid membranes. *Hydrometallurgy* 3:153.
14. Ho, W.S.W., Poddar, T.K. 2001. New Membrane Technology for removal and recovery of chromium for waste waters. *Environ. Prog.* 20:44.
15. Gabelman, A., Hwang, S.-T. 1999. Hollow-fiber membrane contactors. *J. Membr. Sci.* 159:61–106.
16. Pabby, A.K., Sastre, A.M. 2013. State-of-the-art review on hollow fiber contactor technology and membrane based extraction processes. *J. Membr. Sci.* 430:263–303.

17. Dixit, S., Chinchale, R., Govalkar, S., Mukhopadhyay, S., Shenoy, K.T., Rao, H., Ghosh, S.K. 2013. A mathematical model for size and number scale up of hollow fiber modules for the recovery of uranium from acidic nuclear waste using the DLM technique. *Sep. Sci. Technol.* 48:2444–2453.

18. Dixit, S., Mukhopadhyay, S., Govalkar, S., Shenoy, K.T., Rao, H., Ghosh S.K. 2012. A mathematical model for pertraction of uranium in hollow fiber contactor using TBP. *Desalin. Water Treat.* 38:195–206.

19. Dixit, S., Mukhopadhyay, S., Shenoy, K.T., Juvekar, V.A. 2017. A process intensified technique of liquid membrane employed in in-house hollow fibre contactor. *Desalin. Water Treat.* 79:40–48.

20. Mukhopadhyay, S., Ghosh, S.K., Juvekar, V.A. 2008. Mathematical model for swelling in a liquid emulsion membrane system. *Desalination* 232:110–127.

21. Geankoplis, C.J. 2002. *Transport Processes and Unit Operations.* Prentice Hall of India, Third edition, 400–412.

22. Davis, W., Mrochek, Jr., J., Judkins, R.R. 1970. Thermodynamics of the two-phase system: Water-uranyl nitrate- tributyl phosphate-Amsco 125–82. *J. Inorg. Nucl. Chem.* 32:1689.

23. Rathore, N.S., Sonawane, J.V., Kumar, A., Venugopalan, A.K., Singh, R.K., Bajpai, D.D., Shukla, J.P. 2001. Hollow fibre supported liquid membrane: A novel technique for separation and recovery of plutonium from aqueous acidic wastes. *J. Membr. Sci.* 189:119.

24. Cussler, E.L. 1984. *Diffusion – Mass Transfer in Fluid Systems.* Cambridge University Press: New York, NY.

25. Al-Enezi, G.A., Abdo, M.S.E. 1991. Mass transfer measurements around a single cylinder in cross flow at low reynolds numbers. *Chem. Ing. Tech.* 63:381–384.

26. Prasad, R., Khare, S., Sengupta, A., Sirkar, K.K. 1990. Novel liquid-in-pore configurations in membrane solvent extraction. *AIChE J.* 36(10):1592–1596.

10 Advances in Membrane Emulsification and Membrane Nanoprecipitation Using Membrane Contactors
State-of-the-Art and Perspectives

E. Piacentini, F. Bazzarelli, E. Drioli, and L. Giorno

CONTENTS

10.1 INTRODUCTION

Membrane emulsification is a type of membrane contactor-based process where a porous membrane controls the contact, the interaction, and the distribution of an immiscible fluid into another. Usually liquid/liquid and gas/liquid phases are used as fluids, with the first type much more common than the latter one. The phases may remain liquid or solidify after dispersion. In the first case, the system is composed of liquid droplets dispersed into a continuous liquid phase, i.e., an emulsion is obtained. In the latter case, the system is composed of solid (or jelly) spheres distributed into a liquid phase, i.e., a suspension is obtained. Post-treatments may apply to both systems ending up in dry spheres. Depending on the size of the obtained spheres, the systems are termed as micro- or nano-emulsion, suspension particles. Different institutions refer to different definitions based on particles size. According to IUPAC classification, particles with at least one dimension in the range 0.1–100 μm and 0.001–0.1 μm, are defined as microparticles and

nanoparticles respectively [1]. Likewise, the European Commission describes a nanomaterial as a synthetic or natural material with external dimensions between 1 and 100 nm size range. The International Organization for Standardization refers to a nanomaterial as a material with an external nanoscale dimension or with an internal nanoscale structure. The British Standards Institution refers to nanoscale as approximately 1–1,000 nm range.

Membrane nanoprecipitation is a new type of membrane contactor where a porous membrane controls the interaction between two miscible phases to implement the formation of nanoscale particles (i.e., spheres or capsules) by nanoprecipitation. Thanks to the fluid dynamics conditions, such as dispersed phase flux and continuous phase axial velocity, it is possible to tune the micromixing of the two phases and govern the nanoprecipitation process for the production of uniform nanoparticles.

In this chapter both membrane emulsification and membrane nanoprecipitation will be discussed, including their fundamentals, applications, and future perspectives.

10.2 MEMBRANE EMULSIFICATION

In membrane emulsification (ME), porous membranes are used as tools for liquid-liquid inter-phase mass transfer to achieve the dispersion of one phase (dispersed phase) within the other (continuous phase).

Different methods have been developed depending upon the properties of the membrane surface and of the two phases used (Figure 10.1).

In *direct membrane emulsification*, membranes unwetted by the dispersed phase are required in order to promote the formation of liquid droplets at the interface between the contacting immiscible phases at the pore opening (Figure 10.1A). The emulsion is formed by a drop-by-drop mechanism on the surface of the membrane controlled by the pore size according to a factor of 2 to 10 between pore size and droplet size. Multiple emulsions are obtained when the dispersed phase is a simple emulsion, while solidified particles can be obtained by a dispersion of a preformed polymer or a polymerization of dispersed monomers by combining the emulsification step with the appropriate secondary reaction. A limitation of this process is the low dispersed phase flux because the membrane is unwetted by the dispersed phase, and to increase the fraction of dispersed phase the recirculation of the emulsion is often required.

In *premix membrane emulsification*, membranes unwetted or wetted by the dispersed phase could be alternatively used to produce a uniform emulsion starting from a pre-emulsion obtained from two immiscible liquids (Figure 10.1B). The emulsion is broken up into smaller droplets when the pre-emulsion is passed through a microporous membrane unwetted by the dispersed phase of the pre-emulsion. If the membrane surface is wetted by the dispersed phase of the pre-emulsion, phase inversion can take place during the permeation through the membrane. In particular, an oil-in-water (O/W) emulsion is obtained if a pre-emulsified water-in-oil (W/O) emulsion is permeated through a hydrophilic membrane, and in a short time a formulation with high O/W ratio is produced starting from a low concentrated pre-emulsified W/O emulsion. The pre-emulsion is repeatedly passed through the same membrane in order to achieve additional droplet size reduction and to improve droplet size uniformity. The emulsion is formed inside of the membrane pore rather than on the surface of the membrane, and the ratio between droplet and pore size is decreased with increasing the pore size and the number of passes. In contrast to direct membrane emulsification, premix emulsification can be used to produce emulsions with a high-dispersed phase flux and high dispersed phase–continuous phase fraction because the membrane is wetted by the continuous phase of the pre-emulsion. A limitation of this process is membrane fouling due to the interaction of the formulation components with the membrane.

A pressure gradient driving force is responsible for the permeation of the dispersed phase through the membrane pores while a shear is usually generated at the membrane surface in contact with the continuous phase to promote droplets detachment. According to the way in which the surface shear is generated, membrane emulsification can operate in cross-flow/stirred/pulsed mode or by using rotating/vibrating/oscillating membranes (Figure 10.2).

When comparing various emulsification methods, the energy consumption of premix-membrane emulsification is usually lower than that for direct membrane emulsification since no high wall shear is needed.

10.2.1 MEMBRANE EMULSIFICATION FUNDAMENTALS

The conventional direct membrane emulsification (ME) is based on a drop-by-drop mechanism in which one liquid (a dispersed phase) is forced through the microporous membrane into another immiscible liquid (the continuous phase) to form droplets. These grow at pore openings until a certain dimension and then detach from the pore border.

The production of droplets is governed by the balance between the detaching and retaining forces that act on the membrane pore level [2, 3]. Figure 10.3 illustrates these forces:

- The static pressure force (F_{SP}), which is due to the pressure difference between the dispersed phase and the continuous phase at the membrane surface;
- The drag force (F_D), which is generated by the continuous phase flowing tangentially to the membrane surface;

A) DIRECT MEMBRANE EMULSIFICATION

B) PREMIX MEMBRANE EMULSIFICATION

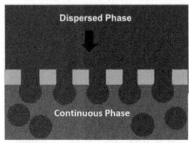

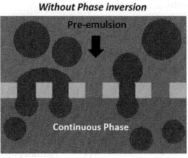
Without Phase inversion

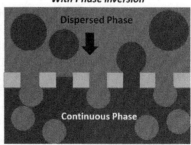
With Phase inversion

FIGURE 10.1 Emulsification based on membrane operations: A) direct membrane emulsification and B) premix membrane emulsification.

SURFACE SHEAR GENERATED BY MOVING THE CONTINUOUS PHASE

CROSS-FLOW

PULSED-FLOW

STIRRED

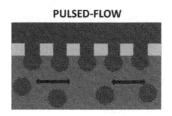 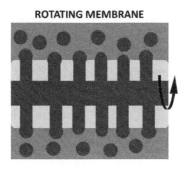 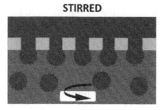

SURFACE SHEAR GENERATED BY MOVING THE MEMBRANE

VIBRATING MEMBRANE

ROTATING MEMBRANE

AZIMUTALLY OSCILLATING MEMBRANE

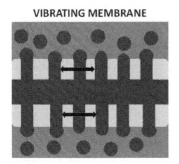 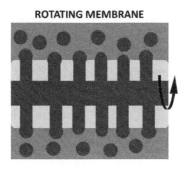 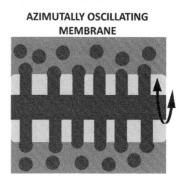

FIGURE 10.2 Membrane emulsification mode of operations.

- The dynamic lift force (F_L), which results from the asymmetric velocity profile of the continuous phase near the droplet;
- The buoyancy force (F_B), which is due to the density difference between the continuous phase and the dispersed phase;

- The inertial force (F_I), which is caused by the dispersed phase flow moving out from the pore opening.

Regarding retaining force, this is mainly represented by interfacial tension force (F_Υ), which is provided by the effects of dispersed phase adhesion around the pore edge.

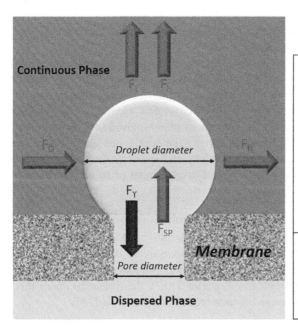

DETACHING FORCES	Static pressure force (F_{SP})	$F_{SP} = \dfrac{\Upsilon}{d_d} \pi d_p^2$
	Drag force (F_D)	$F_D = \dfrac{3}{2} k_x \pi \tau_{c,s} d_d^2$
	Dynamic lift force (F_L)	$F_L = 0.761 \dfrac{\tau_{c,s}^{1.5} \rho_c^{0.5}}{\mu_c} d_d^3$
	Buoyancy force (F_B)	$F_B = \dfrac{1}{6} \pi g \Delta \rho d_d^3$
	Inertial force (F_I)	$F_I = \rho_d \left(\dfrac{J_d}{\varepsilon}\right) A_N$
RETAINING FORCES	Interfacial tension force (F_Υ)	$F_\Upsilon = \pi d_p \Upsilon$

FIGURE 10.3 The detaching and retaining forces acting on a droplet at the membrane pore level (d_d: droplet diameter; d_p: membrane pore diameter; γ: dynamic interfacial tension; K_x: equal to 1.7 and takes into account the wall correction factor for a single sphere touching an impermeable wall; $\tau_{c,s}$: wall shear stress; ρ_c: density of the continuous phase; ρ_d: density of dispersed phase; μ_c: viscosity of the continuous phase; g: acceleration due to gravity; $\Delta\rho$: difference between continuous and dispersed phase density; J_d: dispersed phase flow; A_N: cross-sectional area of the droplet neck; ε: membrane porosity).

In membrane emulsification where the shear is generated by the movement of the membrane, the droplet detachment from the pore outlets is promoted by rotation or vibration of the membrane within the stationary continuous phase.

For rotating membrane emulsification, the droplets' detachment is due to an additional force such as centrifugal force and the tangential velocity differences between the droplet and the continuous phase. In the case of the vibrating membrane, an additional drag force and an inertial force are provided by transversal membrane excitation.

Equations for the generation of shear stress in different membrane emulsification configurations are reported as follows [4–6]:

- In cross-flow membrane emulsification, the shear stress (τ) is constant along the membrane surface and is correlated with the tangential velocity of the flowing continuous phase (ν), the density of the continuous phase (ρ) and the friction factor (λ). The shear stress is given by:

$$\tau = \frac{\lambda \rho \nu^2}{2} \quad (10.1)$$

- In stirred membrane emulsification, the shear stress is not constant at the membrane surface because it varies with the radial distance (r) from the stirrer axis. The shear stress (τ) depends on the angular velocity (ω) of the stirrer and can be estimated as follows:

$$\tau = 0.825\mu\omega r \frac{1}{\delta} \text{ For } r < r_{trans} \quad (10.2)$$

$$\tau = 0.825\mu\omega r_{trans} \left(\frac{r_{trans}}{r}\right)^{0.6} \frac{1}{\delta} \text{ For } r > r_{trans} \quad (10.3)$$

where μ is the continuous phase viscosity, r_{trans} is the transitional radius and δ is the boundary layer thickness given from the Landau–Lifshitz equation:

$$\delta = \sqrt{\frac{\mu}{\omega\rho}} \quad (10.4)$$

where ω is the angular velocity, μ and ρ are the viscosity and the density of the continuous phase, respectively.

- In rotating membrane emulsification, the shear stress (τ) is a function of the membrane rotational rate (n) and the width of the annular gap between the rotating membrane and the stationary vessel. The shear at the surface of the membrane is given by:

$$\tau = \frac{\pi R_1^2 n}{15\left(R_2^2 - R_1^2\right)} \quad (10.5)$$

where R_1 is the outer radius of the membrane and R_2 is the inner radius of the vessel.

- In pulsed, vibrating and azimuthally oscillating membrane emulsification, the maximum shear stress (τ_{max}) is a function of the frequency and amplitude of oscillation of the continuous phase or the membrane according to the following equation:

$$\tau_{max} = 2a\left(\pi f\right)^{\frac{3}{2}} \left(\mu\rho\right)^{\frac{1}{2}} \quad (10.6)$$

Where μ is the viscosity and ρ is the density of the continuous phase, a and f are the amplitude and frequency, respectively.

The particle diameter and particle size distribution are affected by different parameters, which comprise: process parameters (wall shear stress and trans-membrane pressure) membrane characteristics (surface wettability, porosity, pore size distribution, and distance between the pores) and phase parameters (interfacial tension, viscosity, and density of the dispersed and continuous phase) [7–9]. These parameters will be described in detail in the following paragraphs.

10.2.2 Process Parameters

10.2.2.1 The Wall Shear Stress

The drop diameter decreases with increasing wall shear stress until reaching a constant particle size for further increase of shear stress. The fast detachment of the droplets at the membrane pore level reduces coalescence phenomena of the droplets. The action of the wall shear stress on decreasing droplet size is more effective for smaller pore size.

Depending on the membrane emulsification configuration, the shear stress may be provided at the membrane surface by: continuous phase flowing tangentially to the membrane surface (cross-flow membrane emulsification); pulsing the continuous phase (pulsed-flow membrane emulsification), or stirring the continuous phase (stirred membrane emulsification); the speed of membrane rotation or the frequency and displacement of membrane oscillation/vibration.

10.2.2.2 Trans-Membrane Pressure

The trans-membrane pressure (TMP) is defined as the difference between the pressure of the dispersed phase and the mean pressure of the continuous phase. According to Darcy's law Equation (10.7), the dispersed phase flux (J_d) through the membrane increases with TMP.

$$J_d = K \frac{\Delta P}{L} \quad (10.7)$$

where ΔP is pressure difference, L is the membrane thickness, K is the permeability coefficient. This can be expressed in terms of solution viscosity, membrane pore size, and porosity as given by the Hagen–Poiseuille equation:

$$K = \frac{\varepsilon r^2}{8\mu\tau}$$ (10.8)

Where ε is the membrane porosity; r is the pore radius, μ is the viscosity, and τ is the tortuosity factor.

Over a certain range, the high flux may lead to droplet coalescence at the membrane surface, as it may cause a jet-like flow of the dispersion phase thus resulting in large droplet size and wide size distribution. Therefore, the membrane emulsification process must be carried out in the dripping regime. The trans-membrane pressure applied must be higher than the critical pressure in order to ensure that the dispersed phase starts flowing through the pores. The critical pressure depends on the membrane pore diameters and oil-water interfacial tension and can be evaluated with the Young–Laplace equation [9]:

$$P_c = \frac{4\gamma\cos\theta}{d_p}$$ (10.9)

Where P_c is the critical pressure, γ is the O/W interfacial tension, θ is the contact angle between the oil droplet and the membrane surface, and d_p is the average pore diameter.

10.2.3 Membrane Properties

10.2.3.1 Surface Wettability

The membrane surface should not be wetted by the dispersed phase, especially at the pore edge where the droplet grows; membranes unwetted by the dispersed phase permit the growth of droplets with their size being controlled by the membrane pore size. Therefore, a hydrophilic membrane is generally required for O/W emulsion and a hydrophobic membrane for W/O emulsion to avoid wetting of the membrane surface. Recently, the use of membranes with asymmetric wettability (i.e., membranes having high interfacial tension with the dispersing phase only at surface and pore edge where the droplets form and grow, whilst being wetted through its thickness) was proven a suitable strategy to improve membrane emulsification performance [10].

10.2.3.2 Porosity, Pore Size Distribution, and Distance between Pores

The droplet diameter can be predicted on the basis of the pore size of the membrane, because, in general a linear relationship between the average droplet diameter and the average membrane pore diameter is observed. Pore size and uniformity of pore size distribution reflect the droplet size and uniformity of droplet size distribution. Membranes specifically designed for membrane emulsification are therefore needed. Membrane porosity is another parameter that affects the production of uniform particle size.

The porosity of the membrane determines the distance between two adjacent pores. This should be larger than the droplet diameter. Controlled pore distance and low membrane porosity are permitted to prevent droplet coalescence at the membrane surface by preventing contact between two neighboring droplets.

10.2.4 Phase Parameters

10.2.4.1 Interfacial Tension, Viscosity, and Density of the Dispersed and Continuous Phase

The formation of droplets between two immiscible phases increases the area at the interface between the two phases, which also increases the free energy. The higher the interfacial tension the more difficult it is to form stable emulsions. Emulsifiers are amphiphilic molecules with hydrophilic and lipophilic properties and can be adsorbed at the interface of immiscible phases. Therefore, during droplet formation, emulsifier molecules can be used to lower the interfacial tension between oil and water as well as to increase the repulsion between the droplets. Furthermore, an emulsifier facilitates droplet detachment from the membrane surface. The faster the emulsifier molecules are adsorbed at the newly formed interface the smaller are the obtained droplets. Emulsifiers present in the continuous phase enhance membrane wetting by the continuous phase, which facilitates droplet formation; as well as they stabilize the droplets against creaming, flocculation, coalescence phenomena.

The viscosity of the dispersed and continuous phase is another key parameter for the formation of uniform droplets by membrane emulsification. The droplet diameter is constant at high dispersed-phase to continuous-phase viscosity ratio, while it increases at lower ratio [11]. The viscosity of the dispersed phase also affects membrane emulsification productivity. In fact, according to Darcy's and Hagen–Poiseuille's laws (**Equations 10.7 and 10.8**), the flux is inversely proportional to the permeating phase viscosity.

Besides small droplet radius, highly viscous continuous phase and low density difference between oil and water phases also contrast the emulsion destabilization such as creaming rate, as given by Stokes' equation:

$$v = \frac{2gr^2(\rho_d - \rho_c)}{9\mu}$$ (10.10)

Where v is the creaming rate, r is the droplet radius, ρ_d is the density of dispersed phase, ρ_c is the density of the continuous phase, μ is the viscosity of the continuous phase, and g is the acceleration due to gravity.

10.2.5 Membranes for Membrane Emulsification

Depending on the type of formulation, a range of membranes has been fabricated to make particles with precisely controlled size and size distributions. Membrane emulsification can be carried out using microporous membranes made with different materials (organic and inorganic materials) and configurations (flat, tubular, and hollow fibers).

The mechanical, thermal, and chemical properties of membranes needed for membrane emulsification are dictated by the type of materials forming the emulsion. The main methods used for fabrication of porous membranes are sintering of powders (ceramic, metallic, and polymeric materials), sol-gel process (ceramic materials), and phase inversion (polymers) [12]. The most commonly used membranes for membrane emulsification are inorganic membranes such as Shirasu Porous Glass (SPG) membranes and microsieve membranes. SPG membranes were obtained from Na_2O–CaO–MgO–Al_2O_3–B_2O_3–SiO_2 mother glass through phase separation by spinodal decomposition. The mother glass is prepared by mixing and melting Shirasu (Japanese volcanic ash), calcium carbonate, and boric acid [13].

After the mother glass is formed into tubes, phase separation of the homogeneous glass melt is induced by thermal treatment (650–750°C). The acid-soluble Na_2O–CaO–MgO–B_2O_3 phase dissolves when the phase-separated glass is immersed into a hydrochloric acid solution resulting in the formation of porous Al_2O_3–SiO_2 skeleton with a porosity of 50 to 60%. SPG membranes have uniform internal structure with no voids or cracks, they are commercially available with a wide range of mean pore sizes (0.050–20 μm) and have a relatively high hydraulic resistance due to the high wall thickness of 700–900 μm. Microsieves are increasingly used for membrane emulsification. Typical microsieves are silicon nitride, nickel, aluminum, and stainless steel. They are ultra-thin membranes with controlled pore geometry (i.e., slotted pores and micro nozzles) and precise spatial arrangement manufactured by semiconductor fabrication method (reactive ion etching, UV-LIGA-lithography, electroplating, molding, and pulsed laser drilling). Other membranes used for membrane emulsification include ceramic membranes (fabricated by sintering of fine inorganic oxide powders) and polymeric membranes (mainly obtained by phase-inversion).

In this section, membranes used for particles production by direct and pre-mix membrane emulsification will be highlighted.

10.2.5.1 Membranes Used in Direct Membrane Emulsification and Related Formulations

Membranes used for the preparation of emulsions should be characterized by uniform pore size distribution, low hydraulic resistance, mechanical, thermal, and chemical resistance. Membrane emulsification can be integrated with physicochemical or chemical processes in order to transform emulsion droplets into solid particles. Membranes used in direct membrane emulsification are listed in Table 10.1 The most widely used membranes are Shirasu Porous Glass (SPG) membranes and microsieve membranes, which are among the membranes specifically designed for membrane emulsification processes. Other inorganic membranes include ceramic membranes. Organic membranes made with hydrophilic or hydrophobic polymers were also investigated for direct membrane emulsification. They share the same trend known for other

membrane processes, i.e., the inorganic membranes are chemically inert and possess ease of cleaning after fouling compared to organic materials, as for other membrane processes; conversely, polymeric membranes are easily fabricated at low cost. Some of them were adapted from membranes initially designed for other processes, such as micro- or ultra-filtration. In these cases, emulsions with larger droplet size distribution are obtained.

The surface wettability of SPG membrane can undergo modifications for the production of different type of formulations. Usually, it is modified by chemical reaction with organosilane compounds, such as chlorosilanes or physical coating with silicone resin for hydrophobized surfaces.

In the following, some examples of formulations using various membranes are illustrated.

Lai et al. [14] used SPG membrane with a pore size of 3.0 μm to formulate an artificial oxygen carrier composed of bovine hemoglobin and bovine serum albumin. The membrane was treated with silicone resin to render its surface hydrophobic. This permitted the production of W/O emulsion using a cross-flow direct membrane emulsification process. The subsequent cross-linking of the emulsions by glutaraldehyde formed stable microspheres (droplets diameter between 3.9 and 4.9 μm) with narrow size distribution and coefficient of variation (CV, ratio of the standard deviation to the average particles size) around 10%.

The performance of hydrophilic SPG membrane (pore size of 1 and 0.1 μm) was investigated for the preparation of polymeric particles by using pulsed back-and-forward cross-flow membrane emulsification in combination with solvent diffusion process [15]. The membrane emulsification process permitted the control of particles production based on polycaprolactone with tuned structural properties under mild shear stress conditions. Polymeric particles were prepared via W/O emulsion in a size range between 2.35 μm and 0.20 μm suitable for injectable administration.

Rotating membrane emulsification was studied to produce water-in-oil-in-water (W/O/W) double emulsions using hydrophilic SPG membrane (pore size of 2.8 μm). The potential of cross-flow membrane emulsification for the production of double emulsions was also investigated. For this configuration hydrophilic SPG membrane with a pore size of 3.9 μm was used. In order to produce the W/O/W emulsion, the primary W/O emulsion was produced by high-shear mixer, then it was used as dispersed phase and pressurized through the membrane pores by compressed air. Minimum droplet size of 8.5 μm and 13μm at highest shear forces was obtained for rotating and cross-flow membrane emulsification, respectively [16].

Membrane emulsification was also studied with metallic membranes. These types of membranes tolerate higher dispersed fluxes and are less prone to fouling compared to the SPG and ceramic membranes. Rectilinear pores in a highly regular array characterize flat or tubular metallic membranes. Morelli et al. [17] prepared W/O emulsion using flat nickel membrane by dispersion cell membrane

TABLE 10.1

Representative Characteristics of Membranes and Related Formulations Produced by Direct Membrane Emulsification

Membrane material	Surface Wettability	(D_p) (µm)	Membrane Configuration	Type of Particles	K= (D_d)/ (D_p)	Particle Size Distribution*	Type of Me Process	Reference
	Membrane Characteristics			**Formulation Characteristics**				
SPG	Hydrophobic	0.2	Tubular	Silver nanoparticles	0.05	N/A	Stirred	22
SPG	Hydrophobic (silicone resin treatment)	3	Tubular	Hemoglobin-bovine serum albumin microspheres	1.3–1.6	N/A	Stirred	14
SPG	Hydrophobic	3.1	Tubular	W/O emulsion	2.3	Span 0.33	Pulsed	23
SPG	Hydrophobic (modification with silane coupling agent)	4.50	Flat	Agarose microspheres	2.26	N/A	Stirred	24
SPG	Hydrophilic	0.1	Tubular	PCL particles	8.35	N/A	Pulsed	15
SPG	Hydrophilic	1			2.1 / 2.35	PDI 0.06		
SPG	Hydrophilic	1	Tubular	O/W emulsion	2	Span 1.2	Rotating	25
SPG	Hydrophilic	1.1	Tubular	Microbubbles	12.1	Span 0.53	Cross-flow	26
SPG	Hydrophilic	1.1	Tubular	Solid lipid particles	3.68	Span 0.45	Pulsed	27
SPG	Hydrophilic	2.8 / 3.9	Tubular	W/O/W (primary W/O emulsion produced by high-shear mixer)	3–14 / 3–10	N/A	Rotating / Cross-flow	16
SPG	Hydrophilic	10	Tubular	W/O/W (primary W/O emulsion produced by dispersing tool)	5.98	Span 0.87–0.90	Cross-flow	28
Silica glass	Treated with silicone resin for surface hydrophobic	0.6	Flat	W/O emulsion	5	N/A	Stirred	29
Silicon nitride	Pre-treated to improve the hydrophilicityby oxygen plasma	2.0 (stationary membrane) / 2.5 (piezoactuated membrane)	Flat	O/W emulsion	19 / 43	N/A	Cross-flow	30
Nickel	Hydrophobic coating with poly(tetrafluoroethylene) / Hydrophobic coating with poly(tetrafluoroethylene)	30	Flat	Gelatin/chitosan blend or pure gelatin microparticles	2.07 / 2.73	N/A	Stirred	17
Nickel	Hydrophilic	10 / 20	Tubular	O/W emulsion	3.00 / 2.55	Span 0.40 / Span 0.55	Oscillatory	6
Stainless steel	Hydrophilic	15	Tubular	O/W emulsion	~1	Span1	Rotating	25
Stainless steel	Hydrophilic	100	Tubular	O/W emulsion	1.06	N/A	Rotating	4

(Continued)

TABLE 10.1 (CONTINUED)

Representative Characteristics of Membranes and Related Formulations Produced by Direct Membrane Emulsification

Membrane material	Surface Wettability	(D_p) (µm)	Membrane Configuration	Type of Particles	K= (D_d)/ (D_p)	Particle Size Distribution*	TYPE OF ME PROCESS	REFERENCE
Zirconia	Hydrophilic modified by hydrophobic protein adsorption	0.05	Tubular	W/O emulsion	142	Span 0.89	Cross-flow	10
Zirconia	Hydrophilic modified by cationic surfactant adsorption	0.16	Tubular	W/O emulsion	6.3–12.5	N/A	Stirred	18
Zirconia	Hydrophilic (pre-filled pore with oil)	0.20	Tubular	W/O emulsion	1	PDI 0.1	Cross-flow	31
	Hydrophilic (pre-filled pore with water)				0.5–1	PDI >0.3		
Zirconia α-Alumina	Hydrophilic	0.1 0.2 0.5	Tubular	O/W emulsion	19 18 12	Span 0.89 Span 0.87 Span 1.51	Cross-flow	32
Alumina	Hydrophilic	0.2 0.8	Tubular	O/W emulsion	14.6 5.95	Span 1.59 Span 2.27	Cross-flow	33
Titania	Hydrophilic	1	Tubular	O/W emulsion	2.5	Span 1.3	Rotating	25
Anodic alumina	Hydrophilic	0.058	Flat	O/W emulsion	2.48	N/A	Stirred	34
Polypropylene	Hydrophobic (pores pre-filled with oil)	0.4	Hollow fibers	W/O emulsion	0.65–0.80	Span 1.1–1.6	Cross-flow	19
Poly(tetrafluoroethylene)	Hydrophobic	1.0	Flat	W/O emulsion Hydrogel	3.12 5.73	N/A N/A	Stirred	35
Epoxy-based polymer	Hydrophobic	9.45	Flat	Agarose microspheres	1.75	N/A	Stirred	24
Polyamide	Hydrophilic conditioned with organic solvents	<0.01 (NMWCO of 10 kDa)	Hollow fibers	O/W emulsion	187	N/A	Cross-flow	20
Polyamide	Hydrophilic conditioned with organic solvents	0.038 (NMWCO of 10 kDa) 0.04 (NMWCO of 50 kDa)	Hollow fibers	O/W emulsion	58 70	N/A	Cross-flow	21
Poly(tetrafluoroethylene)	Hydrophilically treated	0.5, 1 and 5	Flat	O/W emulsion	9.85	N/A	Stirred	36
Polycarbonate	Hydrophilic	10	Flat	O/W emulsion	2	N/A	Stirred	37

ME: membrane emulsification; O/W: oil-in-water; W/O: water-in-oil; W/O/W: water-in-oil-in-water; PCL: polycaprolactone; NMWCO: nominal molecular weight cut-off.

* The particle size distribution is reported in terms of the Span and PDI (polydispersity index). The span value is calculated as Span = d90-d10/d50, where d10, d50, and d90 correspond to the equivalent volume diameters at 10%, 50%, and 90% of the cumulative volume.

emulsification device. The hydrophilic nickel membrane was covered with PTFE (polytetrafluoroethylene) or FAS (fluoro alkyl silane) to hydrophobize the surface. Microparticles based on blended chitosan with gelatin or pure gelatin were produced for encapsulation and release of yeast cells in the intestine-colon tract. The membrane emulsification device was integrated with the drop solidification process that consisted of thermal gelation and/or ionic cross-linking. PTFE coated membrane was characterized by a higher degree of hydrophobicity and gave smaller droplets than the FAS coated membrane. This approach was effective for preparing monodispersed particles (CV below 25%) with hydrophobic membranes of 30 μm pore diameter and 200 μm pore spacing. The PTFE coating was less favorable for repeated use of the membrane after cleaning, while a simple coating with FAS allowed the repeated use of the same membrane. Other inorganic membranes such as alumina (Al_2O_3), zirconia (ZrO_2), and titania (TiO_2) ceramic membranes are attractive because of their availability in large-scale modules and high flux. Furthermore, these ceramic membranes are economically advantageous compared to SPG membrane. Piacentini E. et al., 2014 [10], studied a strategy to modify the wettability properties of hydrophilic ceramic membrane by adsorption of hydrophobic proteins such as lipase on the lumen side of the membrane. A cross-flow membrane system was used. Results showed that lipase-modified membranes led to the production of W/O emulsions with 60% smaller size compared to the unmodified hydrophilic membrane. In addition, a significant increase of dispersed phase flux was reached compared to the hydrophobic membrane. In another work, monodispersed W/O emulsions were prepared by hydrophilic ZrO_2 ceramic membrane (nominal pore size of 0.16 μm) by adding a cationic surfactant into the dispersed phase. This strategy permitted the preparation of mono dispersed W/O emulsions with a droplet size of 1 to 2 μm [18]. In addition to inorganic membranes, several authors studied the potential of polymeric membranes in flat and hollow fiber configuration for the membrane emulsification process.

Vladisavljević [19] produced W/O emulsion using pretreated hydrophobic hollow fibers made of polypropylene (pore size of 0.4 μm). The pre-treatment based on pores pre-filled with oil permitted the production of emulsions with droplet sizes smaller than pore size (droplets size of 0.26 to 0.32 μm). Giorno et al. [20, 21] studied a pretreatment procedure for hydrophilic polyamide hollow fiber membranes having nominal molecular weight cutoff (NMWCO) of 10 kDa. The procedure was based on contacting the membrane with a gradient of solvents and solvent mixtures of decreasing polarity in order to permit the passage, at low trans-membrane pressure, of nonpolar solvent through polar nanopores. Conditioned hollow fiber membranes were investigated for the preparation of O/W emulsions by cross-flow membrane emulsification. Isooctane-in-water droplets with a size of 1.87 μm (±0.58) were obtained.

10.2.5.2 Membranes Used in Premix Membrane Emulsification and Related Formulations

Membrane materials used for premix membrane emulsification have been described (Table 10.2). Few studies are specifically focused on the influence of membrane properties in premix membrane emulsification [38–40]. Also for this type of process the most commonly used membranes are SPG, even though the use of polymeric membranes is also reported.

The ratio between droplet diameter (D_d) and pore diameter (D_p) decreased with increasing pore size and it is in the range between 0.2 and 1 when the extrusion step is repeated in several cycles. The D_p distribution does not affect the D_d when using premix membrane emulsification with repeated extrusion cycles [38]. To assist the successful preparation of simple O/W or W/O/W emulsions by hydrophilic membrane, it was necessary to wet the membrane with the continuous phase. In particular, the use of different hydrophilic membrane materials led to different emulsion qualities depending on the emulsifier used for emulsion formulations [40]. The thickness of the membrane also influence the D_p; thicker membranes resulted in more uniform and smaller emulsion when the number of passes through the membrane is controlled [38]. In addition, the membrane structures have an influence on the emulsion properties [39]. The track-etched membranes (such as polycarbonate, polyester), which exhibited pore openings of uniform size running straight through the membrane, allowed the production of smaller emulsion droplets compared to the extrusion processes with membranes of a more irregular structure (i.e., sponge-like internal structure). This is correlated with the droplet break-up mechanism.

10.3 MEMBRANE NANOPRECIPITATION

Membrane contactors play a significant role also in assisting nanoprecipitation. In particular, they represent a suitable and reproducible system to scale up the process. So far, the scale-up of nanoprecipitation is among the challenges to exploit the process for industrial production.

Nanoprecipitation occurs when two miscible solvents, such as a polar organic solvent (i.e., acetone) containing macromolecules (i.e., polymers or phospholipids) and water are used. As soon as the miscible polar organic solvent and water are in contact, solvent displacement and non-solvent induced phase separation occur. If the polymer concentration in the organic solvent is below the dilute or semi-dilute regime, nanoparticles are formed and amorphous precipitation is avoided. The formation of nanoparticles has been explained in terms of super saturation, followed by nucleation, growth, and aggregation. The aggregation stops as soon as the colloidal stability is reached. The nanoprecipitation process has also been described in terms of difference of surface tensions, which causes interfacial turbulence and diffusion stranding. The interfacial turbulence causes vortices between the two non-equilibrated liquid phases where a

TABLE 10.2

Representative Characteristics of Membranes and Related Formulations Produced by Premix Membrane Emulsification

	Membrane characteristics			Formulation characteristics			
Membrane Material	Surface Wettability	$(D_p)(\mu m)$	Membrane Configuration	Type of Particles	$K= (D_d)/(D_p)$	Span (–)	Reference
SPG	Hydrophilic	2.7 and 4.2	Tubular	O/W emulsion	1.4–2.1	0.4–0.62	41
SPG	Hydrophilic	1.1	Tubular	S/O/W emulsion	1	N/A	42
SPG	Hydrophilic	10.7	Tubular	W/O/W emulsion	0.41–1.2	0.28–0.6	43
SPG	Hydrophilic	5.4–20.3	Tubular	W/O/W emulsion	0.37–1.2	0.28–0.93	44
SPG	Hydrophilic	8	Tubular	O/W emulsion	0.5–1.4	0.33–0.77	45
SPG	Hydrophilic	8	Tubular	W/O/W emulsion	0.2–0.29	N/A	46
SPG	Hydrophilic	5.2	Tubular	PELA microspheres	0.19	N/A	47
SPG	Hydrophilic	1.4	Tubular	PLA microspheres	0.2	N/A	48
SPG	Hydrophilic	25.9	Tubular	Chitosan-coated alginate particles	9.1	N/A	49
SPG	Hydrophobic	10.2	Tubular	Agarose microbeads	1	N/A	50
Glass filter	Hydrophilic	1	Flat	PLA microspheres	1	0.7	51
Glass filter	Hydrophilic	1	Flat	PLA microspheres	0.35–5	0.7–1.5	52
Glass Filter	Hydrophilic	1	Flat	PLA microspheres	1	N/A	53
α-Alumina	Hydrophilic	1.5	Tubular	O/W emulsion	1.5–1.8	1–1.2	54
α-Alumina	Hydrophilic	0.2	Tubular	O/W emulsion	1.25	N/A	40
Nickel	Hydrophilic	13.2	Flat	O/W emulsion	0.45	1	55
Nickel	Hydrophilic	10	Flat	O/W emulsion	0.6	N/A	56
Poly(tetrafluoroethylene)	Hydrophilic	1	Flat	O/W emulsion	2–4.1	N/A	57
Poly(tetrafluoroethylene)	Hydrophobic	1	Flat	W/O emulsion	2.8–4.0	N/A	58
Cellulose acetate	Hydrophilic	0.2-3	Flat	W/O/W emulsion	1–3.5	N/A	46
Cellulose acetate	Hydrophilic	0.2	Flat	O/W emulsion	2.5	0.6	40
Polycarbonate	Hydrophilic	0.2	Flat	O/W emulsion	1	0.6	40
Polycarbonate	Hydrophilic	0.33–1	Flat	O/W emulsion	1.6	0.6	59
Polycarbonate	Hydrophilic	0.7–2.5	Flat	W/O/W emulsion	1	N/A	60
Polycarbonate	Hydrophilic	1	Flat	O/W emulsion	1–12	N/A	61

O/W: oil-in-water; S/O/W: solid-in-oil-in-water; W/O/W: water-in-oil-in-water; PELA: poly(lactide-co-ethylene glycol); PLA: Polylactide.

change of the physicochemical properties appears in order to compensate discrepancies in free energy. The diffusion-stranding mechanism results in polymer partition into the aqueous non-solvent phase, where it aggregates into colloidal polymer particles due to solvent displacement [62].

In membrane nanoprecipitation, the membrane is settled between the two miscible phases (an organic phase containing the polymer and an aqueous phase) and works as an intensified high throughput system. The organic phase flows through the membrane pores and meets the non-solvent aqueous phase at the pore edge on the other side of the membrane, along which the aqueous phase flows tangentially. The fluid dynamics of the cross-flow system promote mixing of the miscible phases at the pore mouth thus leading to solvent displacement and polymer nanoprecipitation (Figure 10.4).

A comprehensive theoretical understanding of the process is still not available. Experimental observation confirmed that in membrane nanoprecipitation, the influence of the surface shear stress on the size of the nanoparticles is

different and much lower with respect to the direct membrane emulsification, and low energy consumption is expected. To some extent the pore size may influence nanoparticle size, in particular when they have an influence on the local (where the two phases meet) polymer concentration, ratio between dispersed and continuous phase, and their mixing. However, particles significantly smaller than the pore size are produced. High-dispersed phase flux is usually obtained due to the fact that the membrane is wetted by the dispersed phase and a high fraction of the dispersed phase to the continuous phase can be reached in a single pass.

10.3.1 MEMBRANES USED IN MEMBRANE NANOPRECIPITATION AND RELATED FORMULATIONS

Metallic membranes (nickel or stainless steel) are the most used for membrane nanoprecipitation (Table 10.3), and particles smaller than the pore size of the membrane were usually obtained [63–65]. The same results were also obtained

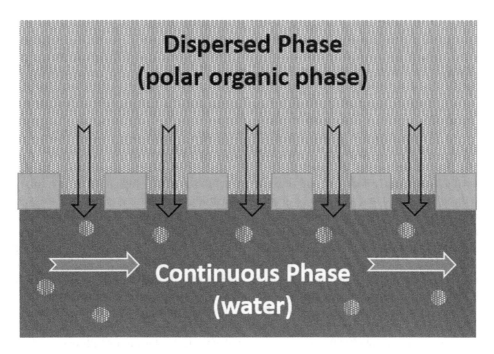

FIGURE 10.4 Membrane nanoprecipitation.

TABLE 10.3

Representative Characteristics of Membranes and Related Formulations Produced by Membrane Nanoprecipitation

Membrane characteristics				Formulation characteristics			
Membrane Material	Surface Wettability	$(D_p)(\mu m)$	Membrane Configuration	Type of Particles	$K= (D_d)/ (D_p)$	Span (-)	Reference
Nickel	Hydrophilic	5	Flat	PEG-PCL micelles	0.02	N/A	63
		10			0.01		
		20			0.006		
		40			0.004		
Nickel	Hydrophilic	5	Flat	Phospholipids	0.02	N/A	64
		10			0.09		
		20			0.005		
		40			0.002		
		10	Flat	PCL nanoparticles	0.022	N/A	65
		20			0.0098		
		40			0.0056		
Stainless steel		10			0.016	N/A	
SPG	Hydrophilic	0.2	Tubular	PCL nanoparticles	0.20	N/A	66
		0.9			0.17		
		10.2			0.02		
SPG	Hydrophilic	0.2	Tubular	PLGA-PEG	1	N/A	67
		1			0.28		
Zirconia	Hydrophilic	0.1	Tubular	PCL nanoparticles	3	N/A	68
		0.06			5		
		0.006			43		
		0.1	Tubular	PCL nanoparticles	4	N/A	69
Polypropylene	Hydrophobic	0.004	Hollow fibers	Phospholipids	2.8	N/A	70

PEG-PCL: poly(ethyleneglycol)-b-polycaprolactone; PCL: polycaprolactone; PLGA-PEG: Poly[(D,L lactide-co-glycolide)-copolyethyleneglycol] diblock.

with SPG membranes [66, 67]. Only a few works reported the use of membranes with a pore diameter equal or lower than 0.1 μm. In these cases, droplets with a size close to the membrane pore diameter membranes and up to five times bigger than the pore size were produced [68, 69]. Pore shade and the interpore distance also have a certain influence on the process. Different particle sizes were obtained using membranes with the same pore size but different pore shape. Funnel-shaped pores of nickel membranes were less efficient in mixing than straight pores of stainless steel membranes due to the lower exit velocity of the organic phase [65]. When the interpore distance is increased, a constant dispersed phase flow rate, the dispersed phase stream, is fragmented into a smaller number of sub-streams and the flow velocity of each sub-stream is higher. Membranes with higher interpore distance were demonstrated to produce smaller particles because the micromixing was more efficient [63, 64]. Only one example is reported on the use of polypropylene (PP) hollow fiber membranes [70]. The PP hollow fibers are hydrophobic and they were treated before their first use by using a water/ethanol (50:50, v/v).

10.4 CONCLUSIONS AND PERSPECTIVES

Membrane contactors offer great flexibility to precisely formulate micro- and nano-particles (mainly in the shape of spheres) through both membrane emulsification and membrane nanoprecipitation processes.

They have unique capability in producing droplets with highly uniform size distribution using quite low shear stress, which make them suitable for formulations involving labile molecules and low energy input. The unique feature of direct membrane emulsification to formulate emulsions drop-by-drop through porous membranes is very useful for emulsions used for formulations in biomedical, pharmaceutical, cosmetic, and food. Besides, fuels, dyes, synthetic textile, bioactive films, catalysts formulations, smart materials, etc., will also benefit from this membrane contactor technology.

Membrane nanoprecipitation is certainly a process that will permit the development of nanoprecipitation at an industrial scale since it can easily guarantee reproducibility and scalability. Uniform micro- and nano-sized particles are required for most application sectors; in fact, size and size uniformity strongly influence parameters such as texture, taste, optical, electronic, dielectric, magnetic, mechanical (such as elastic modulus, hardness, stress and strain, interfacial adhesion, and friction) properties, thermal conductivity, reactivity, and wettability.

Functionalized nano-capsule membranes carrying payload represent a promising strategy for cancer therapy. In fact, thanks to the low size, they can penetrate tissues through cells' fenestrae thus reaching tumors and increasing drug transport and retention. Furthermore, nano-capsules grafted with ligands specific for target receptors present on tumor cells permit the release of a drug only in tumor sites where the drug is needed, avoiding serious side effects. Nanoparticles may also provide improvements in the bioavailability of drugs

to the brain to treat central nervous system disorders as they could be designed to penetrate the blood brain barrier.

All of these promising perspectives can be reached in a short to medium term scale provided that significantly increased research effort will be promoted. Research advances are needed in order to (i) fully understand the fundamentals of the basic mechanisms and properties governing the droplets and particles formation, functionalization, and stabilization; (ii) design and develop novel membranes tuned for membrane emulsification and nanoprecipitation in order to achieve high precision as well as high productivity; (iii) optimize, develop and proof the robustness of the technology; (iv) test and proof the unique properties of formulations with break through properties; and (v) innovate production lines using such clean processes and low carbon footprint technology.

Research papers and patents focusing on the technology increased in the last decades. Initially, they mainly dealt with the development of membranes, membrane contactors configurations, and operating conditions, then focused also on the improved properties of formulations. Nevertheless, the potential of the technology is not fully exploited yet. Where present, data on production scale are proprietary and not available in the open literature. Although membrane contactors for niche emulsions applications have been developed on a productive scale, the technology is still at an emerging stage of development and the gap between academia and industry needs to be overcome. The advancement of the technology to a developing and then to a mature stage will permit translation of the scientific results and be of huge benefit to the society contributing to solve challenges in strategic areas including health, food, and environment.

REFERENCES

1. Vert, M., Doi, Y., Hellwich, K. H., Hess, M., Hodge, P., Kubisa, P., Rinaudo, M., Schué, F. 2012. Terminology for biorelated polymers and (IUPAC recommendations 2012). *Pure and Applied Chemistry* 84: 377–410.

2. De Luca, G., Di Maio, F. P., Di Renzo, A., Drioli, E. 2008. Droplet detachment in cross-flow membrane emulsification: Comparison among torque-and force-based models. *Chemical Engineering and Processing: Process Intensification* 47: 1150–1158.

3. Schröder, V., Behrend, O., Schubert, H. 1998. Effect of dynamic interfacial tension on the emulsification process using microporous, ceramic membranes. *Journal of Colloid and Interface Science* 202: 334–340.

4. Vladisavljević, G. T., Williams, R. A. 2006. Manufacture of large uniform droplets using rotating membrane emulsification. *Journal of Colloid and Interface Science* 299: 396–402.

5. Laouini, A., Charcosset, C., Fessi, H., Holdich, R. G., Vladisavljević, G. T. 2013. Preparation of liposomes: A novel application of microengineered membranes–From laboratory scale to large scale. *Colloids and Surfaces B: Biointerfaces* 112: 272–278.

6. Holdich, R. G., Dragosavac, M. M., Vladisavljević, G. T., Piacentini, E. 2013. Continuous membrane emulsification with pulsed (oscillatory) flow. *Industrial & Engineering Chemistry Research* 52: 507–515.

7. Joscelyne, S. M., Trägårdh, G. 2000. Membrane emulsification–A literature review. *Journal of Membrane Science* 169: 107–117.

8. Charcosset, C., Limayem, I., Fessi, H. 2004. The membrane emulsification process–A review. *Journal of Chemical Technology & Biotechnology: International Research in Process, Environmental & Clean Technology* 79: 209–218.

9. Vladisavljević, G. T. 2015. Structured microparticles with tailored properties produced by membrane emulsification. *Advances in Colloid and Interface Science* 225: 53–87.

10. Piacentini, E., Imbrogno, A., Drioli, E., Giorno, L. 2014. Membranes with tailored wettability properties for the generation of uniform emulsion droplets with high efficiency. *Journal of Membrane Science* 459: 96–103.

11. Kukizaki, M., Goto, M. 2006. Effects of interfacial tension and viscosities of oil and water phases on monodispersed droplet formation using a Shirasu-porous-glass (SPG) membrane. *Membrane* 31: 215–220.

12. Strathmann, H., Giorno, L., Drioli, E. 2006. *Introduction to Membrane Science and Technology*. Roma, Italy: CNR Publisher.

13. Nakashima, T., Shimizu, M., Kukizaki, M. 2000. Particle control of emulsion by membrane emulsification and its applications. *Advanced Drug Delivery Reviews* 45: 47–56.

14. Lai, Y. T., Sato, M., Ohta, S., Akamatsu, K., Nakao, S.-I., Sakai, Y., Ito, T. 2015. Preparation of uniform-sized hemoglobin–albumin microspheres as oxygen carriers by Shirasu porous glass membrane emulsification technique. *Colloids and Surfaces B: Biointerfaces* 127: 1–7.

15. Imbrogno, A., Piacentini, E., Drioli, E., Giorno, L. 2014. Micro and nano polycaprolactone particles preparation by pulsed back-and-forward cross-flow batch membrane emulsification for parenteral administration. *International Journal of Pharmaceutics* 477: 344–350.

16. Pawlik, A. K., Norton, I. T. 2012. Encapsulation stability of duplex emulsions prepared with SPG cross-flow membrane, SPG rotating membrane and rotor-stator techniques – A comparison. *Journal of Membrane Science* 415: 459–468.

17. Morelli, S., Holdich, R. G., Dragosavac, M. M. 2017. Microparticles for cell encapsulation and colonic delivery produced by membrane emulsification. *Journal of Membrane Science* 524: 377–388.

18. Jing, W., Wu, J., Jin, W., Xing, W., Xu, N. 2006. Monodispersed W/O emulsion prepared by hydrophilic ceramic membrane emulsification. *Desalination* 191: 219–222.

19. Vladisavljević, G. T., Tesch, S., Schubert, H. 2002. Preparation of water-in-oil emulsions using microporous polypropylene hollow fibers: Influence of some operating parameters on droplet size distribution. *Chemical Engineering and Processing: Process Intensification* 41: 231–238.

20. Giorno, L., Li, N., Drioli, E. 2003. Preparation of oil-in-water emulsions using polyamide 10 kDa hollow fiber membrane. *Journal of Membrane Science* 217: 173–180.

21. Giorno, L., Mazzei, R., Oriolo, M., De Luca, G., Davoli, M., Drioli, E. 2005. Effects of organic solvents on ultrafiltration polyamide membranes for the preparation of oil-in-water emulsions. *Journal of Colloid and Interface Science* 287: 612–623.

22. Kakazu, E., Murakami, T., Akamatsu, K., Sugawara, T., Kikuchi, R., Nakao, S. I. 2019. Preparation of silver nanoparticles using the SPG membrane emulsification technique. *Journal of Membrane Science* 354: 1–5.

23. Piacentini, E., Poerio, T., Bazzarelli, F., Giorno, L. 2016. Microencapsulation by membrane emulsification of biophenols recovered from olive mill wastewaters. *Membranes* 6: 25.

24. Mi, Y., Zhou, W., Li, Q., Gong, F., Zhang, R., Ma, G., Su, Z. 2015. Preparation of water-in-oil emulsions using a hydrophobic polymer membrane with 3D bicontinuous skeleton structure. *Journal of Membrane Science* 490: 113–119.

25. Hancocks, R. D., Spyropoulos, F., Norton, I. T. 2016. The effects of membrane composition and morphology on the rotating membrane emulsification technique for food grade emulsions. *Journal of Membrane Science* 497: 29–35.

26. Melich, R., Valour, J. P., Urbaniak, S., Padilla, F., Charcosset C. 2019. Preparation and characterization of perfluorocarbon microbubbles using Shirasu porous glass (SPG) membranes. *Colloids and Surfaces A: Physicochemical and Engineering Aspects* 560: 233–243.

27. Bazzarelli, F., Piacentini, E., Giorno, L. 2017. Biophenols-loaded solid lipid particles (SLPs) development by membrane emulsification. *Journal of Membrane Science* 541: 587–594.

28. Matos, M., Gutiérrez, G., Iglesias, O., Coca, J., Pazos, C. 2017. Enhancing encapsulation efficiency of food-grade double emulsions containing resveratrol or vitamin B12 by membrane emulsification. *Journal of Food Engineering* 166: 212–220.

29. Fuchigami, T., Toki, M., Nakanishi, K. 2000. Membrane emulsification using sol-gel derived macroporous silica glass. *Journal of Sol-Gel Science and Technology* 19: 337–341.

30. Zhu, J., Barrow, D. 2005. Analysis of droplet size during crossflow membrane emulsification using stationary and vibrating micromachined silicon nitride membranes. *Journal of Membrane Science* 261: 136–144.

31. De los Reyes, J. S., Charcosset, C. 2010. Preparation of water-in-oil and ethanol-in-oil emulsions by membrane emulsification. *Fuel* 89.11: 3482–3488.

32. Joscelyne, S. M., Trägårdh, G. 1999. Food emulsions using membrane emulsification: Conditions for producing small droplets. *Journal of Food Engineering* 39: 59–64.

33. Zanatta, V., Rezzadori, K., Penha, F. M., Zin, G., Lemos-Senna, E., Petrus, J. C. C., Di Luccio, M. 2017. Stability of oil-in-water emulsions produced by membrane emulsification with microporous ceramic membranes. *Journal of Food Engineering* 195: 73–84.

34. Medina-Llamas, M., Mattia, D. 2017. Production of nanoemulsions using anodic alumina membranes in a stirred-cell setup. *Industrial & Engineering Chemistry Research* 56: 7541–7550.

35. Yamazaki, N., Naganuma, K., Nagai, M., Ma, G. H., Omi, S. 2003. Preparation of W/O (water-in-oil) emulsions using a PTFE (polytetrafluoroethylene) membrane—A new emulsification device. *Journal of Dispersion Science and Technology* 24: 249–257.

36. Yamazaki, N., Yuyama, H., Nagai, M., Ma, G. H., Omi, S. 2002. A comparison of membrane emulsification obtained using SPG (Shirasu porous Glass) and PTFE [poly (tetrafluoroethylene)] membranes. *Journal of Dispersion Science and Technology* 23: 279–292.

37. Kobayashi, I., Yasuno, M., Iwamoto, S., Shono, A., Satoh, K., Nakajima, M. 2002. Microscopic observation of emulsion droplet formation from a polycarbonate membrane. *Colloids and Surfaces A: Physicochemical and Engineering Aspects* 207: 185–196.

38. Zhou, Q., Ma, G., Su, Z. 2009. Effect of membrane parameters on the size and uniformity in preparing agarose beads by premix membrane emulsification. *Journal of Membrane Science* 326: 694–700.

39. Joseph, S., Bunjes, H. 2013. Influence of membrane structure on the preparation of colloidal lipid dispersions by premix membrane emulsification. *International Journal of Pharmaceutics* 446: 59–62.

40. Gehrmann, S., Bunjes, H. 2018. Influence of membrane material on the production of colloidal emulsions by premix membrane emulsification. *European Journal of Pharmaceutics and Biopharmaceutics* 126: 140–148.

41. Suzuki, K., Shuto, I., Hagura, Y. 1996. Characteristics of the membrane emulsification method combined with preliminary emulsification for preparing corn oil-in-water emulsions. *Food Science and Technology International* 2: 43–47.

42. Toorisaka, E., Ono, H., Arimori, K., Kamiya, N., Goto, M. 2003. Hypoglycemic effect of surfactant-coated insulin solubilized in a novel solid-in-oil-in-water (S/O/W) emulsion. *International Journal of Pharmaceutics* 252: 271–274.

43. Vladisavljević, G. T., Shimizu, M., Nakashima, T. 2004. Preparation of monodisperse multiple emulsions at high production rates by multi-stage premix membrane emulsification. *Journal of Membrane Science* 244: 97–106.

44. Vladisavljević, G. T., Shimizu, M., Nakashima, T. 2006. Production of multiple emulsions for drug delivery systems by repeated SPG membrane homogenization: Influence of mean pore size, interfacial tension and continuous phase viscosity. *Journal of Membrane Science* 284: 373–383.

45. Vladisavljević, G. T., Surh, J., McClements, J. D. 2006. Effect of emulsifier type on droplet disruption in repeated Shirasu porous glass membrane homogenization. *Langmuir* 22: 4526–4533.

46. Surh, J., Vladisavljevic, G. T., Mun, S., McClements, D. J. 2007. Preparation and characterization of water/oil and water/oil/water emulsions containing biopolymer-gelled water droplets. *Journal of Agricultural and Food Chemistry* 55: 175–184.

47. Wei, Q., Wei, W., Tian, R., Wang, L. Y., Su, Z. G., Ma, G. H. 2008. Preparation of uniform-sized PELA microspheres with high encapsulation efficiency of antigen by premix membrane emulsification. *Journal of Colloid and Interface Science* 323: 267–273.

48. Wei, Q., Wei, W., Lai, B., Wang, L.-Y., Wang, Y.-X., Su, Z.-G., Ma, G.-H. 2008. Uniform-sized PLA nanoparticles: Preparation by premix membrane emulsification. *International Journal of Pharmaceutics* 359: 294–297.

49. Nan, F. F., Wu, J., Qi, F., Fan, Q., Ma, G., Ngai, T. 2014. Preparation of uniform-sized colloidosomes based on chitosan-coated alginate particles and its application for oral insulin delivery. *Journal of Materials Chemistry B* 2: 7403–7409.

50. Zhou, Q. Z., Wang, L. Y., Ma, G. H., Su, Z. G. 2008. Multi-stage premix membrane emulsification for preparation of agarose microbeads with uniform size. *Journal of Membrane Science* 322: 98–104.

51. Sawalha, H., Purwanti, N., Rinzema, A., Schroën, K., Boom, R. 2008. Polylactide microspheres prepared by premix membrane emulsification—Effects of solvent removal rate. *Journal of Membrane Science* 310: 484–493.

52. Sawalha, H., Fan, Y., Schroën, K., Boom, R. 2008. Preparation of hollow polylactide microcapsules through

premix membrane emulsification – Effects of nonsolvent properties. *Journal of Membrane Science* 325: 665–671.

53. Kooiman, K., Böhmer, M. R., Emmer, M., Vos, H. J., Chlon, C., Shi, W. T., Hall, C. S., de Winter, S. H. P. M., Schroën, K., Versluis, M., de Jong, N., van Wamel, A. 2009. Oil-filled polymer microcapsules for ultrasound-mediated delivery of lipophilic drugs. *Journal Controlled Release* 133: 109–118.

54. Jing, W. H., Wu, J., Xing, W. H., Jin, W. Q., Xu, N. P. 2005. Emulsions prepared by two-stage ceramic membrane jet-flow emulsification. *AIChE Journal* 51: 1339–1345.

55. Nazir, A., Schroën, K., Boom, R. 2011. High-throughput premix membrane emulsification using nickel sieves having straight-through pores. *Journal of Membrane Science* 383: 116–123.

56. Santos, J., Vladisavljević, G. T., Holdich, R. G., Dragosavac, M. M., Muñoz, J. 2015. Controlled production of eco-friendly emulsions using direct and premix membrane emulsification. *Chemical Engineering Research and Design* 98: 59–69.

57. Suzuki, K., Fujiki, I., Hagura, Y. 1998. Preparation of corn oil/water and water/corn oil emulsions using PTFE membranes. *Food Science and Technology International* 4: 164–167.

58. Suzuki, K., Hayakawa, K., Hagura, Y. 1999. Preparation of high concentration O/W and W/O emulsions by the membrane phase inversion emulsification using PTFE membranes. *Food Science and Technology Research* 5: 234–238.

59. Park, S. H., Yamaguchi, T., Nakao, S. 2001. Transport mechanism of deformable droplets in microfiltration of emulsions. *Chemical Engineering Science* 56: 3539–3548.

60. Yafei, W., Tao, Z., Gang, H. 2006. Structural evolution of polymer-stabilized double emulsions. *Langmuir* 22: 67–73.

61. Trentin, A., Ferrando, M., López, F., Güell, C. 2009. Premix membrane O/W emulsification: Effect of fouling when using BSA as emulsifier. *Desalination* 245: 388–395.

62. Schubert, S., Delaney, Jr., J. T., Schubert, U. S. 2011. Nanoprecipitation and nanoformulation of polymers: From history to powerful possibilities beyond poly(lactic acid). *Soft Matter* 7: 1581–1588.

63. Laouini, A., Charcosset, C., Fessi, H., Holdich, R. G., Vladisavljević, G. T. 2013. Preparation of liposomes: A novel application of microengineered membranes – Investigation of the process parameters and application to the encapsulation of vitamin E. *RSC Advances* 3: 4985.

64. Laouini, A., Koutroumanis, K. P., Charcosset, C., Georgiadou, S., Fessi, H., Holdich, R. G., Vladisavljević, G. T. 2013. pH-sensitive micelles for targeted drug delivery prepared using a novel membrane contactor method. *ACS Applied Materials & Interfaces* 5: 8939–8947.

65. Othman, R., Vladisavljevic, G. T., Shahmohamadi, H., Nagy, Z. K., Holdich, R. G. 2016. Formation of size-tuneable biodegradable polymeric nanoparticles by solvent displacement method using micro-engineered membranes fabricated by laser drilling and electroforming. *Chemical Engineering Journal* 304: 703–713.

66. Khayata, N., Abdelwahed, W., Chehna, M. F., Charcosset, C., Fessi, H. 2012. Preparation of vitamin E loaded nanocapsules by the nanoprecipitation method: From laboratory scale to large scale using a membrane contactor. *International Journal of Pharmaceutics* 423: 419–427.

67. Albisa, A., Piacentini E., Sebastian, V., Arruebo, M., Santamaria, J., Giorno, L. 2017. Preparation of drug-loaded PLGA-PEG nanoparticles by membrane-assisted

nanoprecipitation. *Pharmaceutical Research* 34: 1296–1308.

68. Charcosset, C., Fessi, H. 2005. Preparation of nanoparticles with a membrane contactor. *Journal of Membrane Science* 266: 115–120.

69. Blouza, I. L., Charcosset, C., Sfar, S., Fessi, H. 2006. Preparation and characterization of spironolactone-loaded nanocapsules for paediatric use. *International Journal of Pharmaceutics* 325: 124–131.

70. Laouini, A., Jaafar-Maalej, C., Sfar, S., Charcosset, C., Fessi, H. 2011. Liposome preparation using a hollow fiber membrane contactor-application to spironolactone encapsulation. *International Journal of Pharmaceutics* 415: 53–61.

Part IV

Chapters on Supported Liquid Membranes

11 Application of Liquid Membrane Technology at Back End of Nuclear Fuel Cycle—Perspective and Challenges

S. Panja, P. S. Dhami, J. S. Yadav, and C. P. Kaushik

CONTENTS

11.1 INTRODUCTION

A closed nuclear fuel cycle option is being practised in many countries, including India, for a sustainable long term nuclear energy option [1]. In this option, reprocessing and waste management play a very important role in maximizing the resource utilization. During the operation of reprocessing and waste management plants large volumes of liquid waste streams of different natures are generated which are conventionally categorized as "low level", "intermediate level", and "high level" liquid wastes. Liquid wastes pose several chemical and radiological hazards to the environment due to the presence of several radionuclides if not managed properly. Public acceptability of nuclear energy also mainly depends on the safe management of radioactive wastes. With the advancement in separation science and technology, some of the liquid waste streams are now

being considered as resources rather than as waste as they contain useful radionuclides such as ^{137}Cs, ^{90}Sr, ^{90}Y, ^{241}Am, etc., which have many applications if they are recovered in a form suitable for societal use. Almost all of the conventional separation methods either alone or in combinations are adopted in reprocessing and waste management areas at the back end.

Membrane techniques are relatively new and have gained considerable importance over the last two decades. They are considered to be the most energy saving and green techniques [2]. Due to the availability of tailor made solvents and membranes of desired characteristics, there has been a rapid growth in the field of membrane separation both in the chemical as well as in the nuclear industry. Depending upon the process driving force, membrane technologies are broadly divided into the following four categories.

i)   Pressure gradient as driving force,
ii)   Electrical potential gradient as driving force,
iii) Concentration gradient as driving force, and
iv)   Temperature gradient as driving force.

Among these, concentration and pressure gradient driven membrane technologies are found to be more suitable at the back end of the nuclear fuel cycle. Liquid membrane techniques use concentration gradient as the driving force for separation and this method allows the use of tailor made expensive solvents for a variety of applications especially from waste streams containing low metal ion concentration. The advantages of SLM over conventional techniques can be listed as follows:

- Simultaneous extraction and stripping,
- Possibility of achieving high separation factors,
- Economic use of expensive extractants,
- Possibility of concentrating the recovered metal ion species during separation itself,
- Lower capital and operating costs,
- Lesser space requirements, and
- Low energy consumption.

Extensive literature is available on the studies of liquid waste treatment and recovery of useful radionuclide for societal applications involving liquid membrane technology. In this chapter, we intend to give an account of the studies carried out using liquid membrane technology at the back end of nuclear fuel cycle using real waste solutions to establish the technology as a mature industrially acceptable technique. Prior to entry into the actual applications of liquid membrane techniques at the back end of nuclear fuel cycle, a brief description of LM techniques, the principle behind the separation, and their categorizations are described.

11.2   LIQUID MEMBRANE SEPARATION

Liquid membrane (LM)-based separation methods work on the principle of solvent extraction [3]. Liquid membranes usually consist of a water immiscible (organic) layer separating a source and a receiving aqueous phase. Liquid membranes may be divided into two categories, *viz.* non-supported liquid membranes and supported liquid membranes. In the case of non-SLMs, the most common types are bulk liquid membranes (BLM) and emulsion liquid membranes (ELM). SLMs again can be categorized into two types as mentioned above, *viz.*, flat sheet supported liquid membrane (FSSLM), and hollow fiber liquid membrane (HFSLM). A short description of these techniques is discussed below.

11.2.1   Non-Supported Liquid Membranes

In case of bulk liquid membrane (BLM) the source phase is separated from the receiving phase by a bulk organic ligand solution. Figure 11.1 gives schematic presentation of BLM. The amount of metal ion transported is determined by its concentration in the receiving phase. So long as the stirrer is not spinning too vigorously, the stability is maintained.

11.2.1.1   Emulsion Liquid Membranes

Emulsion liquid membranes (ELM) have a very thin membrane and a large surface area per unit source phase volume, which enhances the transport rate of this membrane. In this type of liquid membrane, the emulsion contains the receiver solution as the internal phase. The organic phase or the emulsion is added to a source phase of the external phase, and mass transfer is determined similar to a two-phase extraction system. The membrane stability is the same as the stability of the emulsion. In order to recover the receiving phase, and in order to replenish the carrier phase, it is required to break the emulsion by a process called demulsification. However, factors affecting the emulsion stability like ionic strength, pH, etc., must be controlled as it is vital for the resultant separation. Schematic presentation of ELM is given in Figure 11.2.

11.2.2   Supported Liquid Membrane

In supported liquid membrane systems, the feed phase and the receiver phase are separated by a polymeric support material impregnated with the organic phase containing a carrier ligand. There are two types of supported liquid membranes, *viz.* flat sheet supported liquid membrane

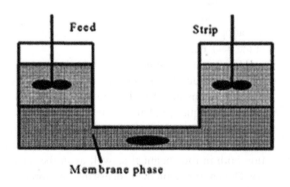

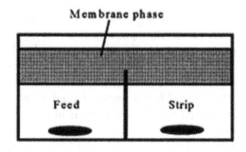

FIGURE 11.1   Schematic presentation of bulk liquid membrane.

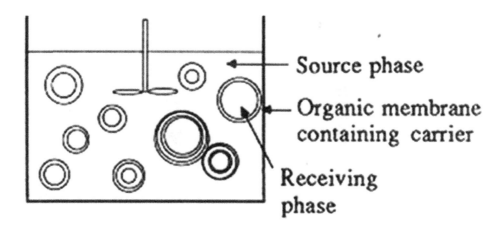

FIGURE 11.2 Schematic presentation of emulsion liquid membrane.

(FSSLM) and hollow fiber supported liquid membrane (HFSLM).

11.2.2.1 Flat Sheet Supported Liquid Membranes

In the case of FSSLMs, the feed and receiver phases are separated by a flat sheet polymeric support material impregnated with the organic extractant solution. A schematic presentation of FSSLM system is given in Figure 11.3. The feed and the receiver phases are constantly stirred using magnetic stirrers in order to minimize the aqueous diffusion layer. FSSLM systems are usually associated with slow transport rates.

11.2.2.2 Hollow Fiber Supported Liquid Membranes

Higher mass transfer can be achieved by increasing the membrane surface area many fold using the hollow fiber polymeric support material in place of the flat sheets. A schematic presentation of the HFSLM is shown in the Figure 11.4. The outer shell consists of nonporous material. Inside that shell, there are many thin fibers running along the length of the shell, all in neat rows. The liquid membrane is created by pumping the carrier solution inside the module low pressure, followed by washing with plenty of water. The feed phase and the receiver phase are circulated around these fibers in countercurrent direction

without mixing. If the feed and the receiver phases are moving along same direction, then it is termed as co-current mode while an opposite flow mode is termed as countercurrent mode. Different configurations of membrane contactors have been utilized to maximize the efficiency of the process with a long-term stability. These include non-dispersive solvent extraction (NDSX), hollow fiber supported liquid membrane (HFSLM), pseudo-emulsion-based hollow-fiber strip dispersion (PEHFSD), hollow fiber contained liquid membrane (HFCLM), and hollow fiber renewal liquid membrane (HFRLM). Among these configurations, the HFRLM has been preferred due to its several advantages over the other methods. Because of the renewal effect of the liquid membrane layer developed at the inner surface of the fibers and the high membrane surface area, the mass transfer rate is higher than the conventional methods in the hollow fibers. In addition, the continuous replenishment of the liquid membrane embedded in the pores of hollow fibers causes the long-time stability of the liquid membrane. Moreover, due to lateral shear forces, no secondary pollution could be created by the emulsification. Finally, the low investment costs due to the low energy consumption and low carrier consumption as well as the easy scale-up of the process and the wide range of operation regions are some other advantages of this process

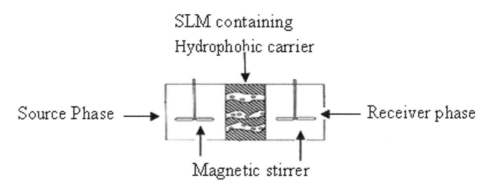

FIGURE 11.3 Schematic presentation of metal ion transport through FSSLM.

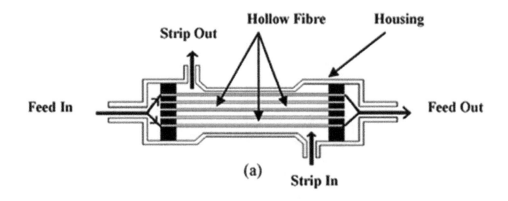

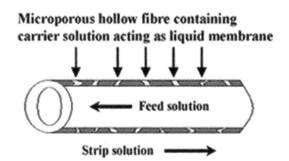

FIGURE 11.4 Schematic presentation of the hollow fiber supported liquid membrane system.

11.3 MECHANISM OF TRANSPORT IN LIQUID MEMBRANES

In membranes, the permeating species can be transported across the membrane against their concentration gradient as a consequence of an existing concentration gradient of a second species present in the system [4]. Furthermore, the transport process may take place in the presence of an extractant or carrier contained within the membrane. Facilitated transport originated in biochemistry where natural carriers contained in cell walls are involved. In 1970, Bloch first proposed the use of organic extractants immobilized on microporous inert supports for the separation of metal ions from a mixture. Subsequently, other researchers observed that the carrier could assist in the transport process (facilitated transport) by reacting competitively with the two species which were being transported across the membrane. Depending on the nature of the extractant, facilitated coupled transport can be of two types:(i) counter, and (ii) co-transport [3]. When the extractant exhibits acidic properties, coupled counter transport takes place and the extraction reaction proceeds *via*:

$$M^+ + HX\left(membrane\right) \rightarrow MX\left(membrane\right) + H^+ \quad (11.1)$$

However, when neutral extractants are used, coupled co-transport takes place according to:

$$M^+ + X^- + E\left(membrane\right) \rightarrow EMX\left(membrane\right) \quad (11.2)$$

Where, pH and counter ion concentration are used as driving forces, respectively, and X is the extractant ligand. Transport mechanism can also be explained schematically as follows in Figure 11.5.

The definition and equations describing membrane parameters can be found in any standard text book of membrane science and technology [5]. For hollow fiber supported liquid membrane, the equation for permeability [17,18] can be expressed as perthe following equation:

$$\ln\frac{C_t}{C_0} = -\frac{A.P}{V}\left(\frac{\varphi}{\varphi+1}\right).t \quad (11.3)$$

where, P is the permeability co-efficient of the metal ion, C_t and C_o are the concentrations of the metal ions at time t and at zero time in the feed solution, respectively, and V is the volume of the feed solution (mL). The total effective surface area of the hollow fiber (cm²) represented by A is calculated by the following equation:

$$A = 2\pi r_i LN\varepsilon \quad (11.4)$$

where, r_i denotes the internal radius of the fiber (cm), L is the length of the fiber (cm), N is thenumber of fibers, andε is the membrane porosity. The parameter ϕfor module containing N number of fibers can be expressed by the given equation:

$$\varphi = \frac{Q_T}{Pr_i LN\pi\varepsilon} \quad (11.5)$$

where, Q_T is the total flow rate of the feed solution (mL/min).

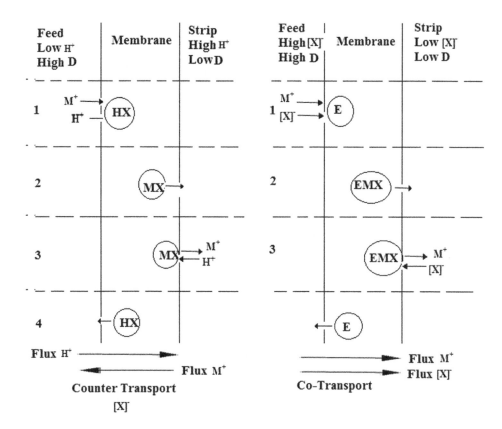

FIGURE 11.5 Schematic presentation of metal ion transport through membrane.

The prime requirement for the successful liquid membrane system is the choice of carrier. Depending on the requirement, a variety of organic solvents has been studied and some of the extractants explored at the back end of nuclear fuel cycle are shown in Figure 11.6.

11.4 APPLICATIONS OF LIQUID MEMBRANE TECHNIQUES AT THE BACK END OF NUCLEAR FUEL CYCLE

11.4.1 Use of FSSLM Partitioning of Actinide from HLLW

Alpha-emitting long-lived nuclides present in high level liquid waste (HLLW) solutions originating from the reprocessing of spent nuclear fuel are of great environmental concern. Considerable work has been carried out all over the world in the last two decades on the partitioning of actinides from acidic waste solutions using solvent extraction and extraction chromatography [6,7]. A variety of novel solvents like octylphenyl -N,N'-diisobutylcarbamoylmethyl phosphine oxide (CMPO) [Figure 11.6a], di-methyl di-butyl tetra decyl malonamide (DMDBTDMA), N,N,N'',N''- tetra alkyl diglycolamide [Figure 11.6b] etc., have been synthesized and studied in solvent extraction, extraction chromatography, and liquid membrane technology [8–14]. Among various reagents used in solvent extraction mode, octyl phenyl-N,N'-diisobutylcarbamoylmethyl phosphine oxide (CMPO) [Figure 11.6a] was studied by us for partitioning of actinide from HLLW because of its ability to extract tri-, tetra-, and hexavalent actinides from acidic wastes without any feed adjustment [15]. CMPO has also been used as a carrier in the supported liquid membrane (SLM) technique for the separation of americium from nitric acid medium [16]. Initially, transport studies were carried out for ^{241}Am under different experimental conditions. The studies also included the transport of other radionuclides, viz., ^{233}U, $^{237+}$^{238}Np, and ^{239}Pu, from 3.0 M nitric acid. The studies were then extended to simulated HLLW originating from the reprocessing of spent fuel from a pressurized heavy water reactor (PHWR-HLLW) as well as actual HLLW solutions originating from the reprocessing of spent fuel from a research reactor to test the applicability of this technique for the separation of actinides from waste solutions.

The composition of the simulated PHWR-HLLW used was based on the fission product inventory of the spent uranium fuel from a pressurized heavy water reactor with a burn up of 6,500 MWd/Te of UO$_2$ and three years of cooling. Inert constituents such as sodium and corrosion products like iron, chromium, and nickel were added in quantities anticipated to occur in the HLLW. The composition of the simulated HLLW used can be found in the literature [16]. U removal prior to liquid membrane transport studies was carried out for both the SHLW as well as the actual HLW by contacting thrice with fresh 30% TBP/n-dodecane solution. The concentration of uranium in both wastes was lowered to less than 10 ppm. Figure 11.7 shows the transport of ^{241}Am from Uranium lean simulated PHWR-HLLW using

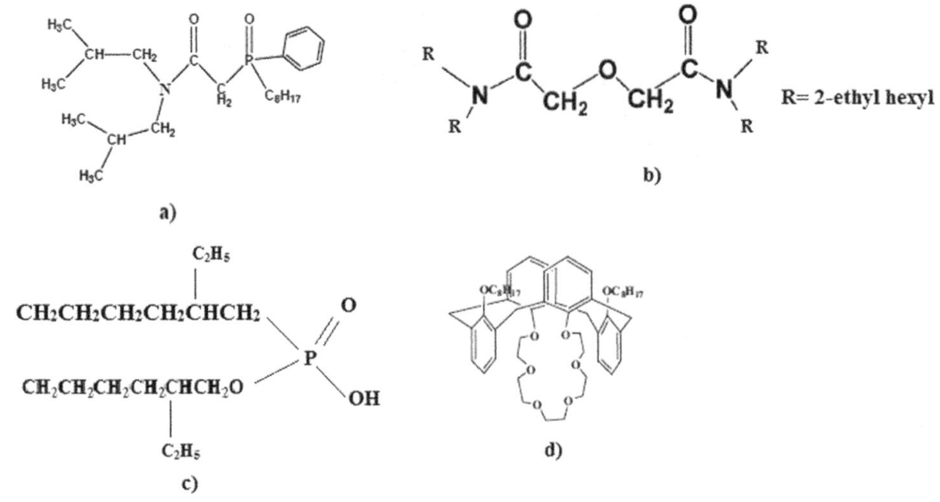

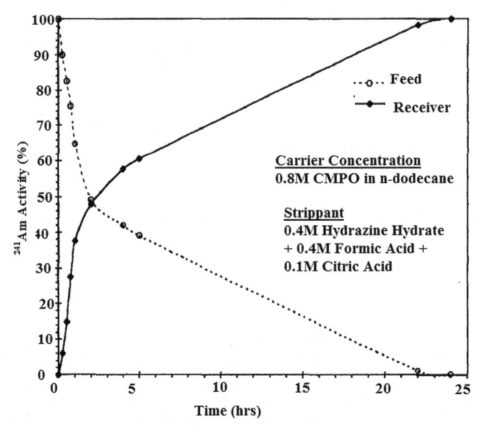

FIGURE 11.6   Structures of different extractants used in liquid membrane technology at the back end of nuclear fuel cycle. a) CMPO, b) Tetra alkyl diglycolamide (TEHDGA), c) 2-Ethylhexyl2-ethylhexyl phosphoric acid (KSM-17/PC88A), d) bis-(octyloxy)calix[4] arene-mono-crown-6 (CMC).

a mixture of hydrazine hydrate, formic acid, and citric acid as the complexing agent. Transport of ^{241}Am increased initially and gradually reached to > 99.96% only after ~24 h.

Figure 11.8 gives the transport of alpha activity from actual uranium lean HLLW of research reactor origin.

Under experimental conditions quantitative transport of alpha activity was observed within a short interval of ~90 min. A lesser time requirement for the partitioning of alpha activity in the case of HLLW originated from the reprocessing of research reactor fuel compared to the SHLW of

FIGURE 11.7   Transport of ^{241}Am from Uranium lean PHWR-HLLW using CMPO as the carried in presence of complexing agent as strippant across SLM [Reprinted from Separation Science and Technology, 34, A. Ramanujam, P. S. Dhami, V. Gopalakrishnan, N. L. Dudwadkar, R. R. Chitnis, J. N. Mathur, Partitioning of Actinides from High Level Waste of PUREX Origin Using Octylpheny l-N,N‘-diisobutylcarbamoylmethyl Phosphine Oxide (CMPO)-Based Supported Liquid Membrane, 1717–1728, Copyright (1999) with permission from Taylor and Frances [16]].

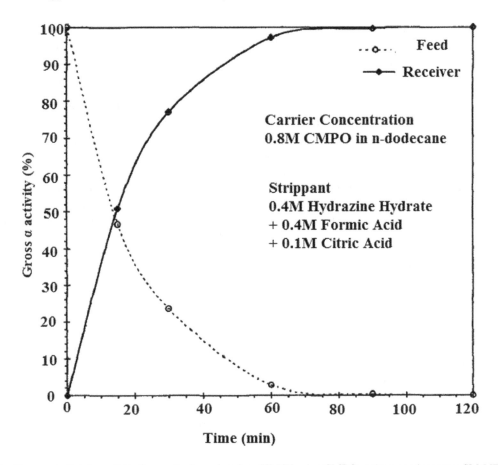

FIGURE 11.8 Transport of alpha activity from actual uranium lean HLLW using CMPO as the α carrier across SLM [Reprinted from Separation Science and Technology, 34, A. Ramanujam, P. S. Dhami, V. Gopalakrishnan, N. L. Dudwadkar, R. R. Chitnis, J. N. Mathur, Partitioning of Actinides from High Level Waste of PUREX Origin Using Octylphenyl-N,N'-diisobutylcarbamoylmethyl Phosphine Oxide (CMPO)-Based Supported Liquid Membrane, 1717–1728, Copyright (1999) with permission from Taylor and Frances [16].

PHWR origin could be attributed to the presence of low concentrations of lanthanides.

Results indicated that CMPO-based supported liquid membranes can be employed for the partitioning of actinides from high level liquid waste solutions. Since the transport rates were found to be very low in the presence of a high concentration of uranium, the present method will have its best results only after the uranium content of the waste solution is lowered. Hollow fiber liquid membrane technology with larger surface area can also be used to increase the transport rate as indicated by the results from various research groups [17–20].

11.4.2 Production of Clinical Grade ⁹⁰Y-Acetate for Therapeutic Applications Using Ultra Pure ⁹⁰Sr Recovered from PUREX HLLW Employing SLM

The utilization of radioisotopes for diagnosis and therapy of various diseases in health care is one of the important objectives of the Indian nuclear program. Several radionuclides are produced using research reactors and particle accelerators for applications in nuclear medicine. HLLW obtained from the reprocessing of spent fuel by PUREX

process is a rich source of several radionuclides *viz.* ^{137}Cs, ^{90}Sr, ^{90}Y, ^{106}Ru, ^{147}Pm, ^{144}Ce, ^{231}Pa, ^{237}Np, ^{241}Am, and ^{252}Cf, etc. Among these, ^{90}Sr is a pure β- emitter and decays to ^{90}Y, a very useful radionuclide in nuclear medicine for cancer therapy. A solvent extraction-based plant is being operated at WIP, Trombay, Mumbai for the management of HLLW using indigenously synthesized solvents. The secondary streams from the plant are being deployed for the recovery of several useful radionuclides. One of the streams from this process is found to be ideal to recover ^{90}Sr(NO$_3$)$_2$ solution of high specific activity.

Yttrium-90, a pure β-emitter (E_{max} = 2.28MeV, $T_{1/2}$ = 64.1h), is a potential therapeutic radionuclide formed by β- decay of ^{90}Sr which can be used as a long lasting source for the generation of carrier-free ^{90}Y. To separate ^{90}Y from ^{90}Sr, several techniques such as extraction chromatography, electrochemical, and supported liquid membrane-based generator systems were studied [21–24]. Among these, a two-stage supported liquid membrane (SLM)-based generator system [25,26] was pursued in our laboratories and found to be convenient for milking carrier-free ^{90}Y. Based on extensive studies, the generator system was found to be suitable for getting high purity ^{90}Y with respect to ^{90}Sr (< 10^{-6} Ci/Ci of ^{90}Y, i.e., 10^{-4}% of total activity) as desired for clinical grade

[90]Y. As per the European Pharmacopeia, the purity requirements with respect to α activity are more stringent (10^{-9} Ci/Ci ^{90}Y) than compared to [90]Sr [27].

Extensive studies were carried out at our laboratory for purification of high specific activity ^{90}Sr recovered from actual HLLW after removal of bulk uranium, cesium, lanthanides, and actinides at Waste Immobilization Plant (WIP), Trombay using a multi stage solvent extraction loop. Since a single separation technique is unable to recover ^{90}Sr of desired radionuclidic purity from such waste solution containing a host of minor actinides, fission products, and other metallic elements, multi separation techniques, *viz.* solvent extraction, ion-exchange, extraction chromatography, and membrane-based methods were employed for purification. The purified [90]Sr $(NO_3)_2$ was used to generate carrier-free ^{90}Y by employing two-stage supported membrane (SLM) generator. Various steps employed during separation, purification, and assaying its quality to make it suitable for clinical grade ^{90}Y generation, are discussed below.

HLLW obtained from the recycling of spent fuel is subjected to three-cycle solvent extraction processes at WIP, Trombay where a multi-cycle solvent extraction process is adopted for the management of HLLW. In the first cycle, depletion of residual uranium and plutonium from the waste is carried out using PUREX solvent. Uranium and plutonium from organic phase are stripped in aqueous phase using dilute HNO_3 and sent back for recycling. The U/Pu lean raffinate phase from the first cycle is subjected to a second solvent extraction cycle wherein indigenously synthesized 1,3-dioctyl oxy calix[4] arene-crown-6 (CC6), in isodecanol (IDA) and n-dodecane is used for selectiverecovery of [137]Cs. Recovered Cs is used for making Cs glass pencils for blood irradiators. Raffinate from the Cs recovery cycle is treated by tetra 2-ethylhexyldiglycolamide (TEHDGA) in IDA and n-dodecane. In this third cycle, entire actinides/lanthanides and ^{90}Sr are extracted quantitatively in the organic phase leaving the raffinate stream amenable to direct dilution and dispersal. Stripping of the organic phase using dilute nitric acid generates an aqueous stream rich in ^{90}Sr activity. This aqueous phase generated

from the stripping run also contains minor actinides/lanthanides and traces of Cs activity. The chemical and radiochemical composition of this aqueous phase used as feed for ^{90}Sr purification is given in Table 11.1.

11.4.2.1   [90]Sr Purification steps

A series of steps were followed to recover and purify ^{90}Sr from PUREX-HLLW which are described below:

STEP-1: In this step, trace impurity of ^{137}Cs is removed using granulated ammonium molybdo phosphate (AMP) column to reduce the man-rem exposure in the subsequent purification steps.

STEP-2: To remove α emitters, *viz.* minor actinides, lanthanides, and traces of U and Pu, 0.2M CMPO +1.2M TBP in n-dodecane was used both in solvent extraction as well as in extraction chromatographic mode. Extraction chromatography was found simpler when compared to liquid–liquid extraction.

STEP-3: Raffinate/effluent containing ^{90}Sr in ~3–4 M HNO_3 from STEP-2 was subjected toSr extraction step using di-(t-butyl cyclohexano)-18-Crown-6 in IDA and n-dodecane, and the extractaed ^{90}Sr was stripped quantitatively using dilute nitric acid.

STEP-4: The resultant strip solution was evaporated to get the concentrated ^{90}Sr product after removal of dissolved organic material, if any, by passing through a glass column containing polymeric resin (XAD). Radiochemical composition of the ^{90}Sr product is shown in Table 11.2.

Further purification of ^{90}Sr from the above feed solution was carried out using modified radiochemical precipitation method [28].

11.4.2.2   Ultra Purification of ^{90}Sr by SLM

In order to achieve the desired decontamination factor for ^{90}Y with respect to the α activity, a SLM technique was followed for further purification of this [90]Sr$(NO_3)_2$ solution using 2-Ethylhexyl 2-ethylhexylphosphonic acid (KSM-17) as carrier (Figure 11.6c). In this technique Sr$(NO_3)_2$ solution, purified by radiochemical precipitation method andadjusted to a pH 1–2 was used as the feed and 4M HNO_3 was usedas the receiver phase. After about every 8 h, the receiver compartment was replaced with fresh 4M HNO_3 for six timeswhile retaining the same [90]Sr$(NO_3)_2$ in the

TABLE 11.1

Composition of the Feed Solution Used for [90]Sr Purification

Constituents	Conc.	Constituents	Conc.
$[HNO_3]$, M	4.00	Cr, mg/L	931.4
Gross α, mCi/L	–	Fe, mg/L	4063
Gross β, Ci/L	20.90	La, mg/L	334
[90]Sr, Ci/L	9.50	Mg, mg/L	7.8
[137]Cs, mCi/L	7.50	Mn, mg/L	292.7
[106]Ru, mCi/L	2.00	Mo, mg/L	550
Al, mg/L	48.2	Na, mg/L	285.4
Ba, mg/L	25.0	Ni, mg/L	467.1
Ca, mg/L	2440	Sr, mg/L	248
Ce, mg/L	566.4	U, mg/L	35.9

TABLE 11.2

Radiochemical Composition of Sr Product

Analyte	Analysis
$[HNO_3]$, M	8.02
[90]Sr, Ci/ L	80–100
α-activity, Ci/L	10^{-5}–10^{-6}
γ-emitters	BDL

Stage-1

Stage-2

FIGURE 11.9 SLM based two stages for generation of carrier-free ^{90}Y.

feed compartment. This purified ^{90}Sr-nitrate solution by SLM technique was used as feed for ^{90}Y generation after the growth of ^{90}Y activity. ^{90}Y (10^{-8}Ci α/Ci) detected earlier in ^{90}Y acetate product could be reduced to less than 10^{-9}Ci α/Ci of ^{90}Y as desired for clinical applications. This ^{90}Sr-nitrate solution was preserved and used for ^{90}Y milking as and when required after allowing it to reach radioactive secular equilibrium.

Recent experiments carried out in our laboratory have indicated the possibility of removal of α-contamination in ^{90}Sr solution using HFSLM containing CMPO as carrier. More than 99.7% of α-activity due to ^{241}Am and Pu from a simulated solution could be removed. The technique will be adopted for purification of ^{90}Sr from HLLW after Cs depletion. This has given further confidence to adopt HFSLM at the back end of nuclear fuel cycle.

11.4.2.3 Milking of Carrier-Free ^{90}Y from Ultra Pure ^{90}Sr(NO$_3$)$_2$ Using Two-Stage SLM Generator System

A two-stage SLM-based generator system developed in-house [25] was used for the separation of carrier-free ^{90}Y, which is principally based on the solvent extraction properties of two ligands, namely KSM-17 and CMPO under optimum conditions. The system was operated in sequential modes with each cell having 5mL capacity. In the first stage, the equilibrium mixture of ^{90}Sr and ^{90}Y adjusted to a pH of 1–2 was used in the feed compartment and the receiver compartment contained 4 M HNO$_3$. KSM-17-based SLM was used for selective transport of ^{90}Y to the receiver phase in about 4 h. The product from this stage was taken out and placed in the feed compartment of the second stage, whereas the ^{90}Y depleted lean ^{90}Sr left out in the feed compartment of the first stage was transferred back to the feed reservoir for next cycle. CMPO-based SLM was used for transport of ^{90}Y in second stage where 1M acetic acid was used as receiver phase. Stage 1 and 2 of the generator system are shown in Figure 11.9. Beta activity transported after 4h are given in Table 11.3.

Thus, carrier-free ^{90}Y product of ~40 Ci/L in acetic acid medium could be generated under optimized conditions and supplied to RMC, Mumbai.

Quality control of the final product with respect to ^{90}Sr was carried out as per "BARC Method for Quality Control of ^{90}Y" which is based on extraction paper chromatography (EPC) [29]. The contamination of ^{90}Sr in all batches of ^{90}Y-acetate product was found within the permissible level. The decay curve of ^{90}Y activity as a function of time was also plotted to further ascertain the purity. The slope of the line plotted between Ln (β- activity) vs time in hour (Figure 11.10) was used found to be $T_{1/2}$ which was found to be 64.17h indicating the purity of the product.

Gross α impurity analyses in six different ^{90}Y-acetate batches (each about 140-160 mCi in 4 mL at the time of supply) were assayed after their complete decay using low background ZnS (Ag) scintillator counting system. The results are given in Table 11.4. These results were also validated using the Solid State Nuclear Track Detector (SSNTD) counting system.

Gross α impurity analyses in six different ^{90}Y-acetate batches (each about 140–160 mCi in 4 mL at the time of supply) were assayed after their complete decay using low background ZnS (Ag) scintillator counting system. The results are given in Table 11.4. These results were also validated using the Solid State Nuclear Track Detector (SSNTD) counting system.

The elemental concentrations *viz.* Al, Ca, Fe, Cu, Zn, Zr, and Pb in six batches of ^{90}Y-acetate products were analyzed by Inductively Coupled Plasma-Optical Emission

TABLE 11.3

Transport of β-Activity After 4 h in Two Stages
Experimental Condition **Cell Volume: 5 mL for Each Compartment, Feed: pH 1.24, Gross β-Activity: 82.96 Ci/L, Receiver Phase in 1st Stage: 4M HNO$_3$, Receiver Phase in 2nd Stage: 1M CH$_3$COOH**

STAGE-1		STAGE-2	
β-activity (Ci/L)		β-activity (Ci/L)	
Feed	Product	Feed	Receiver
41.48	40.86	0.52	39.64
Yield ~98.5%		Yield ~97.01%	

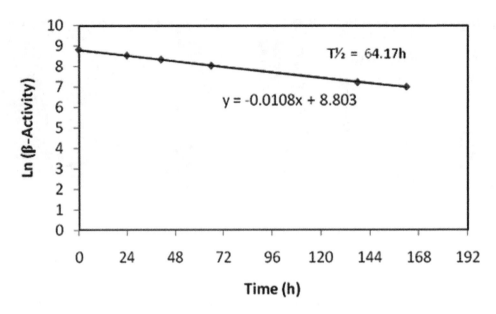

FIGURE 11.10   Ln (β-activity) vs time plot for product in acetic acid medium.

TABLE 11.4

Alpha Assay of Decayed ^{90}Y Product [Permissible Limit-α: <1 × 10^{-9} Ci α /Ci ^{90}Y, ^{90}Sr: < 10^{-6} Ci/Ci of ^{90}Y]

Batch No.	^{90}Y β- activity in separated product (Ci/L)	α activity in decayed ^{90}Y-acetate solution (Ci α /Ci ^{90}Y)	^{90}Sr activity (Ci ^{90}Sr/Ci ^{90}Y)
1	38.90	1.65 × 10^{-9}	1.5 × 10^{-7}
2	38.56	0.55 × 10^{-9}	1.4 × 10^{-7}
3	37.48	1.07 × 10^{-9}	1.2 × 10^{-7}
4	37.60	0.66 × 10^{-9}	1.4 × 10^{-7}
5	35.46	0.34 × 10^{-9}	1.3 × 10^{-7}
6	34.96	0.15 × 10^{-9}	1.4 × 10^{-7}

^{90}Sr: β counting after its separation from ^{90}Y by KSM-17-Based Extraction Paper Chromatography)

TABLE 11.5

Chemical and Radiochemical Analysis and Quality of ^{90}Y-Acetate

Description	Specifications
Activity Lots	~140–160 mCi
Form	^{90}Y(CH$_3$COO)$_3$
Clarity and Color	Clear &Colorless
pH	2.2-2.5
Extraction Yield	>90%
Radiochemical Purity	>99%
Radionuclide γ impurity	<0.001%
Radionuclide Impurity (β-content due to ^{90}Sr)	≤ 2.5μCi of^{90}Sr /Ci ^{90}Y
Radionuclide Gross α Impurity	≤ 1nCi α /Ciof ^{90}Y
Fe Content	<5μg/Ci
Pb Content	<10μg/Ci
Zn, Cu, Al, Cd Content	<30μg/Ci each

Spectrometry (ICP-OES) after near complete decay of the beta activity and found more insignificant than that in reagent blank (1M acetic acid) indicating no metallic impurities in the ^{90}Y-aceate product. The chemical, radiochemical, and other analyses are given in Table 11.5. The values reported are the average values of six ^{90}Y-acetate batches separated using a two-stage SLM-based ^{90}Sr–^{90}Y generator system.

Presently, carrier-free ^{90}Y-acetate products having specific activity of ~ 40Ci/L are being supplied for radiopharmaceutical applications in one of the radiation medicine centers. 140–160 mCi activity lots are being milked nowadays using the generator system. To meet the rising demand for ^{90}Y activity, multiple generator systems and use of HFSLM-based techniques are being looked at.

11.4.3 ^{137}Cs Separation

The fission product nuclide, ^{137}Cs ($t_{1/2}$ ~30 y), which is a major constituent of HLLW, has large heat output (~0.42 W/g) and its removal from HLLW is important from a waste management point of view as it may create deformations in the vitrified waste product resulting in the release of the vitrified radionuclides into the environment [30]. Secondly, the separation of the radiotoxic nuclides including ^{137}Cs will reduce the waste volume significantly. Long half-life and high energy gamma ray (661 keV) emitted from ^{137}Cs make this radionuclide a viable alternative source for gamma irradiators to replace the commonly used ^{60}Co ($t_{1/2}$ = 5.2 y, γ = 1173 keV and 1332 keV) for the sterilization of medical accessories, food preservation, sewage sludge treatment, etc. [31]. In view of these, it is required to separate the radio-cesium from the HLLW for its subsequent use.

Depending on the nature of the waste solution, various types of solvent extraction and ion exchange methods have been studied. From alkaline wastes Resorcinol formaldehyde poly-condensate resin (RFPR resin) and sodium tetra phenyl borate are studied [32,33] for removal of [137]Cs. Ammonium molybdophosphate (AMP), cobalt dicarbolides, crown ether, calix-crown-6, etc., have beenused for ^{137}Cs removal from acidic waste solutions. The calix-crown reagents have been extensively studied in various research groups using SLM, PIM, and HFSLM techniques with promising results [34–37]. Mohapatra etal. [38,39] have studied various calix[4]-bis-crown-6 ligands for their efficacy in transporting ^{137}Cs from nitric acid medium employing HFSLM technique. They have reported near quantitative transport (> 99%) of ^{137}Cs in 6h time duration from 3M HNO_3 when the strippant was distilled water. The carrier solvent was composed of 1 mM calix[4]-bis-2,3-naphtho-crown-6 (CNC) in 20% (v/v) dodecane with 80% (v/v) NPOE and 0.4% alamine 336 [38]. According to them, the high decontamination factor with respect to other metal ions present in HLW and long term stability of the system shows that the possible application of the HFSLM system in real samplesthough radiation stability of the system is still a challenge which needs to be overcome. They further studied Bis-octyloxy-calix[4]arene-mono-crown-6 (CMC) as a carrier in HFSLM mode to understand its behavior for separation of ^{137}Cs from pure nitric acid medium as well as from synthetic high level waste (SHLW) solution containing ~300 mg/L of Cs [39]. A solvent composition of 1×10^{-2} M CMC in 40% IDA/n-dodecane was found to result in near quantitative transport (> 99%) of ^{137}Cs at tracer level within 2h duration in presence of 4M HNO_3 whereas in presence of SHLW condition it required 6h for ~ 95% transport. They found out that the stability of the membrane was highly satisfactory for a continuous operation of 12 days. Recently Chaudhry et al. [40] have reported an electro driven cation transport though HFSLM containing chlorinated cobalt dicarbollide as carrier for selective Cs separation. They applied a working potential of 3.0 V to achieve a significantly faster transport rate of Cs simultaneously achieving a large separation factor with respect to Na. The stability of the membrane was reported to be excellent with an added advantage of elimination of the stripping agent. Most of the studies including the ones mentioned above have been carried out from nitric acid solution and synthetic high level waste solution. Very few reports are available on the application of liquid membrane technology for actual waste solution. In our laboratory, a hollow fiber-based supported liquid membrane study was carried out to understand the feasibility of ^{137}Cs removal from actual high level waste solution from reprocessing origin albeit in diluted form in 300 mL scale [41]. 0.01M bis-octyl- benzo-calix[4]arene-mono-crown-6 (CMC)(Figure 11.6d) in 40% iso-decanol and 60% dodecane was used as the extractant. It was observed that more than 90% of ^{137}Cs could be selectively separated from high level liquid waste stream of reprocessing origin diluted to 100 times employing only 35 fibers. The contractor modules were kept in contact with the diluted HLLW for 50 days and the second run carried out after 50 days showed excellent reproducibility suggesting the radiation tolerance of the fibers upto a certain radiation level. The experimental set up is shown in Figure 11.11. The gamma spectrum of the actual waste solution as well as the product solution from the HFSLM experiment are given in Figure 11.12 indicating the selectivity of the CMC and possibility to obtain highly pure ^{137}Cs product. Further studies using actual waste are being planned.

11.4.4 Actinide/Lanthanide Group Separation

Actinide/Lanthanide group separation is a key step in the "Partitioning and Transmutation" process for management of HLW at the back end of nuclear fuel cycle. Lanthanides with high neutron absorption cross sections have adverse effects on the transmutation of long lived actinides and thus require to be separated from them. Similar charge and ionic radii of the minor actinides and lanthanides make the separation among them a difficult and challenging task. But due to higher spatial distribution of the '5f' valence orbitals compared to '4f' valence orbitals, actinides form stronger complexes with soft donor ligands like N and S. Dithiophosphinic acidextractants with 'S' donor atoms like bis(2,4,4-trimethylpentyl)dithiophosphinic acid(Cyanex-301)andbis-(chlorophenyl)dithiophosphinicacid havebeenstudiedasselectiveextractan tsforthetrivalent actinides.

Bhattacharyya et al. [42] studied Cyanex-301 as carrier molecule in polypropylene (PP) hollow fiber-based supported liquid membrane for separation of trivalent actinides, *viz.* Am^{3+} and lanthanides, *viz.* La^{3+}, Eu^{3+}, Tb^{3+}, Ho^{3+}, Yb^{3+}, and Lu^{3+}. They observed much faster transport rates of Am^{3+} as compared to those of the trivalent lanthanide ions investigated in the study with the transport rates being affected by the concentration of Cyanex-301. Positive results with respect to high decontamination factors (DF) as well as throughputs suggested that this method may be applied for real waste solutions. They further studied [43] the effect of various neutral auxiliary donor ligands, *viz.* tri-n-butyl phosphate (TBP), N,N,-dihexyl octanamide (DHOA) and 2,20 -bipyridyl (Bipy) as synergists on the selective recovery of Am(III) over lanthanides. The presence of synergists showed the possibility of Am(III) recovery at lower pH values with improved separation factors. At pH 2, the separation factor value for different synergists followed the order: bipyridyl (350) > DHOA (50) > TBP (8). They also investigated the possibility of applying an aqueous soluble actinide selective ligand phenyl sulphonic acid functionalized bis-triazinyl pyridine (SO_3-Ph-BTP) [44] using a

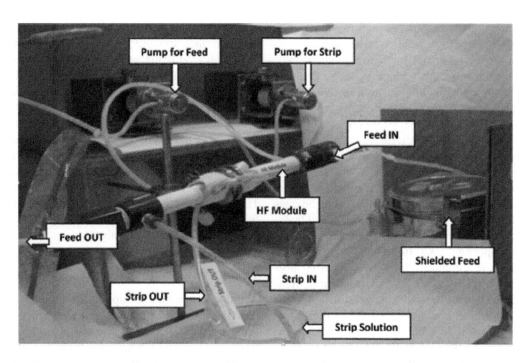

FIGURE 11.11 Experimental set up for the recovery of ^{137}Cs using Cs selective calix crown based HFLSM Technique [Reprinted from Journal of Environmental Chemical Engineering, 5, P. Jagasia; P.S. Dhami; P.K. Mohapatra; S.A. Ansari; S.Y. Jadhav; G.K. Kalyankar; P.M. Gandhi and U.K. Kharul, Recovery of radio-cesium from actual high level liquid waste using solvents containing calix[4]arene-crown-6 ligands, 4134–4140, Copyright (2017) with permission from Elsevier [41]].

TODGA-based HFSLM technique for separation of Am^{3+} from Eu^{3+}. They showed that selective transport of Eu^{3+} from a mixture of Am^{3+} and Eu^{3+} using TODGA as the carrier from a feed solution of 1M HNO_3 containing 10 mM ofSO_3-Ph-BTP was possible with a DF > 100.

11.4.5 ALPHA DECONTAMINATION OF OTHER RADIOACTIVE WASTE STREAMS

11.4.5.1 Decontamination of α-Bearing Radioactive Stream from Metallurgical Waste Treatment Plant

A hollow fiberrenewable liquid membrane, using N,N,N,N-tetra-2-ethylhexyl diglycolamide (TEHDGA) (Figure 11.1b where R_1, R_2, R_3, R_4 = 2-ethyl hexyl) as carrier, was studied extensively for alpha decontamination of aqueous waste stream containing high salt concentrations [45] mainly nitrates of Al, Mg, Ca and Na (> 7M nitrate) having alpha emitters [241]Am and Pu. Gross alpha activity in the aqueous waste stream used as feed was about 228456 Bq/mL in about 1.2M HNO_3. In this study, a mixture of complexing reagents *viz.* diethylene triamine pentacaetic acid (DTPA) and lactic acid (LA) was used as the receiving phase. The method could reduce the gross alpha activity to less than 37 Bq/mL in three cycle of operation. The decontaminated solution showed presence of [241]Am alone. Pu alpha activity was found to be practically undetectable. The experimental set up used at laboratory scale is given in Figure 11.13. The method can be scaled up and employed to treat such waste at engineering scale.

11.4.5.2 Decontamination of α-Bearing Stream from Spent Resin Treatment

An ion-exchange method is widely employed for purification of plutonium from uranium and americium in reprocessing plants. In this method, Pu solution is conditioned with respect to its oxidation state and acidity. Pu is converted to its tetravalent state using NO_2- and free nitric acid concentration of about 6.9–7.2M is maintained in feed solutions. Under such conditions, Pu^{4+} forms its anionic complex *viz.* $Pu(NO_3)_6^{2-}$[46] which gets loaded onto the anion exchange preconditioned column with ~7M nitric acid. After washing the column with ~7M nitric acid, it is eluted with 0.5M HNO_3. Beside purification, the method has the additional advantage of getting Pu in higher concentrations. On prolonged use, the resin often gets degraded due to the prevailing conditions of acidity and alpha radiation requiring resin replacement. To manage the spent resin it needs decontamination as it contains significant alpha activity. Chemical treatment using ammonium carbonate was found to be suitable for resin decontamination with respect to alpha activity. The aqueous carbonate stream thus obtained was decontaminated using TEHDGA-based HFSLM technique after its neutralization and acidification to about 2M HNO_3. The hollow fiber supported liquid membrane set up used for α-decontamination was similar to that as described for the decontamination of α-bearing radioactive stream from metallurgical waste treatment plant (Figure 11.13). Alpha activity of the decontaminated stream could be brought down to less than 1 Bq/mL.

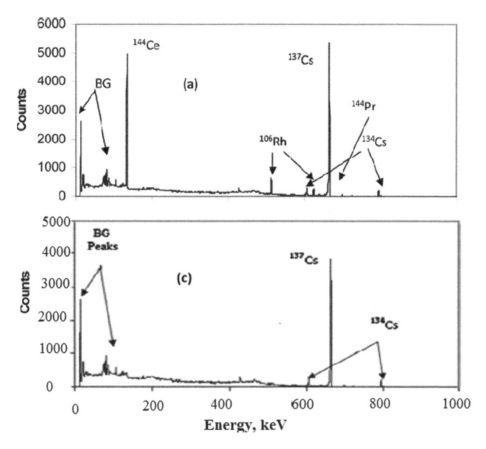

FIGURE 11.12 Gamma spectrum of actual HLLW and the product obtained from HFSLM. Experiment [Reprinted from Journal of Environmental Chemical Engineering, 5, P. Jagasia; P.S. Dhami; P.K. Mohapatra; S.A. Ansari; S.Y. Jadhav; G.K. Kalyankar; P.M. Gandhi and U.K. Kharul, Recovery of radio-cesium from actual high level liquid waste using solvents containing calix[4]arene-crown-6 ligands, Copyright (2017) with permission from Elsevier [41]].

FIGURE 11.13 Hollow Fiber Renewable Liquid Membrane set up used for α-decontamination.

11.4.5.3 Recovery of Pu and U from Laboratory Analytical Waste

Any radiochemical laboratory during its routine operations generates large volumes of secondary waste containing significant quantities of U, Pu, Am, etc. Recovery of such radionuclides from the waste makes it easier to manage the large volume of the waste as well as the recovery of various

important radio-elements for further applications. Rathore et al., [47] have demonstrated the transport feasibility of U and Pu from acidic waste solution via HFSLM employing TBP in n-dodecane as carrier. They have reported the feasibility of more than 99% recovery of U and Pu from process effluent in the presence of fission products employing this technique. Taking this literature report one step further, Ansari et al., [48] applied the same HFSLM technique with 30% TBP/n-dodecane as carrier for recovery of U and Pu from real laboratory waste solution. They recovered pure fractions of U and Pu from analytical waste solutions employing two-stage operations. As per the report, they have been successful in recovering ~50 g of Pu having purity > 99% from 10L waste solution using the HFSLM method which demonstrates the excellent possibility for the application of HFSLM in nuclear industry.

11.5 CONCLUSIONS

The studies carried out at the backend of nuclear fuel cycle, as mentioned above, have clearly illustrated that liquid membrane techniques are very useful for treatment of various process streams including actinide partitioning, actinide/lanthanide group separation, purification of ^{90}Sr to achieve the high purity level, milking of ultrapure carrier free ^{90}Y, separation of ^{137}Cs, etc. Various other secondary streams contaminated with alpha activity could also be decontaminated for better management of such waste and thus protecting the environment. Based on the experiments carried out, it can be concluded that hollow fiber liquid membrane is the configuration most suited for applications with real nuclear waste solutions at higher scales due to large throughput. The requirement of solvents is much less compared to solvent extraction and thus makes it an attractive option to study the costly, exotic solvents for target oriented applications. But challenges are multifold to establish this method as a mature industrial scale technique. Stability of the membrane is a major issue which needs to be overcome. Radiation resistant fibers as well as sealant materials are the prime requirement for the application of this technique in actual waste streams. The new developments in HF contactors are trying to address the current challenges, andrecent studies on radiation effects on polypropylene membrane indicated that these membrane can withstand radiation up to 0.5 Mrad (low level waste can easily be processed).

REFERENCES

1. H.J. Bhabha and N.B. Prasad, A study of the contribution of atomic energy to a power programme in India, Proc. 2nd Int. Conf. on Peaceful Uses of Atomic Energy, Geneva, Vol. 1 (1958) 89.

2. L. Boyadzhiev and Z. Lazarova, *Membrane Separations Technology. Principles and Applications*, R.D.Noble and S.A.Stern (Eds.), Elsevier Science B.V. (1995), p. 283.

3. P.K. Mohapatra and V.K. Manchanda, Liquid membrane based separations of actinides and fission products, *Indian J. Chem.* 42 (2003) 2925–2938.

4. P.R. Danesi, E.P. Horwitz and P.G. Rickert, Rate and mechanism of facilitated americium(III) transport through a supported liquid membrane containing a bifunctional organophosphorus mobile carrier, *J. Phys. Chem.* 87 (1983) 4708.

5. L. Boyadzheiv, Liquid pertraction or liquid membranes: State of the art, *Sep. Sci. Technol.* 25 (1990) 187.

6. J.N. Mathur, M.S. Murali and K.L. Nash, Actinide partitioning-A review, *Solvent Extr. Ion Exch.* 19 (2001) 357.

7. S.A. Ansari, P.N. Pathak, P.K. Mohapatra and V.K. Manchanda, Aqueous partitioning of minor actinides by different processes, *Sep. Purif. Rev.* 40 (2011) 43–76.

8. E.P. Horwitz and D.G. Kalina, The extraction of Am(III) from nitric acid by octyl(phenyl)-N,N-diisobutylcarbamoylmethylphosphine oxide-tri-n-butyl phosphate mixture, *Solvent Extr. Ion Exch.* 2 (1984) 179.

9. W.W. Schulz and E.P. Horwitz, The TRUEX process and the management of liquid TRU waste, *Sep. Sci. Technol.* 23 (1988) 1191.

10. J.N. Mathur, M.S. Murali, R.H. Iyer, A. Ramanujam, P.S. Dhami, V. Gopalakrishnan, L.P. Badheka and A. Bannerji, Extraction chromatographic separation of minor actinides from PUREX high-level wastes using CMPO, *Nucl. Technol.* 109 (1995) 216.

11. S. Sriram and V.K. Manchanda, Transport of metal ions across a supported liquid membrane (SLM) using dimethyldibutyltetradecyl-1,3-malonamide (DMDBT-DMA) as the carrier, *Solvent Extr. Ion Exch.* 20 (2002) 97–114.

12. S.A. Ansari, P.N. Pathak, M. Hussain, A.K. Prasad, V.S. Parmar and V.K. Man-chanda, N,N,N',N' tetraoctyl diglycolamide (TODGA): A promising extractant for actinide-partitioning from high-level waste (HLW), *Solvent Extr. Ion Exch.* 23 (2005) 463.

13. G. Modolo, H. Asp, H. Vijgen, R. Malmbeck, D. Magnusson and C. Sorel, Demonstration of a TODGA-based continuous counter-current extraction process for the partitioning of actinides from a simulated PUREX raffinate. Part II. Centrifugal contactor runs, *Solvent Extr. Ion Exch.* 26 (2008) 62–76.

14. S. Panja, R. Ruhela, S.K. Misra, J.N. Sharma, S.C. Tripathi and A. Dakshinamoorthy, Facilitated transport of Am(III) through a flat-sheet supported liquid membrane (FSSLM) containing tetra(2-ethyl hexyl) diglycolamide (TEHDGA) as carrier, *J. Membr. Sci.* 325 (2008) 158–165.

15. J.N. Mathur, M.S. Murali, P.R. Natarajan, L.P. Badheka, A. Bannerji, A. Ramanujam, P.S. Dhami, V. Gopalakrishnan, R.K. Dhumwad and M.K. Rao, Partitioning of actinides from high-level waste streams of PUREX process using mixtures of CMPO, *Waste Manag.* 13 (1993) 317.

16. A. Ramanujam, P.S. Dhami, V. Gopalakrishnan, N.L. Dudwadkar, R.R. Chitnis and J.N. Mathur, Partitioning of actinides from high level waste of PUREX origin using octylphenyl-N,N'-diisobutylcarbamoylmethyl phosphine oxide (CMPO)-based supported liquid membrane, *Sep. Sci. Technol.* 34 (1999) 1717–1728.

17. P.R. Danesi, Permeation of metal ions through hollow-fibre supported liquid membranes: Concentration equations for once-through and recycling module arrangements, *Solvent Extr. Ion Exch.* 2 (1984) 115–120.

18. A.M. Sastre, A. Kumar, J.P. Shukla and R.K. Singh, Improved techniques in liquid membrane separations: An overview, *Sep. Purif. Methods* 27 (1998) 213–298.

19. A.S. Ansari, P.K. Mohapatra and V.K. Manchanda, Recovery of actinides and lanthanides from high-level waste using hollow-fiber supported liquid membrane with TODGA as the carrier, *Ind. Eng. Chem. Res.* 48 (2009) 8605–8612.

20. S.A. Ansari, P.K. Mohapatra, D.R. Raut, M. Kumar, B. Rajeswari and V.K. Manchanda, Performance of some extractants used for 'actinide partitioning' in a comparative hollow fibre supported liquid membrane transport study using simulated high level nuclear waste, *J. Membr. Sci.* 337 (2009) 304–309.

21. R. Chakravarty, U. Pandey, R.B. Manolkar, A. Dash, M. Venkatesh and M.R.A. Pillai, Development of an electrochemical [90]Sr-^{90}Y generator for separation of ^{90}Y suitable for targeted therapy, *Nucl. Med. Biol.* 35 (2008) 245–253.

22. P.V. Achuthan, P.S. Dhami, R. Kannan, V. Gopalakrishnan and A. Ramanujam, Separation of carrier-free ^{90}Y from high level waste by extraction chromatographic technique using 2-ethyl hexyl-2-ethylhexyl phosphonic acid (KSM-17), *Sep. Sci. Technol.* 35 (2000) 261.

23. A. Ramanujam, P.V. Achuthan, P.S. Dhami, R. Kannan, V. Gopalakrishnan, V.P. Kansra, R.H. Iyer and K. Balu, Separation of carrier free 90Y from high level waste by supported liquid membrane using KSM-17, *J. Radioanal. Nucl. Chem.* 247 (2001) 185–191.

24. B.T. Hsieh, G. Ting, H.T. Hsieh and L.H. Shen, Preparation of carrier-free Yttrium-90 for medical applications by solvent extraction chromatography, *Appl. Radiat. Isot.* 44 (1993) 1473.

25. P.S. Dhami, P.W. Naik, N. Dudwadkar, R. Kannan, P.V. Achuthan, A.D. Moorthy, U. Jambunathan, S.K. Munshi, P.K. Dey, U. Pandey and M. Venkatesh, Studies on the development of a two stage SLM system for the separation of carrier-free ^{90}Y using KSM-17 and CMPO as carriers, *Sep. Sci. Technol.* 42 (2007) 1107–1121.

26. P.W. Naik, P. Jagasia, P.S. Dhami, P.V. Achuthan, S.C. Tripathi, S.K. Munshi, P.K. Dey and M. Venkatesh, Separation of carrier-free ^{90}Y from ^{90}Sr by SLM technique using D2EHPA in N-dodecane as carrier, *Sep. Sci. Technol.* 45 (2010) 554–561.

27. European Pharmacopoeia monographs (Ed 8.2) (2014).

28. A. Ramanujam, P.S. Dhami, R.R. Chitnis, P.V. Achuthan, R. Kannan, V. Gopalakrishnan and K. Balu, *Separation of ^{90}Sr from PUREX High level waste and development of a ^{90}Sr-^{90}Y generator, Report* No. BARC/2000/E/09 (2000).

29. U. Pandey, P.S. Dhami, P. Jagasia, M. Venkatesh and M.R.A. Pillai, Extraction paper chromatography technique for the radionuclidic purity evaluation of ^{90}Y for clinical use, *Anal. Chem.* 80 (2008) 801.

30. L.L. Hench, D.E. Clark and J. Campbell, High level waste immobilization forms, *Nucl. Chem. Waste Manag.* 5 (1984) 149–173.

31. W.W. Schulz and L.A. Bray, Solvent extraction recovery of by product ^{137}Cs and ^{90}Sr from HNO$_3$ solutions-A technology review and assessment, *Sep. Sci. Technol.* 22 (1987) 191–214.

32. A. Dash, R. Ram, Y.A. Pamale, A.S. Deodhar and M. Venkatesh, Recovery of ^{137}Cs from laboratory waste using solvent extraction with sodium tetraphenylboron (TPB), *Sep. Sci. Technol.* 47 (2012) 81–88.

33. D. Banerjee, M.A. Rao, J. Gabriel and S.K. Samanta, Recovery of purified radio-cesium from acidic solution using ammonium molybdophosphate and resor-cinol formaldehyde polycondensate resin, *Desalination* 232 (2008) 172–180.

34. D.R. Raut, P.K. Mohapatra, S.A. Ansari, A. Sarkar and V.K. Manchanda, Selective transport of radio-cesium by supported liquid membrane containing calix[4]-bis-crown-6 ligands as the mobile carrier, *Desalination* 232 (2008) 262–271.

35. P. Kandwal, S.A. Ansari and P.K. Mohapatra, Transport of cesium using hollow fiber supported liquid membrane containing calix[4]arene-bis(2,3-naphtho)crown-6 as the carrier extractant: II. Recovery from simulated high level waste and mass transfer modeling, *J. Membr. Sci.* 384 (2011) 37–43.

36. J.S. Kim, S.K. Kim, J.W. Ko, E.T. Kim, S.H. Yu, M.H. Cho, S.G. Kwon and E.H. Lee, Selective transport of cesium ion in polymeric CTA membrane containing calixcrown ethers, *Talanta* 52 (2000) 1143–1148.

37. D.R. Raut, P.K. Mohapatra, S.A. Ansari and V.K. Manchanda, Evaluation of a calix[4]-bis-crown-6 ionophore-based supported liquid membrane system for selective ^{137}Cs transport from acidic solutions, *J. Membr. Sci.* 310 (2008) 229–236.

38. P. Kandwal, P.K. Mohapatra, S.A. Ansari and V.K. Manchanda, Selective cesium transport using Hollow-fiber

39. P. Jagasia, S.A. Ansari, D.R. Raut, P.S. Dhami, P.M. Gandhi, A. Kumar and P.K. Mohapatra, Hollow fiber supported liquid membrane studies using a process compatible solvent containing calix[4]arene-mono-crown-6 for the recovery of radio-cesium from nuclear waste, *Sep. Purif. Technol.* 170 (2016) 208–216.

40. S. Chaudhury, A. Bhattacharyya, S.A. Ansari and A. Goswami, A new approach for selective Cs$^+$ separation from simulated nuclear waste solution using electro driven cation transport through hollow fiber supported liquid membranes, *J. Membr. Sci.* 545 (2018) 75–80.

41. P. Jagasia, P.S. Dhami, P.K. Mohapatra, S.A. Ansari, S.Y. Jadhav, G.K. Kalyankar, P.M. Gandhi and U.K. Kharul, Recovery of radio-cesium from actual high level liquid waste using solvents containing calix[4]arene-crown-6 ligands, *J. Environ. Chem. Eng.* 5 (2017) 4134–4140.

42. A. Bhattacharyya, P.K. Mohapatra, S.A. Ansari, D.R. Raut and V.K. Manchanda, Separation of trivalent actinides from lanthanides using hollow fiber supported liquid membrane containing Cyanex-301 as the carrier, *J. Membr. Sci.* 312 (2008) 1–5.

43. A. Bhattacharyya, S.A. Ansari, P. Kandwal, P.K. Mohapatra and V.K. Manchanda, Selective recovery of Am(III) over Eu(III) by hollow fiber supported liquid membrane using cyanex 301 in the presence of synergists as the carrier, *Sep. Sci. Technol.* 46 (2011) 205–214.

44. A. Bhattacharyya, S.A. Ansari, D.R. Prabhu, D. Kumar and P.K. Mohapatra, Highly efficient separation of Am^{3+} and Eu^{3+} using an aqueous soluble sulfonated BTP derivative by hollow-fiber supported liquid membrane containing TODGA, *Sep. Sci. Technol.* 54 (2019) 1512–1520.

45. C.S. Kedari, B.K. Kharwandikar and K. Banerjee, Technology development for the alpha decontamination of raffinate of metallurgical waste by hollow fiber renewable liquid membrane process, *BARC Newsl* July–Aug (2016): 6–9.

46. D.K. Veirs, C.A. Smith, J.M. Berg, B.D. Zwick, S.F. Marsh, P. Allen and S.D. Conradson, Characterization of the nitrate complexes of Pu(IV) using absorption spectroscopy, ^{15}N NMR and EXAFS, *J. Alloys Compd.* 213/214 (1994) 328–332.

47. N.S. Rathore, J.V. Sonawane, S.K. Gupta, A.K. Pabby, A.K. Venugopalan, R.D. Changrani and P.K. Dey, Separation of uranium and plutonium from aqueous acidic wastes using a hollow fiber supported liquid membrane, *Sep. Sci. Technol.* 39 (2004) 1295–1319.

48. S.A. Ansari, S. Chaudhury, P.K. Mohapatra, S.K. Aggarwal and V.K. Manchanda, Recovery of plutonium from analytical laboratory waste using hollow fiber supported liquid membrane technique, *Sep. Sci. Technol.* 48 (2013) 208–214.

12 Hollow Fiber Membrane-Based Analytical Techniques
Recent Advances

Anil K. Pabby, B. Swain, V. K. Mittal, N. L. Sonar,
T. P Valsala, D. B. Sathe, R. B. Bhatt, and A. M. Sastre

CONTENTS

12.1 INTRODUCTION

The field of analysis of chemicals, pharmaceuticals in biological matrices, and toxic metallic species in water, food, and in environmental samples has undergone a major renewal during the last decades. There has been a massive focus on improvement of analytical instrumentation for separation and detection of the analytes of interest [1, 2].

Problems of chemical analysis almost always involve two steps; separation of the desired constituent and measurement of the amount or concentration of this constituent. Under the scope of these two steps, sample preparation, proposed analytical technique, and instrumention coupled with analytical system are of paramount importance. Among various analytical techniques, membrane extraction has gained great attention due to its several merits as compared to conventional methods. In all types of membrane extraction, the membrane separates the sample phase (often called donor or feed solution) from the acceptor or strip phase, and the analyte molecules pass through the membrane from the donor to the acceptor. The membrane extraction techniques can be divided into porous and nonporous membrane techniques. Another distinction is between one-, two-, and three-phase membrane extraction techniques. Figure 12.1 is a schematic diagram of the various phase arrangements as shown.

12.1.1 WHY HOLLOW FIBER DEVICES?

The novel geometry of the hollow fiber capillary (high area per unit volume) makes this membrane a desirable candidate for application in the analytical field. In membrane extraction devices, porous polypropylene hollow fibers are generally used, often in a disposable way, which minimizes carry-over problems and reduces costs [2]. On the other hand, manual manipulations are needed, limiting the possibility for automation. With these devices, the extraction can be carried out in a static mode, either in large sample volumes, where the extraction is not intended to be complete, or in small volumes aiming for complete extraction. Usually stirring is applied to increase the speed of mass transfer. Further, hollow fibers can be connected in flow systems [2, 3].

The next important component is sample preparation. This step is still seen as crucial and critical in any analytical determination and can be viewed as the rate determining step in an analytical procedure [4, 5]. Currently, focus is on the search for simple, cheap, and environmentally friendly sample preparation techniques. As the simplest overview, sample preparation techniques can generally be categorized into the solid-phase and liquid-phase methods. Liquid–liquid extraction (LLE) and solid-phase extraction (SPE) are the two most well known traditional modes of sample

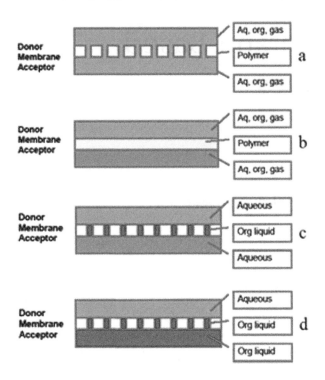

FIGURE 12.1 Schematic diagram of different membrane extraction phase systems: (a) one-phase membrane extraction (dialysis), (b) three-phase polymeric membrane extraction, (c) three-phase SLM extraction, and (d) two-phase SLM extraction system (MMLLE). Copyright (2012) with permission from Elsevier [9].

preparation that have been used extensively for many years. However, these conventional methods are usually time consuming, laborious, entailing several consecutive steps, and requiring excessive quantities of organic solvents. These issues have eclipsed many research studies toward developing more environmentally friendly, simpler, and faster extraction techniques [3, 6]. Consequently, a great variety of miniaturized sample preparation methods have been proposed as an effective substitution for the conventional methods. Some of them are the single-drop microextraction (SDME), dispersive liquid–liquid microextraction (DLLME), and membrane-based liquid-phase microextraction techniques [7]. On the other side, for chromatographic and electrophoretic analysis of trace amounts of ionizable analytes liquid phase microextractions (LPME) in which extraction, clean-up, and pre-concentration are all integrated into a single step are now favored over SPE-based procedures [4, 8, 9]. Under the thrust area, among the recently developed microextraction techniques, membrane-based liquid-phase microextraction methods have gained popularity owing to their advantages over the other similar modes of microextraction, comprising much more effective sample clean-up, more stable extraction procedures, greater preconcentration factors, and negligible consumptions of organic solvents [10]. Membrane-based liquid–liquid–liquid microextraction (LLLME), generally known as hollow fiber-based liquid-phase microextraction (HF-LPME), was first introduced by Pedersen-Bjergaard in 1999 to address the drawbacks associated with the SDME method [11]. In

SDME, proposed by Liu and Dasgupta as well as Jeannot and Cantwell, a single drop of organic solvent is suspended in the aqueous sample solution by a microsyringe and the extraction is based on the distribution of analyte between two organic and aqueous phases [12,13]. However, an apparent disadvantage of this method is the instability of the single drop during the extraction such that the drop can be easily disappeared in the sample solution [14]. Therefore, HF-LPME was initiated as a protected mode of LPME to provide much more stable extractions. In HF-LPME, the organic solvent is immobilized and sustained in the pores of a porous hydrophobic hollow fiber by capillary forces, and the acceptor phase is introduced into the lumen of the hollow fiber by a microsyringe. Then, the system is located in the sample solution and improved mass transfer is also accomplished by a magnetic stirrer in the sample container. Shortly after the inception of this method, it was developed rapidly and became a fascinating method of microextraction in the analytical community due to its inherent merits over conventional analytical methods.

12.1.2 Two-Phase and Three-Phase HF-LPME Methods

These methods can be put into two main classes based on the number of exploited liquid phases. These categories are three-phase HF-LPME and two-phase HF-LPME methods. The three-phase method can be simply described as an aqueous-organic-aqueous system in which the immobilized organic solvent or supported liquid membrane (SLM) is exposed to two aqueous phases of sample solution and the aqueous acceptor phase located inside the hollow fiber. In contrast, in the two-phase HF-LPME, the acceptor phase solution is also the same as the organic solvent of SLM and both are immiscible in water [15]. A schematic illustration of both configurations is shown in Figure 12.2A and 12.2B.

Compared to three-phase HF-LPME, two-phase HF-LPME is capable of extracting uncharged hydrophobic analytes, which cannot be efficiently extracted by a three-phase module. What is more important is that, in the two-phase methods, the final organic extract can be directly injected to the gas chromatographic (GC) instrument. Nevertheless, in the two-phase HF-LPME methods, only the partition coefficient determines the maximum enrichment. In contrast, in the three-phase mode, other degrees of freedom exist, which can be used to tune the extraction conditions, e.g., the pH gradient between the sample solution and the aqueous phase. Three-phase HF-LPME also provides much better clean up than that of the two-phase mode [2].

This chapter begins with a brief introduction to HF-LPME techniques, basic principles, theoretical aspects, possible setups, and applications in the field of environmental, chemical, food and beverage, and biological samples in addition to a comprehensive literature survey. Also, this article covers advances in the implementation of HF-LPME techniques and important parameters used in optimising

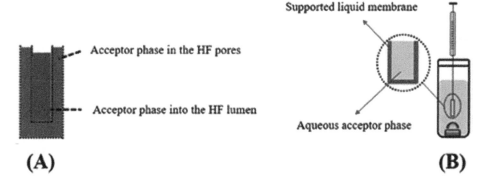

FIGURE 12.2 The schematic of basic configurations of (A) two-phase HF-LPME (rod-like setup), and (B) three-phase HF-LPME. Copyright (2013) with permission from Elsevier [3].

analyte extraction and the role of enrichment, selectivity, and clean up prior to coupling to an on-line or off-line analytical instrument.

12.2 PRINCIPLES OF TWO-PHASE AND THREE-PHASE HF-LPME

The mechanism of the extraction is presented by **Equation 12.1**, which scrutinizes and investigates the two-phase HF-LPME system from a theoretical point of view. As seen in this reaction, the system exists in equilibrium between the concentrations of an analyte in the aqueous sample (A_{sample}) and in the final organic acceptor phase ($A_{acceptor}$) [16].

$$A_{sample} \leftrightarrow A_{acceptor} \qquad (12.1)$$

Passive diffusion is the main driving force in the two-phase HF-LPME system, which deals with direct extraction of analytes from the sample solution to the organic acceptor phase. Thus, the distribution coefficient of analyte can be expressed below as shown in **Equation 12.2**:

$$D_{acceptor/sample} = \frac{C_{eq.acceptor}}{C_{eq.sample}} = \alpha_D . K_{acceptor/sample} \quad (12.2)$$

where $C_{eq,acceptor}$, and $C_{eq,sample}$ are the total concentration of analyte in the organic acceptor phase and the concentration of analyte in the sample solution, respectively. α_D and $K_{acceptor/sample}$ are the extractable fraction of total concentration of A and the partition coefficient of analyte between the two phases. The magnitude of $K_{acceptor/sample}$ is essentially dependent upon the selected organic solvent as the acceptor phase and it is a key parameter in the two-phase HF-LPME methods. Additionally, a proper organic solvent for the specific application should have certain characteristics comprising immiscibility with water, low volatility, compatibility with polypropylene HF, and low viscosity to provide high diffusion coefficients. Commonly, toluene, undecane, 1-octanol, and dihexylether, fulfill all these requirements and are served as the organic solvent in the two-phase HF-LPME [17]. In the same concept, α is a parameter that refers to a fraction of analyte that can be extracted by the two-phase

HF-LPME. Only a fraction of neutral analyte which is not ionized is extractable and, therefore, the pH adjustment for the conversion of the analyte to a neutral form is another crucial factor on the extraction efficiency of the two-phase HF-LPME. According to **Equation 12.3**, α can be calculated based on the pH value and pK_a of the analyte.

$$\alpha = \frac{1}{1 + 10^{S}(pH - pK_a)} \qquad (12.3)$$

The above equation only applies to monoprotic acids or bases, and the S value in the given equation will be –1 and 1 for basic and acidic compounds, respectively. Moreover, the extraction conditions in practise are normally adjusted in such a manner that nearly all quantitative amounts of analyte are neutral and a values approach 1. However, the efficiency (E) and preconcentration factor (PF) in the two-phase HF-LPME methods can be estimated by **Equations 12.4 and 12.5**, respectively [18].

$$E = \frac{D_{acceptor/sample} . V_{acceptor}}{D_{acceptor/sample} . V_{org} + V_{acceptor}} \qquad (12.4)$$

$$P.F = \frac{C_{acceptor.end}}{C_{initial}} = \frac{V_{sample} . E}{V_{org}} \qquad (12.5)$$

In addition, the kinetics of extraction in the two-phase HF-LPME can be expressed through **Equation 12.6**.

$$C_{acceptor} = C_{eq.acceptor}\left(1 - e^{-kt}\right) \qquad (12.6)$$

where, $C_{acceptor}$ and $C_{eq, acceptor}$ are the analyte concentration in the acceptor phase at time t, and concentration of analyte at the equilibrium respectively. On the other hand, k is the rate constant (s^{-1}), which is calculated by **Equation 12.7** given below:

$$k = \frac{A_i}{V_{org}} \beta_0 \left(D_{acceptor/sample} . \frac{V_{org}}{V_{sample}} + 1 \right) \qquad (12.7)$$

In **Equation 12.7**, A_i, and β_0 are the contact area between the sample solution and the organic phase and the mass transfer coefficient, respectively. In order to get faster extractions, A_i and $_{\beta 0}$ should be maximized whereas V_{sample} should be minimized. To accomplish this, high stirring rates should be acquired to maximize the value of $_{\beta 0}$. Therefore, it is expected that the analyte concentration in the acceptor phase increases up to a certain time and then levels off and remains constant over time [18]. Finally, under the present conditions, the influential parameters for achieving a fast and efficient extraction in the two-phase HF-LPME are pH of the sample solution, stirring rate, length of the hollow fiber, volume of the sample solution, and type of the organic solvent.

12.3 VARIATIONS IN CONFIGURATIONS

The importance of different configurations was realized by researchers across the world when trace and ultratrace level detection become challenges and regulatory bodies became very strict with respect to pollutant or contaminant concentration in environment, food, and biological samples. Therefore, new configurations like one-phase and two-phase HF-LPME have gained importance. Since the inception of the two-phase HF-LPME, diverse configurations and strategies have been evaluated to make this method much more efficacious. The method developed in the earlier stage and the most famous configuration of this method is the rod-like setup (Figure 12.2B) in which the organic solvent is immobilized in the pores of a hollow fiber and the fiber is directly attached to the needle of a microsyringe containing a certain amount of the same organic solvent. Then, the organic acceptor phase is introduced into the lumen of HF and the same microsyringe is used for the withdrawal of the acceptor phase and the subsequent injection to the analytical instrument. The bottom of the hollow fiber can be open or sealed and the setup is immersed into the sample solution [19]. Another common technical configuration is a U-shaped one. In this setup, both ends of the HF are connected to the needles of two medical syringes to maintain the organic solvent throughout the extraction. After the extraction, one end of the HF is disconnected to flush the final acceptor phase by excess air and collect it into a microtube. The U-shaped setup is a simple configuration, which is dispensed with any clamp, stand, or even a microsyringe. The other geometry involves the immobilization of SLM, introduction of the same organic acceptor phase, sealing the HF in both ends, and immersion of the sealed HF in the sample solution [19]. This configuration is called solvent bar microextraction (SBME). The scheme of the U-shaped configuration and SBME are illustrated in Figure 12.3A and B, respectively [3, 20].

It is interesting to note that the two-phase HF-LPME can be accomplished either in static or dynamic modes. In the static mode, as described before, a static acceptor phase is used during the extraction and some type of agitation or stirring is responsible for an appropriate mass transfer. On the contrary, a programmable syringe pump is used in the dynamic mode to move the acceptor phase up and down. To perform this, a definite amount of the same immobilized organic solvent is withdrawn from the pores of the HF into a microsyringe located on the syringe pump. As is depicted in Figure 12.3B, the microsyringe is connected to the HF and the other end of the fiber is placed in the sample solution. The syringe plunger continually pulls the sample in and out. When the sample is moved up into the fiber, the thin film of SLM efficiently extracts the analytes and when the sample is pulled out, the extracted analytes are combined with the bulk of the organic phase. Compared to the static mode, it is claimed that the dynamic mode provides much higher efficiencies and repeatability [21].

Similarly to some other microextraction methods, the two-phase HF-LPME is implemented in both direct immersion and headspace modes. While the majority of the reported studies of HF-LPME in the literature have applied the direct immersion mode, as discussed before, there are also some reports related to the utilization of headspace HF-LPME (HS-HF-LPME). In this mode, the fiber containing the organic acceptor phase is located in a space between the cap of the container and the sample to accomplish the extraction of analyte vapor in the headspace of the container. In the case of volatiles and semi-volatile compounds that exist in very complicated matrices, this methodology has a high potential for application [22]. There are also some other variations of the two-phase HF-LPME. A method called microporous membrane LLE (MMLLE) was proposed as a modified version of the two-phase HF-LPME. In this modification, the analytes are extracted to the organic solvent of SLM and then are desorbed by another organic liquid for further analysis [23]. In another similar modification, Montes et al. used a dry polypropylene microporous membrane, which was not impregnated with the SLM solvent. In their methodology, after pH adjustment, a certain amount of 1-octanol was added to the homogenized sample. A stainless steel rod was introduced into the HF along its length and the whole system was magnetically stirred. During the extraction, 1-octanol containing the target analytes diffused into the pores of HF, and the procedure was followed by the subsequent desorption using a mixture of toluene-hexane [24].

12.3.1 AUTOMATION OF PROPOSED TECHNIQUES

Automated sample preparation, which often results in larger sample throughput, is generally expected to be less labor intensive and has other benefits such as minimized contact with hazardous chemicals, reduced sample contamination, and improved overall accuracy and precision. Therefore, various semi-automated and automated setups of HF-LPME have been reported in the literature. The earliest report on the semi-automation of the two-phase HF-LPME goes back to 2003, when HF was attached to a funnel-shaped injection guide [24]. The injection guide also consisted of a dent to hold the other end of the HF unsealed and open for an automated injection to GC. The loop-shaped HF was then placed into a vial containing the sample, and after extraction, the vial was placed into a GC autosampler. In another attempt to automate the two-phase

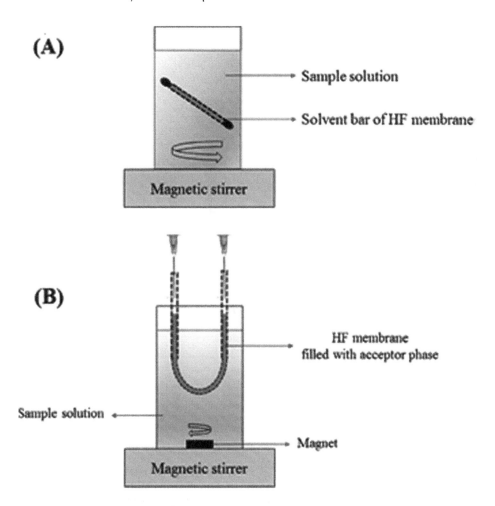

FIGURE 12.3   An illustration of typical setups of (A) solvent bar microextraction (SBME), and (C) U-shaped two-phase HF-LPME. Copyright (2018) with permission from Elsevier [3].

HF-LPME, Barri et al. developed an extracting syringe (ESy), which was a sandwiched device comprising a membrane sheet and two channels for the sample flow and the stagnant organic acceptor phase. The ESy syringe also acted as a GC autosampler that combined both extraction and injection processes in a single step [25]. The other configuration of two-phase HF-LPME was proposed in 2007 by Salafranca et al. with a higher degree of automation [26]. In this work, the dynamic two-phase HF-LPME was accomplished for the simultaneous processing of up to eight samples. A multisyringe pump was used for a continuous flow of the fresh organic solvent inside the HF. In addition, a double three-way valve was exploited to withdraw the solvent flow and flush the final acceptor solvent into a vial for further analysis by GC. Pawliszyn et al. suggested another setup for a fully automated two-phase HF-LPME, which entailed several steps, all accomplished by a CTC autosampler [27]. In 2014, Luan et al. used this automated device for mapping the spacial distribution of organophosphate esters (OPEs) in the Pearl River estuaries [28].

In an interesting study, Yamini et al. reported a specially designed TT-extractor tube, which was used to perform the extractions in a flow system. In this sense, a piece of HF was mounted in a stainless-steel tube and the organic solvent was introduced into the HF by a syringe pump. In addition, the aqueous sample solution was smoothly recycled around the HF by another pump [29]. A scheme of this configuration is shown in Figure 12.4A. Finally, a user-friendly, fully automated setup of HF-LPME was proposed by the same research group in 2012. In this innovative instrument, four different solvent containers filled with an SLM solvent, acceptor phases, and an eluting solvent (usually water) were used. These solvents were injected and withdrawn into the hollow fiber via a syringe-driven pump. The on-line injection of the acceptor phase was carried out by a six-port injection valve. A sensor was utilized in the device, which precisely switched the six-port injection valve from the load to the inject position and vice versa. The extraction parameters including the extraction time, stirring rate, and HF-LPME mode, which could be two- or three-phase, were controlled through a touch screen panel in front of the instrument. A bottle containing the sample solution was moved up for the HF immersion and moved down when the extraction was accomplished via a tiny elevator [30]. Although in the first application of this instrument, a three-phase HF-LPME was carried out, this device is also capable of performing two-phase HF-LPME. Figure 12.4B illustrates the schematic diagram of the fully automated HF-LPME instrument.

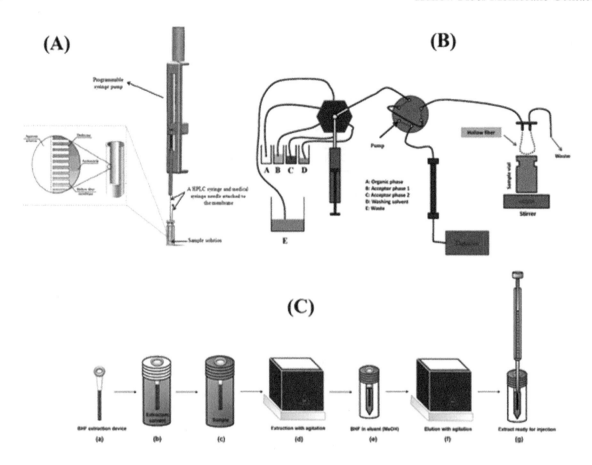

FIGURE 12.4 The schematic of automated instruments (A) T-T extractor tube configuration. Copyright (2011) with permission from John Wiley and Sons [29], (B) fully automated instrument proposed by Yamini et al. (2012). Copyright (2012) with permission from Elsevier [30], and (C) automated BHF-LPME. Copyright (2018) with permission from from Elsevier [31].

Recently, Lee et al. developed another form of automated HF-LPME, which was called automated arrays of bundled hollow fiber liquid-phase microextraction (BHF-LPME). To prepare the bundles of HF, the fibers were heat-sealed at one end while the other end was tightly fitted into a pipette tip. A metal ring was also attached to the micropipette to provide the extraction unit. A tray plate containing 32 sample vials equipped with a micro-editor was used to obtain the array of extraction units. The instrument was programmed to immerse the BHFs into dibutyl ether as the SLM solvent and then immediately placed BHFs into an aqueous sample solution. The procedure continued by the immersion of BHF assembly in methanol for analyte elution and subsequent injection to the analytical instrument [31]. The extraction procedure is depicted in Figure 12.4C. Shortly after the original demonstration, the procedure was modified by the same research group, where the desorption step was assisted by ultrasonication. The modified method has been used for the determination of estrogens in aqueous samples [32].

12.3.2 CONCENTRATION ENRICHMENT AND SELECTIVITY

Concentration enrichment and selectivity are the two parameters that are the integral part of the desired analytical techniques to be developed for specific application.

In simple words, one of the main purposes of membrane extraction in sample preparation is to enrich the analyte, i.e., to increase the concentration of the analyte in order to permit determination of low concentrations. Plotting the concentration of analyte in the acceptor (C_A) either directly as determined by analysis of the acceptor phase or as a concentration enrichment factor Ee (C_A/C_S – where C_S is the initial concentration in the sample) versus time will typically produce a curve which initially raises approximately linearly and asymptotically eventually reaches a steady equilibrium value. See Figure 12.5 [33].

Assuming that the rate of mass transfer is proportional to the concentration difference over the membrane, and noting that in static extraction, the concentration in the donor phase C_D decreases as analyte is transferred over the membrane, we get the following differential equation:

$$\frac{\partial C_A}{\partial t} = k \cdot \left(C_S - \left(\frac{V_A}{V_S} - \frac{1}{D} \right) \cdot C_A \right) \qquad (12.8)$$

where V_A and V_S are the volumes of the acceptor (strip) phase and the extracted sample, respectively, C_S is the initial concentration in the sample ($C_S V_S = C_D V_S + C_A V_A$). D is the equilibrium distribution coefficient and k is a rate constant. The general solution expressed as a concentration enrichment factor Ee is:

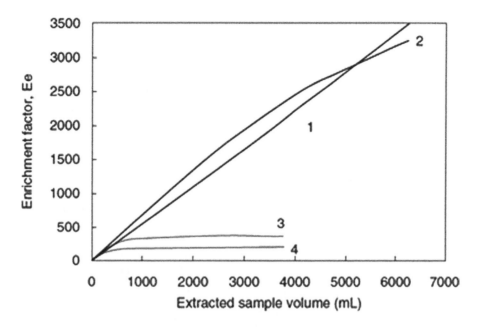

FIGURE 12.5 Enrichment factors for various aniline compounds with different pK_a values and $pH_A = 0$ ($\alpha_D = 1$). Extracted compounds: aniline (1), 3-chloro4-methylaniline (2), 3,5-dichloroaniline (3), and 3-methyl-5-nitroaniline (4). Copyright (1998) with permission from American Chemical Society [33].

$$Ee = \frac{C_A}{C_S} = \frac{1}{\left(\dfrac{V_A}{V_S} + \dfrac{1}{D}\right)} \cdot \left(1 - \exp(-k \cdot \left(\dfrac{V_A}{V_S} + \dfrac{1}{D}\right) \cdot t\right) \quad (12.9)$$

The rate of mass transfer through the membrane is proportional to the gradient of this curve, and thus decreases from an initial value to practically zero. The concentration enrichment factor value at equilibrium will be:

$$Ee_{eq} = \frac{1}{\left(\dfrac{V_A}{V_S} + \dfrac{1}{D}\right)} \quad (12.10)$$

Equations 12.8 to 12.10 are basically valid for static extraction. In the case of extraction in a flow system with flowing donor and stagnant acceptor, the phase ratio (V_A/V_S) is zero, as depletion of the sample is not possible.

There are several limiting cases for **Equations 12.9 to 12.10**.

1. With a large equilibrium distribution coefficient D ("complete trapping"), and a large phase ratio Ee will increase linearly up to large values. In Figure 12.5, which refers to an SLM flow system experiment, D in curve 1 is approximately 40 000 and the enrichment factor is linear at least up to at least 6,000 times. In many cases, especially in flow systems, the extraction is not allowed to go to equilibrium, and an extraction efficiency E is defined as the fraction of the total amount of analyte that is transferred to the acceptor. Thus:

$$E = \frac{C_A}{C_S} \cdot \frac{V_A}{V_S}; \quad Ee = E \cdot \frac{V_S}{V_A} \quad (12.11)$$

The extraction efficiency is related to the slope of the extraction curve, so if the extraction is linear, E is constant and < 100%. By careful calibration and keeping the experimental parameters constant so the system is kinetically stable, repeatable values for Ee and E can be obtained. This will give reproducible quantitative results, and most of the applications of membrane extraction to practical analyses are in fact based on this principle.

2. With a finite sample volume in static extraction mode under complete trapping conditions, the sample will eventually be depleted with regard to analyte, so a decrease of mass transfer is due to the decrease of analyte concentration and the equilibrium occurs after virtually all analyte has been transferred to the acceptor. **Equation 12.10** then leads to:

$$Ee_{eq} = \frac{V_S}{V_A}; \quad C_A = C_S \cdot \frac{V_S}{V_A} \quad (12.12)$$

i.e., the enrichment factor is determined by the phase ratio.

Thus, if the extraction is continued until equilibrium so that the sample is totally depleted with analyte, very simple and straightforward calculations of the enrichment factor are possible, according to **Equation 12.12**, which contains no other parameters than the phase volumes. This can be the basis for accurate quantitative determinations.

3. The third limiting case of **Equation 12.10** is the situation where equilibrium is attained without depletion of the analyte in the sample.

To obtain a high selectivity, i.e., discrimination between the analytes and various unwanted matrix compounds, membrane extraction has a clear advantage over other sample preparation techniques, as all compounds that reach the analytical instrument must travel through the membrane. There is no direct connection and possibility for transferring compounds into the analytical instrument in other ways. This is not the case with other extraction techniques. With solid phase extraction (SPE), solid phase microextraction (SPME), etc, there is a definite possibility that matrix components are absorbed on the sorbing phase and are subsequently being eluted into the extract. With liquid–liquid extraction (LLE) such transfer is less probable and it is generally considered that extracts after LLE are cleaner than after SPE. The possible and common problem of the formation of emulsions at the phase interface with LLE, which is avoided with all types of membrane extraction, is a source of contamination across the phase border.

Further, with SLM extraction, the pH of the donor solution and the acceptor phase can easily be fine-tuned to obtain extraction which is selective for certain groups of compounds, as described above. Alternatively, a carrier can be added in the membrane to increase the selectivity and/or mass transfer of the compounds of interest. When the primary aim of adding a carrier is to increase the mass transfer of the analytes by analyte–carrier interactions in the membrane, it is important to choose other conditions carefully so that selectivity is still retained.

Membrane extraction is especially effective in discriminating towards macromolecules. In environmental analysis, macromolecular humic acids are ubiquitous and usually very efficiently removed. Megersa et al. [34] compared SLM and SPE extraction of some triazine herbicides spiked in river water. As indicated by results, the difference is dramatic. Also, there are a number of examples of how various drugs can be determined in blood plasma and also urine without matrix interferences [35, 36]. In another study, low-cost and convenient ballpoint tip-protected liquid-phase microextraction (BT-LPME) was studied for analysis of organic molecules in water samples [37]. In this study, magnetic field-induced BT spinning was utilized to accelerate the extraction process. Through adhesion effect and sheath protection, the bullet-shaped BT finely confined the organic solvent within its hollow cavity, allowing long extraction under high spinning speeds (>1,000 rpm). Five polycyclic aromatic hydrocarbons (PAHs) were employed to evaluate the extraction performance of BT-LPME, followed by GC-MS analysis. Compared to SDME, higher extraction efficiency was obtained. Good linearities with correlation coefficients, as well as low LODs (0.002–0.011 mg L^{-1}) and LOQs (0.007–0.023 mg L^{-1}), were achieved. Three real samples (river, surface, and tap water) were analyzed by the proposed method, and no contamination was observed. Quantitative spiking experiments showed the high accuracy of this method. Furthermore, successful extraction of organophosphorus compounds, organochlorines, and triazines from water samples demonstrated its wide application range in environmental water sample analysis.

12.4 RECENT DEVELOPMENTS OF THE TWO-PHASE HF-LPME

12.4.1 APPLICATION TO THE ANALYSIS OF ENVIRONMENTAL SAMPLES

Regarding the environmental applications of the two-phase HF-LPME, Yu et al. developed a method for the determination of earthy-musty odorous compounds in water samples. They also investigated the effect of HF type on the extraction efficiency. For this purpose, they compared the applicability of various types of HFs with different thicknesses and porosities including polypropylene (PP), in-outside microfiltration polyvinyldene fluoride, namely MIF-1a-PVDF, and outside-in microfiltration polyvinyldene fluoride, namely MOF-1b-PVDF. According to their obtained results, MOF-1b-PVDF provided higher extraction efficiencies thanks to its highly porous structure. What is more interesting is that the solvent distribution and impregnation occurred much faster in MOF-1b-PVDF than in MIF-1a-PVDF. They used a simple rod-like setup for the extraction and o-xylene was served as the organic solvent in the acceptor phase. By performing the analysis on a GC-MS instrument, the limits of detection (LODs) were lower than 2.0 ng L^{-1} [38]. Following the same approach, an n-octanol impregnated PVDF hollow fiber was used by Cai et al. for the determination of organochlorine pesticides in ecological textiles [39]. In a reported work on the determination of formaldehyde in water samples, Ganjikhah et al. utilized a U-shaped HF-LPME setup. Their work was accomplished through the derivatization of formaldehyde by acetylacetone to convert the analyte to a colored extractable form. 1-Octanol was used as the extracting phase and the procedure was followed by spectrophotometric measurement of the formaldehyde derivative [40]. Recently, a method was validated for the quantification of phthalic acid esters in water samples. In this study, 1-octanol was utilized as the organic solvent in both acceptor phase and SLM. The extraction was pursued by back extraction of the analytes to cyclohexane and further evaporation-reconstitution [41].

A new methodology named dynamic single-interface HF-LPME was introduced by Varanusupakul et al. for the calorimetric detection of Cr(VI). Their colorimetric detection method was based on the formation of a violet complex of chromium and diphenyl carbazide (DCP) and further analysis by a fiber-optic UV-Visible spectrophotometer. To perform this, they used an ionic liquid (Aliquat 336) as the SLM solvent and the sample solution was continually passed through the lumen of the membrane by a peristaltic pump. After the extraction, a DCP solution was fed into the HF to elute the chromium and form the colored complex [42]. In addition to the exploitation of ionic liquids in the HF-LPME methods, the use of surfactants has been repeatedly reported in the literature as a strategy to increase the efficiency of HF-LPME. To refer to some, Sarafraz Yazdi et al. proposed a surfactant enhanced hollow fiber liquid-phase microextraction

(SE-HF-LPME) for the melamine determination in soil samples based on formation of a hydrophobic ion-pair between sodium dodecyl sulfate and protonated melamine [43]. In another research, acidic, basic, and amphiprotic pollutants were successfully co-extracted using supramolecular vesicles of decanoic acid in 1-decanol. This method was called supramolecular nanosolvent based-hollow fiber liquid-phase microextraction (SSHF-LPME) [44]. The latest applications of two-phase and three-phase HF-LPME for quantitative analysis of different analytes in environmental samples are shown in Table 12.1.

12.4.2 Analysis of Biological Samples

As an example, automation of the method is described for extraction of trace amounts of anticancer drugs (exemestane, letrozole, and paclitaxel) from urine [50]. In this setup, the three phases consist of the aqueous, diluted urine sample, n-dodecane with 10% trioctylphosphine oxide (TOPO) as the organic solvent within the membrane, and acetonitrile as the acceptor phase, respectively. The extraction setup was automated and coupled directly to HPLC-UV via a six-port injection valve. Different extraction parameters like type of acceptor phase, SLM composition, sample pH, extraction time, length of hollow fiber, and influence of ionic strength were optimized. With this method, low limits of detections (LODs) were obtained from diluted urine (0.3–0.6 mg/L), and with preconcentration factors in the range 152–411 from 20 mL sample. The repeatability of the method was excellent (below 4.5% (n = 6)) for all three compounds, probably facilitated by the automation of the method. Zhang et al. assembled a U-shaped setup for the two-phase HF-LPME to investigate the pharmacokinetics of echinacoside in Parkinson disease rat plasma after oral administration. They used 1-octanol as the acceptor phase and the echinacoside pharmacokinetics of the normal rats were compared against the Parkinson disease model ones, which showed a difference due to the possible induced physiological changes caused by the Parkinson diseases [51]. By the same setup, the authors used a simple rod-like two-phase HF-LPME setup with n-heptanol as the extracting solvent for the determination of imperatorin and its metabolites in rat plasma [51]. Similarly, Cardeal et al. developed a two-phase HF-LPME method for the determination of bisphenol A and metabolites of plasticizers [52]. Another automated setup including HF-LPME was recently described for determination of four acidic drugs (ketoprofen, ibuprofen, diclofenac, and naproxen) in urine as a proof-of-concept [53]. In this work, an LPME microextraction chamber was created by a 3D printer and coupled inline between syringe pumps and HPLC, allowing continuous flow-extraction of the drugs from a flowing stream donor phase (pH 1.7), through a membrane impregnated with dispersed carbon nanofibers (CNF) in dihexyl ether, and further into a stagnant, alkaline acceptor phase. The membrane was regenerated between each extraction, thereby requiring no need for replacement of the hollow fiber. Different nanomaterials

were tested as CNFs and compared to unmodified polypropylene fibers. The extraction efficiencies were significantly improved by the addition of CNFs into the dihexyl ether membrane. In addition, the composition of organic solvents, donor, and acceptor phases were optimized, so also the configuration of the 3D-printed microextraction chamber. With optimized conditions, LODs in the range 1.6–4.3 mg/L were obtained, and with preconcentration factors within 43.2–96.8. All RSD values were below 6.1% (n = 5), also here demonstrating that automation of the sample preparation method is favorable with respect to the repeatability. In another work, a mixture of organic solvents including 1-octanol, chloroform, and toluene was used as the organic acceptor phase in the two-phase HF-LPME for the extraction of naloxone, buprenorphine, and norbuprenorphine from plasma samples. Both ends of the HF were sealed and the HF was immersed in the sample solution. Not only did the use of three solvents significantly increase the efficiency, but the HF impregnation time was reduced to 10 s and the less volatility of the acceptor composition resulted from adding 1-octanol [54]. HF-LPME could also be used for determination of compounds in teeth [55]. Exposure to benzene, toluene, ethylbenzene, and xylenes (BTEX) from the environment is associated with great health risk for human beings. For the first time, HF-LPME was used for the determination of the hydrophilic BTEX metabolites (mandelic acid (MA), hippuric acid (HA), and 4-methylhippuric acid (4mHA)) as biomarkers in teeth. The analytes were first extracted from the teeth by ultrasonification of the teeth placed in 10 mL 1 M sodium hydroxide (NaOH). Thereafter, water and concentrated HCl were added to the solution to adjust the pH to 2 for optimal HF-LPME conditions. The type of organic solvent was optimized and the relatively polar n-decanol was found to be the best one compared with other pure solvents and with the addition of carriers to the membrane, respectively. Different extraction times, stirring speeds, and the composition of the acceptor phase were also studied. With optimized conditions, MA and HA was quantified in real teeth in the range of 1–11 mg/g (MA) and 70 mg/g (HA). A two-phase HF-LPME method using vesicular coacervates of decanoic acid was proposed for the extraction of benzodiazepines from human urine and plasma, fruit juice, and water samples. In this method, the sole supramolecular solvent, which is a green alternative to the typical organic solvents, was used as both the acceptor phase and SLM. Additionally, it was claimed that the solubilization capability of vesicles provides desirable partitioning of the analytes [56]. In another work, Hadjmohammadi et al. developed a new two-phase HF-LPME based on the reverse micelle for the extraction of quercetin from human plasma, anion, and tomato samples. In this work, cetyltrimethylammonium bromide (CTAB) was used as a cationic surfactant and the extraction was successfully accomplished on the basis of analyte solubilization in the aqueous center of the reverse micelle. CTAB was added to 1-octanol at the concentration of 7 mmol L_1 and the mixture was utilized as the extracting solvent [57].

TABLE 12.1

Environmental Applications of Membrane Extraction. LPME2 = Two-phase liquid phase micro extraction in hollow fibers (aq/org); LPME3 = Three-phase liquid phase micro extraction in hollow fibers (aq/org/aq)

Analytes	Matrices	Membrane Technique	Analytical Technique	Ref
2-methylisoborneol	Water	LPME2	GC-MS	38
2-isopropyl-3-methoxy pyrazine	Water	LPME2	GC-MS	38
2,4,6-trichloroanisole	Water	LPME2	GC-MS	38
2,3,6-trichloroanisole	Water	LPME2	GC-MS	38
geosmin	Water	LPME2	GC-MS	38
Tri-iso-propyl phosphate	Water	LPME2	GC-MS	28
Tributyl phosphate	Water	LPME2	GC-MS	28
Tris (2-chloroethyl) phosphate	Water	LPME2	GC-MS	28
Tris (chloropropyl) phosphate	Water	LPME2	GC-MS	28
Tris (dichloropropyl) phosphate	Water	LPME2	GC-MS	28
Triphenyl phosphate	Water	LPME2	GC-MS	28
2-Ethylhexyl diphenyl phosphate	Water	LPME2	GC-MS	28
Tris (2-ethylhexyl) phosphate	Water	LPME2	GC-MS	28
Triphenylphosphine oxide	Water	LPME2	GC-MS	28
heptachlor	Acetone	LPME2	GC-MS	39
hexachlorobenzene	Acetone	LPME2	GC-MS	39
aldrin	Acetone	LPME2	GC-MS	39
cis-chlordane	Acetone	LPME2	GC-MS	39
trans-chlordane	Acetone	LPME2	GC-MS	39
dieldrin	Acetone	LPME2	GC-MS	39
endrin	Acetone	LPME2	GC-MS	39
o,p-DDT	Acetone	LPME2	GC-MS	39
p,p-DDT	Acetone	LPME2	GC-MS	39
mirex	Acetone	LPME2	GC-MS	39
Melamine	Soil	LPME2	HPLC-UV	43
formaldehyde	Water	LPME2	UV-Vis	40
dipropyl phthalate	Water	LPME2	GC-MS/MS	41
dibutyl phthalate	Water	LPME2	GC-MS/MS	41
di-isobutyl phthalate	Water	LPME2	GC-MS/MS	41
bis-isopentyl phthalate	Water	LPME2	GC-MS/MS	41
bis-2-ethoxyethyl ester	Water	LPME2	GC-MS/MS	41
bis-n-pentyl ester	Water	LPME2	GC-MS/MS	41
butyl benzyl phthalate	Water	LPME2	GC-MS/MS	41
bis-2-n-butoxyethyl ester	Water	LPME2	GC-MS/MS	41
dicyclohexyl phthalate	Water	LPME2	GC-MS/MS	41
4-nitrophenol	Water	LPME2	HPLC-DAD	44
3-nitroaniline	Water	LPME2	HPLC-DAD	44
1-amino-2-naphthol	Water	LPME2	HPLC-DAD	44
bisphenol A	Water	LPME2	GC-MS	46
mono-methyl phthalate	Water	LPME2	GC-MS	46
mono-isobutyl phthalate	Water	LPME2	GC-MS	46
mono-butyl phthalate	Water	LPME2	GC-MS	46
mono-cyclohexyl phthalate	Water	LPME2	GC-MS	46
mono-(ethylhexyl) phthalate	Water	LPME2	GC-MS	46
mono-cyclohexyl phthalate	Water	LPME2	GC-MS	46
mono-(ethylhexyl) phthalate	Water	LPME2	GC-MS	46
monoisononyl phthalate	Water	LPME2	GC-MS	46
mono-benzyl phthalate	Water	LPME2	GC-MS	46
Phenolics	Water, nutrient solutions	LPME3	HPLC	47
Drugs	Water	LPME3	CE, LC/MS	49–50
	Sewage sludge			

HF-LPME, or variants of the technique, is also widely used for sample preparation of food [58–60] and waste water [61]. Zhou et al. introduced a modified version of the two-phase HFLPME, which was based on using Fe_3O_4 in the sample solution as a magnetofluid powder. They also compared the magnetofluidic HF-LPME with the conventional HF-LPME. According to the reported results, the preconcentration factors of mangnetofluidic mode were greater and the extraction time was significantly lower than the conventional mode. They attributed this observation to the uniform magnetic field generated by the exploited magnetofluid powder [62]. In Table 12.2, the applications of two-phase and three-phase HF-LPME for quantitative analysis of biological samples are shown.

TABLE 12.2
Applications of Membrane Extraction to Biological Samples. Abbreviations as in Table 12.1

Analytes	Matrices	Membrane Technique	Analytical Technique	Ref
Antibiotics	Blood plasma	LMPE3	HPLC	63
Exemestane	Water, urine	LMPE3	HPLC-UV	50
letrozole	Water, urine	LMPE3	HPLC-UV	50
paclitaxel	Water, urine	LMPE3	HPLC-UV	50
Ketoprofen	urine	LMPE3	HPLC-DAD	53
ibuprofen,	urine	LMPE3	HPLC-DAD	53
diclofenac	urine	LMPE3	HPLC-DAD	53
naproxen	urine	LMPE3	HPLC-DAD	53
Benzene	Teeth	LMPE3	HPLC-DAD	55
toluene	Teeth	LMPE3	HPLC-DAD	55
ethylbenzene	Teeth	LMPE3	HPLC-DAD	55
xylenes	Teeth	LMPE3	HPLC-DAD	55
metabolites	Teeth	LMPE3	HPLC-DAD	55
estron	Water	LMPE2	LC-MS/MS	32
estriol	Water	LMPE2	LC-MS/MS	32
17β-estradiol	Water	LMPE2	LC-MS/MS	32
17α-ethylestradiol	Water	LMPE2	LC-MS/MS	32
echinacoside	Water	LMPE2	LC-MS/MS	32
imperatorin	2%NaCl/plasma	LMPE2	LC-ESI/MS	51
xanthotoxol	2%NaCl/plasma	LMPE2	LC-ESI/MS	51
echinacoside	Rat plasma	LMPE2	HPLC/DAD	62
tubuloside B	Rat plasma	LMPE2	HPLC/DAD	62
acteoside	Rat plasma	LMPE2	HPLC/DAD	62
isoacteoside	Rat plasma	LMPE2	HPLC/DAD	62
Various drugs	Urine	LPME3	MS	64
Trimethazidine	Blood plasma	LPME3	HPLC	65
Osthol	Blood plasma	LPME3	HPLC	66
Methimazole	Various biol. Samples	LPME3 (ion pair)	HPLC	67
alprazolam	Human urine	LMPE2	HPLC/UV	56
clonazepam	Human urine	LMPE2	HPLC/UV	56
nitrazepam	Human urine	LMPE2	HPLC/UV	56
diazepam	Human urine	LMPE2	HPLC/UV	56
midazolam	Human urine	LMPE2	HPLC/UV	56
naloxone Static/magnetofluid	Human plasma	LMPE2	LC-MS/MS	54
buprenorphine	Human plasma	LMPE2	LC-MS/MS	54
norbuprenorphine	Human plasma	LMPE2	LC-MS/MS	54
Triazine herbicides	Environmental water, honey, tomatoes	LMPE3	CE-DAD	58
Chromium (Cr (VI))	Water	LMPE3	Calorimetric detection	60
Salicylic acid	Estuarine and riverine water	LMPE3	CE-UV	61

12.4.3 Analysis of Chemical, Food, and Beverage Samples

Two-phase hollow fiber-liquid microextraction based on reverse micelle was investigated by Ranjbar et al. [68] for the determination of quercetin in human plasma and vegetable samples. In another study, an innovative method for analysis of Pb (II) in rice, milk, and water samples based on TiO$_2$ reinforced caprylic acid hollow fiber solid/liquid phase microextraction was developed by Bahar et al. [69]. Considering the reported applications of the two-phase HF-LPME in food analysis, Radriguez-Delgado et al. applied the conventional rod-like two-phase HF-LPME to the determination of estrogens in milk samples. They employed 1-octanol as the extracting organic phase and followed the procedure by analyte desorption by acetonitrile [70]. This research group used the same approach for the measurement of oestergens in yogurt and cheese samples [71]. In a interesting peace of work, Mlunguza et al. [72] described a simple and sensitive method for the simultaneous isolation, enrichment, identification, and quantitation of selected antiretroviral drugs; emtricitabine, tenofovir disoproxil, and efavirenz in aqueous samples and plants. The analytical method was based on microwave extraction and hollow fiber liquid phase microextraction technique coupled with ultra-high pressure liquid chromatography-high resolution mass spectrometry. The optimal enrichment factors for emtricitabine, tenofovir disoproxil, and efavirenz from aqueous phase were 78, 111, and 24, respectively. The analytical method yielded recoveries in the range of 86 to 111%, and quantitation limits for emtricitabine, tenofovir disoproxil, and efavirenz in wastewater were 0.033, 0.10, and 0.53 µg L^{-1}, respectively. The drugs were detected in most samples with concentrations up to 37.6 µg L^{-1} recorded for efavirenz in waste water effluent. Roots of the water hyacinth plant had higher concentrations of the investigated drugs ranging from 7.4 to 29.6 µg kg^{-1}. In another investigation, oestrogenic compounds were determined in milk samples by a vortex-assisted two-phase HF-LPME (VA-HF-LPME) method. Nonanoic acid was impregnated in the pores of HF and was also filled into the lumen of HF. Both ends of the fiber were sealed and immersed in the sample solution like a U-shaped solvent bar. The convection of sample was accomplished by the vortex mixing. Finally, the acceptor phase was flushed by 200 mL of methanol for the subsequent injection [73]. A combination of HF-LPME and DLLME was implemented by Carasek et al. for the quantification of aflatoxins in soybean juice [74]. Another fascinating methodology was proposed by Huang et al.; their method was a reverse two-phase HF-LPME in which the acceptor phase and SLM solvent were aquatic, and edible oil samples were chosen as the matrix of interest. For this purpose, they modified the PVDF fibers by polydopamine and polyet hyleneimine to change the surface characteristics of the fibers from hydrophobic to hydrophilic. The method was employed for the determination of aflatoxins (AFTs) in edible oils. To perform this, the modified fiber was immersed into the water to form the SLM solvent, and the lumen was filled with a phosphate buffer solution containing anti-AFTs antibody [75]. A schematic diagram of their modification is illustrated in Figure 12.6. Table 12.3 covers the applications of two- and three-phase HF-LPME for quantitative analysis of different analytes in food samples.

12.5 FUTURE DIRECTIONS AND CONCLUDING REMARKS

In this chapter, the state of the art of the two-phase and three-phase HF-LPME techniques applicable to a wide

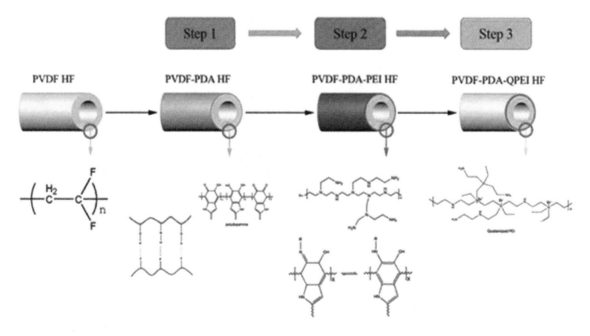

FIGURE 12.6 The utilized modification on static two-phase HF-LPME procedure by Huang et al. (2017). Copyright (2017) with permission from from Royal Society of Chemistry [75].

variety of analytes and connected to several analytical instruments are presented. An overview is provided regarding HF-LPME basic principles, theoretical aspects, possible setups, and applications. It is apparent that in the past two decades, after the introduction of these methods, numerous developments and diverse configurations of these methods have been proposed by analytical chemistry researchers around the world. The obvious advantages of one-phase and two-phase HF-LPME techniques are connected with demand for lower volumes of solvents and less laborious operations in comparison to other conventional techniques as well as the lower generation of wastes. Microvolumes of extracting solvents give considerably less penalty points in the Analytical Eco-Scale than even a thousand higher volumes of solvents used in LLE. In addition, the heavy routine makes it absolutely necessary to choose dissolution and clean up techniques that are similar to the most common analytical techniques, in order to minimize errors and develop expert systems that work without staff supervision. The acceptance by regulatory bodies of up-to-date sample pretreatment techniques would surely give a substantial boost to their widespread use in chemical, environmental, biological, and food analysis. This wide acceptance and interest can be explained by the striking features and inherent advantages of this method including its simplicity, environmentally friendly nature, superior clean up, and noticeable efficiency. More importantly, the investigations of this method are continuing and demonstrate its excellent

TABLE 12.3

Various Applications of Membrane Extraction in Chemical Analysis of Industrial and Food Samples. Abbreviations as in Table 12.1.

Analytes	Matrices	Membrane Techniques	Analytical Techniques	Ref
Pb(II)	Rice, milk	LMPE3	Flame atomic absorption spectrometry	69
quercetin	Human Plasma and vegetable matrix	LPME2	LC-UV	68
17 β-estradiol	Milk	LPME2	HPLC-DAD	70
estrone	Milk	LPME2	HPLC-DAD	70
diethylstilbestrol	Milk	LPME2	HPLC-DAD	70
Cr (VI)	Water	LPME2	UV-Vis	42
2-hydroxyoestradiol	Dairy product	LPME2	HPLC-DAD/FD	70
17α-estradiol	Dairy product	LPME2	HPLC-DAD/FD	70
oestrone	Dairy product	LPME2	HPLC-DAD/FD	70
17α-ethynylestradiol	Dairy product	LPME2	HPLC-DAD/FD	70
17β-estradiol	Dairy product	LPME2	HPLC-DAD/FD	70
diethylstilbestrol	Dairy product	LPME2	HPLC-DAD/FD	70
dienestrol	Dairy product	LPME2	HPLC-DAD/FD	70
hexestrol	Dairy product	LPME2	HPLC-DAD/FD	70
Pesticides	Fruit, water	LPME2	HPLC	76
Pesticides	Juice	LPME2	LC-MS/MS	77
Melamine	Milk	LPME3	HPLC	78
aflatoxin B1	Soybean juice	LPME2	HPLC-FD	74
aflatoxin B2	Soybean juice	LPME2	HPLC-FD	74
aflatoxin G1	Soybean juice	LPME2	HPLC-FD	74
aflatoxin G2	Soybean juice	LPME2	HPLC-FD	74
2-Hydroxyestradiol	Milk	LPME2	HPLC-DAD/FD	70
estriol	Milk	LPME2	HPLC-DAD/FD	70
17b-estradiol	Milk	LPME2	HPLC-DAD/FD	70
17a-estradiol	Milk	LPME2	HPLC-DAD/FD	70
estrone	Milk	LPME2	HPLC-DAD/FD	70
17a-ethynylestradiol	Milk	LPME2	HPLC-DAD/FD	70
diethylstilbestrol	Milk	LPME2	HPLC-DAD/FD	70
dienestrol	Milk	LPME2	HPLC-DAD/FD	70
hexestrol	Milk	LPME2	HPLC-DAD/FD	70
aflatoxin B1	Edible oil	LPME2	LC-MS/MS	75
aflatoxin B2	Edible oil	LPME2	LC-MS/MS	75
aflatoxin G1	Edible oil	LPME2	LC-MS/MS	75
aflatoxin G2	Edible oil	LPME2	LC-MS/MS	75

potential for further developments. It seems that further investigations are needed to develop an automated instrument capable of performing the HF-LPME procedure at a competitive cost. There are several necessary features that should be met in an automated setup that is to be well-suited for mass production and ubiquitous use in laboratories, which include the capability for parallel analysis of large number of different analytes, feasibility of programming, easy controlling, and a miniaturized instrument design.

REFERENCES

1. Veronika, P., Sultani, M., Skoglund A. K., Lucie, N., Stig, P., Gjelstad, A., 2017. One-step extraction of polar drugs from plasma Byparallel artificial liquid membrane extraction, *J. Chromatogr. B Anal. Technol. Biomed Life Sci.* 1043: 25–32.
2. Jönsson, J. Å., 2015. Membrane extraction in preconcentration, sampling and trace analysis. In *Handbook of Membrane Separations: Chemical, Pharmaceutical, Food and Biotechnological Applications*, A. M. Sastre, A. K. Pabby, S. S. H. Rizvi, Eds., Boca Raton, FL: CRC Press, pp. 345–369.
3. Esrafili, A., Baharfar, M., Tajik, M., Yamini, Y., Ghambarian, M., 2018. Two-phase hollow fiber liquid-phase microextraction, *TrAC Trends Anal. Chem.* 108: 314–322.
4. Gjelstad, A., 2019. Three-phase hollow fiber liquid-phase microextraction and parallel artificial liquid membrane extraction, *Trends Anal. Chem.* 113: 25–31.
5. Chimuka, L., Cukrowska, E., Michel, M., Buszewski, B., 2011. Advances in sample preparation using membrane-based liquid-phase microextraction techniques, *TrAC Trends Anal. Chem.* 30: 1781–1792.
6. Yamini, Y., Seidi, S., Rezazadeh, M., 2014. Electrical field-induced extraction and separation techniques: Promising trends in analytical chemistry–A review, *Anal. Chim. Acta* 814: 1–22.
7. He, Y., 2012. Liquid-based micro-extraction techniques for environmental analysis, 835–862. Cambridge, MA: Academic Press.
8. Lin, H., Wang, J., Zeng, L., Li, G., Sha, Y., Wu, D., Liu, B., 2013. Development of solvent micro-extraction combined with derivatization, *J. Chromatogr. A* 1296: 235–242.
9. Jönsson, J. A., 2012. Membrane extraction: General overview and basic techniques. In Pawliszyn, J., Lord, H., Eds., *Comprehensive sampling and sample preparation*. Oxford, UK: Elsevier, pp. 461–474.
10. Carasek, E., Merib, J., 2015. Membrane-based microextraction techniques in analytical chemistry: A review, *Anal. Chim. Acta* 880: 8–25.
11. Pedersen-Bjergaard, S., Rasmussen, K. E., 1999. Liquid-liquid-liquid micro-extractionfor sample preparation of biological fluids prior to capillary electrophoresis, *Anal. Chem.* 71: 2650–2656.
12. Jeannot, M. A., Cantwell, F. F., 1996. Solvent micro-extraction into a single drop, *Anal. Chem.* 68: 2236–2240.
13. Liu, H., Dasgupta, P. K., 1996. Analytical chemistry in a drop. Solvent extraction in a micro-drop, *Anal. Chem.* 68: 1817–1821.
14. Pawliszyn, J., Pedersen-Bjergaard, S., 2006. Analytical microextraction: Current statusand future trends, *J. Chromatogr. Sci.* 44: 291–307.
15. Asensio-Ramos, M., Ravelo-Perez, L. M., Gonzalez-Curbelo, M. A., Hernandez-Borges, J., 2011. Liquid phase microextraction applications in food analysis, *J. Chromatogr. A* 1218: 7415–7437.
16. Rasmussen, K. E., Pedersen-Bjergaard, S., 2004. Developments in hollow fibre-based, liquid-phase micro-extraction, *TrAC Trends Anal. Chem.* 23: 1–10.
17. Fernandez, E., Vidal, L., 2014. Liquid-phase extraction and micro-extraction. In De Los Ríos, A.P., Fernández, F.J.H., Eds., *Ionic Liquids in Separation Technology*. Elsevier, 107–152.
18. Ghambarian, M., Yamini, Y., Esrafili, A., 2012. Developments in hollow fiber based liquid-phase micro-extraction: Principles and applications, *Microchim. Acta* 177: 271–294.
19. Zhao, L., Lee, H. K., 2002. Liquid-phase microextraction combined with hollow fiber as a sample preparation technique prior to gas chromatography/mass spectrometry, *Anal. Chem.* 74: 2486–2492.
20. Kokosa, J. M., 2013. Advances in solvent-microextraction techniques, *TrAC Trends Anal. Chem.* 43: 2–13.
21. Dadfarnia, S., Shabani, A. M. H., 2010. Recent development in liquid phase micro-extraction for determination of trace level concentration of metals–A review, *Anal. Chim. Acta* 658: 107–119.
22. Ghasemi, E., Sillanpää, M., Najafi, N. M., 2011. Headspace hollow fiber protected liquid-phase micro-extraction combined with gas chromatography mass spectroscopy for speciation and determination of volatile organic compounds of selenium in environmental and biological samples, *J. Chromatogr. A* 1218: 380–386.
23. Romero-Gonzalez, R., Frenich, A. G., Vidal, J. L. M., Aguilera-Luiz, M. M., 2010. Determination of ochratoxin A and T-2 toxin in alcoholic beverages by hollow fiber liquid phase microextraction and ultra high-pressure liquid chromatography coupled to tandem mass spectrometry, *Talanta* 82: 171–176.
24. Montes, R., Rodríguez, I., Rubi, E., Ramil, M., Cela, R., 2008. Suitability of polypropylene micro-porous membranes for liquid-and solid-phase extraction of halogenated anisoles from water samples, *J. Chromatogr. A* 1198: 21–26.
25. Barri, T., Bergström, S., Hussen, A., Norberg, J., 2006. Extracting syringe for determination of organochlorine pesticides in leachate water and soilwater slurry: A novel technology for environmental analysis, *J. Chromatogr. A* 1111(1): 11–20.
26. Pezo, D., Salafranca, J., Nerín, C., 2007. Development of an automatic multiple dynamic hollow fibre liquid-phase microextraction procedure for specific migration analysis of new active food packagings containing essential oils, *J. Chromatogr. A* 1174: 85–94.
27. Ouyang, G., Zhao, W., Pawliszyn, J., 2007. Automation and optimization of liquid phase micro-extraction by gas chromatography, *J. Chromatogr. A* 1138: 47–54.
28. Wang, X., He, Y., Lin, L., Zeng, F., Luan, T., 2014. Application of fully automatic hollow fiber liquid phase micro-extraction to assess the distribution of organophosphate esters in the Pearl River Estuaries, *Sci. Total Environ.* 470: 263–269.
29. Esrafili, A., Yamini, Y., Ghambarian, M., Moradi, M., 2011. Dynamic three-phase hollow fiber micro-extraction based on two immiscible organic solvents with automated movement of the acceptor phase, *J. Sep. Sci.* 34: 98–106.

30. Esrafili, A., Yamini, Y., Ghambarian, M., Ebrahimpour, B., 2012. Automated preconcentration and analysis of organic compounds by on-line hollow fiber liquid-phase micro-extraction high performance liquid chromatography, *J. Chromatogr. A* 1262: 27–33.

31. Goh, S. X. L., Lee, H. K., 2018. Automated bundled hollow fiber array-liquid-phase micro-extraction with liquid chromatography tandem mass spectrometric analysis of perfluorinated compounds in aqueous media, *Anal. Chim. Acta* 1019: 74–83.

32. Goh, S. X. L., Lee, H. K., 2017. An alternative perspective of hollow fiber-mediated extraction: Bundled hollow fiber array-liquid-phase micro-extraction with sonication-assisted desorption and liquid chromatography tandem massspectrometry for determination of estrogens in aqueous matrices, *J. Chromatogr. A* 1488: 26–36.

33. Chimuka, L., Megersa, N., Norberg, J., Mathiasson, L., Jönsson, J. A., 1998. Incomplete trapping in supported liquid membrane extraction with a stagnant acceptor for weak bases, *Anal Chem.* 70: 3906–3911.

34. Megersa, N., Solomon, T., Jönsson, J. A., 1999. Supported liquid membraneextraction for sample work-up and preconcentration of methoxy-s-triazine herbicides in a flow system, *J. Chromatogr. A* 830: 203–210.

35. Lindegard, B., Björk, H., Jönsson, J. A., Mathiasson, L., Olsson, A.-M., 1994. Automatic column liquid chromatographic determination of a basic drug in blood plasma using the supported liquid membrane technique for sample pretreatment, *Anal Chem.* 66: 4490–4497.

36. Jönsson, J. A., Andersson, M., Melander, C., Norberg, J., Thordarson, E., Mathiasson, L., 2000. Automated liquid membrane extraction for HPLC determination of Ropi vacaine metabolites in urine, *J. Chromatogr A* 870: 151–157.

37. Ji, B., Xia, B., Fu, X., Lei, S., Ye, Y., Zhou, Y., 2018. Low-cost and convenient ballpoint tip-protected liquid-phase microextraction for sensitive analysis of organic molecules in water, samples, *Anal. Chim. Acta* 1006: 42–48.

38. Yu, S., Xiao, Q., Zhu, B., Zhong, X., Xu, Y., Su, G., Chen, M., 2014. Gas chromatography-mass spectrometry determination of earthyemusty odorous compounds in waters by two phase hollow-fiber liquid-phase micro-extraction using polyvinylidene fluoride fibers, *J. Chromatogr. A* 1329: 45–51.

39. Cai, J., Chen, G., Qiu, J., Jiang, R., Zeng, F., Zhu, F., Ouyang, G., 2016. Hollow fiber basedliquid phase micro-extraction for the determination of organochlorine pesticidesin ecological textiles by gas chromatography mass spectrometry, *Talanta* 146: 375–380.

40. Ganjikhah, M., Shariati, S., Bozorgzadeh, E., 2017. Preconcentration and spectro photometric determination of trace amount of formaldehyde using hollow fiber liquid-phase micro-extraction based on derivatization by Hantzsch reaction, *J. Iran. Chem. Soc.* 14: 763–769.

41. Gonzalez-Salamo, J., Gonzalez-Curbelo, M. A., Socas-Rodríguez, B., Hernandez-Borges, J., Rodríguez-Delgado, M. A., 2018. Determination of phthalic acid esters inwater samples by hollow fiber liquid-phase micro-extraction prior to gas chromatography tandem mass spectrometry, *Chemosphere* 201: 254–261.

42. Pimparu, R., Nitiyanontakit, S., Miro, M., Varanusupakul, P., 2016. Dynamic single interface hollow fiber liquid phase micro-extraction of Cr (VI) using ionic liquid containing supported liquid membrane, *Talanta* 161: 730–734.

43. Yazdi, A. S., Yazdinezhad, S. R., Heidari, T., 2015. Determination of melamine in soilsamples using surfactant-enhanced hollow fiber liquid phase micro-extraction followed by HPLC-UV using experimental design, *J. Adv. Res.* 6: 957–966.

44. Asgharinezhad, A. A., Ebrahimzadeh, H., 2016. Supramolecular nano solvent-based hollow fiber liquid phase micro-extraction as a novel method for simultaneous preconcentration of acidic, basic and amphiprotic pollutants, *RSC Adv.* 6: 41825–41834.

45. Socas-Rodríguez, B., Asensio-Ramos, M., Hernández-Borges, J., Rodríguez-Delgado, M. A., 2014. Analysis of oestrogenic compounds in dairy products by hollow-fibre liquid-phase micro-extraction coupled to liquid chromatography, *Food Chem.* 149: 319–325.

46. Fernandez, M. A. M., Andre, L. C., de Cardeal, Z. L., 2017. Hollow fiber liquid-phasemicro-extraction-gas chromatography-mass spectrometry method to analyze bisphenol A and other plasticizer metabolites, *J. Chromatogr. A* 1481: 31–36.

47. Feng, Y. D., Tan, Z. Q., Liu, J. F., 2011. Development of a static and exhaustive extraction procedure for field passive preconcentration of chlorophenols in environmental waters with hollow fiber-supported liquid membrane, *J. Sep. Sci.* 34: 965–970.

48. Sagrista, E., Larsson, E., Ezoddin, M., Hidalgo, M., Salvado, V., Jönsson, J. Å., 2010. Determination of non-steroidal anti-inflammatory drugs in sewage sludge by direct hollow fiber supported liquid membrane extraction and liquid chromatography-mass spectrometry, *J. Chromatogr. A* 1217: 6153–6158.

49. Saleh, A., Larsson, E., Yamini, Y., Jönsson J. Å., 2011. Hollow fiber liquid phasemicroextraction as a preconcentration and clean-up step after pressurized hot water extraction for the determination of non-steroidal anti-inflammatory drugs in sewage sludge, *J. Chromatogr. A* 1218: 1331–1339.

50. Nazaripour, A., Yamini, Y., Bagheri, H., 2018. Extraction and determination of trace amounts of three anticancer pharmaceuticals in urine by three-phase hollow fiber liquid-phase micro-extraction based on two immiscible organic solvents followed by high-Performance liquid chromatography, *J. Sep. Sci.* 41: 3113–3120.

51. Zhang, J., Zhang, M., Fu, S., Li, T., Wang, S., Zhao, M., Ding, W., Wang, C., Wang, Q., 2014. Simultaneous determination of imperator in and its metabolite xanthotoxol in rat plasma by using HPLC-ESI-MS coupled with hollow fiber liquid phase micro-extraction, *J. Chromatogr. B* 945: 185–192.

52. Fernandez, M. A. M., Andre, L. C., de Cardeal, Z. L., 2017. Hollow fiber liquid-phase micro-extraction gas-chromatography-mass spectrometry method to analyze bisphenol A and other plasticizer metabolites, *J. Chromatogr. A* 1481: 3136.

53. Worawit, C., Cocovi-Solberg, D. J., Varanusupakul, P., Miro, M., 2018. In-line carbonnano-fiber reinforced hollow fiber-mediated liquid phase micro-extraction using a 3D printed extraction platform as a front end to liquid chromatography for automatic sample preparation and analysis: A proof of concept study, *Talanta* 185: 611–619.

54. Sun, W., Qu, S., Du, Z., 2014. Hollow fiber liquid-phase micro-extraction combined with ultra-high performance liquid chromatography tandem mass spectrometry for the simultaneous determination of naloxone, buprenorphine and norbuprenorphine in human plasma, *J. Chromatogr. B* 951: 157–163.

55. Gonzalez, J. L., Pell, A., Lopez-Mesas, M., Valiente, M., 2018. Hollow fibre supported liquid membrane extraction for BTEX metabolites analysis in human teeth asbiomarkers, *Sci. Total Environ.* 630: 323–330.

56. Rezaei, F., Yamini, Y., Moradi, M., Daraei, B., 2013. Supramolecular solvent-based hollow fiber liquid phase micro-extraction of benzodiazepines, *Anal. Chim. Acta* 804: 135–142.

57. Banforuzi, S. R., Hadjmohammadi, M. R., 2017. Two-phase hollow fiber-liquid micro-extraction based on reverse micelle for the determination of quercetin inhuman plasma and vegetables samples, *Talanta* 173: 14–21.

58. Yang, Q., Chen, B., He, M., Hu, B., 2018. Sensitive determination of seven triazineherbicide in honey, tomato and environmental water samples by hollow fiber based liquid-liquid-liquid micro-extraction combined with sweeping micellar electrokinetic capillary chromatography, *Talanta* 186: 88–96.

59. Campillo, N., Lopez-García, I., Hernandez-Cordoba, M., Vinas, P., 2018. Food and beverage applications of liquid-phase microextraction, *Trac. Trends Anal. Chem.* 109: 1161–1123.

60. Alahmad, W., Tungkijananisn, N., Kaneta, T., Varanusupakul, P., 2018. A colorimetric paper-based analytical device coupled with hollow fiber membrane liquid phase micro-extraction (HF-LPME) for highly sensitive detection of hexavalent chromium in water samples, *Talanta* 190: 78–84.

61. da Silva, G. S., Lima, D. L. D., Esteves, V. I., 2017. Salicylic acid determination in estuarine and riverine waters using hollow fiber liquid-phase microextraction and capillary zone electrophoresis, *Environ. Sci. Pollut. Res.* 24: 15748–15755.

62. Zhou, J., Zhang, Q., Sun, J. B., Sun, X. L., Zeng, P., 2014. Two-phase hollow fiber liquid phase micro-extraction based on magneto fluid for simultaneous determination of echinacoside, tubuloside B, acteoside and isoacteoside in rat plasmaafter oral administration of Cistanche salsa extract by high performance liquid chromatography, *J. Pharmaceut. Biomed. Anal.* 94: 30–35.

63. Esrafili, A., Yamini, Y., Ghambarian, M., Shariati, S., Moradi, M., 2012. Measurement of fluoroquinolone antibiotics from human plasma using hollow fiber liquid-phase microextraction based on carrier mediated transport, *J. Liq. Chromatogr. Rel. Technol.* 35: 343–354.

64. Thunig, J., Flo, L., Pedersen-Bjergaard, S., Hansen, S. H., Janfelt, C., 2012. Liquid-phase microextraction and desorption electrospray ionization mass spectrometry for identification and quantification of basic drugs in human urine, *Rapid Commun. Mass Spectrom* 26: 133–140.

65. Lv, J., Zhao, X., Ye, J., Liu, D., Chen, X., Bi, K., 2013. Hollow fiber-based liquid membrane microextraction combined with high-performance liquid chromatography for extraction and determination of trimetazidine in human plasma, *Biomed. Chromatogr.* 27: 292–298.

66. Zhou, J., Zeng, P., Cheng, Z. H., Liu, J., Wang, F. Q., Qian, R. J., 2011 Application of hollow fiber liquid phase micro-extraction coupled with high-performance liquid chromatography for the study of the osthole pharmacokinetics in cerebral ischemia hypoperfusion rat plasma, *J. Chromatogr. B* 879: 2304–2310.

67. Ebrahimzadeh, H., Asgharinezhad, A. A., Adlnasab, L., Shekari, N., 2012. Optimization of ion-pair based hollow fiber liquid phase microextraction combined with HPLC-UV for the determination of methimazole in biological samples and animal feed, *J. Sep. Sci.* 35: 2040–2047.

68. Banforuzi, S. R., Hadjmohammadi, M. R., 2017. Two-phase hollow fiber-liquid microextraction based on reverse micelle for the determination of quercetin in human plasma and vegetables samples, *Talanta* 173: 14–21.

69. Bahar, S., Es'haghi, Z., Nezhadali, A., Banaei, A., Bohlooli, S., 2017. An innovative methodfor analysis of Pb (II) in rice, milk and water samples based on TiO$_2$ reinforce dcaprylic acid hollow fiber solid/liquid phase micro-extraction, *Food Chem.* 221: 1904–1910.

70. Socas-Rodríguez, B., Asensio-Ramos, M., Hernández-Borges, J., Rodríguez-Delgado, M. A., 2013. Hollow-fiber liquid-phase micro-extraction for the determination of natural and synthetic estrogens in milk samples, *J. Chromatogr. A* 1313: 175–184.

71. Socas-Rodríguez, B., Asensio-Ramos, M., Hernández-Borges, J., Rodríguez-Delgado, M. A., 2014. Analysis of oestrogenic compounds in dairy products by hollow-fibre liquid-phase micro-extraction coupled to liquid chromatography, *Food Chem.* 149: 319–325.

72. Mlunguza, N. Y., Ncube, S., Mahlambi, P. N., Chimuka, L., Madikizela, L. M., 2020. Determination of selected antiretroviral drugs in wastewater, surface water and aquatic plants using hollow fibre liquid phase microextraction and liquid chromatography – Tandem mass spectrometry, *J. Hazard. Mater.* 382: 121067.

73. Wang, P., Xiao, Y., Liu, W., Wang, J., Yang, Y., 2015. Vortex-assisted hollow fibre liquid phase micro-extraction technique combined with high performance liquid chromatography-diode array detection for the determination of oestrogens in milk samples, *Food Chem.* 172: 385–390.

74. Simao, V., Merib, J., Dias, A. N., Carasek, E., 2016. Novel analytical procedure using a combination of hollow fiber supported liquid membrane and dispersive liquid-liquid micro-extraction for the determination of aflatoxins in soybean juice by high performance liquid chromatography fluorescence detector, *Food Chem.* 196: 292–300.

75. Huang, S., Chen, X., Wang, Y., Zhu, F., Jiang, R., Ouyang, G., 2017. High enrichment and ultra-trace analysis of aflatoxins in edible oils by a modified hollow-fiber liquid-phase micro-extraction technique, *Chem. Commun.* 53: 8988–8991.

76. Zhao, G., Wang, C., Wu, Q., Wang, Z., 2011. Determination of carbamate pesticides in water and fruit samples using carbon nanotube reinforced hollow fiber liquid-phase microextraction followed by high performance liquid chromatography, *Anal. Methods* 3: 1410–1417.

77. Bedendo, G. C., Jardim, I. C. S. F, Carasek, E., 2012. Multiresidue determination of pesticides in industrial and fresh orange juice by hollow fiber microporous membrane liquid-liquid extraction and detection by liquid chromatography-electrospray-tandem mass spectrometry, *Talanta* 88: 573–580.

78. Gao, L., Jönsson, J. Å., 2012. Determination of melamine in fresh milk with single hollow fiber supported liquid membrane extraction based on ion pair mechanism combined with high performance liquid chromatography, *Anal. Lett.* 45: 2310–2323.

Part V

Chapters on Supported Gas Membranes

13 Hollow Fiber Membrane Contactors for Dehumidification of Air

M. Madhumala, T. Nagamani, and S. Sridhar

CONTENTS

13.1 INTRODUCTION TO AIR DEHUMIDIFICATION

Humidification and dehumidification processes have been primarily used in various chemical industries, space craft, and air conditioning applications. Gas humidification for proton exchange membrane-based fuel cell stack has drawn great attention worldwide, where the humidity of the reactant gas stream is increased during the humidification process before it enters the fuel cell chamber [1]. The conventional techniques adapted for humidification and dehumidification are considered disadvantageous due to problems concerning high pressure drop, liquid entrainment, low contact area, and high capital investments. This chapter deals with the application of membrane contactor systems to dehumidification of air and industrial gas mixtures. Air dehumidification has attained great interest in recent years due to its widespread applications for buildings with high outdoor air facility [2], air conditioning [3], spacecraft mission, [4], etc. Research is ongoing to improvise the present technologies to attain higher performance. The removal of atmospheric moisture or humid air from ambient air is termed as dehumidification. Air dehumidification can be possible either by (a) compressing the ambient air, (b) introducing a desiccant or drying agent into the air stream, or (c) cooling the ambient air below its dew point [5]. The conventional compressor-based air conditioning units consume enormous energy to condense moisture from air [6]. Further problems concerning bacterial growth over the wetting surface can degrade indoor air quality. Desiccant-based dehumidifiers seem to be feasible alternatives over conventional compressor-based air conditioning units due

to advantages of lower energy consumption and operational costs [7]. The low water vapor pressure near the desiccant surface results in a vapor pressure gradient which enables extraction of water vapor molecules from ambient air. Liquid or solid desiccants can be used to facilitate removal of moisture from air. The absorption of water vapor onto the surface of liquid desiccants undergoes chemical or physical changes while the water vapor adsorption onto solid desiccants does not change its chemical or physical structure [8]. The process of moisture sorption and desorption from and into the air and its subsequent cooling during absorption and regeneration processes are depicted in Figure 13.1 [9].

Liquid desiccants have found increasing popularity because of their enhanced ability to absorb moisture from air streams as well as their ease of utilization during operation. Highly concentrated inorganic halide salt solutions made of lithium chloride [7], calcium chloride [10], lithium bromide [11], magnesium chloride [12], etc., and glycols [13] have been majorly used as desiccant materials in industrial dehumidifiers. The use of chilled water as a desiccant solution was first reported in literature to control air humidity in spacecraft applications [4,14,15].

The significant downside of utilizing fluid desiccants for dehumidification is that they get drawn into the conditioned space alongside dehumidified air, while strong desiccants experience issues such as their spillage among wet and dry air streams. The reuse of desiccants after the process is essential to lower the operational costs involved. However, the regeneration of solid desiccants is considered disadvantageous as it consumes greater energy. The use of renewable energy sources for regeneration would reduce running costs to a considerable extent. Therefore, the use of

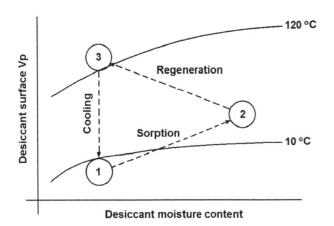

FIGURE 13.1 Dehumidification and regeneration cycle (Reprinted from Thermal Comfort in Buildings, ed. Matthew Hall, Desiccant materials for moisture control in buildings. In *Materials for Energy Efficiency and Thermal Comfort in Buildings*, 365–383, UK, Copyright (2010) with permission Woodhead Publishing, UK [9].)

a non-contacting dehumidifier where problems pertaining to the desiccant's corrosive nature and complexity of design can be avoided is desired. Recently, a membrane-based air dehumidification system combined with desiccants has emerged as a new alternative technology in comparison to traditional dehumidifiers with desiccants as the packing material. The use of selective membrane material prevents the cross flow of the desiccant solution over to the processed air, while permitting the transport of heat and moisture between the solution and the air.

13.2 BASIC PRINCIPLE OF GAS–LIQUID MEMBRANE CONTACTORS

13.2.1 HOLLOW FIBER MEMBRANE-BASED AIR DEHUMIDIFICATION

Hollow fiber membrane contactors (HFMC) for air dehumidification have attained greater importance when compared to other membrane module configurations, e.g. flat sheet, plate and frame, and tubular types. HFMC provides

high packing density, interfacial contact area, self mechanical support, as well as higher heat and mass transfer capability, which thereby enhance the overall efficiency of the dehumidification process [16]. These devices are able to create vapor/liquid interface near the membrane pores that offer efficient mass and heat transfer between the phases. In addition, the hollow fiber modular configuration enables easy scale-up and handling which can be widely useful for household and industrial applications. Figure 13.2 represents the schematic of a basic liquid desiccant HFMC-based air dehumidification system.

This contactor arrangement facilitates selective contact between the gaseous and liquid streams avoiding the penetration of the absorbent into the gaseous phase.

The removal of water vapor from the air is accomplished by passing the gaseous and liquid phases on either side of porous hydrophobic membranes. Porous membranes experience pore-wetting phenomena by the liquid absorbent. To prevent the liquid absorbent from permeating into the gaseous stream, the trans-membrane pressure drop across the membrane should be carefully adjusted. Furthermore, the membrane surface morphology or pore radius (r), its contact angle (θ), the surface tension (σ) of the wetting absorbent solution, and the fluid flow pattern play a major role in membrane pore-wetting (Figure 13.3).

The critical breakthrough trans-membrane pressure (ΔP) can be estimated using Laplace equation [17]:

$$\Delta P = \frac{2\sigma\cos\theta}{r} \qquad (13.1)$$

It was reported that the use of hydrophobic microporous membrane materials such as Polypropylene (PP), Polytetrafluoroethylene (PTFE), Polyvinylidene fluoride (PVDF), etc., coated with a silicone skin could overcome the problems concerning air leakage and intermixing of air/liquid absorbent systems [18].

Studies on use of dense hydrophilic polymer membranes made of Nafion [19], cellulose triacetate [20], polyether sulfone [14], polyurethane [21], etc., for air drying have also been reported in the literature. Hydrophilic membranes

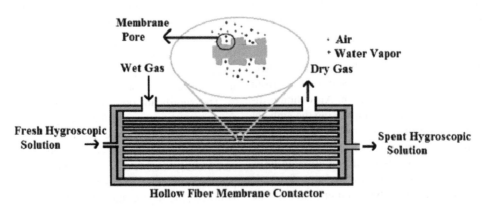

FIGURE 13.2 Schematic of basic liquid desiccant HFMC based air dehumidification system (Reprinted from *Journal of Industrial & Engineering Chemistry Research* 34, Experimental investigation of gas dehumidification by tri-ethylene glycol in hollow fiber membrane contactors. 390–396, Copyright (2016) with permission from Elsevier [25].)

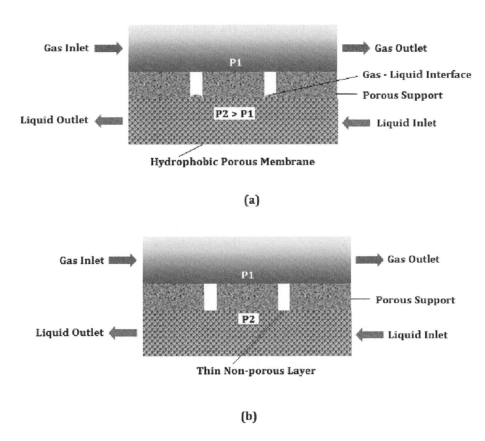

FIGURE 13.3 Mass transfer in a gas / liquid membrane contactor system through (a) porous and (b) composite hydrophobic membranes.

enable higher moisture absorption with less or no permeation with respect to air [22].

13.3 MECHANISMS OF WATER VAPOR TRANSPORT THROUGH THE MEMBRANE

Many industrial applications are established on the countercurrent flow pattern of gas and liquid streams with gas flowing through the shell side of the hollow fiber membrane contactor, while the liquid absorbent flows inside the lumen side [23–25]. The advantage of this flow pattern reduces the pressure drop of gaseous stream flowing inside the shell side which thereby directly improves the overall performance of the contactor. The transport mechanism of fluid (gas/liquid) flow within the membrane contactor system has been widely reported by several investigators. Both heat and mass transfer occur simultaneously during the membrane facilitated dehumidification process. Heat gets transferred from the air side through the membrane to the desiccant solution side as sensible and latent heat, while heat transferred through the membrane is by conduction.

The sensible (q_S) and latent (q_L) heat transfer rates are given by [26]:

$$q_S = h_o A_s \left(t_i - t_a \right) \tag{13.2}$$

$$q_L = h_m A_s \left(W_i - W_a \right) h_v \tag{13.3}$$

where, t_a is the dry-bulb temperature of air (K), t_i is the wetted surface temperature (K), W_a is the humidity ratio of air (kg/kg), W_i is the humidity ratio of saturated air at t_i (kg/kg), h_o and h_m are the convective heat (W/m²K) and mass transfer (kg/m²) coefficients, and h_v is the latent heat of vaporization (J/kg).

13.3.1 HEAT TRANSFER IN A GAS–LIQUID MEMBRANE CONTACTOR

Heat transfer across the membrane is by conduction, whereas the bulk liquid experiences convective heat flow pattern [27]. Several investigations are available in the literature for estimation of the heat transfer coefficients. The governing correlations for hollow fiber membrane modules are similar to shell and tube heat exchangers.

Shell side convective heat transfer coefficient can be expressed in terms of Nusselt number [3]:

$$Nu = \frac{h d_h}{\tau} \tag{13.4}$$

where h is the convective heat transfer coefficient (k Wm⁻² K⁻¹), d_h is the hydrodynamic diameter (m), and τ is the heat conductivity ((k Wm⁻¹ K⁻¹) of the solution.

The Hausenand Sieder–Tate equations are the commonly recommended heat transfer correlations for estimation of Nusselt (Nu) number [4].

$$Nu = 3.685 + \frac{0.085\left(\dfrac{Re.Pr.d_h}{L}\right)}{1 + 0.047\left(\dfrac{Re.Pr.d_h}{L}\right)^{0.67}} \quad (For\ Re.Pr.d_H\,/\,L < 100)$$

$$(13.5)$$

$$Nu = 1.86\left(\frac{Re.Pr.d_h}{L}\right)^{0.33}\left(For\ Re.Pr.d_H\,/\,L < 100\right) \quad (13.6)$$

where, Re and Pr are Reynolds and Prandtl numbers, respectively, and L is the fiber length.

The convective heat transfer correlation for turbulent flow pattern is given by Dittus–Boelter equation [4]:

$$Nu = 0.023Re^{0.8}Pr^{c} \quad (13.7)$$

where, the value of constant, c is 0.4 for hot fluid and 0.3 for cold fluid.

The correlation for Nusselt number considering the packing density (ϕ) of hollow fibers in the shell side and Reynolds number at the air side is described by the following equation [4]:

$$Nu = \left(-0.191\phi^2 + 0.517\emptyset + 0.318\right)\times(0.243Re^{0.8}$$
$$+10.323)Pr^{0.33}\left(For\,\phi \leq 0.2\ and\ Re\,250 \geq 100\right) \quad (13.8)$$

The overall heat transfer coefficient, h_{tot} (W/m^{2}K), from solution to air is calculated by the following equation:

$$\frac{1}{h_{tot}} = \frac{1}{h_{inner}}\left(\frac{d_{outer}}{d_{inner}}\right) + \frac{\delta}{\lambda_{mem}}\left(\frac{d_{outer}}{\overline{d}}\right) + \frac{1}{h_{outer}} \quad (13.9)$$

where $\overline{d}$ is the arithmetic mean diameter of a fiber, i.e., (0.5*(Outer diameter of fiber (d_{outer}) + Inner diameter of fiber (d_{inner})), and λ is the heat conductivity (W/mK) of the membrane.

Convective heat transfer in a randomly packed hollow fiber module is defined by the following equation:

$$\rho C_p Q \frac{dT}{dX} = hN\pi d_o\left(T_w - T_{in}\right) \quad (13.10)$$

where C_p is the specific heat (kJ kg^{-1} K^{-1}), Q is the volumetric flow rate (m^3/s), d_o is the outer diameter of the hollow fiber, T_w and T_{in} are the temperatures of liquid on the wall and at the inlet, respectively.

Various mathematical models have been developed to study the heat transfer profile within a hollow fiber membrane module. Correlations established on heat balance using a free surface model is presented below:

Absorption of water vapor by the liquid desiccant releases heat on the membrane surface at the solution side.

The heat balance equation in terms of heat of absorption (h_{abs}) and moisture diffusivity (D_{va}) in air is given by [3]:

$$\lambda_a\frac{\partial T}{\partial n}\bigg|_{r=r_o} + \rho_a D_{va}h_{abs}\frac{\partial\omega_a}{\partial n}\bigg|_{r=r_o} = \lambda_s\frac{\partial T_s}{\partial n}\bigg|_{r=r_i} \quad (13.11)$$

where ω is humidity ratio, λ is heat conductivity, and ρ is the density of the air stream.

Equation 13.11 can be normalized as follows [3]:

$$\lambda^*\frac{\partial\theta_a}{\partial n}\bigg|_{r=r_o} + h_{abs}^*\frac{\partial\xi_a}{\partial n}\bigg|_{r=r_o} = \lambda_s\frac{\partial\theta_s}{\partial n}\bigg|_{r=r_i} \quad (13.12)$$

where, the dimensionless absorption heat (h_{abs}^*) and the dimensionless heat conductivity (λ^*) are defined by [3]:

$$h_{abs}^* = \frac{\rho_a D_{va}h_{abs}}{\lambda_s}\left(\frac{\omega_{si} - \omega_{ai}}{T_{si} - T_{ai}}\right) \quad (13.13)$$

$$\lambda^* = \frac{\lambda_a}{\lambda_s} \quad (13.14)$$

The basic equation for determination of total heat transfer rate (Q)is expressed as follows [28]:

$$Q = UA\Delta T_{ln} \quad (13.15)$$

$$U = \left(\frac{1}{h_{air}} + \frac{\delta}{k_{mem}} + \frac{1}{h_{sol}}\right)^{-1} \quad (13.16)$$

where U is the overall heat transfer coefficient (Wm^2/K), A is membrane surface area (m^2), h_{air} and h_{sol} are the air and solution convective heat transfer coefficients (Wm^2/K), δ is membrane thickness (m), and k_{mem} is membrane thermal conductivity (W/mK).

The amount of heat exchanged by the air within the membrane module is expressed as[28]:

$$Q = m_a c_{pa}\left(t_1 - t_2\right) \quad (13.17)$$

Combining **Equations 13.15 and 13.16**, the following expression is obtained:

$$U = \frac{m_a c_{pa}\left(t_1 - t_2\right)}{A\Delta T_{ln}} \quad (13.18)$$

The log mean temperature difference (ΔT_{ln}) is given as:

$$\Delta T_{ln} = \left(\frac{t_{a1} - t_{a2}}{\ln\left(\dfrac{t_{a1} - t_{w1}}{t_{a2} - t_{w1}}\right)}\right) \quad (13.19)$$

where, m_a is the mass flow rate of air (kg/s), C_{pa} is the specific heat of air (kJ/kg.K), t_{a1} and t_{a2} are inlet and outlet temperatures (K) of air, and t_{w1} is the inlet temperature of the liquid desiccant.

13.3.2 MASS TRANSFER IN A GAS–LIQUID MEMBRANE CONTACTOR

The mass transfer involved during removal of water vapour from air by a liquid absorbent in a gas/liquid membrane contactor is described in the literature [24]. The water vapour flux (J) from the air stream to the surface of the liquid absorbent is represented by the following expression [29]:

$$J = \frac{\Delta\omega_{lm}}{K_T} = D_{eff}\left(\frac{\omega_m}{z}\right) \quad (13.20)$$

where, $1/K_T$ is the overall mass transfer resistance (m^2s/kg) from the air stream to the liquid absorbent, $\Delta\omega_{lm}$ is the logarithmic mean humidity difference between the air stream and the solution surface, D_{eff} is the effective diffusivity (m^2/s) of water vapour in the membrane, ω_m is the mean humidity across the membrane, and Z is the membrane thickness (μm).

At steady state, the overall mass transfer resistance ($1/K_T$) is expressed as a function of individual mass transfer resistances in the gas or air phase, the membrane, and in the liquid phase. Considering resistances-in-series theory for a contactor we obtain [30]:

$$\frac{1}{K_T} = \frac{Hd_i}{k_g d_o} + \frac{Hd_i}{k_m d_{lm}} + \frac{1}{k_l} \quad (13.21)$$

where, k_l, k_m, and k_g are the overall mass transfer coefficients (m/s) in the liquid boundary layer, membrane, and gas boundary layer whereas d_i, d_o, and d_{lm} are the diameters (m) of the hollow fiber inside, outside, and the log mean value, respectively with H being the Henry's law coefficient.

For a composite membrane structure, the resistance offered by the membrane is given by the following equation:

$$\frac{1}{k_m} = \frac{1}{k_{ms}} + \frac{1}{k_{mc}} \quad (13.22)$$

where, $1/k_{ms}$ and $1/k_{mc}$ represent the individual mass transfer resistances of the porous support and dense skin layer, respectively.

The logarithmic mean humidity represented in **Equation 13.20** can be calculated using the following relationship [29]:

$$\Delta\omega_{lm} = \left| \frac{\omega_i - \omega_o}{\ln\left(\frac{\omega_L - \omega_o}{\omega_L - \omega_i}\right)} \right| \quad (13.23)$$

where, ω_i, ω_o, and ω_L represent humidity ratios at the inlet, outlet, and the liquid stream, respectively.

The diffusion of water vapor from the bulk of the air stream through the hydrophobic membrane into the liquid desiccant stream is based on Knudsen's diffusion, Poiseuille flow, molecular or transitional diffusion [29]. The smaller kinetic diameter of the water molecule enables its easier transport through the membrane compared to gases.

When the membrane pore size is smaller than the mean free path of the permeating molecules, collisions between the molecules and the pore walls dominate over intermolecular collisions, indicating that the mass transfer is governed by Knudsen's model [31, 32]. If the membrane pore size is greater than the mean free path of the permeating molecules, the intermolecular collisions preside over molecule-wall collisions and mass transfer is governed by Poiseuille flow mechanism (viscous flow). When molecules collide with each other and diffuse through the air gap, then mass transfer is explained by the Transition model. In most of the global air conditioning systems, the mass transfer of water vapor through microporous membranes is controlled by a combination of Knudsen's diffusion and ordinary diffusion. The molecular diffusion (D_M) and Knudsen's diffusion (D_K) coefficients of water vapour (V) in air (a) are given by the following correlations [29]:

$$D_M = \frac{3.203 \times 10^{-4} T^{1.75}}{P_m\left(v_V^{1/3} + v_a^{1/3}\right)^2} \sqrt{\frac{1}{M_V} + \frac{1}{M_a}} \quad (13.24)$$

$$D_K = \frac{d_p}{3}\sqrt{\frac{8RT}{\pi M_V}} \quad (13.25)$$

where, ν is the molecular diffusion volume (m^3), M is the molecular weight (kg/mol), P_m is the atmospheric pressure (Pa), d_p is the pore diameter, T is the temperature (K), and R is the universal gas constant (J/mol.K).

The effective diffusion coefficient (D_{eff}) in the membrane is calculated as [29]:

$$D_{eff} = \frac{\epsilon}{\tau}\left(\frac{D_K D_m}{D_K + D_m}\right) \quad (13.26)$$

where, ϵ is the porosity and τ is the pore tortuosity.

13.4 PERFORMANCE EVALUATION OF MEMBRANE-BASED AIR DEHUMIDIFIERS

There are limited reports available on the application of membrane based gas / liquid contactors for air dehumidification. Usachov et al. (2007) investigated the performance of apolydimethylsiloxane (PDMS) dense membrane-based counter-current recycle contactor system with triethylene glycol used as a desiccant solution [33]. The designed system allowed steady air drying up to dew point temperature (T_d) of −30°C. At high gas to liquid flow rate ratio (G/L) of 1,000, the device exhibited reduction in air humidity from 40% RH (initial) to 10–12% RH (final). The overall efficiency of the device could be improved by raising the gas side pressure (Figure 13.4).

Kneifel et al. (2006) predicted the performance of a hollow fiber module with transversal flow configuration [24]. A composite hollow fiber membrane module coated with a thin layer of PDMS on the lumen side of porous polyetherimide (PEI) support was prepared to study the water

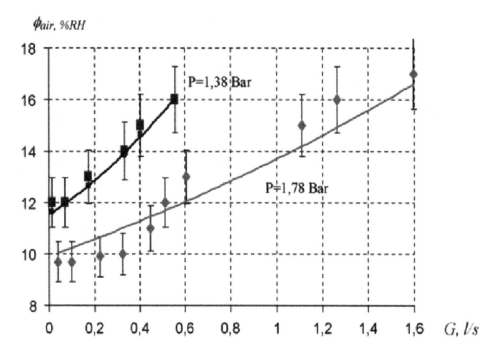

FIGURE 13.4 Air humidity dependence on feed gas flow in spiral wound membrane contactor (Reprinted from *Separation and Purification Technology* 57, Usachov, V.V., V.V. Teplyakov, A. Yu. Okunev, and N.I. Laguntsov. Membrane contactor air conditioning system: Experience and prospects, 502–506, Copyright (2007) with permission from Elsevier [33].)

vapour permeance into LiCl absorbent solution. Figure 13.5 reveals the scanning electron microscope (SEM)images of hollow fiber composite membrane with respect to: (a) inner wall surface, (b) cross-section, (c) outer surface, and (d) PDMS coated layer at the inner surface.

The cross-section of the membrane showed a finger-like porous structure extended over a dense PDMS skin while the outer surface of membrane consisted of a highly porous configuration. Membrane with an enhanced shell side surface porosity exhibited the highest water vapor permeance of 0.64 $gm^{-2}h^{-1}Pa^{-1}$. A porous polypropylene hollow fiber membrane based gas / liquid contactor utilizing dry TEG solution as a desiccant enabled reduction in the dew point of air below −30°C from 96% humidity level at standard conditions [34]. Annadurai et al. (2018) studied the performance of adiabatic and internally cooled dehumidifiers for moisture removal using porous PVDF membrane and LiCl solution as desiccant [35]. The performance of internally cooled adiabatic dehumidifiers for different inlet desiccant temperatures is compared in Figure 13.6.

It was reported that the internally cooled dehumidifier arrangement exhibited improved mass transfer when compared to an adiabatic dehumidifier. The pre-cooling of desiccant with chilled water in an adiabatic dehumidifier improves moisture absorption from air. Experimental investigation was carried out by Fakharnezhad and Keshavarz (2016) using tri-ethylene glycol as absorbent in Polyvinylidene fluoride and Polypropylene hollow fiber membrane contactors [25]. The study revealed an improvement in water vapour absorption efficiency as well as the outlet water dew point with a change in gas flow rate when operated under a counter-current mode of flow. The

performance of a hollow fiber based air-liquid membrane contactor was studied by Das and Jain (2013) using LiCl as desiccant solution [18]. The system exhibited maximum water vapor flux of 1,295 g/m^2h with the dehumidifying effectiveness varying from 23% to 45%. Further, the study showed improved heat and mass transfer characteristics which could be pressed to advantage for dehumidification application. Zhao et al. (2015) developed Polyacrylonitrile (PAN) / PDMS composite hollow fiber membrane modules [23]. A water vapor permeance ranging from 800 to 3,700 GPU (i.e., about 65% of humidity removal) was achieved under a low vacuum maintained in the lumen side. It was reported that the pilot scale air dehumidifier incorporated with PDMS composite membrane exhibited energy savings of up to 26.2% when compared to conventional air conditioning units (Figure 13.7).

Yuan et al. (2015) performed a simulation study for a novel environmental control system in an aircraft based on the membrane dehumidification process to evaluate the cooling performance [36]. The simulation results revealed greater feasibility when compared to the currently employed four-wheel high pressure de-watering systems. Zhang et al. (2008) fabricated a novel polyethersulfone (PES) / polyvinylalcohol (PVA) composite membrane to investigate water vapour permeability [22]. LiCl salt was added to the PVA solution to increase the hydrophilicity of the composite membrane. The composite membrane doped with 2wt% LiCl exhibited an extent of moisture absorption that was 4.7 times greater than the membrane containing no additive (Figure 13.8).

TiO_2 supported PVA/LiCl composite membranes fabricated by Bui et al. (2016) for air dehumidification showed

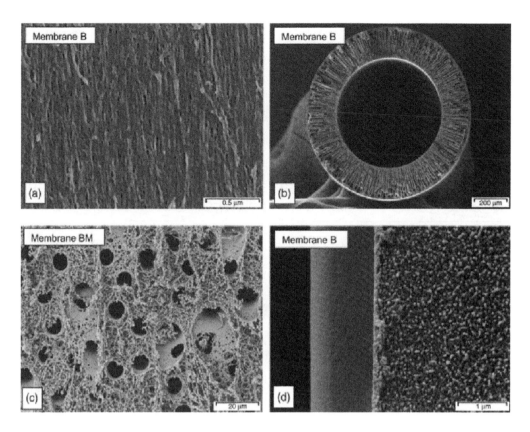

FIGURE 13.5   SEM images of hollow fiber composite membrane (a) inner wall surface, (b) cross-section, (c) outer surface, (d) PDMS layer on inner surface (Reprinted from *Journal of Membrane Science* 276, Kneifel, K., S. Nowak, W. Albrecht, R. Hilke, R. Just, and K.V. Peinemann, Hollow fiber membrane contactor for air humidity control: Modules and membranes, 241–251, Copyright (2006) with permission from Elsevier [24].)

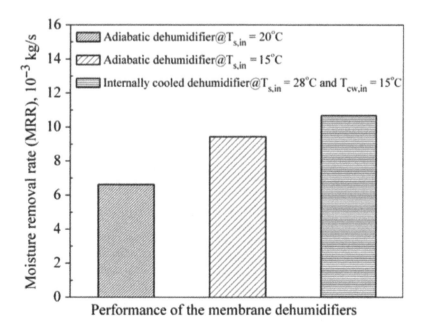

FIGURE 13.6   Performance comparison of the membrane air dehumidifiers (Reprinted from *Journal of Low-Carbon Technologies* 13(3), Annadurai, G., S. Tiwari, and M.P. Maiya, Experimental performance comparison of adiabatic and internally-cooled membrane dehumidifiers. 240–249, Copyright (2018) with permission from Oxford University Press [35].)

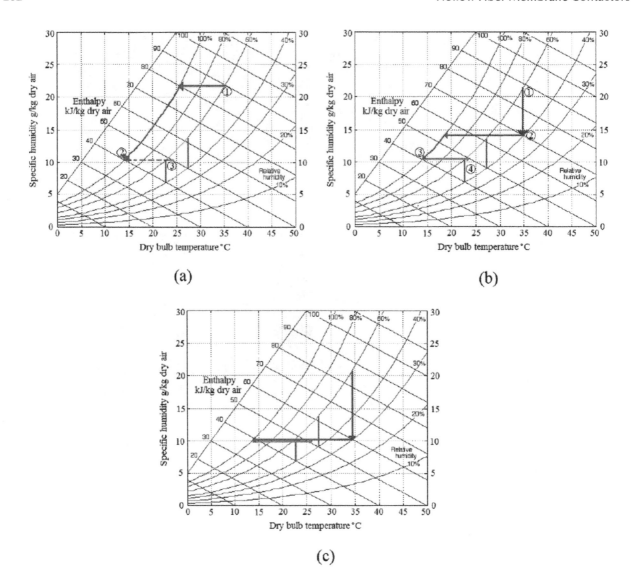

FIGURE 13.7 Comparison of energy consumption of (a) traditional air conditioning system without dehumidification; (b) air conditioning using PAN/PDMS hollow fiber system; and (c) air conditioning using an ideal membrane system (Reprinted from *Membranes* 5, Zhao, B., N. Peng, C. Liang, W.F. Yong, and T.S. Chung. Hollow Fiber Membrane Dehumidification Device for Air Conditioning System, 722–738 Copyright (2015) with permission from MDP (open access journal) [23].)

improved water vapor permeance with higher selectivity as shown in Figure 13.9 [37].

Dehumidification experiments were carried out at 24°C and 70% RH. The water vapor selectivity was found to increase with the number of polymer dips made to increase membrane thickness. Membranes with five dips exhibited a water vapour permeability of 300,000 barrer with a selectivity of 2,800 indicating potential for utilization in air dehumidification applications. Kudasheva et al. (2016) performed air dehumidification experiments by liquid membranes using hygroscopic ionic liquid desiccants [38]. Ionic liquids showed higher water recovery rates with permeability ranging from 26,000 to 46,000 barrer, in contrast to TEG desiccant solution. A water recovery rate of 10 g/h was achieved from a humid inlet air containing 80% RH (Figure 13.10).

13.5 APPLICATIONS OF MEMBRANE CONTACTOR SYSTEMS FOR REMOVAL OF WATER VAPOR FROM GASEOUS MIXTURES

13.5.1 NATURAL GAS

The demand for production of natural gas has been increasing in recent years due to its advantages of lower CO_2 emissions over coal and petroleum-based fuels for electricity generation. Natural gas comprises primarily of methane with low concentrations of higher hydrocarbon gases, CO_2, N_2, and H_2S. It is considered to be one of the important fuels for both domestic and industrial applications. In order to meet the requirement of end users, the natural gas should be free of water content and other entrained liquids. Hollow fiber

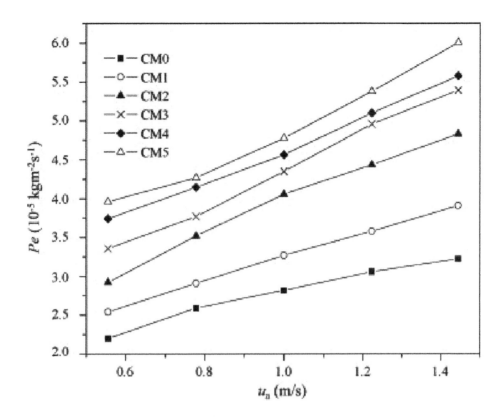

FIGURE 13.8 Effects of LiCl content on vapour permeation rates (Reprinted from *Journal of Membrane Science* 308, Zhang, L.Z., Y.Y. Wang, C.L. Wang, and H. Xiang, Synthesis and characterization of a PVA/LiCl blend membrane for air dehumidification, 198–206, Copyright (2008) with permission from Elsevier [22].)

membrane technology has been a cost-effective method for removal of undesirable components from gaseous streams. During the membrane dehydration process the wet gaseous stream is pumped through a membrane contactor using a gas compressor. The gas–liquid interface at the membrane surface facilities the separation of water vapor from other gaseous components. There are only a few studies reported on natural gas dehumidification using membrane contactor systems. PRISM (Air Products) company established a polydimethylsiloxane polymer based contactor unit to process 600,000 Nm^3/h of natural gas at Shell, Nigeria [39]. The membrane effectively removed CO_2, H_2S, and water vapor in a single step. Dalane et al. (2018) experimentally tested dehumidification of subsea natural gas by PTFE based hollow fibers using TEG desiccant [40]. The results predicted from the model showed error value of 3 to 7% compared to experimental data. High pressure and low temperature are considered favourable conditions for water vapor removal due to increased water solubility in the glycol solution.

13.5.2 Flue Gas

Coal fired power plants generate electric power with emission of large volumes of flue gas. The removal of water vapor from the flue gas prevents condensation and provides water for cooling towers in power plants. Sijbesma et al. (2008) explored the possibility of removing water vapor using

polyether-*block*-amide (PEBAX) and sulfonated poly(ether ether ketone) (SPEEK) membranes [41]. Composite hollow fibers with SPEEK as the upper layer exhibited high water vapor permeability with low permeation of non-condensable inert gases. Membranes enabled removal of 0.6–1 kg/m²h of water vapor during an operation time of 150 h using artificial flue gas. Macedonio et al. (2012) employed commercial porous PVDF hollow fibers supplied by MEMBRANA GmbH (Germany) for water vapor removal from flue gas stream [42]. Synthetic flue gas mixture of 78% nitrogen, 17% carbon dioxide, and 5% oxygen with 100% RH was fed on the shell side of the dehumidifier at a temperature of 53–55.2 °C. The membrane effectively retained water in the shell side while the dehydrated gaseous stream was recovered from the lumen side of the module. The patent by Wang et al. (2012) was based on an advanced transport membrane condenser (TMC) technology for water vapour removal using nanoporous ceramic membranes [43]. A two-stage TMC unit was built and its performance was validated using simulated coal fired flue gas. The technology has proven to be more beneficial for coal fire power plants for removing moisture and inert gases, as well as recovery of heat energy for reuse in power plants. A schematic of the two-stage TMC unit for recovering both heat energy and water is shown in Figure 13.11.

The recovered steam condensate and water vapor from stage I of the TMC unit is sent through a de-aerator before making up

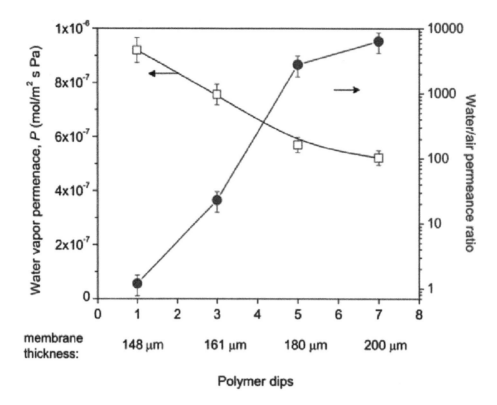

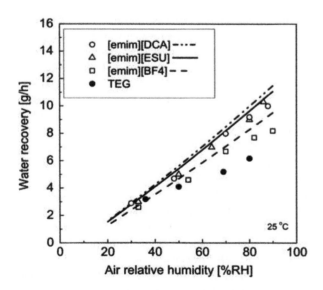

FIGURE 13.9 Water vapor permeance, water vapor permeance/air permeance ratio of the membranes, and membrane thickness with different number of polymer dips and PVA:LiCl ratio of 3:1 (Reprinted from *Journal of Membrane Science* 498, Bui, D.T., A. Nida, K.C. Ng, and K.J. Chua. 2016. Water vapour permeation and dehumidification performance of poly (vinyl alcohol)/lithium chloride composite membranes, 254–262, Copyright (2016) with permission from Elsevier [37].)

FIGURE 13.10 The dehumidification performance of the liquid membranes with ionic liquids (Reprinted from *Journal of Membrane Science*, 499, Kudasheva, A., T. Kamiya, Y. Hirota, A. Ito. Dehumidification of air using liquid membranes with ionic liquids, 379–385, Copyright (2016) with permission from Elsevier [38].)

the level of boiler feed water, while the outlet water from stage II of the TMC is routed back to the cooling water stream.

13.6 CONCLUSIONS AND FUTURE PERSPECTIVES

Air dehumidification has attained considerable significance due to its widespread application in aircrafts, and industries, as well as households. Among the available technologies, membrane-based dehumidification has proven to be an effective method in resolving the problems associated with traditional air/ liquid contacting systems. Details pertaining to hollow fiber membrane contactors and water vapor transport mechanisms through non-wetted membranes along with mathematical correlations for estimation of heat and mass transfer coefficients have been discussed. Additionally, a detailed review on use of polymeric, composite, and liquid membranes for air dehumidification is also presented. It is reported that the use of nonporous membranes can avoid the risk of desiccant permeation into the air stream. Recent literature has revealed membrane modification through incorporation of hydrophilic additives to be beneficial in enhancing

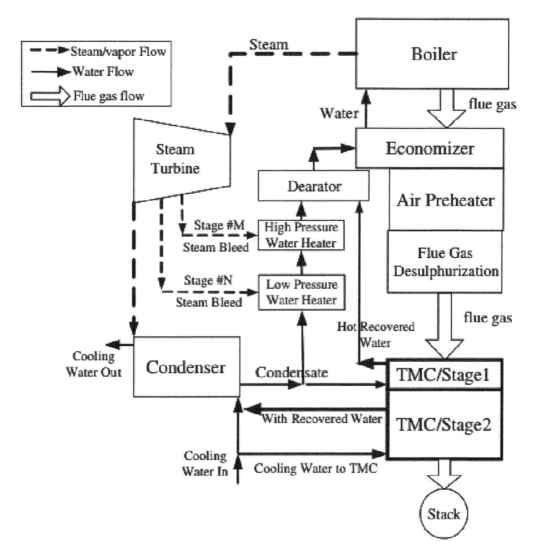

FIGURE 13.11 Schematic of two-stage transport membrane condenser (Reprinted from *Applied Energy* 91, Wang, D., A. Bao, W. Kunc, and W. Liss, Coal power plant flue gas waste heat and water recovery. 341–348 Copyright (2012) with permission from Elsevier [43].)

he water permeation rate. Further, the study demonstrated the application of membrane contactors for dehydration of gaseous mixtures including natural gas and flue gas. Liquid desiccant based membrane contactors were found to be economical and efficient for removal of water vapor in coal fired power plants. The recovered water can be recycled to cooling towers in industries. The design of multistage membrane air dehumidifiers could provide a competitive edge to this technology in the near future.

REFERENCES

1. Bakeri, G. 2019. A comparative study on the application of porous PES and PEI hollow fiber membranes in gas humidification process. *Journal of Membrane Science and Research* 5:11–19.
2. Pang, S.C., H.H. Masjuki, M.A. Kalam, and M.A. Hazrat. 2013. Liquid absorption and solid adsorption system for household, industrial and automobile applications: A review. *Renewable and Sustainable Energy Reviews* 28:836–847.
3. Zhang, L.Z., S.M. Huang, and L.X. Pei. 2012. Conjugate heat and mass transfer in a cross-flow hollow fiber membrane contactor for liquid desiccant air dehumidification. *International Journal of Heat and Mass Transfer* 55:8061–8072.
4. Yang, B., W. Yuan, X. He, and K. Ren. 2015. Air dehumidification by hollow fibre membrane with chilled water for spacecraft applications. *Indoor and Built Environment* 25(5):1–14.
5. El-Ghonemy, A.M.K. 2012. Fresh water production from/by atmospheric air for arid regions, using solar energy: Review. *Renewable and Sustainable Energy Reviews* 16:6384–6422.
6. Pérez-Lombard, L., J. Ortiz, and C. Pout. 2008. A review on buildings energy consumption information. *Energy and Buildings* 40(3):394–398.
7. Fumo, N., and D.Y. Goswami. 2002. Study of an aqueous lithium chloride desiccant system: Air dehumidification and desiccant regeneration. *Solar Energy* 72(4):351–361.
8. Roland, V., and N. Wahlgre. 2001. Atmospheric water vapour processor designs for potable water production: A review. *Water Research* 35(1):1–22.

9. Warwicker, B. 2010. Desiccant materials for moisture control in buildings. In *Materials for Energy Efficiency and Thermal Comfort in Buildings*, ed. M. Hall, 365–383. Cambridge, UK: Woodhead Publishing.

10. Seenivasan, D., V. Selladurai, and T.V. Arjunan. 2018. Experimental studies on the performance of dehumidifier using calcium chloride as a liquid desiccant. *International Journal of Energy Technology and Policy*1 4(1):49–63.

11. Wang, Z., F. Wang, J. Peng, F. Liu, X. Luo, and Z. Gu. 2015. Study on LiBr-liquid desiccant system driven by heat pump. *Procedia Engineering* 121:2068–2074.

12. Lychnos, G., and P.A. Davies. 2012. Modelling and experimental verification of a solar-powered liquid desiccant cooling system for greenhouse food production in hot climates. *Energy* 40(1):116–130.

13. Zurigat, Y., M. Abu-Arabi, and S. Abdul-wahab. 2004. Air dehumidification by triethylene glycol desiccant in a packed column. *Energy Conversion and Management* 45(1):141–155.

14. Scovazzo, P., A. Hoehn, and P. Todd. 2000. Membrane porosity and hydrophilic membrane based dehumidification performance. *Journal of Membrane Science* 167:217–225.

15. Scovazzo, P., J. Burgos, A. Hoehn, and P. Todd. 1998. Hydrophilic membrane-based humidity control. *Journal of Membrane Science* 149:69–81.

16. Bazhenov, S.D., A.V. Bildyukevich, and A.V. Volkov. 2018. Gas-liquid hollow fiber membrane contactors for different applications. *Fibers* 6:1–41.

17. Kim, B.S., and P. Harriott. 1987. Critical entry pressure for liquids in hydrophobic membranes. *Journal of Colloid and Interface Science* 115(1):1–8.

18. Das, R.S., and S. Jain. 2013. Experimental performance of indirect air-liquid membrane contactors for liquid desiccant cooling systems. *Energy* 57:319–325.

19. Ye, X.H., and M.D. Levan. 2003. Water transport properties of Nafion membranes: Part I. Single-tube membrane module for air drying. *Journal of Membrane Science*221:147–161.

20. Pan, C.Y., C.D. Jensen, C. Bielech, and H.W. Habgood. 1978. Permeation of water vapour through cellulose triacetate membranes in hollow fiber form. *Journal of Applied Polymer Science* 22:2307–2323.

21. Dilandro, L., M. Pegoraro, and L. Bordogna. 1991. Interaction of polyetherpolyurethane with water vapour and water–methane separation selectivity. *Journal of Membrane Science* 64:229–236.

22. Zhang, L.Z., Y.Y. Wang, C.L. Wang, and H. Xiang. 2008. Synthesis and characterization of a PVA/LiCl blend membrane for air dehumidification. *Journal of Membrane Science* 308:198–206.

23. Zhao, B., N. Peng, C. Liang, W.F. Yong, and T.S. Chung. 2015. Hollow fiber membrane dehumidification device for air conditioning system. *Membranes* 5:722–738.

24. Kneifel, K., S. Nowak, W. Albrecht, R. Hilke, R. Just, and K.V. Peinemann. 2006. Hollow fiber membrane contactor for air humidity control: Modules and membranes. *Journal of Membrane Science* 276:241–251.

25. Fakharnezhad, A., and P. Keshavarz. 2016. Experimental investigation of gas dehumidification by tri-ethylene glycol in hollow fiber membrane contactors. *Journal of Industrial and Engineering Chemistry* 34:390–396.

26. Khamis Mansour, M., and M. Hassab. 2012. Thermal design of cooling and dehumidifying coils. In *Heat Exchangers – Basics Design Applications*, ed. J. Mitrovic, 367–394. Rijeka, Croatia: InTechOpen.

27. Yang, B., W. Yuan, F. Gao, and B. Guo. 2013. A review of membrane-based air dehumidification. *Indoor and Built Environment* 24(1):11–26.

28. Englart, S. 2018. Comparison heat and mass transfer coefficients in the shell side of the hollow fiber membrane module. *E3S Web of Conferences* 44(00040):1–8.

29. Zhang, L.Z. 2006. Mass diffusion in a hydrophobic membrane humidification/dehumidification process: The effects of membrane characteristics. *Separation Science and Technology*41:1565–1582.

30. Ahmadi, H., S.A. Hashemifard, and A.F. Ismail. 2017. A research on CO_2 removal via hollow fiber membrane contactor: The effect of heat treatment. *Chemical Engineering Research and Design* 120:218–230.

31. Nagy, E. 2018. *Basic Equations of Mass Transport through a Membrane Layer*. Amsterdam, Netherlands: Elsevier.

32. Li, N.N., A.G. Fane, W.S.W. Ho, and T. Matsuura. 2011. *Advanced Membrane Technology and Applications*. Hoboken, NJ: John Wiley & Sons.

33. Usachov, V.V., V.V. Teplyakov, A.Y. Okunev, and N.I. Laguntsov. 2007. Membrane contactor air conditioning system: Experience and prospects. *Separation and Purification Technology* 57:502–506.

34. Petukhov, D.I., A.A. Eliseev, A.A. Poyarkov, A.V. Lukashin, and A.A. Eliseev. 2017. Porous polypropylene membrane contactors for dehumidification of gases. *Nanosystems: Physics, Chemistry, Mathematics* 8(6):798–803.

35. Annadurai, G., S. Tiwari, and M.P. Maiya. 2018. Experimental performance comparison of adiabatic and internally-cooled membrane dehumidifiers. *International Journal of Low-Carbon Technologies* 13(3): 240–249.

36. Yuan, W., B. Yang, B. Guo, X. Li, Y. Zuo, and W. Hu. 2015. A novel environmental control system based on membrane dehumidification. *Chinese Journal of Aeronautics* 28(3):712–719.

37. Bui, D.T., A. Nida, K.C. Ng, and K.J. Chua. 2016. Water vapour permeation and dehumidification performance of poly (vinyl alcohol)/lithium chloride composite membranes. *Journal of Membrane Science* 498:254–262.

38. Kudasheva, A., T. Kamiya, Y. Hirota, A. Ito. 2016. Dehumidification of air using liquid membranes with ionic liquids. *Journal of Membrane Science* 499:379–385.

39. Hoek, E.M., and V.V. Tarabara. 2013. *Encyclopedia of Membrane Science and Technology*. Hoboken, NJ: John Wiley & Sons.

40. Dalane, K., H.F. Svendsen, M. Hillestad, and L. Deng. 2018. Membrane contactor for subsea natural gas dehydration: Model development and sensitivity study. *Journal of Membrane Science* 556:263–276.

41. Sijbesma, H., K. Nymeijer, R. Marwijk, R. Heijboer, J. Potreck, and M. Wessling. 2008. Flue gas dehydration using polymer membranes. *Journal of Membrane Science* 313:263–276.

42. Macedonio, F., A. Brunetti, G. Barbieri, and E. Drioli. 2012. Membrane condenser as a new technology for water recovery from humidified "waste" gaseous streams. *Industrial and Engineering Chemistry Research* 52:1160–1167.

43. Wang, D., A. Bao, W. Kunc, and W. Liss. 2012. Coal power plant flue gas waste heat and water recovery. *Applied Energy* 91:341–348.

14 Gas Filled Membrane Pores – Fundamentals and Applications

A. Sengupta and S. Ranil Wickramasinghe

CONTENTS

14.1 INTRODUCTION

Gas filled membranes are a type of membrane contactor, where a porous membrane is a physical boundary between feed side and permeate side [1–4]. The pores are filled with gas. Depending upon the application, the gas might have a special affinity for the elements of interest to be separated or it can be inert. In general, mass transfers across the membrane consists of three steps. In the first step, the species that transfers from the feed side vaporizes. This could be a volatile solute or the solvent. The species then diffuses through the membrane pores. In the final step, the vapor condenses in the liquid phase on the permeate side.

For membrane contactors consisting of gas filled membrane pores there are three main unit operations: membrane distillation, osmotic distillation, and gas absorption/stripping. In the case of membrane distillation, the feed is maintained at a higher temperature than the permeate. This imposed temperature gradient results in a vapor pressure gradient across the membrane which is the driving force for mass transfer. Non-isothermal transport of water vapor across the hydrophobic membrane was employed as a common membrane-based separation technique since the mid-1960s and is called "membrane distillation" [5–8].

Osmotic distillation and gas absorption are considered "isothermal" processes. Here it is the difference in activity of the species being transferred from the feed to the permeate, that leads to differences in vapor pressure across that membrane, that drives mass transfer. In reality, since vaporization and condensation of the species being transferred across the membrane occur in the feed and permeate sides, respectively; the process cannot be truly isothermal. Blood oxygenation used during open-heart surgery is the best-known commercialized example of membrane-based gas absorption. This chapter focuses on isothermal operations, i.e., gas absorption and osmotic distillation.

In the case of gas absorption, Figure 14.1 summarizes the steps involved for transfer of a solute species from the feed to the permeate or strip solution. In these applications it is typically a volatile dilute solute that is transferred to the permeate stream.

1. Transfer of solute species to the membrane surface through the concentration boundary layer, that exists in the feed solution.
2. Phase equilibration between feed solution and gas filled membrane pores.

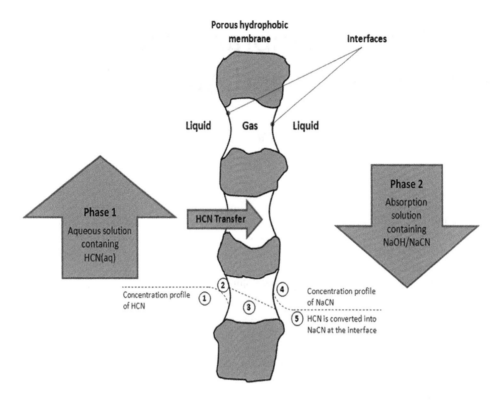

FIGURE 14.1 Schematic representation of transfer of a solute from the bulk feed to the bulk permeate in a gas absorption/stripping [Reprinted from *J. Membr. Sci.*, 466, Humberto Estay, Elizabeth Troncoso, Julio Romero, Design and cost estimation of a gas-filled membrane absorption (GFMA) process as alternative for cyanide recovery in gold mining, 253–264, Copyright (2014) with permission from Elsevier [1]].

3. Mass transfer across the membrane mainly by diffusion through gas in the pores.
4. Phase equilibration between gas in membrane pores and permeate.
5. Transfer from the membrane surface through the concentration boundary layer, that exists in the permeate solution.

While osmotic distillation is related to gas absorption it is often the solvent (i.e., water) that is transferred across the membrane thus concentrating nonvolatile species in the feed. For example; the permeate stream could consist of a high concentration salt solution [9, 10]. Given the fact that elevated temperatures are not required, osmotic distillation is highly favorable for concentration of different streams, especially for bioactive and agro-food products, that are temperature sensitive. Figure 14.2 is a schematic representation of the steps involved in transfer of water during osmotic distillation.

14.2 MEMBRANE MATERIALS AND MEMBRANE MODULE

The desired characteristics for the membrane materials of membrane contactors with gas filled membrane pores are as follows.

1. The membrane should be hydrophobic in nature, so that only water vapor/gas phase mass transfer across the membrane is possible. It should be

noted that if organic feed and permeate solutions are used, a hydrophilic membrane is required. However, there are fewer examples of using gas filled membrane pores with organic solvent. Thus, this chapter focuses in the aqueous feed streams.

2. The membrane should be highly porous (almost 60–80 %) with average pore size 0.1–1 μm range, as the permeability increases with the pore size of the membrane.

3. The membrane should be as thin as practical because the permeability decreases with the thickness of the membrane. However, sufficient mechanical stability must be retained.

4. The membrane material must conduct heat efficiently in order to minimize heat transfer requirements to maintain isothermal conditions.

Organic polymeric porous membranes, such as polypropylene (PP), polytetrafluoroethylene (PTFE), polyvinylidene difluoride (PVDF), ethylenechlorotrifluroethylene (ECTFE) are the most commonly used hydrophobic materials for gas filled membranes [11–16]. Fluoropolymers are found to have wide applications as gas filled membranes, due to the combination of unique properties like excellent chemical stability, lower surface tension, and good mechanical and thermal stability. Significantly, lower polarizability, and higher electronegativity of F atoms on the polymer backbones, result in strong C–F bond with lower Vander Waal's radius. Ease of fabrication and even fabrication of

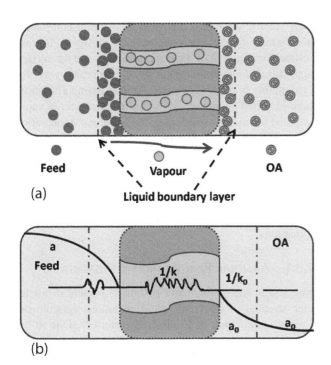

Feed **Vapour** **OA**

(a)

Liquid boundary layer

(b)

FIGURE 14.2 Schematic of osmotic distillation and water activity profile with mass transfer resistance.

copolymer membranes to achieve desired properties are advantages of fluoropolymers. Molar flux through a pore can be quantified in terms of different membrane parameters and are given by **Equation 14.1** as follows [17].

$$N = \frac{\langle r^\alpha \rangle \varepsilon}{\tau \delta}. \qquad (14.1)$$

where, $\langle r^\alpha \rangle$ is the average pore size for Knudsen diffusion ($\alpha = 1$) or for the viscous flux ($\alpha = 2$). ε and τ are the membrane porosity and tortuosity, respectively, while δ is the thickness. Membrane hydrophobicity is one of the important criteria. Generally, the surface hydrophilicity or hydrophobicity of a membrane can be evaluated in terms of the water contact angle. However, the method of estimation differs from that for smooth dense material surface. In accordance with the Laplace equation, the liquid entry pressure for a membrane can be quantified by Laplace equation [18]:

$$\Delta P_{entry} = \frac{-2\beta\chi_L Cos\theta}{\gamma_{max}} \qquad (14.2)$$

where, ΔP_{entry} refers to liquid entry pressure, β refers to the geometric factor, χ_L refers to the liquid surface tension, θ refers to the liquid–solid contact angle, and γ_{max} refers to the maximum pore radius (m). In a situation when the pressure drop across the vapor–liquid interface exceeds the penetration pressure, the feed solution starts penetrating through the pores of the membrane resulting in "pore wetting", which is highly undesirable. The water contact angle on the membrane surface increases with increasing polarity difference between water and the membrane material.

A higher contact angle indicates a greater polarity difference and hence a lower chance of pore wetting. Generally, membrane surfaces with water contact angles greater than 90°, refer to the hydrophobic membrane surface and more hydrophobicity leads to better wetting resistance characteristic of the membrane surface.

The membrane pore size significantly influences the wetting pressure of the membrane. It is reported that the pore size exhibits an inversely proportional relationship with the wetting pressure. Therefore, small pore size is better to avoid the wetting. The surface tension of the liquid, which is highly dependent on the mutual interaction like H bonding, dispersive interaction, and dipole-dipole interaction also shows a detrimental effect on the wetting property of the hydrophobic membrane. Therefore, the liquid with the higher surface tension is desired to avoid the wetting. The surface energy of the polymeric material can also play a crucial role in deciding the wetting characteristic of the membrane. The lower the surface energy the better the wetting resistance. The surface tension of some common liquids (a) and the surface energy of commonly used hydrophobic materials (b) are given in Figure 14.3.

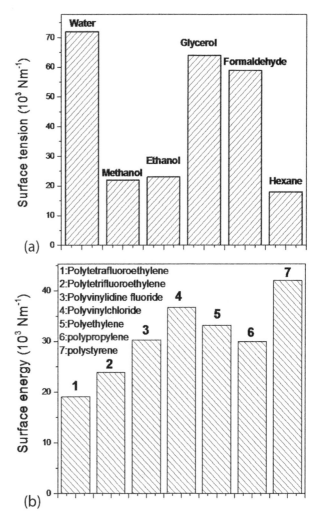

FIGURE 14.3 (a) Surface tension of common liquids and (b) surface energy of common polymeric membrane materials.

Several membrane configurations have been reported: flat sheet membranes in stirred cells, plate and frame, tubular, spiral wound, and hollow fiber. Stirred cells are mainly used to study the laboratory scale separations due to their simplicity. The hollow fiber configuration exhibits several advantages associated with larger surface area and high volumetric efficiency. However, the circulation of feed solution in hollow fiber modules is associated with laminar flow at low Reynolds numbers resulting in lower mass transfer across the membrane due to the existence of concentration boundary layers on both sides of the membrane. This is one of the serious limitations of this configuration. On the other hand, the flat sheet configuration was found to be more versatile compared to hollow fiber and tubular configurations in pilot scale operations due to ease of cleaning.

14.3 APPLICATIONS

In gas filled membranes, either permeate or the feed/retentate stream can be the product depending upon the application. Analogous to a concentration of certain bioactive compounds from a diluted agro-byproduct, the retentate is important. This application not only demands the mass transfer of the water molecules in the form of vapor, but also the retention of bioactivity is very important. The quality of retentate determines the effectiveness of the process.

On the other hand, treatment of industrial/municipal wastewater by gas filled membranes is concerned with the quality of the permeate. In such applications, large volumes of wastewater are required to be processed, and the permeate should be as pure as possible. Another interesting application of gas filled membranes is the preferential separation of the volatile organic compounds from multi-component aqueous mixtures, which is encountered mostly in the case of effluent arising out of different industries and the surface water. Broadly, the utility of gas filled membranes was projected mainly in two directions: (i) concentrating a solution, where gas filled membrane works for mass transfer of vapor from feed solution – wastewater treatment and desalination are the major domain; (ii) removal of a volatile component from a multi-component mixture – if the separation is based on the vapor-liquid equilibrium, the volatile component will be the major component of the permeate. The hollow fiber module and the capillary module were found to be advantageous compared to conventional distillation and small-scale applications.

14.3.1 AMMONIA REMOVAL FROM WASTEWATER

Ammonia is one of the major pollutants present in wastewater streams of different origin including agricultural, domestic, municipal, and industrial resources. Due to the extreme toxicity to aquatic species, mainly fish and bio-oxidation organisms of undesired nitrides and nitrates, stringent specification in wastewater has been enforced in order to have safe disposal of these wastewater streams. The removal of ammonia using gas pore filled membrane contactors has been reported using PP and PTFE membranes in hollow fiber configuration employing diluted sulphuric acid in the permeate side [1,19]. The whole separation process is highly influenced and equilibrated to the forward direction by the following chemical reaction of ammonia with the sulphuric acid, generating ammonium sulfate. The arrangement of ammonia removal by gas filled membrane has been presented in Figure 14.4. The feed solution was recycled

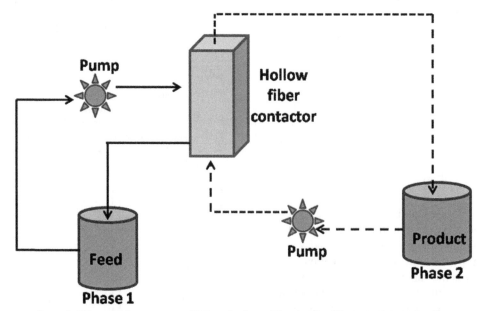

Case 1: Phase 1 is aqueous NH_3 solution, Phase 2 is dilute H_2SO_4 solution

Case 2: Phase 1 is aqueous HCN solution, Phase 2 is KOH solution

FIGURE 14.4 The schematics of experimental set up of ammonia removal by gas filled membrane.

through the shell side of the membrane. The concentration of ammonia was determined by ion-selective electrode.

$$2NH_{3g} + H_2SO_{4aq} = (NH_4)_2SO_{4aq} \qquad (14.3)$$

14.3.2 BLOOD OXYGENATION

Blood oxygenators are utilized to provide cardiopulmonary bypass during open-heart surgery. Gas filled membranes have been employed for the separation of blood and gas, where blood flows on one side of the membrane, while the perfusion gases flow on the other side [20–22]. The blood side concentration boundary layer is the major resistance to oxygen transfer to the blood and carbon dioxide transfer from the blood. Minimizing the priming volume of the oxygenator is critical for minimizing transfusion requirements. Based on hematological and immune responses during surgery, a reduction in membrane surface area as well as contact time is highly desirable.

Mass transfer and friction factor correlations for blood oxygenation have been developed for hollow fiber modules by Wickramasinghe et al. [20]. The influence of the kinematic viscosity of blood, diffusion coefficient of oxygen in blood, on mass transfer on the Sherwood number, and Schmidt numbers were evaluated. The driving force for fluid flow (F) of such system can be expressed as:

$$F = \zeta \left[\frac{\pi}{4} \left(D_0^2 - D_i^2 \right) \right] \Delta P \qquad (14.4)$$

where, ζ denotes the void fraction, D_0 and D_i denote the inside diameter of the module casing, and the outside diameter of the inner core, respectively. ΔP is the pressure drop. The friction factor is defined as the ratio between the force required to move a section of pipe and the vertical contact force applied by the pipe on the seabed. For laminar flow, the head loss is proportional to velocity rather than velocity squared, thus the friction factor is inversely proportional to velocity. The Darcy friction factor for laminar (slow) flow is a consequence of Poiseuille's law, that is given by following equations: Laminar flow: Re < 2000. The transport of momentum is described by a friction factor, f, which is defined differently to the overall mass transfer coefficient.

$$F = A_w \left(0.5 \rho \upsilon^2 \right) f \qquad (14.5)$$

where, A_w is wetted surface, ρ and υ are the density and velocity of the fluid, respectively.

The friction factor has been expressed as

$$f = \frac{\zeta \left(D_0^2 - D_i^2 \right) \Delta P}{4 d_0 L_f \left(\frac{1}{2} \right)^2_{\rho v}} \qquad (14.6)$$

$$f = \frac{d_e \Delta P}{4 L_0 \left(\frac{1}{2} \right)^2_{\rho v}} \qquad (14.7)$$

where, L_f is the total length of fiber, d_e is the equivalent diameter and L_0 is the length of the module. For laminar flow of Newtonian fluids in round tubes the friction factor is $16/R_e$. The friction factors for the membrane flow across the hollow fiber was evaluated as $260 R_e^{-1.1}$, when Re is in between 0.1 and 5, while it is $100 R_e^{-0.5}$, when Re is in between 5 and 100. The friction factors were obtained at the inlet of the header and channel lengths using the actual channel and header geometry. The fitted friction factors between laminar and turbulent values are also reasonable. It is seen that, the flow is in the mixed zone in the channels, and turbulent in the headers. Actually, the friction factor varies along the channel due to varying conditions along the channel. The friction factor of channel is 0.1, while that for header is 0.2. For flow in a rectangular duct, where B is the average channel thickness and W is the width, the friction factor can be expressed as follows:

$$f = \frac{16}{R_e \left[\frac{2}{3} + \frac{11}{24} \frac{2B}{W} \left(2 - \frac{2B}{W} \right) \right]} \qquad (14.8)$$

In case 2B<<W; f becomes $\frac{24}{R_e}$ Figure 14.5 is showing the Moody diagram relating Re, f, and relative roughness of the pore wall. It also depicts the regions of laminar, and turbulent flow of gas within the pores of the membranes

The Stanton number for mass transfer (St_{mass}) can quantitatively be expressed in terms of Chilton–Colburn factor for mass transfer (j_D) and Schmidt number (S_c) as follows:

$$St_{mass} = \frac{j_D}{S_c^{2/3}} \qquad (14.9)$$

A schematic representation of the hollow fiber and the experimental set up for blood oxygenation are depicted in Figure 14.6.

14.3.3 SEPARATION OF SULPHUR DIOXIDE FROM AQUEOUS SOLUTION

The removal of sulfur dioxide from aqueous feed streams, mainly in the food and wine industries, is based on the pH dependent conversation of aqueous HSO_3^- to volatile SO_2, subsequent mass transfer in the form of SO_2 through gas filled membrane, followed by capturing volatile SO_2 in permeate side by chemical reaction [24]. In general, acidic feed was utilized for the conversion to volatile SO_2 in feed side, where as NaOH was used in permeate side to capture SO_2 in the form of Na_2SO_3 using the chemical reaction as follows.

$$HSO_{3aq}^- + H_{aq}^+ = SO_{2g} + H_2O_{aq} \qquad (14.10)$$

$$SO_2 + 2NaOH_{aq} = Na_2SO_{3aq} + H_2O_{aq} \qquad (14.11)$$

In this case also hollow fiber configuration has been employed. A typical instrumental arrangement for SO_2 removal from red wine has been shown in Figure 14.7.

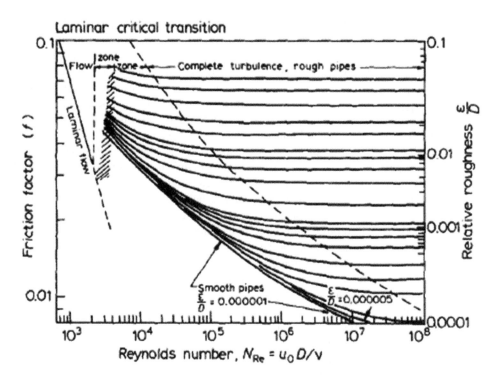

FIGURE 14.5 Moody diagram for relating friction factor with Reynolds number [Reprinted from *Cryogenics*, 23 (5), Wu Peiyi, W.A. Little, Measurement of friction factors for the flow of gases in very fine channels used for microminiature Joule–Thomson refrigerators, 273–277, Copyright (1983) with permission from Elsevier [23]].

14.3.4 SEPARATION OF CYANIDE FROM WASTEWATER

Gas filled membranes have been used to separate the volatile HCN from wastewater, followed by its conversion to KCN. KOH was generally employed in order to trap the recovered HCN using the equation as shown below [25–28]:

$$HCN_g + KOH_{aq} = KCN_{aq} + H_2O \qquad (14.12)$$

Cyanide removal from wastewater streams has been demonstrated at pilot scale by Han et al. in batch mode with a capacity of processing a feed of 1,000 L [27]. A total of ten hollow fiber modules having 180 m² surface area have been used continuously for two months. In this demonstration, 10% NaOH was employed on the permeate side as the strip solution. A reduction in the overall mass transfer coefficient was attributed to the membrane fouling mainly due to the presence of particulate matter in the wastewater. Out of different cleaning solutions, dilute acid was found to serve the purpose of regeneration of fouled membrane effectively.

The overall mass transfer coefficient (K) is a contribution of feed side mass transfer coefficient (k_F), strip side (k_S), and membrane mass transfer coefficient (k_M):

$$\frac{1}{K} = \frac{1}{k_F} + \frac{1}{k_s} + \frac{1}{k_M} \qquad (14.13)$$

$$\frac{k_F d}{D_F} = 1.64 \left(\frac{d^2 v_F}{l D_F} \right) \qquad (14.14)$$

d and l denote the inside diameter and length of hollow fibers. v_F denotes the velocity of liquid phase, D_F is the diffusion coefficient of volatile components.

$$k_M = \frac{\varepsilon D_M H}{\delta \tau} \qquad (14.15)$$

where, ε, δ, τ, D_M, and H denote void fraction, thickness, tortuosity, diffusion coefficient, and partition coefficient, respectively. Due to the presence of excess NaOH in the strip side, k_S is large and, hence, total mass transfer coefficient for cyanide has been evaluated by the simplified equation as follows:

$$\frac{1}{K} = \frac{1}{k_F} + \frac{1}{k_M} \qquad (14.16)$$

A schematic diagram of the membrane unit and the cyanide removal process are given in Figure 14.8.

Estay et al. have carried out a comparative cost evaluation of membrane gas absorption for cyanide recovery from gold mining in hollow fiber modules with batch and continuous operating modes with the commercially available methods for cyanide recovery from gold mining: acidification, volatilization, and re-neutralization (AVR) [29]; and suphidization, acidification, recycling, and thickening (SART) [30]. They revealed the requirement of a large area of mass transfer, with circuits ranging from 156 to 176 membrane modules for the treatment of 240 m³h⁻¹ of

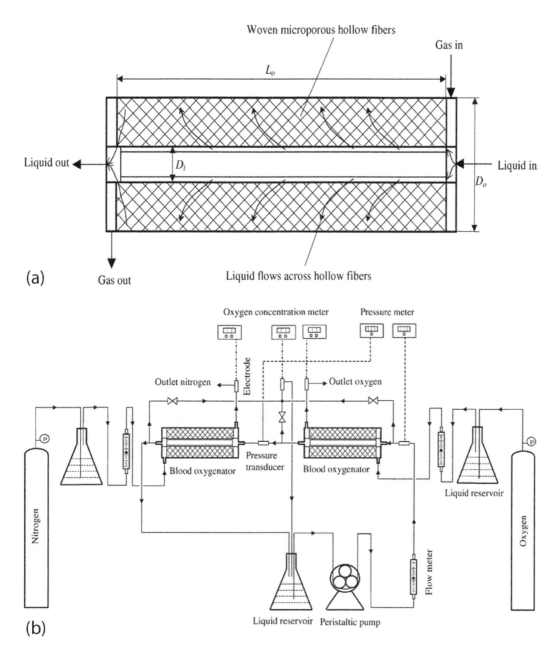

FIGURE 14.6 Gas filled membrane for blood oxygenation (a) hollow fiber module; (b) experimental set up [Reprinted from *Chem. Eng. Sci.*, 57 (11), A. R. Goerke, J. Leung, S. R. Wickramasinghe, Mass and momentum transfer in blood oxygenators, 2035–2046, Copyright (2002) with permission from Elsevier [22]].

cyanide solution having cyanide concentration 1200 mgL^{-1}. The continuous mode of operation was found to be beneficial with respect to cost effectiveness as the batch mode requires a larger number of membrane modules and higher energy requirements than the former [31]. However, the energy requirements for the continuous mode of gas filled membrane absorption is lower compared to AVR and SART processes. Moreover, their calculations revealed that the continuous gas filled membrane absorption process has the lowest capital cost (56.5 kUS$/(m³h^{-1})) compared to other industrial processes. Table 14.1 summarizes the cost effectiveness of the gas filled membrane absorption processes.

14.3.5 Separation of CO from a Mixture of CO–N$_2$

An investigation involving hollow fiber membrane modules with polypropylene as membrane material was reported in the literature for the selective absorption and separation of CO from a mixture of N$_2$ and CO using NH$_4$Cl/CuCl as effective stripping agent, using the following chemical equation [32, 33]:

$$CO + CuCl - NH_4Cl = [Cu(NH_4Cl)_3CO]^+Cl^- \qquad (14.17)$$

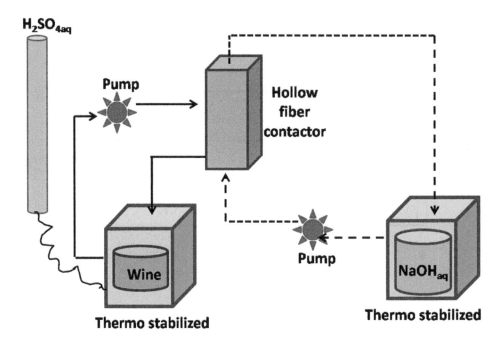

FIGURE 14.7   The schematics of experimental set up of SO_2 removal by gas filled membrane

14.3.6 H_2S SEPARATION

Due to its corrosive nature and toxicity, H_2S gas, which is one of the impurities present in coal gas, natural gas, and refinery gas must be removed. Gas filled membranes have been applied widely for separation of H_2S mainly using hollow fiber modules by application of effective stripping solutions; NaOH, monoethanolamine, sodium carbonate, etc., in the permeate side imposing a high degree of selectivity [34–37]. Even ionic liquids have been employed as stripping agent in order to induce selectivity.

14.3.7   SEPARATION OF Hg VAPOR

Hollow fiber membranes in transverse and shell-tube configurations have been efficiently employed for the separation of Hg vapor from industrial gaseous waste streams using H_2O_2/H_2SO_4, $K_2Cr_2O_7$, $K_2S_2O_8$, $KMnO_4$, $NaClO_4$ as stripping solution in the permeate side [38]. Due to the high toxicity of Hg vapor, it has to be separated or immobilized prior to the release of gaseous waste streams from industries to the environment.

14.3.8   CONCENTRATION OF BIOACTIVE COMPOUND FROM FOOD AND AGRO-PRODUCT

Gas filled membranes may be utilized for the concentration of bioactive compounds, typically polyphenols, anthocyanins, polysaccharides; having medicinal and physiological activity, from food and agro-byproducts including juice, pomace, solid residue after juice extraction, etc. [39–42]. This membrane-based technique is particularly advantageous, as it is operated at mild conditions in order to retain the original texture, taste, and other desired quality of the target compounds. Typically,

a concentrated brine solution is used on the permeate side. Figure 14.9 represents a typical process.

14.4   MASS TRANSFER PHENOMENA

The mass transfer can be quantitatively expressed in terms of the flux (J) as [44]:

$$J = K\Delta P \tag{14.18}$$

where, K is overall mass transfer coefficient expressed in kg m^{-2} hr^{-1} Pa^{-1} and associated to all the resistance including feed(k_f), membrane (k_m), and osmotic agent (k_{OA}) as follows:

$$\frac{1}{K} = \frac{1}{k_f} + \frac{1}{k_m} + \frac{1}{k_{OA}} \tag{14.19}$$

Diffusion mediated transport of vapor across the porous membrane is the origin of mass transfer for gas filled membrane guided either by Knudsen diffusion or molecular diffusion as characterized by the Knudsen number (K_n), which is influenced by mean molecular free path (λ) in µm, and the radius of the pore (r) in m as follows [45]:

$$K_n = \frac{\lambda}{2r} \tag{14.20}$$

$$\lambda = \frac{m}{\sqrt{2}\,\pi\rho d^2} \tag{14.21}$$

where, ρ represents the density in kgm^{-3}, d represents the diameter of the molecule in m, and m represents the molar mass in kg.

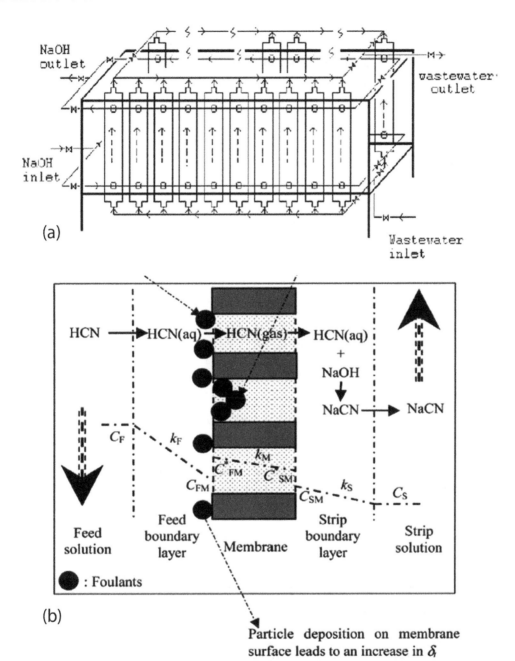

FIGURE 14.8 The schematic diagram of (a) membrane unit and (b) cyanide removal process [Reprinted from *Separation Science and Technology*, 40 (6), Binbing Han, Zhisong Shen, et al, Fouling and Cleaning of Gas-Filled Membranes for Cyanide Removal, 1169–1189, Copyright (2005) with permission from Taylor & Francis [28]].

14.4.1 KNUDSEN DIFFUSION

In the situation when λ is greater than r, $K_n \geq 10$, the collision frequency of the molecules with the pore wall becomes prominent and the Knudsen diffusion mechanism is predominantly operative [46,47]. Knudsen flux (J_K) is highly influenced by the porosity of membrane (ε), tortuosity of the membrane (χ), thickness (δ), and molar mass (M) as follows:

$$J_K = \left[1.064\left(\frac{r\varepsilon}{\chi^\delta}\right)\left(\frac{M}{RT}\right)^{1/2}\right]\Delta P \qquad (14.22)$$

where, R denotes the universal gas constant, T denotes the absolute temperature and ΔP denotes the vapor pressure difference between the feed and permeate side. The membrane characteristics and the driving force both are incorporated in calculating the flux value. In the Knudsen diffusion regime, there is no mutual interaction between the molecules, hence they move uninterruptedly on the pore channel surface.

Self-diffusivity is a measure of the translational mobility of individual molecules. Under thermodynamic equilibrium, a molecule is tagged, and its trajectory followed over a long time. In the case of diffusive motion, and a medium

TABLE 14.1

The Comparative Estimates of the Capital Cost Based on m³h⁻¹ Treated Feed Solution [1]

Item	Unit	Batch GFMA (MR every 5 years)	Batch GFMA (MR every 2 years)	Continuous GFMA (MR every 5 years)	Continuous GFMA (MR every 2 years)	AVR	SART
Labur	US$/m³	0.27	0.27	0.27	0.27	0.27	0.52
Acid cost	US$/ton NaCN	160	160	160	160	160	152
CaO cost	US$/ton NaCN	143	143	143	143	143	135
NaOH cost	US$/ton NaCN	544	544	544	544	544	10
NaHS cost	US$/ton NaCN	0	0	0	0	0	388
Electric energy cost	US$/m³	0.82	0.82	0.26	0.26	0.56	0.29
Metallurgical control	US$/m³	0.002	0.002	0.002	0.002	0.002	0.002
Maintenance	US$/m³	0.24	0.24	0.19	0.19	0.24	0.31
Contactors replacement	US$/m³	0.11	0.42	0.1	0.39	0	0
Packing replacement	US$/m³	0	0	0	0	0.024	0
Contingencies	US$/m³	0.16	0.17	0.13	0.14	0.14	0.13
Total operational cost	US$/m³	3.32	3.65	2.68	2.98	2.97	2.72
Unitary operational costs	US$/ton NaCN	1631	1793	1317	1466	1457	1266
Incomes	US$/ton NaCN	2200	2200	2200	2200	2200	2434

MR is Membrane replacement

Source: Adapted from *Journal of Membrane Science*, 466, Humberto Estay, Elizabeth Troncoso, Julio Romero, Design and cost estimation of a gas-filled membrane absorption (GFMA) process as alternative for cyanide recovery in gold mining, 253–264, Copyright (2014) with permission from Elsevier [1]]

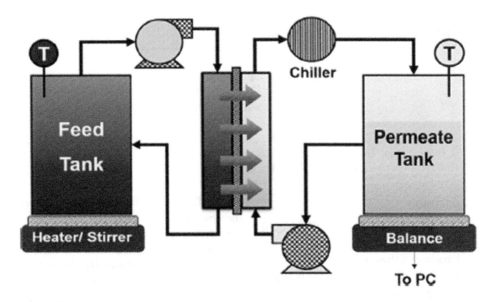

FIGURE 14.9 Schematic diagram for the concentration of bioactive compound from food and agro-byproducts [Reprinted from *Separation and Purification Technology*, 224, Zahra Anari, Arijit Sengupta, Kamyar Sardari, S. Ranil Wickramasinghe, Surface modification of PVDF membranes for treating produced waters by direct contact membrane distillation, 388–396, Copyright (2019) with permission from Elsevier [43]].

having no long-range correlations, the squared displacement of the molecule from its original position will grow linearly with time as suggested by Einstein's equation. To reduce statistical errors in simulations, the self-diffusivity of a species is defined from ensemble averaging Einstein's equation over a large enough number of molecules. Knudsen diffusion is applicable only to gases, since the mean free path for molecules in the liquid state is too small and near the diameter of the molecule itself. Figure 14.10 illustrates the different diffusion mechanisms.

For molecular sieve applications, membrane pores should have in between diameters of the gas molecules of interest. For effective separation between two competitive species, the smaller one would pass through the membrane pores, while the larger one retains and hence selectivity in separation is achieved. However, for a membrane though of average pore size or molecular weight cut-off is reported; practically, a pore size distribution does exist. Depending upon the quality of the membrane, the pore size distribution can be sharper. Hence the permeability is highly influenced by a combination of transport mechanisms. The average pore size and porosity must be optimized to achieve an efficient membrane. The separation can also be influenced by partial condensation of some components in the pores, and its transport across the pores. There is also the possibility of strong and selective adsorption of the component on the pore surface of the membrane and its subsequent surface diffusion across the pores. The selectivity in separation governed by Knudsen diffusion has the following characteristics:

I. Generally low selectivity of separation in mixture of gases.
II. Moderately high selectivity and high permeability for the smaller components in the mixture.

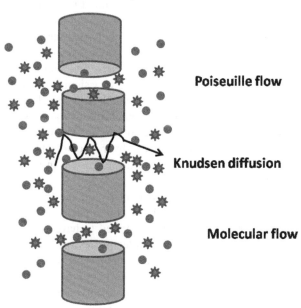

FIGURE 14.10 Diffusion mechanisms for gases in pores for gas filled membranes.

III. Membranes with meso-porous size range (diameter > 30 A) are desirable. The condensation partial pressure of the component, pore size of the membrane, and the geometry of the membrane highly influence the removal efficiency of the component from the mixture.
IV. The separation selectivity is determined by the preferential adsorption of certain components on the surface of the membrane pores, and the subsequent diffusion of the absorbed molecules.

In case of meso- and micro-porous membranes, the enhancement in relative pressure leads to the permeation of both adsorbed and capillary condensed species together. The porous medium acts as a semi-permeable barrier for the free flowing of the permeates, whereas the weakly adsorbed species will be blocked. Hence, both the pore size of the membrane and the physico–chemical nature of pore surface exhibit vital roles in achieving desired selectivity.

14.4.2 MOLECULAR DIFFUSION

In the case of membranes with large pore sizes, collisions between the vapor molecules themselves are more important and the mass transfer is governed by a molecular diffusion mechanism [46,47]. The kinetic theory of gases accounts for the forces of attraction and repulsion between the gas molecules when they diffuse. Lennard–Jones potential has been employed to evaluate the influence of their molecular interactions, the equation for the diffusion coefficient for non-polar, non-reacting gaseous molecules A and B can be expressed as:

$$D_{AB} = \frac{0.001858 T^{3/2} \left(\frac{1}{M_A} + \frac{1}{M_B} \right)^{1/2}}{P_t \sigma_{AB}^2 \Omega_D} \quad (14.23)$$

where, D_{AB} denotes the diffusivity of A through B; M_A and M_B imply the molecular weights of A and B, respectively; σ_{AB} denotes Lennard–Jones parameter; and Ω_D denotes the collision integral. In that situation the $K_n << 0.01$. In that case, the permeate flux (J_M) can be expressed as follows:

$$J_m = \left[\left(\frac{D\varepsilon}{\chi^\delta} \right) \left(\frac{1}{Y_{ln}} \right) \right] \left(\frac{n}{RT} \right) \Delta P \quad (14.24)$$

where, D denotes the diameter of the membrane pore and Y_{ln} denotes the mole fraction of air, respectively. Though, moderate to high partial pressure of air predicts the mass transfer across the gas filled membrane very well, it is limited to lower partial pressure.

14.4.3 DUSTY GAS MODEL

The dusty gas model is another well-known model for understanding vapor transport through gas filled membranes using the kinetic theory of gases [48, 49]. In accordance with the theory, porous membranes were assumed as homogeneously distributed dust. There are mainly six special conditions for which the dusty gas model is preferred:

I. All the particles are at stationary phase and uniformly distributed.

II. An external force is exerted to the particles to make them mobile under pressure gradient.

III. The porosity and tortuosity effects of the porous membrane have been accounted for the dusty gas model.

IV. The total density and total pressure are not the actual density and pressure, as the stationary dust particles are included in the counting.

V. Due to the static phase of the dust particle, it does not influence the λ and η of the system.

VI. Total force on the slab is the same, regardless of the existence of holes.

Based on the dusty gas model and the simultaneous existence of Knudsen diffusion and molecular diffusion mechanism, the permeate flux can be expressed as:

$$J_{K/M} = -\frac{1}{RT}\left[\frac{1}{D_{we}^K} + \frac{P_{air}}{D_{w-air}^0}\right]^{-1}\nabla P_w \qquad (14.25)$$

where, P_{air} denotes partial pressure of air, D_{w-air}^0 and D_{we}^K are the effective diffusivity for molecular diffusion and Knudsen diffusion mechanism, respectively.

$$D_{we}^K = K_0\left(\frac{8RT}{\pi M}\right)^{1/2} \qquad (14.26)$$

$$K_0 = \frac{\varepsilon d_p}{3\chi} \qquad (14.27)$$

$$D_{w-air}^0 = K_1 P D_{w-air} \qquad (14.28)$$

$$K_1 = \frac{\varepsilon}{\chi} \qquad (14.29)$$

where, K_0 and K_1 are the constants depend on pore geometry. For water-air systems, PD_{w-air} can be evaluated by the following expression [50]:

$$PD_{w-air} = 4.46 \times 10^{-6} T^{2.33} \qquad (14.30)$$

14.4.4 Influence of Boundary Layer on Mass Transfer Across the Membrane

The influence of boundary layers on overall mass transfer should also be incorporated to model the actual performance of gas filled membranes as proposed by Mengual et

al. Accordingly, the Sherwood number (S_h), Reynolds number (R_e), and Schmidt number (S_c), are correlated as follows [51–53]:

$$S_h = b_1 R_e^{b_2} S_c^{b_3} \qquad (14.31)$$

$$S_h = \frac{K_l L}{D_w} \qquad (14.32)$$

$$R_e = \frac{\rho u L}{\mu} \qquad (14.33)$$

$$S_c = \frac{\mu}{\rho D_w} \qquad (14.34)$$

where, D_w, K_l, L, u, ρ, μ denote water diffusion co-efficient, liquid mass transfer coefficient, length of membrane, velocity of fluid, density, and viscosity, respectively, while b_1, b_2, and b_3 are the constants. Table 14.2 summarizes the correlation of S_h for different membrane configurations.

14.5 OSMOTIC DISTILLATION

The basic characteristics of a reagent to be used as osmotic agent are as follows: non-volatile, high osmotic activity to maintain a lower vapor pressure, and thermally stable to allow re-concentration of diluted stripping solution by evaporation. Apart from that, the solubility, toxicity, corrosivity, and cost of such reagents are also considered for their commercial applications.

14.6 MEMBRANE FOULING

Since, osmotic pressure or vapor pressure difference across the membrane is the driving force for most of the gas filled membrane applications, it is most likely that the problem of membrane fouling in such applications is less than compared to the other membrane processes, and also most of the membrane materials do not have very significant deviation in membrane performance [59–63]. However, there are several reports in the literature demonstrating the deviation in membrane performance due to fouling. The hydrophobicity of the membrane induces hydrophobic interaction with the organic nonpolar constituents present in the feed side [64–67].

The introduction of hydrophobicity on the membrane surface either by grafting hydrophilic polymer brushes or surface oxidation was demonstrated to have very high antifouling characteristics, while processing produced water having a high total organic content as well as very high salinity. Kamaz et al. and Anari et al. have reported that, the growth of poly(ionic liquid brush) on hydrophobic membranes reduced the membrane fouling compared to the conventional hydrophilic polymer (2-Hydroxyethyl methacrylate, ethylene diamine, acrylic acid, etc.) grafting, while

TABLE 14.2

Comparison of Mass Transfer Coefficient for Different Membrane Module Configurations

Correlation	Configuration	Observation	Reference
$S_h = \alpha \left(\dfrac{d_{in}}{L} R_e S_c \right)^{0.33}$	Lumen side	The value of coefficient a can be 1.86, 1.64 (empirical), or 1.62 (theoretical). Characteristic length is d_{in}	[54]
$S_h = 1.25 \left(\dfrac{d_h}{L} R_e \right)^{0.93} S_c^{0.33}$	Shell side, parallel flow	$0 < R_e < 500$; $0 < \phi < 0.26$. Characteristic length is d_h	[54]
$S_h = 0.022 R_e^{0.6} S_c^{0.33}$	Shell side, parallel flow	Characteristic length is d_h	[55]
$S_h = \beta (1-\varphi) \left(\dfrac{d_h}{L} \right) R_e^{0.66} S_c^{0.33}$	Shell side, parallel flow	$\beta = 5.85$ for hydrophobic membranes and 6.1 for hydrophilic membranes $0 < Re < 500$; $0.04 < \varphi < 0.4$, Characteristic length is d_h	[56]
	Shell side, parallel flow	$0 < R_e < 100$; $0.25 < \varphi < 0.48$, Characteristic length is d_h	[57,58]
$S_h = 17.4 (1-\varphi) \left(\dfrac{d_h}{L} \right) R_e^{0.66} S_c^{0.33}$			
$S_h = 0.9 R_e^{0.4} S_c^{0.33}$	Shell side, cross flow	$1 < R_e < 25$, $\varphi = 0.03$, Characteristic length is d_{out}	[54]

processing the produced water through membrane distillation [68–70].

Fouling may be suppressed by optimizing the surface characteristics of the membrane in such a way that the adsorption of foulants becomes thermodynamically unfavorable. Several reports were also found in the literature on the development of omniophobic surface of the membranes, membranes with super-hydrophobic characteristics, and membrane surface having much lower surface tensions than water in order to reduce the membrane fouling in commercially available membranes applied for gas filled membrane applications. Development of economically viable commercially adoptable surface modification strategies for such types of membranes would definitely be challenging.

In the case of fouling prone feed, sometimes applications of suitable hydrodynamics (bubbling of inert gas, Ritz vortex, etc.) in designing membrane modules or pretreatment (e.g., electrocoagulation) of the feed itself can be better options compared to the modification of surface properties of membranes. Spin coating and plasma grafting are more easily adoptable techniques compared to chemical modification for large scale applications. The formation of the coating layer can be optimized by adjusting the operating conditions and experimental parameters. The coating with an appropriate material can lead to formation of an active layer incorporating selectivity or reducing fouling in gas filled membrane applications. Development of more fouling resistant membranes results in the enhancement of membrane lifetimes as well as lower pretreatment requirements.

Figure 14.11 shows the surface modification of polypropylene membranes for imposing antifouling characteristics (a), and their effect on treatment of real produced water by gas filled membrane application (b).

14.7 CONCLUSION AND FUTURE PERSPECTIVE

This chapter summarizes the basics of gas filled membranes, the feasibility of different applications, and associated mass and energy transfer. The major applications for gas filled membranes are found in food-agricultural and biomedical separation without compromising the quality of the end products. Though gas filled membrane applications were demonstrated to be highly efficient and selective in laboratory scale applications, only a few reports are available in the literature demonstrating pilot scale.

Gas filled membrane applications can be a cost-effective substitute for lyophilization, where thermally sensitive products like enzymes/proteins, natural food colors, etc., are associated. Delocalization of beverages (wine or beer) is another popular industrial application for such membranes to selectively remove ethanol from beverages without compromising the taste, odor, or mouth feel. In recent years, the literature reports demand the need for integrated membrane processes by coupling gas filled membrane application as one of the components of entire membrane-based separation systems [72–75]. The research demonstrates in pilot scale that, the incorporation of micro/ultrafiltration membrane procedure with osmotic distillation can provide fruit juice to intermediate concentration with high quality flavor. Nagaraj et al. demonstrated the large-scale feasibility of UF-RO-OD integrated membrane separation process in increasing the productivity of pineapple juice [75]. Johnson et al reported a three-stage integrated membrane separation process involving gas filled membrane application as one of the components to concentrate the water-alcohol extracts of immuno-stimulant plant, *Echinacea* to be suitable as a commercial product in the form of a capsule [76]. To achieve the stage of using gas filled membranes as a commercially viable option, it is desired to fabricate membranes from suitable materials having better diffusional properties, improved pore geometry for easy mass transfer, and stability of the membrane material for long term applications. Incorporation of all of these factors in one system requires to be cost effective. Figure 14.12 gives an integrated membrane-based separation scheme employing gas filled membrane application as one of the components for the concentration of juice.

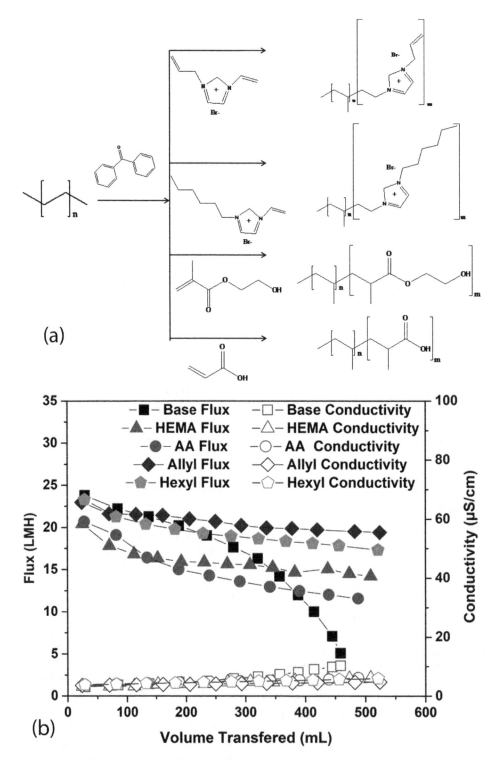

FIGURE 14.11 (a) Modification of polypropylene membranes for improving antifouling properties and (b) their implication on gas filled membrane performance [Reprinted from *Separation Science and Technology*, 54 (17), Tharaka Gamage, Arijit Sengupta, et al., Surface modified polypropylene membranes for treating hydraulic fracturing produced waters by membrane distillation, 2921–2932, Copyright (2019) with permission from Taylor & Francis [71]].

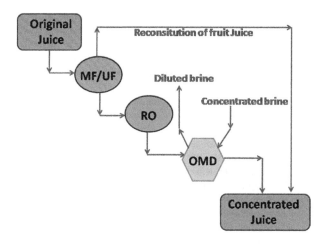

FIGURE 14.12 Integrated membrane-based separation scheme employing gas filled membrane application as one of the components for the concentration of juice.

REFERENCES

1. H. Estay, E. Troncoso, J. Romero, Design and cost estimation of a gas-filled membrane absorption (GFMA) process as alternative for cyanide recovery in gold mining, *J Membr Sci.* 466 (2014) 253–264.
2. A. Hasanoglu, J. Romero, A. Plaza, W. Silva, Gas-filled membrane absorption: A review of three different applications to describe the mass transfer by means of a unified approach, *Desal Water Treat.* 51 (2013) 5649–566.
3. P. Bernardo, E. Drioli, G. Golemme, Membrane gas separation: A review/state of the art, *Ind Eng Chem Res.* 48 (10) (2009) 4638–4663.
4. D. F. Mohshim, H. Mukhtar, Z. Man, R. Nasir, Latest development on membrane fabrication for natural gas purification: A review, *J Eng.* 101746 (2013) 1–7.
5. T. Y. Cath, D. Adams, A. E. Childress, Membrane contactor processes for wastewater reclamation in space: Part I. Direct osmotic concentration as pretreatment for reverse osmosis, *J Membr Sci.* 257 (2005) 111–119.
6. B. R. Babu, N. K. Rastogi, K. S. M. S. Raghavarao, Mass transfer in osmotic membrane distillation of phycocyanin colorant and sweet-lime juice, *J Membr Sci.* 272 (2006) 58–69.
7. Q. Ge, P. Wang, C. Wan, T.-S. Chung, Polyelectrolyte-promoted forward osmosis–membrane distillation (FO–MD) hybrid process for dye wastewater treatment, *Environ Sci Technol.* 46 (11) (2012) 6236–6243.
8. B. Ravindra Babu, N. K. Rastogi, K. S. M. S. Raghavarao, Concentration and temperature polarization effects during osmotic membrane distillation, *J Membr Sci.* 322 (2008) 146–153.
9. J. I. Mengual, J. M. Ortiz de Z'rate, L. Peiia, A. Vela'zquez, Osmotic distillation through porous hydrophobic membranes, *J Membr Sci.* 82 (1993) 129–140.
10. J.-P. Mericq, S. Laborie, C. Cabassud, Vacuum membrane distillation of seawater reverse osmosis brines, *Water Res.* 44 (2010) 5260–5273.
11. N. Nagaraj, B. S. Patil, P. M. Biradar, Osmotic membrane distillation–A brief review, *Int J Food Eng.* 2 (2) (2006), 1–22.
12. S. Atchariyawut, C. Feng, R. Wang, R. Jiraratananon, D. T. Liang, Effect of membrane structure on mass-transfer in the membrane gas–liquid contacting process using microporous PVDF hollow fibers, *J Membr Sci.* 285 (2006) 272–281.
13. H. H. Park, B. R. Deshwal, I. W. Kim, H. K. Lee, Absorption of SO_2 from flue gas using PVDF hollow fiber membranes in a gas–liquid contactor, *J Membr Sci.* 319 (2008) 29–37.
14. Y. Lv, X. Yu, S.-T. Tu, J. Yan, E. Dahlquist, Wetting of polypropylene hollow fiber membrane contactors, *J Membr Sci.* 362 (2010) 444–452.
15. V. Y. Dindore, D. W. F. Brilman, F. H. Geuzebroek, G. F. Versteeg, Membrane–solvent selection for CO_2 removal using membrane gas–liquid contactors, *Sep Purif Technol.* 40 (2004) 133–145.
16. H.-Y. Zhang, R. Wang, D. T. Liang, J. H. Tay, Theoretical and experimental studies of membrane wetting in the membrane gas–liquid contacting process for CO_2 absorption, *J Membr Sci.* 308 (2008) 162–170.
17. M. Mulder, *Basic Principles of Membrane Technology*, Kluwer Academic Publishers (1996).
18. B.-S. Kim, P. Harriott, Critical entry pressure for liquids in hydrophobic membranes, *J Colloid Inter Sci.* 115 (1987) 1–8.
19. A. Hasanoglu, J. Romero, B. Perez, A. Plaza, Ammonia removal and recovery from wastewater using natural zeolite: An integrated system for regeneration by air stripping followed ion exchange, *Chem Eng J.* 160 (2) (2010) 530–537.
20. S. R. Wickramasinghe, B. Han, J. D. Garcia, R. Specht, Microporous membrane blood oxygenators, *AIChE J.* 51 (2005) 6656–6670.
21. S. R. Wickramasinghe, B. Han, Designing microporous hollow fibre blood oxygenators, *Trans IChemE, Part A, Chem Eng Res Des.* 83 (A3) (2005) 256–267.
22. A. R. Goerke, J. Leung, S. R. Wickramasinghe, Mass and momentum transfer in blood oxygenators, *Chem Eng Sci.* 57 (2002) 2035–2046.
23. W. Peiyi, W. A. Little, Measurement of friction factors for the flow of gases in very fine channels used for microminiature Joule-Thomson refrigerators, *Cryogenics* 23 (1983) 273–277.
24. Z. Shen, J. Huang, G. Qian, Recovery of cyanide from wastewater using gas-filled membrane absorption, *Water Environ Res.* 69 (1997) 363–397.
25. Z. Xu, L. Li, Z. Shen, Treatment of praziquantel wastewater using the integrated process of coagulation and gas membrane absorption, *Water Res.* 39 (2005) 2189–2195.
26. B. Han, Z. Shen, S. R. Wickramasinghe, Cyanide removal from industrial wastewaters using gas membranes, *J Membr Sci.* 257 (2005) 171–181.
27. Z. Shen, B. Han, S. R. Wickramasinghe, Cyanide removal from industrial praziquantel wastewater using integrated coagulation–gas-filled membrane absorption, *Desalination.* 195 (2006) 40–50.
28. B. Han, Z. Shen, S. R. Wickramasinghe, Fouling and cleaning of gas-filled membranes for cyanide removal, *Sep Sci Technol.* 40 (2005) 1169–1189.
29. C. A. Fleming, Cyanide recover, *Dev Miner Process.* 15 (2005) 703–727.
30. H. Estay, F. Arriagada, S. Bustos, Design, development and challenges of the SART process, HydroProcess 2010, In 3th International Workshop on Process Hydrometallurgy, Santiago, Chile (2010).
31. H. Estay, M. Ortiz, J. Romero, A novel process based on gas filled membrane absorption to recover cyanide in gold mining, *Hydrometallurgy* 134–135 (2013) 166–176.

32. A. Ghosh, S. Borthakur, N. Dutta, Absorption of carbon monoxide in hollow fiber membranes, *J Membr Sci.* 96 (1994) 183–192.

33. J. R. Figueroa, H. E. Cuenca, Membrane gas absorption processes: Applications, design and perspectives. Osmotically driven processes, Approach, development and current status, Edited by H Du, A. Thompson, X Wang, InTech (2018), Rijeka, Croatia.

34. K. Huang, D. Cai, Y. Chen, T. Wu, X. Hu, Z. Zhang, Thermodynamic validation of 1-alkyl-3-methylimidazolium carboxylates as task-specific ionic liquids for H_2S absorption, *AIChE J.* 59 (6) (2013) 2227–2235.

35. G. Vallée, P. Mougin, S. Jullian, W. Furst, Representation of CO_2 and H_2S absorption by aqueous solutions of diethanolamine using an electrolyte equation of state, *Ind Eng Chem Res.* 38 (9) (1999) 3473–3480.

36. H. Wubs, A. Beenackers, Kinetics of H_2S absorption into aqueous ferric solutions of edta and hedta, *AIChE J.* 40 (3) (1994) 433–444.

37. N. Boucif, E. Favre, D. Roizard, Hollow fiber membrane contactor for hydrogen sulfide odor control, *AIChE J.* 54 (1) (2008) 122–131.

38. R. van der Vaart, J. Akkerhuis, P. Feron, B. Jansen, Membrane gas absorption processes: Applications, design and perspectives, *J Membr Sci.* 187 (2001) 151–159.

39. K. V. Kotsanopoulos, I. S. Arvanitoyannis, Membrane processing technology in the food industry: Food processing, wastewater treatment, and effects on physical, microbiological, organoleptic, and nutritional properties of foods, *Crit Rev Food Sci Nutr.* 55 (2015) 1147–1175.

40. M. Celere, C. Gostoli, Heat and mass transfer in osmotic distillation with brines, glycerol and glycerol–salt mixtures, *J Membr Sci.* 257 (2005) 99–110.

41. M. Courel, M. Dornier, J. M. Herry, G. M. Rios, M. Reynes. Effect of operating conditions on water transport during the concentration of sucrose solutions by osmotic distillation, *J Membr Sci.* 170 (2000) 281–289.

42. M. Courel, M. Dornier, J. M. Herry, G. M. Rios, M. Reynes, Modelling of water transport in osmotic distillation using asymmetric membrane, *J Membr Sci.* 173 (2000) 107–122.

43. Z. Anari, A. Sengupta, K. Sardari, S. R. Wickramasinghe, Surface modification of PVDF membranes for treating produced waters by direct contact membrane distillation, *Sep Purif Technol.* 224 (2019) 388–396.

44. V. D. Alves, I. M. Coelhoso, Effect of membrane characteristics on mass and heat transfer in the osmotic evaporation process, *J Membr Sci.* 228 (2004) 159–167.

45. C. J. Geankoplis. Principals of mass transfer. In *Transport Processes and Unit Operations*, Prentice-Hall, 3rd ed., London, UK (1993).

46. T. K. Sherwood, R. L. Pigford, C. R. Wilke. *Mass Transfer*, 1st ed., New York, NY: McGraw-Hill (1975).

47. R. W. Schofield, A. G. Fane, C. J. D. Fell, Heat and mass transfer in membrane distillation, *J Membr Sci.* 33 (1987) 299–313.

48. B. Ravindra Babu, N. K. Rastogi, N. K. M. S. Raghavarao, Mass transfer in osmotic membrane distillation of phycocyanin colorant and sweet-lime juice, *J Membr Sci.* 272 (2006) 58–69.

49. E. A. Mason, A. P. Malinauskas, R. B. Evans, Flow and diffusion of gases in porous media, *J Chem Phys.* 46 (1967) 3199.

50. R. B. Bird, W. E. Stewart, E. N. Lightfoot. *Transport Phenomenon*, 2nd ed., New York, NY: Wiley (1960).

51. V. A. Alves, I. M. Coelhoso, Mass transfer in osmotic evaporation: Effect of process parameters, *J Membr Sci.* 208 (2002) 171–179.

52. M. Courel, M. Dornier, J. M. Herry, G. M. Rios, M. Reynes, Modelling of water transport in osmotic distillation using asymmetric membrane, *J Membr Sci.* 173 (2000) 107–122.

53. N. Nagaraj, G. Patil, B. Ravindra Babu, U. Hebbar, K. S. M. S. Raghavarao, S. Nene, Mass transfer in osmotic membrane distillation, *J Membr Sci.* 268 (2006) 48–56.

54. M. Yang, E. Cussler, Designing hollow-fiber contactors, *AIChE J.* 32(11) (1986) 1910–1916.

55. J. Knudsen, D. Katz, *Fluid Dynamics and Heat Transfer*, New York, NY: McGraw Hill (1958).

56. R. Prasad, K. Sirkar, Dispersion-free solvent extraction with microporous hollow-fiber modules, *AIChE J.* 34(2) (1988) 177–188.

57. R. Basu, R. Prasad, K. Sirkar, Non-dispersive membrane solvent back extraction of phenol, *AIChE J.* 36(3) (1990) 450–460.

58. C. Yun, R. Prasad, A. Guha, K. Sirkar, Hollow fiber solvent extraction removal of toxic heavy metals from aqueous waste streams, *Ind Eng Chem Res.* 32 (1993) 1186–1195.

59. C. Gostoli, Thermal effects in osmotic distillation, *J Membr Sci.* 163 (1999) 75–91.

60. M. Celere, C. Gostoli, The heat and mass transfer phenomena in osmotic membrane distillation, *J Membr Sci.* 147 (2002) 133–138.

61. P. Onsekizoglu. Membrane distillation: Principle, advances, limitations and future prospects in food industry, distillation. In *Advances from Modeling to Applications*, ed., Dr. S. Zereshki, InTech, Edirne, Turkey (2012). Available from: http://www.intechopen.com/books/distillation-advances-from-modeling-toapplications/ membrane - distillation-principle-advances-limitations-and-future-prospects-in-food-industry

62. A. Burgoyne, M. M. Vahdati, Direct contact membrane distillation, *Sep. Sci Technol.* 35 (2000) 257–1284.

63. F. Lagana, G. Barbieri, E. Drioli, Direct contact membrane distillation: Modeling and concentration experiments, *J Membr Sci.* 166 (2000) 1–11.

64. J. Phattaranawik, R. Jiraratananon, A. G. Fane, Heat transport and membrane distillation coefficients in direct contact membrane distillation, *J Membr Sci.* 212 (2003) 177–193.

65. M. S. Khayet, T. Matsuura, T. *Membrane Distillation: Principles and Applications*, London, UK: Elsevier (2011).

66. K. W. Lawson, D. R. Lloyd, Membrane distillation, *J Membr Sci.* 124 (1997) 1–25.

67. J. Phattaranawik, R. Jiraratananon, Direct contact membrane distillation: Effect of mass transfer on heat transfer, *J Membr Sci.* 188 (2001) 137–143.

68. Z. Anari, C. Mai, A. Sengupta, L. Howard, C. Brownmiller, S. R. Wickramsinghe, Combined osmotic and membrane distillation for concentration of anthocyanin from muscadine pomace, *J Food Sci.* 84(8) (2019) 2199–2208.

69. M. Kamaz, A. Sengupta, A. Gutierrez, R. Wickramasinghe, Surface modification of PVDF membranes for treating produced waters by direct contact membrane distillation, *Int J Environ Res Public Health.* 16(5) (2019) 685.

70. Z. Anari, A. Sengupta, S. R. Wickramasinghe, Surface oxidation of ethylenechlorotrifluoroethylene (ECTFE) membrane for the treatment of real produced water by membrane distillation, *Int J Environ Res Pub Health.* 15 (2018) 1561.

71. T. Gamage, A. Sengupta, S. R. Wickramasinghe, Surface modified polypropylene membranes for treating hydraulic fracturing produced waters by membrane distillation, *Sep Sci Technol.* 54(17) (2019) 2921–2932.

72. V. D. Alves, I. M. Coelhoso, Orange juice concentration by osmotic evaporation and membrane distillation: A comparative study, *J Food Eng.* 74 (2006) 125–133.

73. P. Jaoquen, B. Lepine, N. Rossignol, R. Royer, F. Quemeneur, Clarification and concentration with membrane technology of a phycocyanin solution extracted from Spirulina platensis, *Biotechnol Tech.* 13 (1999) 877.

74. P. M. Biradar, Osmotic membrane distillation for the processing of biomolecules, M. Tech. thesis, VTU Belgaum, India (2005).

75. N. Naveen, *Integrated Biotechnological Approaches for the Purification and Concentration of Liquid Foods, Proteins and Food Colors*, Ph.D. thesis, University of Mysore, India (2004).

76. R. A. Johnson, J. C. Sun, J. Sun, A pervaporation–microfiltration–osmotic distillation hybrid process for the concentration of ethanol–water extracts of the Echinacea plant, *J Membr Sci.* 209 (2002) 221–232.

15 Ammonium Valorization from Urban Wastewater as Liquid Fertilizers by Using Liquid–Liquid Membrane Contactors

A. Mayor, X. Vecino, M. Reig, N. de Arespacochaga,
C. Valderrama, and J. L. Cortina

CONTENTS

15.1 INTRODUCTION

Nitrogen (N) is an essential nutrient for all living forms and, additionally, a key component of fertilizer production; indeed, it is critical for agriculture. A dramatic increase in the production of artificial nitrogen fertilizers over the last 50 years has led to increasing nutrient pollution of water and air. Excess nitrogen in water can induce excessive algae growth and a consequent reduction of the dissolved oxygen; thus nutrient overloading in surface water receiving bodies, namely eutrophication, occurs [1].

In the last decades, the reduction of ammonium concentrations from wastewater effluents by means of conventional activated sludge (CAS) schemes, in order to avoid the consequences of the eutrophication phenomenon, has become an environmental challenge. A nitrogen treatment step is essential in wastewater treatment plants (WWTPs), due to the required low levels of ammonium nitrogen (N-NH_4^+) in the discharged wastewater effluents, which should be lower than 1 mgN-NH_4^+/L [2]. In WWTPs, biological nitrification-denitrification processes are used as conventional methods of nitrogen removal, where ammonium (NH_4^+) is transformed into nitrogen (N_2) gas [3]. A promising alternative to the conventional methods is the anaerobic ammonium oxidation process, namely anammox. This oxidation process does not need an external carbon source for denitrification and also presents low energy consumption [4,5]. However, anammox technology still presents relevant limitations, for instance long start-up periods, strong sensitivity to variations in operational conditions, and a high susceptibility to reactor limitations [6, 7]. According to the regulated discharge values, in the future these biological processes will probably not be able to achieve the required nitrogen removal rates [8].

Figure 15.1 shows a conventional WWTP scheme in which the activated sludge aeration tank, which is used after the primary settling, can work as a CAS or a high-rate activated sludge (HRAS) treatment. As shown in Figure 15.1, anammox would be used as a secondary process in order to obtain a treated effluent.

With the above mentioned treatments, N is removed in WWTPs instead of recovered, even though it is well known that N, as well as phosphorus (P), is an important added-value product in the agriculture and chemical industries. About 60 to 65% of the N present in wastewater influent is returned to the atmosphere, mainly as dinitrogen gas ($N_{2(g)}$), and 15 to 20% ends up as organic nitrogen in the sludge [9]. On the other hand, around 90% of the P is retained in CAS sludge [10]; and in the EU27, 38% of this sludge is disposed of through land spreading [11], where this P is not readily bio-available. In this sense, since wastewater is an important source of nutrients, and taking into account that fertilizer prices are increasing, any alternative that considers nutrients recovery to obtain value-added products and increases the final effluent water quality must be considered. Therefore, there has been renewed interest in the recovery of nutrients from waste streams to tackle economic resource recovery and environmental concerns synergistically [12, 13].

In this sense, society must consider wastewater effluents as a solution to the problem of scarcity of natural resources, where energy and nutrients can be recovered in an economical way [14–17]. Table 15.1 summarizes the potential recovery of resources from municipal wastewater, excluding heat recovery, assuming that chemical energy is recovered in the form of CH_4 by means of an anaerobic digestion (AD) process and that the organic carbon remaining after this treatment is used as organic fertilizer [14].

The purpose of fully recovering energy and nutrient resources from wastewater effluents represents a paradigm shift in how WWTPs become facilities for the recovery of resources in a self-sufficient and sustainable way in which value-added products (e.g., nutrients and energy) are recovered [14, 23, 24].

This chapter presents one of the promising technologies, liquid–liquid membrane contactors (LLMCs), which allow the recovery of nitrogen from wastewater effluents (Figure 15.2). This recovery process involves the reaction of ammonia with an acid stripping solution to produce ammonium salts that can be used as liquid fertilizers. The fundaments of the technique, ammonia removal mechanisms, mass transfer coefficients, influence of the types of acid solutions, and limitations of water transport are discussed. Furthermore, a case study of ammonia recovery from zeolite regeneration streams is presented. Finally, the production of fertilizer formulations and new perspectives are also shown.

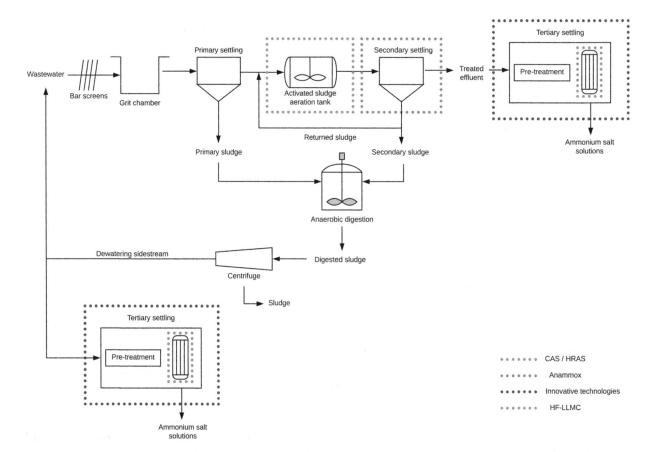

FIGURE 15.1 WWTP scheme with conventional and innovative treatments

TABLE 15.1

Potential Resources from Municipal Wastewater. Prices Are Based on the Market Values of Comparable Products [14]

Potential Recovery	Per m³ of Sewage	Current Market Prices	Total (€)
Water	1 m³	0.22 €/m³ [18]	0.22
Nitrogen (N)	0.05 kg	0.23 €/kg [19]	0.01
Methane (CH₄)	ᵃ 0.14 m³	0.29 €/m³ [20]	0.04
Organic fertilizer	ᵇ 0.10 kg	0.44 €/kg [21]	0.04
Phosphorous (P)	0.01 kg	0.88 €/kg [22]	0.01
		Total	0.32

ᵃ Based on 80% recovery of organic matter in the form of biogas and 0.35 m³/kg chemical oxygen demand (COD) removal.

ᵇ Based on 20% organic matter remaining after AD.

15.2 NITROGEN RECOVERY SOURCES IN WASTEWATER TREATMENT PLANTS (WWTPs)

Nowadays, one of the most commonly used processes for the production of nitrogen fertilizers is the Haber–Bosch process, although it is considered as an expensive and energy-intensive technology [25]. Thus, the current challenge is to recover inorganic forms of nitrogen as valuable components for fertilizer production for agricultural uses while reducing the amount of nitrogen released into the environment in an effective, economical, and sustainable way [26, 27].

WWTPs generate loads of sludge from anaerobic digestion. For this reason, subsequent sludge treatments are required (e.g., dewatering and drying). These streams are nitrogen-rich (15 to 20% of the overall incoming nitrogen load of a WWTP) [28]; therefore, recovery of N after the sludge aeration has emerged as one of the points of interest towards which more effort has been directed in recent years. Moreover, ammonium concentration values vary depending on the WWTP source, ranging from 0.6 to 1.5 g/L in anaerobic side-streams, from 10 to 20 mg/L in CAS, and up to 40 to 60 mg/L in HRAS without an N removal stage [29]. Therefore, to reduce this N-load, new treatment technologies

N recovery from wastewater treatment plants: resource recovery based on a circular economy scheme

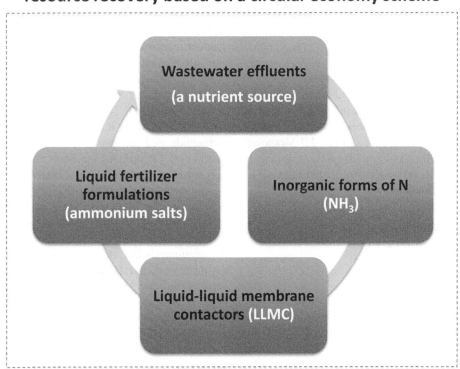

FIGURE 15.2 N recovery from WWTPs: resource recovery based on a circular economy scheme

have been recently installed in WWTPs in order to pre-treat the ammonia-rich sludge produced. Most of these technologies are based on biological processes, mainly denitrification, which releases $N_{2(g)}$ into the atmosphere from the inorganic and organic nitrogen. However, the inorganic form of nitrogen, a potential source of fertilizer, is lost.

Different techniques have been proposed to recover the ammonia from WWTP streams. According to the technology used, different by-products could be achieved, such as NH_3 aqueous solutions after evaporation, ammonium salts after ammonia removal by using stripping technologies, and so on. However, the most common process for the selective recovery of ammonia is its extraction by conversion to gaseous form as $NH_{3(g)}$ by using stripping techniques in three main approaches: (i) ammonia absorption in a strong acid using a wet scrubber [30]; (ii) ammonia air stripping by using a hollow fiber (HF) gas-liquid membrane contactor (GLMC) [31]; and (iii) ammonia recovery by flat-sheet membrane or HF-LLMC by adsorption in strong acid solutions [32–34].

On the other hand, more limited efforts have been reported to recover ammonium (NH_4^+) from the main streams generated in WWTPs. Bio-sorption as the carbon redirection process followed by ion-exchange using zeolites as the nitrogen recovery system has been proposed [29]. This integration of technologies is required, since streams treated by biosorption processes have a low ammonium content (40–60 mg/L) and a pre-concentration step using zeolites (synthetic or natural) has shown a higher ammonium ion selectivity in comparison to other competitive ions (e.g., sodium, potassium, calcium, and magnesium) [35–38] with low cost (0.2 €/kg) [39–41]. Moreover, this two-stage process (sorption and membranes), where ammonium ions are selectively concentrated in the zeolite steps (at basic pH), is conditioned by the chemical oxygen demand (COD) reduction. In this case, after zeolite treatment, the high ammonium concentration (at basic pH) and low COD content in the WWTP stream reduce the LLMC membrane operation problems associated with organic fouling. For this reason, LLMC could be used as a membrane technique that could be included as a tertiary treatment in a WWTP in order to valorize the ammonium as ammonium salts, as shown in Figure 15.1 [13, 42]. Moreover, since LLMC allows only gas transport ($NH_{3(g)}$), its use to recover the ammonia as liquid fertilizer could improve its purity and, as a result, could increase its market value.

15.3 LIQUID–LIQUID MEMBRANE CONTACTORS (LLMCS)

15.3.1 FUNDAMENTALS

The use of LLMCs is a novel and eco-friendly technique where two liquid phases are separated by a membrane and the target specie is only transported by diffusion phenomena because the phases do not mix [43]. Therefore, this technique allows ammonia recovery from wastewater effluents using an acid as a stripping solution in several industrial applications [44, 45]. Then, by using an LLMC it is possible to transform the ammonia present in wastewater

into ammonium salts which could be used as liquid fertilizers, such as NH_4NO_3, $(NH_4)_2SO_4$, $(NH_4)_2HPO_4$, and $(NH_4)H_2PO_4$, among others [13, 42, 46, 47].

Hasanoğlu et al. [48] proposed that gaseous species (like ammonia gas) from an aqueous stream can be transferred through a hydrophobic membrane in five steps as follows: (i) ammonia is transferred from the feed solution through the boundary layer via membrane pores; (ii) equilibrium is reached between the ammonia solution and the gas (air) present in the membrane pores; (iii) ammonia gas is transferred through the air filled pores; (iv) ammonia gas reacts with the receiving component (acid) in the stripping solution; (v) ammonium salt is transferred through the boundary layer into the acid stripping solution. Figure 15.3 summarizes the above mentioned steps.

Membrane contactors are based on hydrophobic or hydrophilic polymeric membranes. The most commercialized ones are hydrophobic membranes made of polyethylene (PE) and polypropylene (PP), which can be wetted by non-polar phases (e.g., non-polar organics) or filled by gas, while the aqueous/polar phase cannot access the pores [43]. Concentration or vapor pressure differences between the two sides of the membrane are the driving force of the process [45]. LLMCs have critical parameters that control membrane wettability, such as the pores' geometry, the membrane's chemical composition, and the surface tension of the absorption liquid or its interactions with the membrane (e.g., contact angle, fouling) [43, 49]. For instance, the contact angle decides whether the membrane presents hydrophobic or hydrophilic properties, with the membrane being hydrophobic when the contact angle is greater than 90°. Otherwise, the wetting phenomenon occurs at 0° [43]. On the other hand, solutions of inorganic compounds in water show high surface tension values ($\gamma \geq$ 72 mN/m), whereas when organic solutes are present, the surface tension decreases, and if the concentration of organic elements becomes too high, the membrane wetting phenomenon happens [50]. Additionally, another criterion that determines the wettability is the surface tension penetration of a liquid, a parameter that indicates how a liquid penetrates a porous medium [43]. Franken et al. [50] observed that for the same aqueous solution (like ethanol, acetic acid, dimethyl formamide, and acetone), the values of surface tension penetration for PP surfaces are lower than those obtained for polyvinylidene fluoride (PVDF) materials.

The membrane can also lose its hydrophobicity when pore wetting occurs, in other words, when the liquid pressure exceeds the breakthrough pressure and its pores get wet [51]. In fact, several parameters can affect the breakthrough pressure. As mentioned before, the contact angle between the fluid and the membrane, interfacial/surface tension, and pore size are some of the critical parameters that control membrane wettability. In this case, Laplace's equation permits the calculation of the breakthrough pressure, which is inversely dependent on the pore radius in asymmetric membranes. For this reason, bigger pores have partial membrane wetting in comparison to smaller ones, which continue to be aqueous/polar phase free. Moreover, membrane hydrophobicity depends on interactions with the phases involved since they imply modifications in the morphology and structure of the membrane [43].

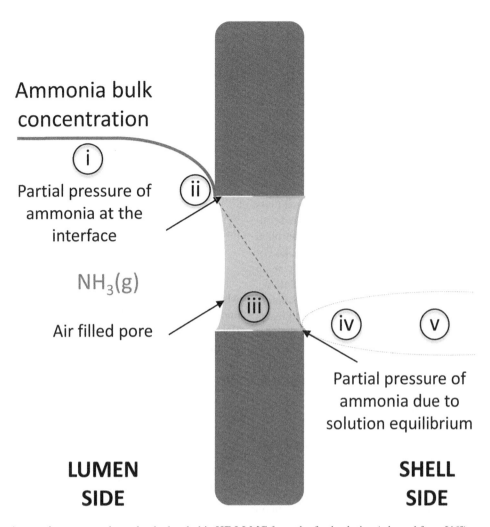

FIGURE 15.3 Ammonia transport through a hydrophobic HF-LLMC from the feed solution (adapted from [46])

15.3.2 Ammonia Recovery Mechanism

The ammonia recovery mechanism from a concentrated basic feed solution (containing $NH_{3(g)}$) by an LLMC is schematically represented in Figure 15.3. The hydrophobic membrane separates the feed solution from the stripping solution. The LLMC system for ammonia recovery, working in a closed-loop configuration, can be described as follows: ammonia-rich wastewater is fed in on the feed side (lumen side) while the acid stripping solution is fed in on the opposite side (shell side). Then, the ammonia recovery process inside the LLMC takes place: ammonia gas diffuses from the feed stream to the feed-membrane interface, volatilizes inside it, and then diffuses through the pores of the membrane to react with the acid stripping solution on the shell side. In this sense, hydrophobic porous membranes provide a safe technology to ensure ammonia recovery from WWTPs since the potential transport of organic and inorganic micro-pollutants on the generated by-product is avoided [46].

Licon et al. (2016) proposed a numerical algorithm to describe the ammonium removal on the lumen side by an HF-LLMC module, incorporating the influence of the chemical equilibrium in the acid stripping collector [46, 52]. This algorithm takes into account an unsteady state and isothermal conditions while considering Henry's law at the feed–membrane interface. When no pore blockage occurs, the feed aqueous solution does not fill the membrane pores, and the reaction of the ammonia with the acid solution is instantaneous. The algorithm was developed considering constant flow rates of ammonia and acid solutions, and operation of both tanks under laminar conditions and perfect mixing mode. The model was evaluated for two acid solutions (monoprotic acid, e.g., HNO_3, and triprotic acid, e.g., H_3PO_4) as stripping agents and, accordingly, the different neutralization reactions were incorporated. Under these conditions, the partial pressure of ammonia gas and free dissolved ammonia in solution can be estimated.

In concentrated NaOH solutions containing ammonia [pH values higher than the $pK_{a(298K)} = 9.3$ (**Equation 15.1**)], >99% of the total ammonia is present as ammonia, which is the gas form ($NH_{3(g)}$) as described by **Equation 15.2**:

$$NH_4^+ \leftrightarrow NH_3 + H^+; K_a = \frac{[NH_3][H^+]}{[NH_4^+]}; pKa_{(298K)} = 9.3$$

(15.1)

$$NH_3 \leftrightarrow NH_{3(g)}; K_{H(298K)}\left(Pa \cdot m^3\Big/mol\right) = 1.62 \quad (15.2)$$

Ammonia gas reacts with both stripping acids (nitric and phosphoric acids) to form ammonium salts as described by **Equations 15.3 and 15.4**, respectively.

$$NH_{3(g)} + HNO_3 \leftrightarrow NH_4NO_3 \quad (15.3)$$

$$NH_{3(g)} + H_3PO_4 \leftrightarrow \left(NH_4\right)_2 HPO_4 \text{ or } \left(NH_4\right)H_2PO_4 \quad (15.4)$$

Since the feed solution is strongly basic, the ammonia mass transfer coefficient can only be affected by the free acid concentration on the shell side of the LLMC and is controlled by neutralization reactions (**Equations 15.3** and **15.4**). For this reason, the free acid stripping concentration is the main driving force of ammonia recovery from the feed solution.

On the other hand, the strong acid stripping solution receives the ammonia gas from the feed solution and reacts immediately to generate ammonium salts. Thus, the free proton concentration diminishes until the acid is neutralized. For this reason, the addition of concentrated acid is needed to maintain a constant pH of the stripping solution.

15.3.2.1   Model Description for Ammonia Removal by LLMC

This section presents a brief description of the mathematical model reported by Licon et al. (2016), which describes the ammonia recovery when treating a basic feed solution by HF-LLMC in a closed-loop configuration (recirculation mode) on the feed side, taking into account the main assumptions previously explained [46]. The concentration of ammonium salts and pH variation on the stripping side are also described. The mass balance equations inside the LLMC fibre and boundary conditions are defined as follows.

Ammonia transport through the lumen is based on convective and diffusion transport according to **Equation 15.5**:

$$\frac{\partial C_j}{\partial t} + \tilde{U} \cdot \nabla C_j = D_j \nabla^2 C_j \quad (15.5)$$

where C_j is the total feed ammonia concentration (mol/m³), $\tilde{U}$ is the velocity vector (m/s), and D_j is the ammonia diffusivity in water (m²/s).

During the experiment, no chemical reactions take place on the feed side (lumen) and its symmetry is supposed to be cylindrical. Thus, **Equation 15.5** could be expressed as follows, since the radial component of advection becomes negligible:

$$\frac{\partial C_j}{\partial t} + U_z \frac{\partial C_j}{\partial Z} = D_j \left[\frac{1}{r}\frac{\partial}{\partial r}\left(r\frac{\partial C_j}{\partial r}\right) + \frac{\partial^2 C_j}{\partial Z^2}\right] \quad (15.6)$$

Besides, assuming laminar flux (ideal condition), the mass transfer inside the pore coefficient is defined as:

$$K_{g,pore} = D_{a,c,pore}\left\{\frac{\varepsilon}{\tau b}\right\} \quad (15.7)$$

where $D_{a,c,pore}$ is the ammonia diffusivity in the pore (m²/s), ε is the membrane porosity, τ is the pore tortuosity (defined as $\tau = 1/\varepsilon^2$), and b is the membrane thickness (m).

The total ammonia concentration in the feed solution (mol/m³) is determined by **Equation 15.8**:

$$C_j = [NH_3] + [NH_4^+] = [NH_3] \cong \quad (15.8)$$

where $[NH_3]$ and $[NH_4^+]$ are the ammonia and ammonium equilibrium concentrations in the feed solution, respectively (mol/m³).

Once the lumen side equations and boundary conditions have been defined, Henry's law describes the interaction at the liquid–gas interface by **Equation 15.9**:

$$P_{a,int}^g = H_a^T \left[NH_3\right]_{int} \quad (15.9)$$

where the liquid and gas phases are assumed to be the feed and the porous phase, respectively. H_a^T is Henry's constant (Pa·m³/mol) and $[NH_3]_{int}$ is the ammonia concentration at the liquid–gas interface (mol/m³).

In this case, Henry's constant depends on the temperature and is determined as follows:

$$H_a^T = \left[\left(\frac{0.2138}{T}\right) * 10^{6.123 - 1825/T}\right] * R_g T \quad (15.10)$$

Moreover, the ammonia mass balance, assuming perfect mixing and recirculation in the feed tank, must take into account **Equation 15.11**:

$$V\frac{dC_{tank}}{dt} = Q\left(C_{j,\,Z=L} - C_{tank}\right) \quad (15.11)$$

where C_{tank} is the ammonia concentration in the feed tank at any moment. At the initial time, $C_0 = C_{tank}$. Finally, Table 15.2 collects the values used for solving the mathematical model.

15.3.3   Ammonia and Water Transport

This section briefly describes how the ammonia mass transfer coefficient ($K_{m(NH3)}$) and water flux (J_W) could be determined experimentally in order to characterize the potential limitations of this technique in terms of the water transport.

15.3.3.1   Experimental Determination of Ammonia Mass Transfer Coefficient ($K_{m(NH3)}$)

The flux of ammonia (J_{NH3} (mol/m²·s)) through the LLMC can be defined by **Equation 15.12**:

$$J_{NH3} = \frac{K_{m(NH3)}\left(p_{NH3,f} - p_{NH3,s}\right)}{RT} \quad (15.12)$$

where $K_{m(NH3)}$ is the ammonia mass transfer coefficient (m/s) and p_{NH3} is the partial pressure in the feed ($_f$) and in the stripping ($_s$) stream (Pa).

Since the pH in the feed stream remains constant during the LLMC experiment, the ammonia feed partial pressure ($p_{NH3,f}$)

TABLE 15.2
HF-LLMC Parameters Used to Solve the Mathematical Model Transport [53]

Parameter	Value (m)	Parameter	Value
Pore diameter, d_{pore}	3×10^{-8}	Number of fibres, N	9950
Membrane thickness, b	3×10^{-5}	Porosity, ε	0.4
Inner lumen diameter, d_i	2.4×10^{-4}	Pore tortuosity, τ	2.25
Outer lumen diameter, d_o	3×10^{-4}	**Parameter**	**Value (m²)**
Distribution tube diameter, d_D	0.22	Active membrane area	1.4
Effective fibre length, L	0.15		
Fibre bundle diameter d_a	0.47		
Shell diameter, d_s	0.63		

and its total feed concentration ($C_{t,NH3}$) can be considered proportional. Additionally, the pH of the stripping solution is also maintained to ensure that the ammonia from the feed tank is transformed into ammonium form, and the ammonia partial pressure in the stripping tank ($p_{NH3,s}$) can be considered to be negligible. For this reason, the ammonia mass balance of **Equation 15.12** can be described as follows [54]:

$$\ln \frac{C_{o(NH3)f}}{C_{t(NH3)f}} = \frac{K_{m(NH3)}A_m}{V_f} t \quad (15.13)$$

where $C_{0(NH3),f}$ and $C_{t(NH3),f}$ are the total ammonia concentrations in the feed tank at the initial time and experimental time (mol/m³), respectively, A_m is the membrane area (m²), V_f is the feed volume (L), and t is the experimental time (h).

The ammonia mass transfer coefficient can be determined experimentally by plotting the left term of **Equation 15.13** over the experimental time. In this case, a linear relationship is obtained, whose slope can determine the ammonia mass transfer coefficient ($K_{m(NH3)}$) in each experiment.

15.3.3.2 Experimental Determination of Water Flux (J_W)

As explained in the section "Fundamentals", an LLMC consists of a hydrophobic porous membrane, which prevents the transport of liquid through its pores. For this reason, only ammonia (gas) is transported from the feed to the stripping solution by diffusion phenomenon [43]. Nevertheless, water transport takes place inside the LLMC depending on the difference in water partial pressures (Δp_w) between the two sides (lumen and shell) and the permeability of the membrane to water vapor (P_w). The water flux (Jw (L/m²·h)) can be defined by **Equation 15.14** [55]:

$$J_W = P_W \Delta p_w = P_W \left(p_{W,f}^o a_{W,f} - p_{W,s}^o a_{W,s} \right) \quad (15.14)$$

where p^o_W is the pressure of water vapour at a given T (Pa), and a_w is the water activity on the feed ($_f$) and stripping sides ($_s$). The water activity can be calculated following Raoult's law taking into account the ionic composition as follows:

$$J_W = P_W \left(p_{W,f}^o \gamma_{W,f} x_{W,f} - p_{W,s}^o \gamma_{W,s} x_{W,s} \right) \quad (15.15)$$

where γ_w is the water activity coefficient and x_w is the mole fraction of water on the feed ($_f$) and stripping sides ($_s$), respectively.

Changes in the composition of the solution control the water activity during ammonia removal by the LLMC, since both streams work under constant temperature and pressure. In this sense, due to the production of ammonia salts and the addition of concentrated acid on the stripping side, the salinity will increase and its activity coefficient will be lower than that on the feed side. Thus, the water transport from the feed to the stripping side can be described by **Equation 15.16**:

$$J_w = \frac{V_w}{A_m \cdot \Delta t} \quad (15.16)$$

where J_w is the water flux (L/m²·h), V_w is the volume of water transported from the feed side to the stripping side (L), A_m is the membrane area (m²), and t is the experimental time (h).

The water volume transported can be quantified by the mass balance in the shell tank, considering its initial and final volumes, and also the ammonia transported from the feed to the stripping side.

15.3.3.3 Impact of Water Transport on LLMC Performance

The transport of water though porous hydrophobic membranes has been used as a transport mechanism in membrane distillation applications, where the main driving force to achieve high water fluxes is typically the use of temperature gradients. However, differences in salinity between the two sides of the membrane is a potential mechanism to increase the differences in vapor pressure between the two sides. The use of salinity gradients in porous hydrophobic membranes, which ensures that membrane pores are occupied by air, has been recognized as an osmotic distillation process [56].

In the case of the LLMC process, as the filtration proceeds, the water flux values achieve a plateau with a steady state behaviour as a consequence of: (i) the salinity increase in the stripping phase due to the transport of ammonia and the addition of strong acids; and (ii) the water transport from the feed phase to the stripping phase, as shown in Figure 15.4. Consequently, water transport in the LLMC has been proven and must be taken into account for the liquid fertilizer formulation.

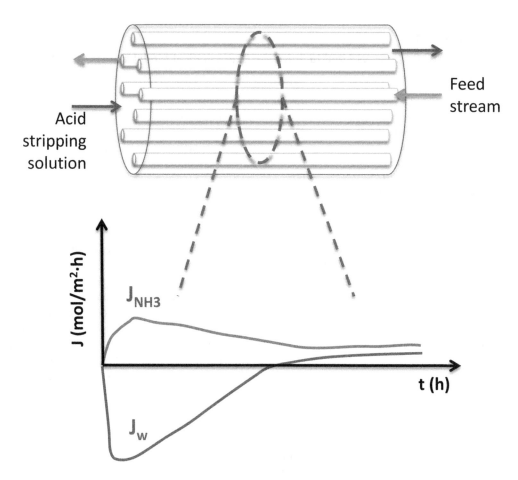

FIGURE 15.4 Ammonia mass transfer and water transport impact through the LLMC

15.3.4 INFLUENCE OF THE TYPE OF ACID STRIPPING SOLUTION

The following sections present the ammonia recovery using two types of acids on the stripping side, HNO_3 and H_3PO_4, the influence of the type of acid in terms of the ammonium salt production.

15.3.4.1 Recovery of Ammonia as Ammonium Nitrate in the Stripping Solution

The variations of the total ammonium concentration and the pH of the nitric acid stripping solution over time for two consecutive LLMC cycles are depicted in Figure 15.5. In the first cycle, after the reduction of the ammonium concentration from 1700 to 20 mg NH_3/L (98% removal) the initial acidity of the 0.5 M HNO_3 stripping solution is reduced to pH = 0.8, due to the neutralization reaction defined by **Equation 15.17**. At the end of the first filtration cycle, a new fresh solution of NH_3 (1700 mg NH_3/L) is introduced in the feed circuit.

$$C_{t(N)} = \left(\left[NO_3^- \right] + 10^{-\log\left[H^+ \right]-14} - 10^{-\log\left[H^+ \right]} \right) \cdot \left(1 + 10^{-\log\left[H^+ \right]-pka} \right)$$

$$(15.17)$$

The theoretical description (plotted as solid lines in Figure 15.5) for the development of the ammonia mass transport algorithm, mentioned in Section 3.2.1, was used to predict

values of free ammonia concentration in the feed and stripping phases as well as the pH of the stripping phase. The modelling parameters used and related to the membrane transport properties are summarized in Table 15.2.

During the second filtration cycle, the reduction of NH_3 follows an exponential decay; however, it stabilizes around a value of 800 mgNH_3/L [C/C_0= 0.5 or $R_{NH3}(\%)$= 50]. During the second cycle, the acidity decreases from an initial pH value of 0.8 to a neutral pH value of 8 at the moment when the ammonia transfer stops, as shown by the fact that the NH_3 concentration in the feed phase is kept constant with time. In this step, the free acid concentration of the stripping phase is insufficient to promote ammonia transfer through the membrane phase and is thus reduced by the partial pressure of NH_3, which appears due to the equilibrium described in **Equations 15.18** and **15.19**.

$$C_{t(N)} = \left[NH_3 \right] + \left[NH_4^+ \right] = \left[NH_3 \right] + \frac{C_{t(N)}}{1 + 10^{-\log\left[H^+ \right]-pka}}$$

$$(15.18)$$

$$\left[H^+ \right] + \left[NH_4^+ \right] = \left[NO_3^- \right] + \left[OH^- \right] \quad (15.19)$$

The continuous operation of the acid circuit could be linked to a simple pH control and nitric acid could be dosed in the stripping circuit using a control loop to assure free acidity

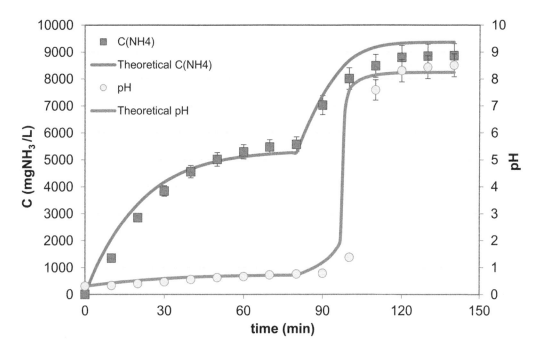

FIGURE 15.5 Ammonium concentrations and pH profiles over time in the stripping phase after two consecutive cycles (cycle 1 from 1 to 80 min and cycle 2 from 80 to 140 min) for 1.7 g/L NH_3 and 8 g/L NaOH concentrates at flow rates of 0.5 L/min and for a 0.5 M HNO_3 as the stripping solution (adapted from [46])

values and pH values below 2. The acid requirements were predicted theoretically as a function of the total amount of ammonia transported. For nitric acid solutions, the stripping phase pH would depend on the total ammoniacal concentration ($C_{t(N)}$) as described by **Equations 15.18** and **15.19**.

Predicted concentrations of free ammonia in the feed solutions were described well by the model, and the prediction error between the experimental and predicted values was below 5%. The free concentration in the stripping phase was also calculated, with values below 0.1 $mgNH_3$/L when the -log[H^+] was lower than 2. The reduction of the free acid concentration (H^+) below 0.01 M implies that the increase of the free ammonia concentration in the stripping phase as a neutralization reaction (**Equation 15.3**) is not favoured. Also the variation of the pH of the stripping phase was calculated and measured and the predicted values showed differences in -log[H^+] below 0.2 units with the exception of values corresponding to non-buffered solutions with stoichiometric ratios of nitric acid and ammonia (between -log[H^+] 4 and pH 8). Then, the variation of pH of the stripping solution is predicted well by **Equation 15.18** and could be used to control the process of production of ammonium nitrate salts as liquid fertilizers.

Additionally, the increase of the ammonium nitrate concentration in the stripping phase with time can be observed in Figure 15.5. After these two LLMC cycles, 9 g/L NH_4NO_3 was produced for a ratio of feed tank volume/stripping tank volume of 5.

15.3.4.2 Recovery of Ammonia as Ammonium-Phosphate in the Stripping Solution

As phosphoric acid has a lower strength than nitric acid, it is supposed that the ammonia removal profiles measured were different from those observed for nitric acid.

The variations of the total ammonium concentration and the pH of the stripping solution with time for two consecutive LLMC cycles are plotted in Figure 15.6. In the first cycle, after the reduction of the ammonium concentration from 1760 to 105 $mgNH_3$/L (94% removal), the initial acidity of the 0.4M H_3PO_4 stripping solution measured as -log[H^+]=1 is reduced and reaches values close to neutral.

The pH values followed a typical acid-base neutralization S shape function and were stabilized (at approximately pH = 7) in two steps according to the acid-base properties of H_3PO_4 as defined by **Equation 15.4**.

During the filtration cycles, the driving force, that is, the difference between the ammonia vapor pressures on the two sides of the fiber, is achieved by the fast reaction that occurs at the gas/liquid phosphoric interface.

Control of the operation of the LLMC could be based on a pH control loop and the required phosphoric acid addition would depend on the total ammonium concentration transported as it is derived from the phosphoric mass and the electroneutrality balance (**Equations 15.20** and **15.21**):

$$\left[H_3PO_4\right]+\left[H_2PO_4^-\right]+\left[HPO_4^{2-}\right]+\left[PO_4^{3-}\right]=C_{t(P)} \quad (15.20)$$

$$\left[H^+\right]+\left[NH_4^+\right]=\left[H_2PO_4^-\right]+2\left[HPO_4^{2-}\right]$$
$$+3\left[PO_4^{3-}\right]+\left[OH^-\right] \quad (15.21)$$

Accordingly, the total phosphoric acid concentration/pH dependence is expressed by **Equation 15.22**:

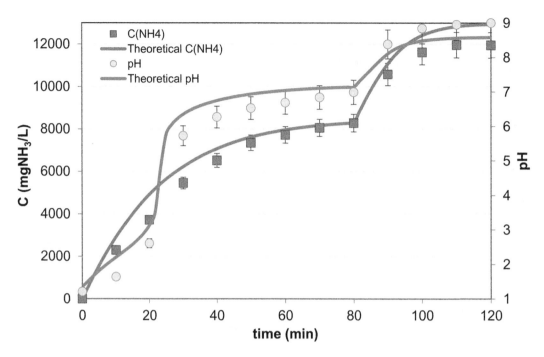

FIGURE 15.6 Ammonium concentration and pH profiles in the stripping phase over time after two consecutive cycles (cycle 1 from 1 to 80 min, and cycle 2 from 80 to 120 min) for 1.7 g/L NH_3 and 8 g/L NaOH concentrates at flow rates of 0.5 L/min and for an initial 0.4 M H_3PO_4 as the stripping solution (adapted from [46])

$$
C_{t(p)} =
$$

$$
\left(1 + 10^{-\log[H^+]-pka}\right) \cdot
$$

$$
\left(
\begin{array}{l}
10^{-\log[H^+]-14} - 10^{+\log[H^+]} + \dfrac{C_{t(p)}}{1 + 10^{-\log[H^+]-pka2} + 10^{pka1+\log[H^+]} + 10^{-2\log[H^+]-pka2-pka3}} + \\[3mm]
\dfrac{2C_{t(p)}}{1 + 10^{pka2+\log[H^+]} + 10^{-\log[H^+]-pka3} + 10^{pka1+pka2+2\log[H^+]}} + \\[3mm]
\dfrac{3C_{t(p)}}{1 + 10^{pka1+pka2+pka3+3\log[H^+]} + 10^{pka2+pka3=2\log[H^+]} + 10^{pka3+\log[H^+]}}
\end{array}
\right)
\qquad (15.22)
$$

where K_{a1}, K_{a2}, and K_{a3} are the acid dissociation constants for the phosphoric system ($H_3PO_4/H_2PO_4^-/HPO_4^{2-}/PO_4^{3-}$).

The theoretical pH was calculated by **Equation 15.22**, and a good prediction of the measured values was obtained, as can be seen in Figure 15.6. However, at the end of the first removal cycle (approximately 80 minutes) ($R_{NH3}(\%) =$ 94%), a new fresh solution of NH_3 (1760 mgNH_3/L) was introduced in the feed stream.

During the second cycle (from 80 to 140 min), the NH_3 concentration was stabilized at a value of around 1060 mgNH_3/L ($C/C_0 = 0.6$). During the second cycle, the acidity increased from an initial pH of approximately 7 with a solution containing a mixture of $(NH_4)H_2PO_4/(NH_4)_2HPO_4$ to pH 9 with a solution of $(NH_4)_2HPO_4$ at this stage. Then, the ammonia transfer stopped as shown by the fact that the NH_3 concentration in the feed phase remained constant

with time, following the same trend as was observed with nitric acid.

The predicted concentrations of free ammonia in the feed solutions were described well by the model since the error between the measured and estimated values was below 5%. The free concentration in the stripping phase was also calculated with values below 0.001 mgNH_3/L when the pH values were lower than 7 (Figure 15.6). The reduction of the free acid concentration below 10^{-3}M represented an increase of the free ammonia concentration in the stripping phase as the neutralization reactions (**Equation 15.4**) are not thermodynamically favoured. The change in the pH of the stripping phase was calculated, and the measured and predicted values showed differences in pH below 0.2 units with the exception of pH values corresponding to buffered solutions with stoichiometric ratios of

$(NH_4)H_2PO_4$ /$(NH_4)_2HPO_4$. Thus, the variation in the pH of the stripping phase is predicted well by **Equation 15.22** and could be used to control the process of production of di-ammonium phosphate salts.

The second cycle starts with a new solution of NH_3 (1.8 g/L) following a typical exponential decay and stabilizing at a C_t/C_o value of about 13 g/L. The evolution of the pH also follows an S shape function and reaches a value of 9 when it is stabilized, corresponding to the stabilization of the ammonium removal. Then when a certain number of moles of ammonia have been transported, the $H_2PO_4^-$ species are converted quantitatively to HPO_4^{2-}, according to **Equation 15.22**, and then the transformation of $(NH_4)H_2PO_4$ to $(NH_4)_2HPO_4$ and the pH of the expected solution can be calculated, taking into account the mixture of NH_4^+ and $H_2PO_4^-$ species. Prediction of the pH of this solution was performed by **Equation 15.22** and an overestimation of pH values can be seen in Figure 15.6 (errors were between 10 and 15%), indicating that additional work should be done, probably on the ammonia mass transfer properties in acidic solutions that are not strong enough.

Lastly, Figure 15.6 shows both experimental (points) and predicted (solid lines) increases of the ammonium phosphate concentration in the stripping phase with time. After two cycles, 9 g/L $(NH_4)_2HPO_4$ was produced at a ratio of feed tank volume/stripping tank volume of 5. Again the model provided a good prediction of the total ammonium phosphate concentration in the stripping phase with an error below 5%.

15.4 AMMONIA RECOVERY FROM ZEOLITE REGENERATION CONCENTRATES: EFFECT OF OPERATIONAL PARAMETERS ON PRODUCT QUALITY (CASE STUDY)

Ammonia concentrated streams generated in the regeneration of zeolites, which is a based sorption stage, from an industrial pilot plant (located in Vilanova i la Geltrú WWTP, Barcelona, Spain) were used to evaluate the performance of HF-LLMC. The different operational conditions studied were: (i) treated volume; (ii) feed concentration; and (iii) type of acid stripping solution. For this purpose, 0.5 L of acid solution [e.g., H_3PO_4 (0.4 M), HNO_3 (0.4 M), or a mixture of both acids, 0.3 M HNO_3 + 0.1 M H_3PO_4] was used as the stripping solution in the HF-LLMC. The pH was kept constant at 6–7 when using phosphoric acid, whereas it was fixed at 2–3 when nitric acid or a mixture of both acids was used as the stripping solution. The feed volumes were 5, 30, and 60 L and the feed concentration solutions were composed of 1.8 and 3.8 g/L of ammonia with a pH of around 12 (to have ammonia in gas form). The experimental setup used an HF-LLMC (2.5 + 8 Liqui-Cel® Membrane Contactor X-50 PP fibre) from 3M Company (USA) with an active membrane area of 1.4 m², working in a closed-loop and counter-current mode (flow rate of 450 mL/min in both circuits). Figure 15.7 shows the HF-LLMC laboratory setup scheme used to evaluate the production of liquid fertilizers from a wastewater ammonia-rich stream. The methodology used was described elsewhere [46].

Samples from the feed tank and from the stripping solution tank were taken over experimental time in order to determine their composition. Additionally, the HF-LLMC efficiency for ammonia recovery was evaluated as follows: (1) ammonia removal on the feed side; and (2) percentage composition of ammonia salts (N from feed ammonia and nitric acid and P_2O_5 from the phosphoric acid) on the stripping side.

The results of HF-LLMC treatment of wastewater containing 1.7 g NH_3/L when different volumes were used as feed solution (5, 30, and 60 L) are summarized in Table 15.3. It can be seen that on increasing the feed volume, the ammonia removal decreased (from 96.3% to 85.0%) as the experimental time increased (from 1.7 to 12.5 h); however, the percentages of N and P_2O_5 obtained were higher.

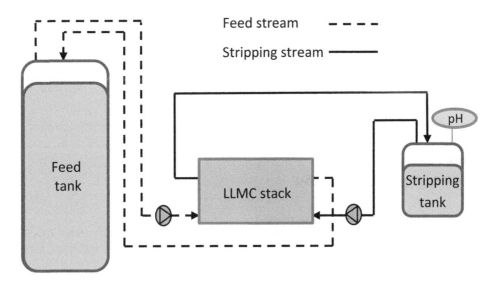

FIGURE 15.7 HF-LLMC laboratory setup scheme

Therefore, a larger treated volume (e.g., 60 L) was needed in order to obtain more liquid fertilizer as ammonia salt (5.1% N and 15.0% P_2O_5).

On the other hand, once the target ammonia recovery (> 90%) has been reached, the exhausted ammonia stream (at pH=12) can be recirculated, taking advantage of the sodium hydroxide (present in the ammonia stream) for successive uses in the sorption/desorption stage of the zeolite regeneration.

The results evaluating the effect of the initial feed concentration (1.5 and 3.7 g/L) for the same treated volume (60 L) and using H_3PO_4 as the stripping acid solution are collected in Table 15.4. It was observed that on increasing the feed concentration (from 1.5 to 3.7 g/L), the ammonia removal decreased (between 85.9 and 76.1) as the experimental time increased; however, the percentages of N and P_2O_5 obtained were higher (7.8% N and 21.6% P_2O_5). Thus, it was concluded that better results can be obtained when working with a higher feed volume (60 L) and higher ammonia feed concentration (3.7 g NH_3/L).

The results evaluating the effect of acid stripping solutions on the treatment of 60 L of wastewater with an initial concentration of ammonia of around 3.8 g/L are collected in Table 15.5. Among the acids used, H_3PO_4 provided higher ammonia removal (76.1%) and higher amounts of N and P_2O_5 (7.8% and 21.6%, respectively).

Optimal operational conditions (e.g., 60 L of wastewater with 3.7 g NH_3/L and using H_3PO_4 as the stripping solution) made it possible to produce 7.8% N and 21.6% P_2O_5, which could be used as liquid fertilizer. Figure 15.8 shows the evolution of ammonia removal under the optimal conditions obtained. Approximately 76% of the ammonia was recovered in 24 h from the feed solution with a mass transfer coefficient (Km_{NH3}) of 8.8× 10^{-7} m/s. Additionally, the NH_3 concentration factor was 26.3on the stripping side [42].

15.5 AMMONIUM SALT PRODUCTION AS LIQUID FERTILIZER FORMULATIONS

The liquid fertilizers obtained by the LLMC process are composed of ammonium salts which are produced in the acid stripping solution. For this reason, depending on the acid used in the process, it is possible to produce specific types of fertilizers: (i) based on one carrier (e.g., N), called single-nutrient fertilizers; and (ii) based on more than one carrier (e.g., N and P_2O_5), called multi-nutrient fertilizers [42, 57].

These N and P_2O_5 amounts must increase in order to approach the target reference values reported by fertilizer producers such as Fertiberia [58], Agralia [59], and Compost [60], among others, which are around 15 to 25% for both nutrients [57, 61]. However, the increase of the

TABLE 15.3
Volume Effect on HF-LLMC

Volume Shell Tank (L)	Feed Concentration (g/L)	Experiment Time (h)	Acid Stripping	Ammonia Removal (%)	N% – P_2O_5%
5	1.7	1.7	HNO_3	96.3	2.2 – 0.0
30		6.0		90.8	4.1 – 0.0
60		12.5	H_3PO_4	85.0	5.1 – 15.0

TABLE 15.4
Effect of Feed Concentration on HF-LLMC

Volume Shell Tank (L)	Feed Concentration (g/L)	Experiment Time (h)	Acid Stripping	Ammonia Removal (%)	N% – P_2O_5%
60	1.5	12.2	H_3PO_4	85.9	5.0 – 14.9
	3.7	23.5		76.1	7.8 – 21.6

TABLE 15.5
Acid Stripping Effect on HF-LLMC

Volume Shell Tank (L)	Feed Concentration (g/L)	Experiment Time (h)	Acid Stripping	Ammonia Removal (%)	N% – P_2O_5%
60	4.0	24.0	HNO_3	69.3	11.4 – 0.0
	3.7	23.5	H_3PO_4	76.1	7.8 – 21.6
	4.0	24.0	$HNO_3 + H_3PO_4$	69.8	10.0 – 0.1

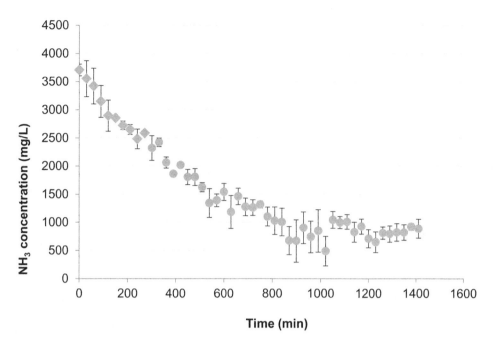

FIGURE 15.8 Evolution of ammonia concentration in the feed tank over time under the optimal conditions

nutrients has only been proved in batch operation cycles, where the stripping phase is re-circulated with continuous addition of free acid in closed-loop configuration as mentioned above. Recently, Vecino et al. [62] proposed the integration of LLMC with electrodialysis (ED) to concentrate the fertilizer solutions from LLMC in order to achieve the nitrogen content required for direct application in agriculture for fertigation. The results showed that it was possible to concentrate these ammonium salts by a factor of 1.6 ± 0.3 using ED with an optimal energy consumption of 0.21 ± 0.08 kWh/kg ammonium salt and $93.1 \pm 4.2\%$ of faradaic yield; giving a liquid fertilizer composed of 15.6% (w/w) nitrogen as NH_4NO_3.

Ammonium nitrate and ammonium phosphate fertilizers are requested formulations in agriculture regulation since they are composed of two key nutrients (N, P). For this reason, more expensive acids such as HNO_3 and H_3PO_4 are preferred over more common ones such as H_2SO_4 [63].

15.6 CONCLUSIONS AND PERSPECTIVES

Liquid–liquid membrane contactors are a novel technology to recover nitrogen from wastewater effluents, transforming it into ammonium salts that could be used as liquid fertilizers. This technology can be a suitable alternative to the conventional methods such as biological (e.g., nitrification-denitrification) or anammox processes since it is possible to increase the nitrogen value. Additionally, this technology makes it possible to formulate different types of fertilizers (like single-nutrient or multi-nutrient fertilizers) depending on the acid used as the stripping solution in the process.

The case study reported in this chapter demonstrated that taking into account the HF-LLMC operational parameters (e.g., treated volume, feed concentration, and acid

stripping solution), it was possible to remove 76.1% of the NH_3 from concentrated ammonia streams generated in the regeneration of zeolites containing 3.7 g/L of ammonia in 60 L of volume and producing an ammonium salt with up to 7.8% of N and 21.6% of P_2O_5.

Finally, water transport phenomena should be studied in more detail since the transport of species through the hollow fibers is only possible in gaseous form (e.g., NH_3); however, water transport has also been reported and quantified. This issue represents a limitation that can impact on the quality of the fertilizer or may not satisfy the requirements of the fertilizer industry. In view of that, with regard to the use of LLMC for ammonia recovery, new perspectives should be addressed in order to obtain a better fertilizer composition and can be listed as follows: (i) treating a higher volume of wastewater with H_3PO_4 as the stripping solution (based on the best conditions of case study); (ii) performing more batch operation cycles; and (iii) decreasing the water transport by testing different temperatures on both sides (feed and stripping solutions).

LIST OF SYMBOLS

A_m: membrane area (m^2)

$a_{w,f}$: water activity in the feed side

$a_{w,s}$: water activity in the stripping side

b: membrane thickness (m)

$C_{0(NH3),f}$: total ammonia concentration in the feed tank at initial time (mol/m^3)

C_j: total feed ammonia concentration (mol/m^3)

$C_{t(NH3),f}$: total ammonia concentration in the feed tank at experimental time (mol/m^3)

C_{tank}: ammonia concentration in the feed tank at any moment (mol/m^3)

d_a: fiber bundle diameter (m)

$D_{a,c,pore}$: ammonia diffusivity in the pore (m²/s)

d_D: distribution tube diameter (m)

d_i: inner lumen diameter (m)

D_j: ammonia diffusivity in water (m²/s)

d_o: outer lumen diameter (m)

d_{pore}: pore diameter (m)

d_s: shell diameter (m)

H^T_a: Henry's constant (Pa·m³/mol)

J_{NH3}: flux of ammonia (mol/m²·s)

J_w: water flux (L/m²·h)

$k_{g,pore}$: mass transfer coefficient inside the pore (m/s)

$K_{m(NH3)}$: ammonia mass transfer coefficient (m/s)

L: effective fibers length (m)

N: number of fibers

$[NH_3]$: ammonia equilibrium concentrations in the feed solution (mol/m³)

$[NH_4^+]$: ammonium equilibrium concentrations in the feed solution (mol/m³)

$[NH_3]_{int}$: ammonia concentration at the liquid–gas interface (mol/m³)

p^o_w: pressure of water vapor at a given T (Pa)

$P^g_{a,int}$: ammonia partial pressure at the interface (Pa)

$p_{NH3,f}$: partial pressure in the feed stream (Pa)

$p_{NH3,s}$: partial pressure in the stripping stream (Pa)

P_w: water vapor membrane permeability (Pa)

Q: flow rate (m³/s)

R_g: universal gas constant (m³·Pa/K·mol)

t: experimental time (h)

T: temperature (K)

$\tilde{U}$: velocity vector (m/s)

V_f: feed volume (L)

V_w: volume of water transported from the feed side to the stripping side (L)

$x_{w,f}$: mole fraction of water in the feed side

$x_{w,s}$: mole fraction of water in the stripping side

Δp_w: water partial pressures between both sides (lumen and shell) (Pa)

γ_w: water activity coefficient

ε: membrane porosity

τ: pore tortuosity

ACKNOWLEDGMENTS

This research was supported by the Waste2Product project (CTM2014-57302-R) and by R2MIT project (CTM2017-85346-R) financed by the Spanish Ministry of Economy and Competitiveness (MINECO), the Catalan Government (2017-SGR-312), and the EU for LifeEnrich project (LIFE16 ENV/ES/000375). As well, Xanel Vecino thanks MINECO for her Juan de la Cierva contracts (ref. FJCI-2014-19732 and ref. IJCI-2016-27445). Authors would also like to acknowledge E. Licon and I. Sancho, for their valuable contribution along their PhD work. Finally, Martin Ulbright (3M) and Jorge Muñoz (Separel, SunChemical) are thanked for their helpful comments and guidance on experimental and modules supply.

REFERENCES

1. H. Blaas, C. Kroeze, Excessive nitrogen and phosphorus in European rivers: 2000–2050, *Ecol. Indic.* 67 (2016) 328–337. doi:10.1016/j.ecolind.2016.03.004

2. M. Salomon, E. Schmid, A. Volkens, C. Hey, K. Holm-Müller, H. Foth, Towards an integrated nitrogen strategy for Germany, *Environ. Sci. Policy.* 55 (2016) 158–166. doi:10.1016/j.envsci.2015.10.003

3. M. Zhang, H. Zhang, D. Xu, L. Han, D. Niu, B. Tian, J. Zhang, L. Zhang, W. Wu, Removal of ammonium from aqueous solutions using zeolite synthesized from fly ash by a fusion method, *Desalination.* 271 (2011) 111–121. doi:10.1016/j.desal.2010.12.021

4. M. Ali, S. Okabe, Anammox-based technologies for nitrogen removal: Advances in process start-up and remaining issues, *Chemosphere.* 141 (2015) 144–153. doi:10.1016/j.chemosphere.2015.06.094

5. B. Ma, S. Wang, S. Cao, Y. Miao, F. Jia, R. Du, Y. Peng, Biological nitrogen removal from sewage via anammox: Recent advances, *Bioresour. Technol.* 200 (2016) 981–990. doi:10.1016/j.biortech.2015.10.074

6. R.C. Jin, G.F. Yang, J.J. Yu, P. Zheng, The inhibition of the Anammox process: A review, *Chem. Eng. J.* 197 (2012) 67–79. doi:10.1016/j.cej.2012.05.014

7. Z. Tian, J. Zhang, Y. Song, Several key factors influencing nitrogen removal performance of anammox process in a bio-filter at ambient temperature, *Environ. Earth Sci.* 73 (2015) 5019–5026. doi:10.1007/s12665-015-4232-y

8. I. Sancho, E. Licon, C. Valderrama, N. de Arespacochaga, S. López-Palau, J.L. Cortina, Recovery of ammonia from domestic wastewater effluents as liquid fertilizers by integration of natural zeolites and hollow fibre membrane contactors, *Sci. Total Environ.* 584–585 (2017) 244–251. doi:10.1016/j.scitotenv.2017.01.123

9. J. Smitshuijzen, J. Pérez, O. Duin, M.C.M. van Loosdrecht, A simple model to describe the performance of highly-loaded aerobic COD removal reactors, *Biochem. Eng. J.* 112 (2016) 94–102. doi:10.1016/j.bej.2016.04.004

10. P. Cornel, C. Schaum, Phosphorus recovery from wastewater: Needs, technologies and costs, *Water Sci. Technol.* 59 (2009) 1069–1076. doi:10.2166/wst.2009.045

11. Milieu Ltd, WRC, RPA, Environmental, economic and social impacts of the use of sewage sludge on land. Final Report. Part II: Report on Options and Impacts, European Commission (2008).

12. D.J. Batstone, T. Hülsen, C.M. Mehta, J. Keller, Platforms for energy and nutrient recovery from domestic wastewater: A review, *Chemosphere.* 140 (2015) 2–11. doi:10.1016/j.chemosphere.2014.10.021

13. I. Sancho, E. Licon, C. Valderrama, N. de Arespacochaga, S. Lopez-Palau, J.L. Cortina, Recovery of ammonia from domestic wastewater effluents as liquid fertilizers by integration of natural zeolites and hollow fibre membrane contactors, *Sci. Total Environ.* 584–585 (2017) 244–251. doi:10.1016/j.scitotenv.2017.01.123

14. W. Verstraete, P. Van de Caveye, V. Diamantis, Maximum use of resources present in domestic "used water," *Bioresour. Technol.* 100 (2009) 5537–5545. doi:10.1016/j.biortech.2009.05.047

15. L.T. Angenent, K. Karim, M.H. Al-Dahhan, B.A. Wrenn, R. Domíguez-Espinosa, Production of bioenergy and biochemicals from industrial and agricultural wastewater, *Trends Biotechnol.* 22 (2004) 477–485. doi:10.1016/j.tibtech.2004.07.001

16. P.M. Sutton, B.E. Rittmann, O.J. Schraa, J.E. Banaszak, A.P. Togna, Wastewater as a resource: A unique approach to achieving energy sustainability, *Water Sci. Technol.* 63 (2011) 2004–2009. doi:10.2166/wst.2011.462

17. R.J. LeBlanc, R.P. Richard, N. Beecher, A review of "global atlas of excreta, wastewater sludge, and biosolids management: Moving forward the sustainable and welcome uses of a global resource," *Proc. Water Environ. Fed.* 2009 (2009) 1202–1208. doi:10.2175/193864709793846402

18. EMATSA, Water rates (2018). https://www.ematsa.cat/en/at-your-service/bill-and-rates/your-rates/ (accessed February 7, 2019).

19. Solid nitrogen suppliers and manufacturers (2019). https://www.alibaba.com/showroom/solid+nitrogen.html?fsb=y&IndexArea=product_en&CatId=&SearchText=solid+nitrogen&isGalleryList=G (accessed February 7, 2019).

20. Methane suppliers (2019). https://tariffe.segugio.it/guide-e-strumenti/domande-frequenti/quanto-costa-uno-standard-metro-cubo-di-gas.aspx (accessed February 7, 2019).

21. Supplier. Product prices (2019). https://www.made-in-china.com/productdirectory.do?word=organic+fertilizer&subaction=hunt&style=b&mode=and&code=0&comProvince=nolimit&order=0&isOpenCorrection=1&log_from=1 (accessed February 7, 2019).

22. Industrial phosphorous suppliers and manufacturers (2019). https://www.alibaba.com/showroom/industrial+phosphorous.html?fsb=y&IndexArea=product_en&CatId=&SearchText=industrial+phosphorous&isGalleryList=G (accessed February 7, 2019).

23. W. Verstraete, S.E. Vlaeminck, ZeroWasteWater: Short-cycling of wastewater resources for sustainable cities of the future, *Int. J. Sustain. Dev. World Ecol.* 18 (2011) 253–264. doi:10.1080/13504509.2011.570804

24. C. Puchongkawarin, C. Gomez-Mont, D.C. Stuckey, B. Chachuat, Optimization-based methodology for the development of wastewater facilities for energy and nutrient recovery, *Chemosphere.* 140 (2015) 150–158. doi:10.1016/j.chemosphere.2014.08.061

25. Y.X. Huo, D.G. Wernick, J.C. Liao, Toward nitrogen neutral biofuel production, *Curr. Opin. Biotechnol.* 23 (2012) 406–413. doi:10.1016/j.copbio.2011.10.005

26. K. Hasler, S. Bröring, S.W.F. Omta, H.W. Olfs, Life cycle assessment (LCA) of different fertilizer product types, *Eur. J. Agron.* 69 (2015) 41–51. doi:10.1016/j.eja.2015.06.001

27. T. Yan, Y. Ye, H. Ma, Y. Zhang, W. Guo, B. Du, Q. Wei, D. Wei, H.H. Ngo, A critical review on membrane hybrid system for nutrient recovery from wastewater, *Chem. Eng. J.* 348 (2018) 143–156. doi:10.1016/j.cej.2018.04.166

28. H. Siegrist, D. Salzgeber, J. Eugster, A. Joss, Anammox brings WWTP closer to energy autarky due to increased biogas production and reduced aeration energy for N-removal, *Water Sci. Technol.* 57 (2008) 383–388. doi:10.2166/wst.2008.048

29. I. Sancho, S. Lopez-Palau, N. Arespacochaga, J.L. Cortina, New concepts on carbon redirection in wastewater treatment plants: A review, *Sci. Total Environ.* 647 (2019) 1373–1384. doi:10.1016/j.scitotenv.2018.08.070

30. L.S. Hadlocon, R.B. Manuzon, L.Y. Zhao, E. Planning, C.R. Act, Optimization of ammonia absorption using acid spray wet scrubbers, *Trans. ASABE.* 57 (2014) 647–659. doi:10.13031/trans.57.10481

31. A. McLeod, P. Buzatu, O. Autin, B. Jefferson, E. McAdam, Controlling shell-side crystal nucleation in a gas-liquid membrane contactor for simultaneous ammonium bicarbonate recovery and biogas upgrading, *J. Membr. Sci.* 473 (2015) 146–156. doi:10.1016/j.memsci.2014.07.063

32. A. Hasanoğlu, J. Romero, B. Pérez, A. Plaza, Ammonia removal from wastewater streams through membrane contactors: Experimental and theoretical analysis of operation parameters and configuration, *Chem. Eng. J.* 160 (2010) 530–537. doi:10.1016/j.cej.2010.03.064

33. E. Licon, M. Reig, P. Villanova, C. Valderrama, O. Gibert, J.L. Cortina, Ammonium removal by liquid–liquid membrane contactors in water purification process for hydrogen production, *Desalin. Water Treat.* 56 (2014) 1–10. doi:10.1080/19443994.2014.974216

34. M. Berdasco, A. Coronas, M. Vallès, Theoretical and experimental study of the ammonia/water absorption process using a flat sheet membrane module, *Appl. Therm. Eng.* 124 (2017) 477–485. doi:10.1016/j.applthermaleng.2017.06.027

35. S.E. Jorgensen, O. Libor, K. Lea Graber, K. Barkacs, Ammonia removal by use of clinoptilolite, *Water Res.* 10 (1976) 213–224. doi:10.1016/0043-1354(76)90130-5

36. A.T. Williams, B.K. Mayer, Advancement in ion exchange processes for municipal wastewater nutrient recovery, *Proc. Water Environ. Fed.* 101 (2013) 6474–6485.

37. E. Ivanova, M. Karsheva, B. Koumanova, Adsorbsion of ammonia ions onto natural zeolite, *J. Univ. Chem. Technol. Metall.* 45 (2010) 295–302.

38. A. Malovanyy, H. Sakalova, Y. Yatchyshyn, E. Plaza, M. Malovanyy, Concentration of ammonium from municipal wastewater using ion exchange process, *Desalination.* 329 (2013) 93–102. doi:10.1016/j.desal.2013.09.009

39. G.J. Millar, A. Winnett, T. Thompson, S.J. Couperthwaite, Equilibrium studies of ammonium exchange with Australian natural zeolites, *J. Water Process Eng.* 9 (2016) 47–57. doi:10.1016/j.jwpe.2015.11.008

40. A. Thornton, P. Pearce, S.A. Parsons, Ammonium removal from solution using ion exchange on to MesoLite, an equilibrium study, *J. Hazard. Mater.* 147 (2007) 883–889. doi:10.1016/j.jhazmat.2007.01.111

41. T. Jorgensen, L. Weatherley, Ammonia removal from wastewater by ion exchange in the presence of organic contaminants, *Water Res.* 37 (2003) 1723–1728. doi:10.1016/S0043-1354(02)00571-7

42. X. Vecino, M. Reig, B. Bhushan, O. Gibert, C. Valderrama, J.L. Cortina, Liquid fertilizer production by ammonia recovery from treated ammonia-rich regenerated streams using liquid-liquid membrane contactors, *Chem. Eng. J.* 360 (2018) 890–899. doi:10.1016/J.CEJ.2018.12.004

43. E. Drioli, A. Criscuoli, E. Curcio, *Membrane Contactors: Fundamentals, Applications and Potentialitie*, Vol. 11, 1st edition, Elsevier Science (2006).

44. S.N. Ashrafizadeh, Z. Khorasani, Ammonia removal from aqueous solutions using hollow-fiber membrane contactors, *Chem. Eng. J.* 162 (2010) 242–249. doi:10.1016/j.cej.2010.05.036

45. M. Darestani, V. Haigh, S.J. Couperthwaite, G.J. Millar, L.D. Nghiem, Hollow fibre membrane contactors for ammonia recovery: Current status and future developments, *J. Environ. Chem. Eng.* 5 (2017) 1349–1359. doi:10.1016/j.jece.2017.02.016

46. E.E. Licon Bernal, C. Maya, C. Valderrama, J.L. Cortina, Valorization of ammonia concentrates from treated urban wastewater using liquid-liquid membrane contactors, *Chem. Eng. J.* 302 (2016) 641–649. doi:10.1016/j.cej.2016.05.094

47. M.A. Boehler, A. Heisele, A. Seyfried, M. Grömping, H. Siegrist, $(NH_4)_2SO_4$ recovery from liquid side streams, *Environ. Sci. Pollut. Res.* 22 (2015) 7295–7305. doi:10.1007/s11356-014-3392-8

48. A. Hasanoğlu, J. Romero, A. Plaza, W. Silva, Gas-filled membrane absorption: A review of three different applications to describe the mass transfer by means of a unified approach, *Desalin. Water Treat.* 51 (2013) 5649–5663. doi:10.1080/19443994.2013.769603

49. A. Zarebska, H. Karring, M.L. Christensen, M. Hjorth, K.V. Christensen, B. Norddahl, Ammonia recovery from pig slurry using a membrane contactor—Influence of slurry pretreatment, *Water. Air. Soil Pollut.* 228 (2017): 1–13. doi:10.1007/s11270-017-3332-6

50. A.C.M. Franken, J.A.M. Nolten, M.H.V. Mulder, D. Bargeman, C.A. Smolders, Wetting criteria for the applicability of membrane distillation, *J. Membr. Sci.* 33 (1987) 315–328. doi:10.1016/S0376-7388(00)80288-4

51. S. Mosadegh-Sedghi, D. Rodrigue, J. Brisson, M.C. Iliuta, Wetting phenomenon in membrane contactors – Causes and prevention, *J. Membr. Sci.* 452 (2014) 332–353. doi:10.1016/j.memsci.2013.09.055

52. E.E. Licon Bernal, A. Alcaraz, S. Casas, C. Valderrama, J.L. Cortina, Trace ammonium removal by liquid–liquid membrane contactors as water polishing step of water electrolysis for hydrogen production from a wastewater treatment plant effluent, *J. Chem. Technol. Biotechnol.* 91 (2016) 2983–2993. doi:10.1002/jctb.4923

53. E.A. Fouad, H.J. Bart, Emulsion liquid membrane extraction of zinc by a hollow-fiber contactor, *J. Membr. Sci.* 307 (2008) 156–168. doi:10.1016/j.memsci.2007.09.043

54. Z. Zhu, Z. Hao, Z. Shen, J. Chen, Modified modeling of the effect of pH and viscosity on the mass transfer in hydrophobic hollow fiber membrane contactors, *J. Membr. Sci.* 250 (2005) 269–276. doi:10.1016/j.memsci.2004.10.031

55. M. Khayet, Membranes and theoretical modeling of membrane distillation: A review, *Adv. Colloid Interface Sci.* 164 (2011) 56–88. doi:10.1016/j.cis.2010.09.005

56. M. Gryta, Osmotic MD and other membrane distillation variants, *J. Membr. Sci.* 246 (2005) 145–156. doi:10.1016/j.memsci.2004.07.029

57. E.A.A. Umweltbundesamt, Production of ammonia, nitric acid, urea and N – fertilizer – industrial processes used (2017).

58. www.fertiberia.com (2018).

59. www.agralia.com (2018).

60. www.compost.com (2018).

61. European Commission, reference document on best available techniques (BAT) for the manufacture of large volume inorganic chemicals – Ammonia, acids and fertilisers (2007).

62. X. Vecino, M. Reig, O. Gibert, C. Valderrama, J.L. Cortina. Integration of liquid-liquid membrane contactors and electrodialysis for ammonium recovery and concentration as a liquid fertilizer. *Chemosphere* 245 (2020): 125606. doi:10.1016/j.chemosphere.2019.125606

63. J.H. Urso, L.M. Gilbertson, Atom conversion efficiency: A new sustainability metric applied to nitrogen and phosphorus use in agriculture, *ACS Sustain. Chem. Eng.* 6 (2018) 4453–4463. doi:10.1021/acssuschemeng.7b03600

Part VI

Chapters on Fluid–Fluid Contacting

16 Hollow Fiber Membrane Contactors in Facilitated Transport-Based Separations – Fundamentals and Applications

M. Fallanza, A. Ortiz, D. Gorri, and I. Ortiz

CONTENTS

16.1 INTRODUCTION

Many industrial gas separations are traditionally carried out by gas–liquid absorption in direct contact equipment like spray towers or packed columns. This type of contact has several disadvantages such as emulsion formation, flooding, unloading, or foaming. In this respect gas–liquid membrane contactors, where the membrane material makes a physical separation between the gas and liquid phases, are an alternative technology that overcomes these disadvantages providing high operational flexibility, large specific area, and the capacity to control the gas as well as the liquid flow rates and pressures independently [1]. Thus, microporous membrane contactors offer a unique alternative to accomplish gas separation processes in a modular, small, robust, and safe way providing at the same time important economic advantages including low investment costs, low energy consumption, and no expensive civil engineering work for their installation [2]. The design of these systems requires the characterization of the mass transport, which is usually limited by the resistance located in the liquid phase or even in the membrane, especially when wetting phenomena take place. These resistances are commonly several orders of magnitude higher than that on the gas phase [3, 4]. This fact has motivated many research efforts aimed at improving the overall kinetics of the gas absorption in G–L membrane contactors. One of the most successful strategies consists of the inclusion of an active carrier in the liquid phase; the carrier acts as a complexing agent that reacts reversibly with the target gas molecules enhancing the mass transfer rate across the membrane and displacing the gas–liquid equilibrium.

The aim of this Chapter is to furnish the reader with the knowledge of the phenomena that control, in membrane contactors, the mass transport of a solute from a gas to a liquid phase; the physico–chemical mechanism will be described by fundamental equations giving rise to useful process design and optimization tools. The analysis reported hereinafter can be extended to any application of membrane contactors where reactive absorption of a gas into a liquid takes place.

16.2 FUNDAMENTALS

A separation process based on gas–liquid absorption involves transfer of one or more species from the gaseous phase to the liquid phase. In general, absorption processes can be categorized as physical absorption or chemical absorption. In the case of physical absorption the gaseous solute is physically dissolved in the liquid phase, whereas in the case of chemical absorption the gaseous solute reacts chemically in the liquid phase [5].

In membrane contactors, the membrane does not play an active role in the separation process; it just acts as a physical barrier to control the interface between two phases.

The gas solute diffuses through the bulk of the gas phase to the gas-membrane interface and then it crosses the membrane to the bulk of the liquid phase [6, 7]. So that there are three main resistances to mass transfer located in the gas phase, membrane, and liquid phase boundary layers that must be considered (Figure 16.1).

As mass transfer phenomena in liquid phase are commonly several orders of magnitude slower than those in gas phase, membrane wetting results in a dramatic increase of the membrane mass transfer resistance and reduces the separation performance. Therefore, membrane wetting is one of the main technical issues in the application of membrane

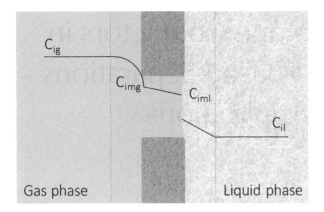

FIGURE 16.1 Concentration profile for the transport of the specie *i* from the gas phase to the liquid phase through a porous membrane.

contactors to G–L separations [8, 9]. The penetration of the liquid phase into the pores of a microporous membrane is based on the capillary forces acting inside the pores of the membrane. This wetting phenomenon is influenced by different factors such as pore size distribution, pressure difference between gas and liquid phases, surface tension of the liquid, and the interaction between the membrane and the liquid phase characterized by the contact angle. The magnitude of the breakthrough pressure, can be calculated by the Young–Laplace equation [10]:

$$\Delta P = -\frac{4 \varnothing \gamma \cos\theta}{d_p} \qquad (16.1)$$

where $\varnothing$ is the shape factor equal to 1 for cylindrical pores, γ is the surface tension of the liquid, θ is the contact angle between the liquid and the membrane, and d_p is the membrane pore diameter which is wetted at a given trans-membrane pressure ΔP.

Therefore, the wetting effects can be minimized by properly choosing the membrane materials (hydrophobic, hydrophilic, or composite) depending on the properties of the liquid phase and by establishing a small overpressure in the gas phase with respect to the liquid phase.

At steady state, the flux, dependent on the local mass transfer coefficients and concentration gradient, can be expressed as:

$$J_i = k_{i,g}\left(C_{ig} - C_{img}\right) = k_{i,m}\left(C_{img} - C_{iml}\right) = k_{il,}\left(C_{iml} - C_{il}\right) \qquad (16.2)$$

In this case, the resistances-in-series approach can be used to describe the mass transfer kinetics. Taking into account that the mass transfer between gas and liquid takes place in the gas–liquid interface, and considering a non-wetted membrane, the overall mass transport resistance can be related to the individual mass transport resistance through the resistances-in-series approach as described in **Equation 16.3** for the liquid phase flowing through the shell side of a hollow fiber membrane contactor:

$$\frac{1}{K_{i,overall}} = \frac{d_{out}}{k_{i,g}d_{in}} + \frac{d_{out}}{k_{i,mg}d_{lm}} + \frac{1}{k_{i,l}H_{di}E_A} \qquad (16.3)$$

Where $K_{i,overall}$ is the overall mass transfer coefficient and $k_{i,g}$, $k_{i,mg}$, $k_{i,l}$ are the individual mass transfer coefficients of the gas phase, the membrane when the pores are filled with gas, and the liquid phase, respectively. H_{di} represents the dimensionless Henry's constant while d_{out}, d_{in}, and d_{lm} are the outer, inner, and logarithmic mean diameters of the fibers, and E_A is the enhancement factor which takes into account the improvement in the overall mass transfer rate due to the chemical reaction.

The local mass transfer coefficient in the lumen side of the fibers can be calculated using different correlations available in the literature. The expression developed by Lévêque in the restricted range of laminar flow inside the fibers has been widely used [11].

$$Sh = 1.62 \times Gz^{0.33} \qquad (16.4)$$

However, the mass transfer rate can be underestimated for low Graetz numbers (Gz); for this reason another correlation proposed by Hausen for laminar flow and low Gz number can be also used [12]:

$$Sh = 3.66 + \frac{0.0668Gz}{1 + 0.04Gz^{2/3}} \qquad (16.5)$$

The mass transport coefficient in a dry microporous membrane depends on the diffusion coefficient of the gas solute through the gas phase inside the pores and on the characteristics of the membrane [13]:

$$k_{i,mg} = \frac{D_{i,g}\varepsilon}{\tau\delta} \qquad (16.6)$$

where ε, ζ, and δ are the porosity, tortuosity, and thickness of the membrane respectively. $D_{i,g}$ is the diffusivity of species in the gas phase inside the pores of the membrane.

In the same way the mass transport coefficient in the shell side can be estimated using different correlations available in literature [14–21]. These correlations take into account the fluid-dynamics of the liquid phase, the geometry of the system, and the rheological properties of the fluid. The general form of these correlations is:

$$Sh = f\left(\varnothing_F\right)\left(Re\right)^\alpha \left(Sc\right)^\beta \qquad (16.7)$$

When the gas solute contacts with the liquid, it is dissolved until gas–liquid equilibrium at the interface is established. The equilibrium concentration is related to the gas partial pressure by the Henry's constant:

$$H_i = \frac{C_i}{p_i} \qquad (16.8)$$

The Henry's constant has an important dependence with temperature which can be described using an Arrhenius-type equation [22]:

$$H = H_0 \cdot e^{\left(-\Delta H_{sol}/RT\right)} \quad (16.9)$$

where the parameters H_0 and $-\Delta H_{sol}$ depend on the gas to be absorbed and the liquid solvent.

The molecular diffusion coefficient of a gas solute in a gas phase $D_{A,b}$ can be calculated by the Fuller's equation [12]:

$$D_{A,b} = \frac{0.01013 \cdot T^{1.75} \left(\frac{1}{M_A} + \frac{1}{M_B}\right)^{0.5}}{P_T \cdot \left[\left(\sum \vartheta_A\right)^{1/3} + \left(\sum \vartheta_B\right)^{1/3}\right]^2} \quad (16.10)$$

where the units of T and P are K and Pa respectively, with the resulting diffusivity in $m^2\ s^{-1}$. M_A and M_B are the molecular weights in $g \cdot mol^{-1}$. $\sum \vartheta_A$ and $\sum \vartheta_B$ are the summation of atomic diffusion volumes for gases A and B.

However, when the membrane pores are filled with the gas phase, the diffusion coefficient of the solute gas A inside the pores of the membrane depends not only on the molecular diffusion, but also on the Knudsen diffusion, which takes into account the fact that the gas molecules frequently collide with the pore wall. Thus, the diffusivity of a gas solute inside the membrane pores filled with gas is a combination between molecular diffusion and Knudsen diffusion that can be calculated as follows:

$$\frac{1}{D_{A,g}} = \frac{1}{D_{A,b}} + \frac{1}{D_{A,Kn}} \quad (16.11)$$

where the Knudsen diffusion depends on the pore diameter, the temperature, and the molecular weight of the gas:

$$D_{Kn} = \frac{1}{3} d_p \sqrt{\frac{8RT}{\pi \cdot M_A}} \quad (16.12)$$

d_p is the pore diameter (m), R is the gas constant ($J\ kmol^{-1}\ K^{-1}$), T is temperature (K), and M_A is the molecular weight of the gas ($g\ mol^{-1}$).

The diffusion coefficient of a gas in a solvent can be estimated using several correlations developed for different absorption media. For instance the Wilke Chang correlation (**Equation 16.13**) can be used for conventional solvents [23] or the one developed by Scovazzo et al. (**Equation 16.14**) for more complex fluids such as ionic liquids [24].

$$D_{A,B} = \frac{7.4 \times 10^{-8} \left(\varnothing_B \cdot M_B\right)^{1/2} \cdot T}{\mu_B \cdot V_A^{0.6}} \quad (16.13)$$

$$D_{AB} = 2.66 \times 10^{-3} \frac{1}{\mu_B^{0.66 \pm 0.03} \cdot V_A^{1.04 \pm 0.08}} \quad (16.14)$$

where A and B refer to the gas and the absorption liquid, respectively. φ_B is the association factor for the solvent, M_B is the molecular weight of the solvent ($g\ mol^{-1}$), T is temperature (K); μ_B is the solvent viscosity (cP), and V_A is the molar volume of the gas at its normal boiling temperature ($cm^3\ mol^{-1}$). Diffusivity is obtained in $cm^2\ s^{-1}$.

Once the gas A is absorbed into the liquid phase, if a reacting agent (C) is present a chemical reaction may take place, giving rise to chemical absorption.

$$A + C \overset{K_{Eq}}{\longleftrightarrow} AC \quad (16.15)$$

Therefore, the chemical equilibrium constant can be defined as follows:

$$K_{Eq} = \frac{[AC]}{[A][C]} \quad (16.16)$$

The dependence of the equilibrium constant with temperature can be described in most cases by the Van't Hoff equation:

$$\frac{dLnK}{d\left(1/T\right)} = -\frac{\Delta H_r}{R} \quad (16.17)$$

where T is the temperature in K, ΔH_r is the enthalpy of reaction ($kJ\ mol^{-1}$), and R is the gas constant ($kJ\ mol^{-1}\ K^{-1}$).

The total concentration of the gas solute in the liquid phase is the sum of the concentration of free dissolved gas in the liquid and the gas molecules which have been bounded by the chemical complexing agent.

$$[A]^T = [A] + [AC] \quad (16.18)$$

A mass balance for the complexing agent can be also established:

$$[C]^T = [C] + [AC] \quad (16.19)$$

Moreover, the chemical reaction that takes place between the gas and the liquid phase can enhance the absorption flux by increasing the overall mass transfer coefficient. This improvement in the absorption rate is characterized by the enhancement factor (E_A), defined as the ratio of the absorption flux of a gas in the liquid in the presence of a chemical reaction to the absorption flux in the absence of a reaction referring these fluxes to the same driving force.

$$E_A = \frac{J_A\ \text{with reaction}}{J_A\ \text{without reaction}} \quad (16.20)$$

Several expressions to calculate the enhancement factor have been proposed in the literature. These expressions are based on the well-known relationship of van Krevelen and Hoftijzer for the film model and depend on the reaction that takes place in the liquid phase. For a second order irreversible reaction DeCoursey proposed the following approximated solution [25]:

$$E_A = -\frac{Ha^2}{2\cdot(E_{A\infty}-1)} + \sqrt{\frac{Ha^4}{4\cdot(E_{A\infty}-1)^2} + \frac{E_{A\infty}\cdot Ha^2}{(E_{A\infty}-1)} + 1} \quad (16.21)$$

In this equation the Hatta number (Ha) relates the maximum conversion in the film, and the maximum mass transport through the film, and $E_{A\infty}$ is the infinite enhancement factor, which can be calculated using the asymptotic solution proposed by Danckwerts [26].

$$Ha^2 = \frac{\dfrac{2}{n+1} k_{1,l} \cdot \left(C_A^{int,L}\right)^{n-1} \cdot C_C^m \cdot D_{A,l}}{k_l^2} \quad (16.22)$$

$$E_{A\infty} = \left(1 + \frac{D_C C_C}{D_{A,l} C_A^{int,L}}\right)\left(\frac{D_{A,l}}{D_C}\right)^{0.5} \quad (16.23)$$

In this section the fundamentals of G–L reactive absorption in membrane contactors have been presented. The effectiveness of the separation process depends on several factors, including the proper selection of membrane materials, system geometry, and physico-chemical properties of the phases involved. However it is well established that in these processes often the controlling step is the mass transfer in the liquid phase. These limitations can be minimized mainly by two different strategies: (i) Improving the fluid-dynamics of the liquid side to physically enhance the mass transport coefficient; and (ii) including in the liquid phase species that can react with the gas to be absorbed; this latter approach enhances the overall mass transfer coefficient while displacing the species equilibrium in the liquid phase contributing to an increase in the concentration gradient between both sides of the membrane.

16.3 APPLICATIONS

The use of membrane contactors for gas reactive absorption has been extensively studied in the literature for the selective absorption of different key components such as of SO_2, CO_2, CO, H_2S, NH_3, VOCs, and olefins from different streams. Among all of these applications, olefin/paraffin separation, CO_2 capture, and SO_2 removal have attracted special attention due to the importance of these separations in the industry and will be the focus of this section.

16.3.1 OLEFIN/PARAFFIN SEPARATION

Light olefins such as ethylene and propylene are the main building blocks for the production of many essential base chemicals and products for industrial and domestic applications. Olefins can be produced by different routes, but are always obtained as a mixture with their corresponding paraffins. Due to the differences in prices and also in the use of each of these compounds, the efficient separation of these

mixtures is a task of primary importance in the petrochemical industry. Traditionally, this separation is performed by distillation operating at high pressures and room temperature or moderate pressures and low temperatures with high reflux ratios. Therefore the process not only requires high capital investment, but also is one of the most highly energy demanding processes in the petrochemical industry.

The reactive absorption of olefins with transition metal cations in aqueous solutions has been extensively studied in the literature. The metal salts commonly used are selected from heavy metal ions such as Cu(I), Ag(I), Au(I), Ni(II), Pt(II), and Pd(II). The interaction is based on the formation of electron donor/acceptor complexes via π-complexation mechanism (Figure 16.2) [27].

Although all of these transition metals selectively bind to olefins, silver salts are particularly preferred since they have been proved to be the most effective ones [28]. A strong dependence has been demonstrated between silver concentration in solution and the olefin absorption capacity of the absorption medium while paraffin solubility remains mostly unaffected [29, 30] as shown in Figure 16.3.

In this system, equilibria of several reactions are possible and the stoichiometry of the complex is highly dependent upon silver and olefin concentrations.

$$Ag^+ + olefin \overset{k_1}{\leftrightarrow} \left[Ag(olefin)\right]^+$$

$$\left[Ag-olefin\right]^+ + Ag^+ \overset{k_2}{\leftrightarrow} \left[Ag_2(olefin)\right]^{2+}$$

$$\left[Ag-olefin\right]^+ + olefin \overset{k_3}{\leftrightarrow} \left[Ag(olefin)_2\right]^+$$

In dilute solutions, the formation of the 1:1 Ag(olefin)+ complex dominates the absorption process. Although at very high concentrations of the silver ion $Ag_2(olefin)^{2+}$ may be formed, it

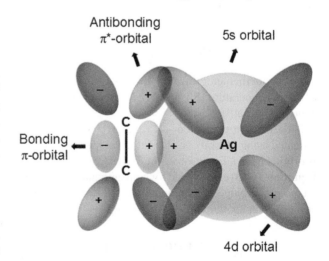

FIGURE 16.2 Dewar-Chatt model of π-bond complexation [Adapted from *Industrial and Engineering Chemistry Research 37*, Safarik, D.J., and R.B. Eldridge. Olefin/paraffin separations by reactive absorption: A review, 2571–2581, Copyright (1998) permission from American Chemical Society) [30]].

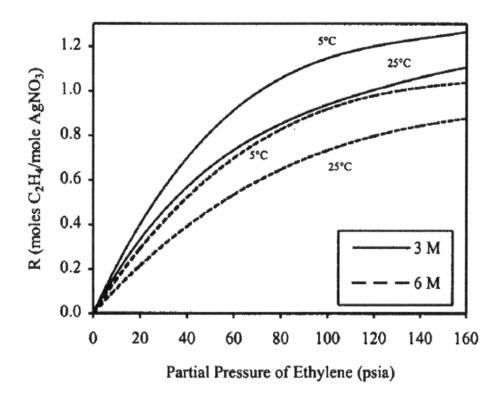

FIGURE 16.3 Ethylene absorption in aqueous $AgNO_3$ at 5°C and 25°C [Reprinted from *Industrial and Engineering Chemistry Research 37*, Safarik, D.J., and R.B. Eldridge, Olefin/paraffin separations by reactive absorption: A review, 2571–2581, Copyright (1998) permission from American Chemical Society) [30]].

can often be ignored since K_2 is typically small. On the other hand it has been reported that at high olefin partial pressures, the concentration of the Ag(olefin)$^+$ complex decreases, and the third reaction starts to occur increasing the concentration of the Ag(olefin)$_2^+$ complex (Figure 16.4) [31].

There are several studies reported in the literature regarding the olefin reactive absorption using silver cations as complexing agent. Some of them using aqueous solutions, while others have evaluated the use of advanced reaction media to improve the absorption capacity and separation selectivity. Ortiz et al. [32] evaluated the enhancement factor due to chemical complexation for propylene using silver cations as complexing agent in ionic liquid as reaction medium (Table 16.1).

Due to the great potential of reactive absorption to carry out the separation of olefin/paraffin gas mixtures some authors have evaluated the experimental implementation of this separation in membrane contactors. Fallanza et al. [33] evaluated the selective absorption of propylene in Ag$^+$-BMImBF$_4$ medium using transverse flow membrane contactors and obtained an improvement of 17.4 times in terms of mass transfer per specific area compared to a conventional stirred tank reactor system (Table 16.2).

Ghasem et al. [27] evaluated the absorption of ethylene using aqueous silver nitrate solutions (1–6M) obtaining absorption fluxes around $2.5 \cdot 10^{-3}$ mol·m^{-2}·s^{-1} and ethylene purities after the desorption stage over 98%. Additionally Faiz et al. [9] evaluated the long-term stability of highly hydrophobic polymeric hollow fibers such as PVDF and PTFE for propylene/propane separation using $AgNO_3$ aqueous solutions

as absorbent. Experimental results revealed that despite the promising results obtained, the process still suffers from Ag$^+$ inherent instability issues that must be overcome before shifting into larger scale applications.

16.3.2 CO_2 Capture

In the last decades there is a growing concern that anthropogenic CO_2 emissions are contributing dramatically to global climate change. In order to mitigate this impact and reduce the CO_2 concentration in the atmosphere, carbon capture is an upcoming technology under development. Chemical absorption, adsorption, cryogenic separation, and membrane based separation are some of the investigated technologies that are already being explored by several researchers to capture CO_2 [34,35]. Although recent studies [36–38] have proposed the use of novel solvents for CO_2 capture in membrane contactors, so far amine solutions have been identified as the most potentially viable absorbents for CO_2 because of cost effective regeneration, high CO_2 capacity, enhanced stability, and high CO_2 absorption rates [39]. For this system the absorption rate is improved to the chemical interaction between dissolved CO_2 and amines. According to some kinetic studies, chemical reactions in CO_2-amine aqueous systems are quite complex [40]; however, commonly the chemical reaction between CO_2 and the amine can be assumed to be pseudo first order reaction due to excess concentration of the amine sorbent in the reaction medium [40,41].

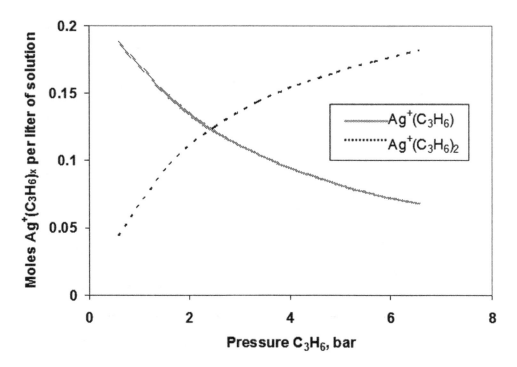

FIGURE 16.4 Distribution of primary and secondary complexes in propylene absorption in $AgBF_4$ / $BMImBF_4$ solutions at 298 K [Reprinted from *Separation and Purification Technology,63,* Ortiz, A., A. Ruiz, D. Gorri, and I. Ortiz. Room temperature ionic liquid with silver salt as efficient reaction media for propylene/propane separation: Absorption equilibrium, 311–318, Copyright (2008) permission from Elsevier) [31].

TABLE 16.1

Experimental E_A for Ag^+ [0.1–0.25 M]- $BMImBF_4$ System at Different Temperatures in a Stirred Cell Reactor at 500 rpm [32]

T (K)	$[AgBF_4]$ (mol L^{-1})	E_A	K_{Eq} (L mol^{-1})
288	0.1	1.81	285.8
	0.25	4.32	
298	0.1	2.45	245.1
	0.25	6.64	
308	0.1	3.05	212.2
	0.25	6.25	

Source: Data from Separation and Purification Technology 73, Ortiz, A., L.M. Galán, D. Gorri, A.B. De Haan, and I. Ortiz. 2010. Kinetics of reactive absorption of propylene in RTIL-Ag^+ media. Separation and Purification Technology, 106–113, Copyright (2010) with permission from Elsevier [32].

Zhang et al. [42] evaluated the implementation of CO_2 capture by DMEA/MEA aqueous solutions in hollow fiber membrane contactors. They studied the performance of different absorbent compositions under different operational conditions and found a synergistic effect between DMEA/MEA obtaining the highest cyclic CO_2 capacities for the 1.0 M DMEA + 1.0 M MEA solution. Hoff et al. [43] evaluated the performance of membrane contactors for CO_2 capture

using MEA as the liquid absorbent. One of the key issues identified was the need to flow the liquid phase on the shell side to improve the fluid dynamics and increase the overall CO_2 mass transfer rate. They conclude that membrane contactors may provide significant improvements in offshore CO_2 capture, both from gas turbine flue gas and in the sweetening of natural gas (Table 16.3).

In addition, several authors [2, 44–47] have studied the performance of CO_2 reactive absorption in G–L membrane contactors in comparison to conventional wet scrubbers. Typically this performance is given by the combination of the overall mass transfer coefficient and the specific area ($K_{overall}$·a). It has been observed that the $K_{overall}$·a values obtained with G–L membrane contactors can be up to 10 times higher than those observed in conventional direct contact equipment (Table 16.4).

16.3.3 SO_2 Removal

Sulfur oxides (SO_x) are the major air pollutant which is emitted from the stationary sources such as power plants, incinerators, and combustors [44, 48]. It is well known that sulfur dioxide emissions cause a direct atmospheric impact leading to acid rain, photochemical smog, and harm to human health and ecosystems [49, 50]. The most effective technology for SO_2 removal is the flue gas desulphurization (FGD) process using limestone slurry as a scrubbing solution. This process has been widely accepted because of its low cost, simple operation, and high SO_2 removal efficiency

TABLE 16.2

Comparison of the Performance of Propylene Reactive Absorption in Different Contactors [33]

	a (m^{2}m^{-3})	$K_{overall}$ (ms^{-1}) *10^6	$K_{overall}*a$ (s^{-1})*10^4
Stirred tank reactor [Ag$^+$= 0.25 M]	212	111	236
Microdyn MD020 TP 2N [Ag$^+$= 0.25 M]	313	3.4	10.7
Liqui-Cell 2.5x8 Extra-Flow [Ag$^+$= 0.25 M]	3500	117	4109

Source: Adapted from Journal of Membrane Science, 385–386, Fallanza, M., Ortiz, A. Gorri, D. and Ortiz, I., Improving the mass transfer rate in G-L membrane contactors with ionic liquids as absorption medium. Recovery of propylene, 217–225, Copyright (2011) with permission from Elsevier [33].

TABLE 16.3

Performance Comparison Between Packed Towers and G–L Membrane Contactors for CO$_2$ Absorption for Post Combustion Processes and Natural Gas Streams [43]

Case	Post Combustion	Post Combustion	NG, Pipeline	NG, Pipeline	NG, LNG
Solvent	30% MEA	30% MEA	30% MDEA	30% MDEA	30% MDEA 5%Pz
Liquid flow mode	Shell side	Tube side	Shell side	Tube side	Shell side
Module width (m)	3.0×1.0*	3.0x1.0*	2.10	2.40	2.40
Fiber length (m)	3.0	3.0	5.0	4.5	5.0
# modules in parallel	25	40	1	2	1
# modules in series	3	5	3	7	5
Pressure drop, gas (kPa)	49.20	5.7	2.32	10.54	1.03
Pressure drop, liq. (kPa)	N/A	37.1	N/A	111.6	N/A
Superficial gas vel. (m s^{-1})	1.43	0.93	0.22	0.07	0.10
Total packed volume (m^3)	675	1800	52	285	113
Volume ratio vs. tower	0.25	0.67	0.25	1.37	0.46

(*Cross flow modules, W=3m, H=3m, D=1m)

Source: Adapted from *Energy Procedia* 37, Hoff, Karl Anders, and Hallvard F Svendsen. 2013. "CO$_2$ absorption with membrane contactors vs. packed absorbers – Challenges and opportunities in post combustion capture and natural gas sweetening, 952–960, Copyright (2013) with permission from Elsevier [43].

compared to other alternatives. However, these processes require large equipment so in the near future are meant to be replaced by more compact, cheap, and modular devices such as hollow fiber membrane contactors [51, 52].

Park et al. [53] evaluated SO$_2$ removal efficiencies and mass transfer coefficients in PVDF hollow fiber membrane contactors. The absorption efficiency of aqueous solutions of various chemicals such as Na$_2$SO$_3$, Na$_2$CO$_3$, NaHCO$_3$, and NaOH were compared, finding that Na$_2$CO$_3$ was the most promising absorbent among them (Figure 16.5).

The chemical reactions involved in the SO$_2$ reactive absorption using different reactant species are as follows:

$$SO_2 + 2NaOH \xrightarrow{k_1} Na_2SO_3 + H_2O$$

$$SO_2 + Na_2SO_3 \xrightarrow{k_2} H_2O + 2NaHSO_3$$

$$SO_2 + Na_2CO_3 \xrightarrow{k_3} Na_2SO_3 + CO_2$$

$$SO_2 + 2NaHCO_3 \xrightarrow{k_4} Na_2SO_3 + H_2O + 2CO_2$$

Jeon et al. [54] studied the separation performance of SO$_2$ using membrane contactors with different membrane materials under different operational conditions. The breakthrough pressure through PTFE membranes was observed to be higher due to its superior hydrophobicity being the most suitable material to prevent membrane wetting, thus providing long-term stability for a real implementation of this technology. Luis et al. [55] evaluated in addition to technical considerations, environmental and economic issues in the use of a ceramic hollow fiber membrane contactor to recover sulfur dioxide from gas streams. They concluded that the use of non-dispersive gas absorption in membrane

TABLE 16.4

Comparison of $K_{overall} \cdot a$ Values Between G–L Membrane Contactors and Conventional Absorption Equipment for CO_2 Capture [57]

Application	Ka Value for Membrane Module	Ka Value for Conventional Absorption Packed Tower
Absorption of CO_2 in water from air	0.12–0.25 s⁻¹	0.01–0.18 s⁻¹
Absorption of CO_2 in monoethanolamine aqueous solution from air	1.3–4.0 kmol/(m³hkPa)	1.1–1.2 kmol/(m³hkPa)
Absorption of CO_2 from flue gas	4.3 s⁻¹	0.47 s⁻¹
Absorption of CO_2 in monoethanolamine aqueous solution from flue gas	8.93×10^{-4} $- 7.53 \times 10^{-3}$ mol/(m³sPa)	2.25×10^{-4} mol/(m³sPa)
Absorption of CO_2 in diethanolamine aqueous solution from air	0.126–0.43 s⁻¹	0.05 s⁻¹

Source: Adapted from Romero and Estay 2018. Membrane gas absorption processes: Applications, design and perspectives. In *Osmotically driven membrane processes – Approach, development and current status*, ed. H. Du, A. Thompson and X. Wang, chapter 12, 255–272 **]57]** IntechOpen.

contactors can increase the competitiveness of this technology that is strongly dependent on the environmental restrictions as well as the SO_2 concentration in the gas stream. Other studies [44, 56] compared the SO_2 absorption efficiency of G–L membrane contactors and conventional absorption equipment, demonstrating the superior performance of membrane contactors in terms of overall mass transfer coefficient per specific area (Table 16.5).

16.4   CONCLUSIONS

In the last decades the implementation of gas separations by reactive absorption in G–L membrane contactors has been successfully explored. The modular concept of membrane devices, together with the high specific surface areas have been combined with the selectivity given by the introduction of a selective complexing agent in the absorption media promoting the development of a new highly efficient, compact, and flexible separation technology that offers great advantages over conventional direct contact equipment. The knowledge acquired so far has provided insights about the mechanisms governing the performance of this technology and has brought full-scale G–L membrane absorbers that are nowadays successfully operating under industrial conditions (Klaassen et al. 2005). Nevertheless, new efforts to increase the competitiveness of this technology are still needed. The development of cheaper membrane materials with improved wetting resistance, the optimization of the

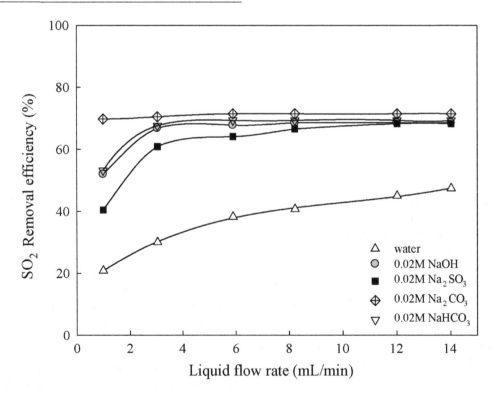

FIGURE 16.5   Effect of absorbent species on SO_2 removal efficiency at different flow rates [Reprinted from *Journal of Membrane Science* 319, Park, H.H., B.R. Deshwal, I.W. Kim, and H.K. Lee. Absorption of SO_2 from flue gas using PVDF hollow fiber membranes in a gas–liquid contactor.29–37, Copyright (2008) permission from Elsevier [53]].

TABLE 16.5

Comparison of $K_{overall} \bullet a$ Values Between G–L Membrane Contactors and Conventional Absorption Equipment for SO_2 Absorption [57]

Application	Ka Value for Membrane Module	Ka Value for Conventional Absorption Packed Tower
Absorption of SO_2 in water from air	0.10–0.13 s^{-1}	0.01–0.04 s^{-1}

Source: Adapted from Romero and Estay 2018. Membrane gas absorption processes: Applications, design and perspectives. In *Osmotically driven membrane processes – Approach, development and current status*, ed. H. Du, A. Thompson and X. Wang, chapter 12, 255–272, [57]. IntechOpen.

reactive absorbent in terms of absorption capacity, complexing agent chemical stability, solvent losses and energetic efficiency for each application, and the optimization of the coupling between absorption and desorption stages is still critical in order to increase the industrial penetration of this technology.

REFERENCES

1. Chabanon, E., and E. Favre. 2017. Membranes contactors for intensified gas–liquid absorption processes. In *Comprehensive Membrane Science and Engineering*, eds. E. Drioli, L. Giorno, and E. Fontananova, chapter 3.9, 249–81. Amsterdam, The Netherlands: Elsevier.
2. Klaassen, R., P.H.M. Feron, and A.E. Jansen. 2005. Membrane contactors in industrial applications. *Chemical Engineering Research and Design* 83: 234–46.
3. Kamo, J., T. Hirai, and K. Kamada. 1992. Solvent-induced morphological change of microporous hollow fiber membranes. *Journal of Membrane Science* 70: 217–24.
4. Barbe, A.M., P.A. Hogan, and R.A. Johnson. 2000. Surface morphology changes during initial usage of hydrophobic, microporous polypropylene membranes. *Journal of Membrane Science* 172: 149–56.
5. Dindore, V.Y., D.W.F. Brilman, and G.F. Versteeg. 2005. Hollow fiber membrane contactor as a gas-liquid model contactor. *Chemical Engineering Science* 60: 467–79.
6. Drioli, E., A. Criscuoli, and E. Curcio. 2006. *Membrane Contactors: Fundamentals, Applications and Potentialities*. Amsterdam, The Netherlands: Elsevier.
7. Sirkar, K.K. 2008. Membranes, phase interfaces, and separations: Novel techniques and membranes–An overview. *Industrial and Engineering Chemistry Research* 47: 5250–266.
8. Naim, R., and S. Abdullah. 2015. Membrane wetting in membrane contactor system: A review. *Journal of Applied Membrane Science & Technology* 17: 9–15.
9. Faiz, R., M. Fallanza, S. Boributh, R. Jiraratananon, I. Ortiz, and K. Li. 2013. Long term stability of propylenene/propane separation using AgNO₃ solutions. *Chemical Engineering Science* 94: 108–19.
10. Pabby, A.K., S.S.H. Rizvi, and A.M. Sastre. 2015. *Handbook of Membrane Separations*. 2nd ed. Boca Raton, FL: CRC Press.
11. Skelland, A.H.P. 1974. *Diffusional Mass Transfer*. New York, NY: Wiley.
12. Perry, R.H., and D.W. Green. 2008. *Perry's Chemical Engineers' Handbook*. 8th ed. New York, NY: McGraw-Hill.
13. Gabelman, A., and S.-T. Hwang. 1999. Hollow fiber membrane contactors. *Journal of Membrane Science* 159 (1–2): 61–106.
14. Shen, S., S.E. Kentish, and G.W. Stevens. 2010. Shell-side mass-transfer performance in hollow-fiber membrane contactors. *Solvent Extraction and Ion Exchange* 28: 817–44.
15. Baudot, A., J. Floury, and H.E. Smorenburg. 2001. Liquid-liquid extraction of aroma compounds with hollow fiber contactor. *AIChE Journal* 47 (8): 1780–793.
16. Fouad, E.A., and H.J. Bart. 2007. Separation of zinc by a non-dispersion solvent extraction process in a hollow fiber contactor. *Solvent Extraction and Ion Exchange* 25: 857–77.
17. Li, K., M.S.L. Tai, and W.K. Teo. 1994. Design of a CO₂ scrubber for self-contained breathing systems using a microporous membrane. *Journal of Membrane Science* 86 (1–2): 119–25.
18. Costello, M.J., A.G. Fane, P.A. Hogan, and R.W. Schofield. 1993. The effect of shell side hydrodynamics on the performance of axial flow hollow fibre modules. *Journal of Membrane Science* 80: 1–11.
19. Wu, J., and V. Chen. 2000. Shell-side mass transfer performance of randomly packed hollow fiber modules. *Journal of Membrane Science* 172: 59–74.
20. Prasad, R., and K.K. Sirkar. 1988. Dispersion-free solvent extraction with microporous hollow-fiber modules. *AIChE Journal* 34: 177–88.
21. Yang, M.-C., and E.L. Cussler. 1986. Designing hollow-fiber contactors. *AIChE Journal* 32: 1910–916.
22. Hansen, K.C., Z. Zhou, C.L. Yaws, and T.M. Aminabhavi. 1993. Determination of Henry's Law constants of organics in dilute aqueous solutions. *Journal of Chemical and Engineering Data* 38: 546–50.
23. Wilke, C.R., and P. Chang. 1955. Correlation of diffusion coefficients in dilute solutions. *AIChE Journal* 1: 264–70.
24. Morgan, D., L. Ferguson, and P. Scovazzo. 2005. Diffusivities of gases in room-temperature ionic liquids: Data and correlations obtained using a lag-time technique. *Industrial and Engineering Chemistry Research* 44: 4815–823.
25. DeCoursey, W.J. 1974. Absorption with chemical reaction: Development of a new relation for the danckwerts model. *Chemical Engineering Science* 29: 1867–872.
26. Danckwerts, P.V. 1970. *Gas-Liquid Reactions*. New York, NY: McGraw-Hill.
27. Ghasem, N., M. Al-Marzouqi, and Z. Ismail. 2014. Gas-liquid membrane contactor for ethylene/ethane separation by aqueous silver nitrate solution. *Separation and Purification Technology* 127: 140–48.
28. Reine, T.A., and R.B. Eldridge. 2005. Absorption equilibrium and kinetics for ethylene-ethane separation with a novel solvent. *Industrial and Engineering Chemistry Research* 44: 7505–510.
29. Keller, G.E., A.E. Marcinkowsky, S.K. Verma, and K.D. Williamson. 1992. Olefin recovery and purification via silver complexation. In *Separation and Purification Technology*, eds. N.N. Li and J. MCalo, 59–83. New York, NY: Marcel Dekker.

30. Safarik, D.J., and R.B. Eldridge. 1998. Olefin/paraffin separations by reactive absorption: A review. *Industrial and Engineering Chemistry Research* 37: 2571–581.

31. Ortiz, A., A. Ruiz, D. Gorri, and I. Ortiz. 2008. Room temperature ionic liquid with silver salt as efficient reaction media for propylene/propane separation: Absorption equilibrium. *Separation and Purification Technology* 63: 311–18.

32. Ortiz, A., L.M. Galán, D. Gorri, A.B. De Haan, and I. Ortiz. 2010. Kinetics of reactive absorption of propylene in RTIL-Ag⁺ media. *Separation and Purification Technology* 73: 106–13.

33. Fallanza, M., A. Ortiz, D. Gorri, and I. Ortiz. 2011. Improving the mass transfer rate in G-L membrane contactors with ionic liquids as absorption medium. Recovery of propylene. *Journal of Membrane Science* 385–386: 217–25.

34. Figueroa, J.D., T. Fout, S. Plasynski, H. McIlvried, and R.D. Srivastava. 2008. Advances in CO_2 capture technology-The U.S. Department of Energy's Carbon Sequestration Program. *International Journal of Greenhouse Gas Control* 2: 9–20.

35. Zhao, S., P.H.M. Feron, L. Deng, E. Favre, E. Chabanon, S. Yan, J. Hou, V. Chen, H. Qi. 2016. Status and progress of membrane contactors in post-combustion carbon capture: A state-of-the-art review of new developments. *Journal of Membrane Science* 511: 180–206.

36. Aghaie, M., N. Rezaei, and S. Zendehboudi. 2018. A systematic review on CO_2 capture with ionic liquids: Current status and future prospects. *Renewable and Sustainable Energy Reviews* 96: 502–25.

37. Zhang, Y., X. Ji, and X. Lu. 2018. Choline-based deep eutectic solvents for CO_2 separation: Review and thermodynamic analysis. *Renewable and Sustainable Energy Reviews* 97: 436–55.

38. Albo, J., P. Luis, and A. Irabien. 2011. Absorption of coal combustion flue gases in ionic liquids using different membrane contactors. *Desalination and Water Treatment* 27 (1–3): 54–9.

39. Kadiwala, S., A.V. Rayer, and A. Henni. 2012. Kinetics of carbon dioxide (CO_2) with ethylenediamine, 3-amino-1-propanol in methanol and ethanol, and with 1-dimethylamino-2-propanol and 3-dimethylamino-1-propanol in water using stopped-flow technique. *Chemical Engineering Journal* 179: 262–71.

40. Sema, T., A. Naami, K. Fu, M. Edali, H. Liu, H. Shi, Z. Liang, R. Idem, and P. Tontiwachwuthikul. 2012. Comprehensive mass transfer and reaction kinetics studies of CO_2 absorption into aqueous solutions of blended MDEA–MEA. *Chemical Engineering Journal* 209: 501–12.

41. Ali, S.H. 2005. Kinetics of the reaction of carbon dioxide with blends of amines in aqueous media using the stopped-flow technique. *International Journal of Chemical Kinetics* 37 (7): 391–405.

42. Zhang, P., R. Xu, H. Li, H. Gao, and Z. Liang. 2019. Mass transfer performance for CO_2 absorption into aqueous blended DMEA/MEA solution with optimized molar ratio in a hollow fiber membrane contactor. *Separation and Purification Technology* 211: 628–36.

43. Hoff, K.A., and H.F. Svendsen. 2013. CO_2 absorption with membrane contactors vs. packed absorbers–Challenges and opportunities in post combustion capture and natural gas sweetening. *Energy Procedia* 37: 952–60.

44. Karoor, S., and K.K. Sirkar. 1993. Gas absorption studies in microporous hollow fiber membrane modules. *Industrial and Engineering Chemistry Research* 32: 674–84.

45. deMontigny, D., P. Tontiwachwuthikul, and A. Chakma. 2005. Comparing the absorption performance of packed columns and membrane contactors. *Industrial & Engineering Chemistry Research* 44: 5726–732.

46. Nishikawa, N., M. Ishibashi, H. Ohta, N. Akutsu, H. Matsumoto, T. Kamata, and H. Kitamura. 1995. CO_2 removal by hollow-fiber gas-liquid contactor. *Energy Conversion and Management* 36: 415–18.

47. Rangwala, H.A. 1996. Absorption of carbon dioxide into aqueous solutions using hollow fiber membrane contactors. *Journal of Membrane Science* 112: 229–40.

48. Qi, Z., and E.L. Cussler. 1985. Microporous hollow fibers for gas absorption. I. Mass transfer in the liquid. *Journal of Membrane Science* 23: 321–32.

49. Rahmani, F., D. Mowla, G. Karimi, A. Golkhar, and B. Rahmatmand. 2014. SO_2 removal from simulated flue gas using various aqueous solutions: Absorption equilibria and operational data in a packed column. *Separation and Purification Technology* 153: 162–69.

50. Bokotko, R.P., J. Hupka, and J.D. Miller. 2005. Flue gas treatment for SO_2 removal with air-sparged hydrocyclone technology. *Environmental Science and Technology* 39: 1184–189.

51. Mansourizadeh, A., and A.F. Ismail. 2009. Hollow fiber gas-liquid membrane contactors for acid gas capture: A review. *Journal of Hazardous Materials* 171: 38–53.

52. Bazhenov, S.D., A.V. Bildyukevich, and A.V. Volkov. 2018. Gas-liquid hollow fiber membrane contactors for different applications. *Fibers* (MDPI) 6: 76.

53. Park, H.H., B.R. Deshwal, I.W. Kim, and H.K. Lee. 2008. Absorption of SO_2 from flue gas using PVDF hollow fiber membranes in a gas-liquid contactor. *Journal of Membrane Science* 319: 29–37.

54. Jeon, H., H. Ahn, I. Song, H.-K. Jeong, Y.L. Lee, and H.-K. Lee. 2008. Absorption of sulfur dioxide by porous hydrophobic membrane contactor. *Desalination* 234 (1–3): 252–60.

55. Luis, P., A. Garea, and A. Irabien. 2012. Environmental and economic evaluation of SO_2 recovery in a ceramic hollow fibre membrane contactor. *Chemical Engineering and Processing: Process Intensification* 52: 151–54.

56. Yu, H., Q. Li, J. Thé, and Z. Tan. 2016. Absorption of sulfur dioxide in a transversal flow hollow fiber membrane contactor. In *Clean Coal Technology and Sustainable Development*, ISCC 2015, eds. G. Yue, and S. Li, 393–99. City of Singapore, Singapore: Springer.

57. Romero, J., and H. Estay. 2018. Membrane gas absorption processes: Applications, design and perspectives. In *Osmotically Driven Membrane Processes – Approach, Development and Current Status*, eds. H. Du, A. Thompson, and X. Wang, chapter 12, 255–72. Rijeka, Coratia: IntechOpen.

17 Membrane Distillation, Membrane Crystallization, and Membrane Condenser

M. Frappa, F. Macedonio, and E. Drioli

CONTENTS

17.1 INTRODUCTION

Membrane contactors are membrane-based systems, in which microporous and hydrophobic membranes are employed for promoting the mass transfer and/or energy transfer between phases. Membrane contactor technology has been demonstrated in several fields, such as in liquid/liquid metal ion separation, as a powerful approach to improve the performance of solvent extraction processes [1,2]. Other systems based on contactor devices and equipped with well-controlled membranes are largely used in semiconductors [3], air and water purification [4,5], pharmaceuticals [6], biotechnology [7,8] and metal mass transfer operations [9]. In this context, membrane operations have the potential to replace conventional energy intensive techniques, accomplish the selective and efficient transport of specific components, improve the performance of reactive processes, and, ultimately, provide reliable options for a sustainable industrial growth [10]. In particular, the reliability of membrane contactor (MC) technology is today increasing in many industrial processes where traditional membrane separation units, such as reverse osmosis (RO), microfiltration (MF), ultrafiltration (UF), nanofiltration

(NF), electrodialysis (ED), pervaporation (PV), etc., are already routinely utilized.

In this chapter three different membrane contactor processes will be discussed: membrane distillation (MD), membrane assisted crystallization (MCr), and membrane assisted condenser (MCo). The driving force of MD process is the difference of vapor pressure existing between the two membrane surfaces. It forces only vapor molecules to pass through a porous hydrophobic membrane. The higher the gradient, the higher the distillate production will be [11–13]. MD is a promising technology, for example, for desalting highly saline waters. MCr is an extension of the previous MD process, in which a simultaneous separation and purification of chemicals species from solution take place. The process has been applied in food, pharmaceutical, chemical, and environmental divisions of process [11,12]. The evaporative mass transfer of volatile solvents through the microporous hydrophobic membranes is utilized to concentrate feed solutions above their saturation limit, thus attaining a supersaturated environment where crystals may nucleate and grow [14]. Membrane condenser is a new and innovative membrane operation for the selective recovery of liquid water from humid waste gaseous streams [15]. Like the two previous processes, it bases its principle of operation in the use of hydrophobic and microporous membranes. The humid gaseous stream, at a certain temperature, is brought into contact with a microporous hydrophobic membrane on which the water vapor and other condensable components can be transformed in the liquid phase at an appropriate lower temperature. The hydrophobic nature of the membrane inhibits the liquid from penetrating into the membrane pores. The liquid water is therefore retained on the retentate side of the membrane [15–18].

The performances of MCs strongly depend on the properties of the used membranes [19]. In general, high hydrophobicity is required in order to prevent wetting and mixing between contacting phases; elevated porosity leads to high fluxes, but might cause bubble coalescence in gas–liquid operations [10].

In this chapter the transport principles and the applications of MD, MCr, and MCo processes will be investigated and analysed.

17.2 MEMBRANE DISTILLATION

17.2.1 MD Process

As earlier described, membrane distillation is a thermally-driven, promising separation process that involves the use of microporous hydrophobic membranes. MD differs from reverse osmosis (RO) where the absolute pressure difference is the driving force for mass transfer. MD process is generally applied for desalination and treatment of water solutions. Theoretically, it can reject 100% of the salts and non-volatile components of the feed. In this process, separation is carried out based mainly on the two-phase changes, evaporation, and condensation. Evaporation occurs at the liquid–vapor interfaces formed at one side of no-wetted pores of a hydrophobic membrane (i.e., feed side), so that only water vapor is allowed to pass through the membrane's pores. Successively, condensation will occur and it can take place inside or outside the membrane module, depending on the MD variant. Other advantages of this process are lower operating temperatures with respect to conventional distillation columns, reduced influence of concentration polarization with respect to pressure-driven membrane processes, and lower operating pressure with respect to reverse osmosis.

The understanding of heat and mass transfer phenomena in MD allows identifying membrane characteristics for enhancing MD performance.

Regardless of the MD configuration used, water and solute (if the solute is volatile) evaporate from the liquid–vapor interface on the feed side of the membrane, diffuse and/or convect across the membrane, and are either condensed or are removed from the membrane module as vapor on the permeate side.

For what concerns heat transfer, heat is first transferred from the heated feed solution of uniform temperature T_f across the thermal boundary layer to the membrane surface at a rate $Q = h_f \cdot \Delta T_f$.

At the surface of the membrane, liquid is vaporized and heat is transferred across the membrane at a rate $Q_V = h_V \cdot \Delta T_m = N \cdot \Delta H_V$ (where N is the rate of mass transfer and ΔH_V is the heat of vaporization). Additionally, heat is conducted through the membrane material and the vapor that fills the pores at a rate $Q_m = h_m \cdot \Delta T_m$ where $h_m = \varepsilon\, h_{mg} + (1-\varepsilon)\, h_{ms}$ (ε is the membrane porosity, h_{mg} and h_{ms} represent the heat transfer coefficients of the vapor within the membrane pores and the solid membrane material, respectively). Conduction is considered a heat loss mechanism, because no corresponding mass transfer takes place. Total heat transfer across the membrane is $Q = Q_V + Q_m$. Finally, as vapor condenses at the liquid–vapor interface, heat is removed from the cold-side membrane surface through the thermal boundary layer at a rate $Q = h_p \cdot \Delta T_p$ (Figure 17.1).

The mass transport model of membrane distillation suggests that the mass transport rate or water flux is proportional to vapor pressure difference across the membrane

$$J = C*(P_F - P_p) = C\Delta P \qquad (17.1)$$

where C is a membrane distillation coefficient which depends on the membrane morphology (e.g., pore size, porosity, membrane thickness, and tortuosity):

$$C \propto \left(r^a \varepsilon\right)/\left(bt\right) \qquad (17.2)$$

r is the nominal pore size of membrane, a is the exponent of r in the range of 1 to 2 (a=1 for Knudsen diffusion, a=2 for viscous flux), t is the pore tortuosity, and b is the membrane thickness.

The terms "temperature polarization coefficient" (TPC) and "concentration polarization coefficient" (CPC) are defined to quantify the heat and mass transport resistance within the boundary layer with respect to the total transfer resistance of the system:

$$TPC = \left(T_{fm} - T_{pm}\right)/\left(T_f - T_p\right) \qquad (17.3)$$

$$CPC = \left(C_{Bm}\right)/\left(C_{Bbf}\right) \qquad (17.4)$$

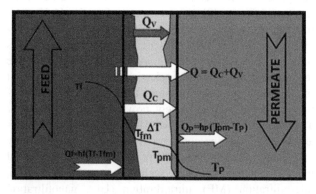

FIGURE 17.1 Heat and mass transfer in MD.

where T_{fm} is the temperature of the feed at the membrane surface, T_{pm} is the temperature of the permeate at the membrane surface, T_f is the temperature of the feed in the bulk, T_p is the temperature of the permeate in the bulk, C_{Bm} is the concentration of the non-volatile solutes at the membrane surface, whereas C_{Bbf} is the concentration of the non-volatile solutes at the bulk feed. As a result of the solvent trans-membrane flux across the membrane, C_{Bm} becomes higher than C_{Bbf} as long as the separation process takes place. The increased concentration of non-volatile compounds next to the membrane surface would have the influence of reducing the trans-membrane flux due to the establishment of a concentration polarization (CP) layer at the feed side that acts as a mass transfer resistance to the volatile molecule species (water).

In MD process, heat transfer across the boundary layers is often the rate limiting steps because a large quantity of heat must be supplied to the membrane surface to vaporize the liquid. Therefore, MD process is generally not limited by mass transfer across the membrane but it is limited by heat transfer through the boundary layers on either side of the membrane.

Polarization being a phenomenon mainly related to fluid dynamics, the only way to reduce it is to generate sufficient mixing and turbulence in the flow thus reducing the thickness of the boundary layer. To do this there are two options: either (1) increasing the flow rates (even if this increase is not to produce an operating pressure above the liquid entry pressure (LEP) for preventing wetting), or (2) including the spacers for promoting turbulence [20].

In addition to the transport resistance offered by boundary layers and membrane, other factors that can induce a reduction in trans-membrane (or even block) flux and/or membrane damage are: membrane wetting, scaling, and fouling.

Membrane wetting occurs when the liquid penetrates into and through the membrane pores. Once a pore has been penetrated it is said to be "wetted" and the membrane must be completely dried and cleaned before the wetted pores can once again support a vapor–liquid interface. The Laplace (Cantor) equation provides the relationship between the membrane's largest allowable pore size (r_{max}) and the related operating conditions [21,22]:

$$LEP = (-2B\, \gamma_L \cos\theta)/r_{max} \qquad (17.5)$$

where B is a geometric factor determined by pore structure, γ_L the liquid surface tension, and θ is the liquid/solid contact angle. When the hydrostatic pressure on the feed side of an MD membrane exceeds LEP, liquid penetrates the pores and is able to pass through the membrane.

For what concerns fouling, it is the deposition and/or an adsorption of certain feed constituents at the membrane surface. Membrane fouling causes a decline in flux over time when all operating parameters, such as pressure, flow rate, temperature, and feed concentration are kept constant. The deterioration of flux due to the fouling layer may be attributed to two effects: first, the reduction of surface area available for vaporization as the fouling layer blocked the pore entrances; second, the fouling layer may decrease the heat transfer driving force. In the MD process membrane fouling is not a relevant problem such as in other membrane separations but fouling particles/molecules attached to the membrane surface might cause plugging of the membrane pore entrances causing first some flux decay and, then, may lead to membrane pore wetting. In fact, the increased deposition of the fouling species at the membrane surface might eventually lead to an increase in the pressure drop to levels at which the hydrostatic pressure may exceed the LEP.

17.2.2 MD Modules and Configuration Process

In MD processes, the hot feed solution is in contact with one side of the membrane. Depending on the nature of the permeate side, the main MD configurations (Figure 17.2) can be identified as follows:

Direct Contact Membrane Distillation (DCMD): a cold, pure water solution is circulated tangentially to the permeate side of the membrane. The trans-membrane temperature difference induces the necessary vapor pressure difference. In this case the volatile molecules of the feed aqueous solution (e.g., water, volatile organic compounds, etc.) evaporate at the hot feed liquid–vapor interface, diffuse and/or convect across the membrane in vapor phase, and condense at the cold liquid–vapor interface inside the membrane module [23]. During the process, the concentration of the volatile solute in the permeate aqueous solution will increase. In general, DCMD is used for production of distilled/potable water using feed aqueous solutions containing non-volatile solutes, e.g., desalination.

Air Gap Membrane Distillation (AGMD): In this case an air gap is interposed between the membrane and a cold, condensing surface placed inside the membrane module. The air gap helps in increasing the conductive heat transfer resistance thus decreasing, with respect to DCMD, the amount of heat lost by conduction through the membrane [24]. However, because the permeate flux has to overcome the air barrier between the membrane and the condensing surface, AGMD fluxes are typically reduced depending on the effective air gap width [25]. Xu et al. [26] showed that, because the air gap width is much larger than the membrane thickness, the effect of membrane thickness and membrane porosity in AGMD can be neglected due to the stagnant air gap representing the predominant resistance to mass transfer. Among the various MD configurations, AGMD is the most versatile and can be applied to almost any applications.

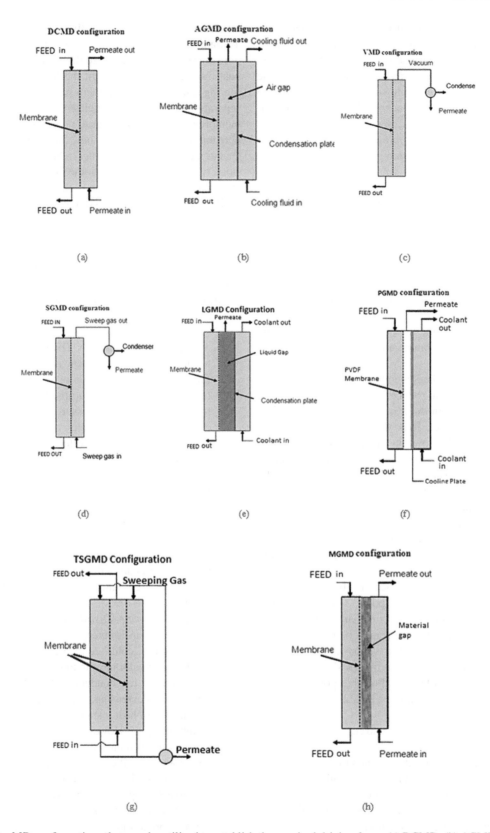

FIGURE 17.2 MD configurations that may be utilized to establish the required driving force: (a) DCMD, (b) AGMD, (c) VMD, (d) SGMD, (e) LGMD, (f) PGMD, (g) TSGMD, (h) MGMD.

Vacuum Membrane Distillation (VMD): In this case a vacuum or a low pressure is applied at the permeate side of the membrane module [12]. The downstream pressure is maintained below the saturation pressure of volatile molecules to be separated from the feed aqueous solution. External condensers are needed to condense the vapor water and collect the permeate [27]. At laboratory scale, nitrogen liquid cold traps are often used when a very low downstream pressure is applied [28]. Membranes having smaller pore size than in the other MD variants are used, providing that in VMD the risk of pore wetting is high. When a dense and selective membrane is used in the membrane module instead of a porous and hydrophobic membrane this process is termed pervaporation [29].

Sweep Gas Membrane Distillation (SGMD): A gas, such as air or nitrogen, sweeps the permeate side of the membrane carrying the evaporated molecules outside the membrane module for condensation instead of the vacuum as VMD [12,30,31]. In this MD variant the gas temperature and its hydrostatic pressure are kept below those of the feed aqueous solution. SGMD is used mostly for distilled/potable water production, concentration of solutes in the feed membrane side, as well as the removal and concentration of contaminants in aqueous solutions [32].

Liquid Gap Membrane Distillation (LGMD): This MD configuration is a combination of DCMD and AGMD configurations. In this case the air gap between the membrane and the cold surface is kept, to be filled by a stagnant, cold liquid solution, frequently the produced distilled water [33].

Permeate Gap Membrane Distillation (PGMD): This is an enhancement of DCMD in which a third channel is introduced by an additional non-permeable foil. One significant advantage of PGMD is the separation of the distillate from the coolant. Therefore, the coolant can be any other liquid, such as cold feed water. In module development this opens the opportunity to integrate an efficient heat recovery system. The presence of the distillate channel reduces sensible heat losses due to an additional heat transfer resistance. An additional effect is the reduction of the effective temperature difference across the membrane, which slightly lowers the permeation rate.

Thermostatic Sweeping Gas Membrane Distillation (TSGMD): This MD variant combines both SGMD and AGMD in order to minimize the temperature of the sweeping gas. This is necessary due to the increases of the temperature of the sweeping gas along the membrane module length because of the heat transferred from the feed side through the membrane to the permeate side [34]. In SGMD, the gas temperature, the heat transfer rate, and the mass transport through the membrane change during the gas progression along the membrane module. The presence of the cold wall in the permeate side reduces the increase of the sweeping gas temperature resulting in an enhancement of the driving force and the water production rate as a consequence. TSGMD can also be applied for distilled/potable water production, concentration of the non-volatile solutes in the feed aqueous solution, and concentration of contaminants in aqueous solutions [34].

Material Gap Membrane Distillation (MGMD): This configuration consists of filling the gap between the membrane and the condensation plate with materials, unlike the AGMD where stagnant air is used. These materials can be of different types with different characteristics such as polyurethane (sponge), polypropylene mesh, sand, and deionized water [35]. MGMD configuration is comparable to the DCMD configuration in terms of heat transfer mainly due to the high heat loss through conduction. The heat developed in the materials at higher temperatures decreases the mass transfer during the MGMD process, a possible reason for the reduction in the percentage flux enhancement [35].

17.2.3 Membrane Characteristics

In general, the membrane used in the MD system should have low resistance to mass transfer and low thermal conductivity to prevent heat loss across the membrane. In addition, the membrane should have good thermal stability in extreme temperatures, and high resistance to chemicals, such as acids and bases. Usually, membranes in polytetrafluoroethylene (PTFE), polypropylene (PP), or polyvinylidene fluoride (PVDF), eventually coated with super-hydrophobic materials, are used in MD processes [36]. Membranes with pore size between 100 nm to 1 μm are usually used in MD systems. The permeate flux increases with increasing membrane pore size and membrane porosity.

Membrane porosity can be determined by gravimetric method [37], measuring the weight of liquid (such as kerosene) contained in the membrane pores. The porosity of the membranes (ε) can be calculated using the following formula:

$$\varepsilon = \frac{\left(W_1 - W_2\right)/\rho_k}{(W_1 - W_2)/\rho_k + W_2/\rho_p} * 100\% \qquad (17.6)$$

where W_1 is the weight of the wet membrane, W_2 is the weight of the dry membrane, ρ_k is the density of the liquid (e.g., in the case of kerosene is 0.82 g/cm^3), and ρ_p is the polymer density.

In general, MD membranes possess porosity as high as 70–80%, thickness of 10–300 μm, and pore sizes of 0.2–1 μm. Moreover, above all:

- the membranes need to be hydrophobic in order to serve their intended function;
- the membranes should be intrinsically hydrophobic and preserve their character over a long period of time;
- the membrane material preferably should have low thermal conductivity to minimize conduction losses and achieve better thermal efficiency;
- the membranes need to have good thermal and chemical stability, as well as sufficient mechanical strength and high liquid entry pressure (LEP); and
- the membranes need to have a low propensity for fouling.

Among the hydrophobic materials applied in the fabrication of membranes for membrane contactors, fluoropolymers constitute a unique class of materials with a combination of interesting properties that attracted significant attention from material researchers over the past few decades [38]. Generally, these polymers have high thermal stability, improved chemical resistance, and lower surface tension because of the low polarizability and the strong electronegativity of the fluorine atom, its small van der Waals radius (1.32°A), and the strong C–F bond (485 kJ mol−1). The important fluoropolymers for membrane operations are poly(vinylidene fluoride) (PVDF), poly(tetrafluoroethylene) (PTFE), poly(ethylene chlorotrifluoroethylene) (ECTFE), poly(chlorotrifluoroethylene) (PCTFE), poly(vinyl fluoride) (PVF), poly(fluorenyl ether), Hyflon® AD, Teflon® AF, and Cytop®.

Robust omniphobic surfaces that repel both water (hydrophobic) and oil (oleophobic) have attracted significant attention in a wide range of technological areas, and in MD, too. Tuteja et al. [39] demonstrated that re-entrant geometries and specific chemical compositions are critical factors in the design of omniphobic surfaces that display excellent wetting resistance to a number of liquids with low surface tension [39]. In the past few decades, a variety of methods has been employed to create re-entrant structures on several substrate materials to impart surface omniphobicity with low surface tension [40–43]. Recently, re-entrant geometries have been successfully created on a variety of membrane substrates by the deposition of silica nanoparticles. The modified membranes exhibited omniphobic properties and were applicable to MD with a low surface tension feed [44–47].

17.2.4 Wetting, Fouling, and Biofouling

The main limits of membrane distillation application in industry and its commercialization are membrane wetting, temperature, and concentration polarization, fouling, and scaling [112]. The wetting of the membrane pores involves a complex of physical and chemical interactions. The non-wetting of a liquid is the result of its high surface tension formed by the liquid in contact with the hydrophobic membrane surface [48]. This contact surface forms a convex meniscus, which prevents the liquid from entering the pore of the membrane. Therefore, this balance remains until the pressure difference resulting from the surface tension of the curved interface balances the pressure drop caused by the partial pressures of vapors and air through the membrane. The pressure caused by the surface tension is called capillary pressure [49]. The primary metric for measuring membrane wettability is liquid entry pressure (LEP) defined through **Equation 17.5**. The latter links the maximum capillary pressure for a hydrophobic membrane with the liquid surface tension, the free surface energy, and the maximum pore size of the membrane. Many membranes and process conditions can impact the LEP through the variables in **Equation 17.5**, including operating temperature, solution composition, surface roughness, surface porosity, and pore shape (i.e., pore radius and fiber radius). Moreover, the utility of **Equation 17.5** for calculating LEP is limited because the contact angle and surface tension of the feed may not be known for the system of interest. Therefore, **Equation 17.5** can be only used to interpret the experimental data [21].

In its simplest form, the wettability of a liquid droplet on a flat, smooth surface is commonly determined by Young's equation:

$$\cos\theta = \left(\gamma_{sv} - \gamma_{sl}\right) / \gamma_{lv} \qquad (17.7)$$

where θ is the contact angle in the Young's model, γ_{sv}, γ_{sl}, γ_{lv}, are the interfacial tensions solid/vapor, solid/liquid and liquid/vapor, respectively. However, in reality, smooth surfaces are rare and some roughness is contained. Therefore, Wenzel's theory [50] was proposed where the roughness of the surface was considered for wettability determination.

$$\cos\theta_w = r\left(\gamma_{sv} - \gamma_{sl}\right) / \gamma_{lv} \qquad (17.8)$$

where θ_w is the apparent contact angle in the Wenzel mode and r is the surface roughness factor as the ratio of the actual solid/liquid contact area to its vertical projection.

Once the wetting takes place, the membrane begins to lose its hydrophobicity locally. Membrane wetting can be distinguished into four degrees: non-wetted, surface-wetted, partially-wetted, and fully-wetted [48]. Surface wetting shifts the interface of liquid/vapor inward of the membrane cross-section. As the feed solution penetrates deeper into the membrane pores, partial wetting can take place. In this case, the MD process can be continued if the majority of the pores are dry. However, partial wetting under certain conditions can reduce the permeate flux due to a reduction of the active surface area for mass transport associated with partial wetting. In the case of full wetting, the MD process no longer acts as a barrier, resulting in a viscous flow of liquid water through membrane pores, incapacitating the MD process [51,52].

Fouling is an extremely important factor in MD processes, since it involves greater energy consumption, longer downtimes due mainly to the continuous request for cleaning or replacing the membrane, and is one of the causes of membrane wetting [53,54]. However, membrane fouling is less of a problem in MD than in other membrane separations because (1) the pores are relatively large compared to the pores in (for example) reverse osmosis and ultrafiltration, (2) the low operating pressure of the MD process (which causes the deposition of foulants on the membrane surface to be less compact and only slightly affect the transport resistance), and (3) there is no feed inside the membrane pores (therefore foulants are deposited only on the membrane surface but not in the membrane pores).

17.2.5 APPLICATIONS AND FUTURE DIRECTIONS

Membrane distillation is an innovative process in continuous evolution that can be used to overcome the continuous demand for clean water in the world. Researchers are still trying to identify new MD process applications, including MD systems integrated with other separation processes. In addition to the already mentioned desalination and treatment of wastewater [55], membrane distillation is a process widely used [56] also for concentration of solutions (brines, fruit juices, acids, proteins, radioactive components, etc.) [57], applications of non-food chemical processes, separation of mixtures [58], recovery of oil and water produced by gas, and removal of heavy metals and dyes [59]. After more than 45 years of difficult and continuous research, MD technology has started to acquire industrial interest promoted by some companies such as Memsys, Memstill, Scarab Development AB, Keppel Seghers, and Fraunhofer ISE [60]. Factors of interest to improve MD research are the reduction of energy consumption, the difficulties faced with long-term operations, and the simultaneous risk of wetting, scaling, and fouling. To improve MD performance, the preparation of new types of membranes or of special coating on commercial membranes is of great interest, with the improvement of the characteristics in terms of pore size, overall porosity, thickness, and hydrophobicity. The preparation of membrane modules designed specifically for MD applications and the optimized coupling of renewable energy systems to MD plants make the MD research of interest. In the literature, new research in these fields has been developed. These include, for example, composite porous hydrophobic/hydrophilic membrane prepared using fluorinated surface modifying macromolecule (SMM) and polyethermide [61]; CF4 plasma surface modification of hydrophilic membranes into hydrophobic ones [62,63]; preparation of nanotube-structured TiO_2 surfaces with superomniphobic characteristics [64]; and fabrication of hybrid graphene/PVDF membranes [62]. Graphene has attracted considerable attention in recent years for water treatment and purification processes as it improves the membrane properties and adds functionalities such as anti-fouling properties [65,66]. Graphene possesses many interesting properties for MD application such as high hydrophobicity, ion selectivity and water vapor transport, good thermal stability, and mechanical properties [67,68].

The possibility of cheaper synthesis of large areas of graphene as recently reported makes it more interesting for modern use [69]. Graphene shows great potential as filler material and ought to be explored. In addition to graphene, two-dimensional (2D) materials of atomic thickness represent the next generation membrane materials with extraordinarily high permeability. 2D membranes with well-defined transport channels and ultra-low thicknesses have demonstrated exceptional performance for liquid and gas separation applications. The unique atomic thickness of the membrane offers ultra-low resistance to mass transport. The potential materials for 2D membranes include zeolites, mixed-organic frameworks, graphene, molybdenum disulfide, etc. The current main challenges for the commercial scale implementation of these membranes include limited available techniques for exfoliating the high aspect ratio and intact nanoporous monolayers from bulk crystals, drilling of the pores with required characteristics (uniform, high-density, large-area, subnanosized) in membrane matrix, and scaling up of these atomic scale membranes into real scale separation devices [70].

17.3 MEMBRANE CRYSTALLIZATION

17.3.1 MEMBRANE CRYSTALLIZATION PROCESS

Membrane crystallization (MCr) is emerging as an interesting candidate to extract additional freshwater and raw materials from high-concentrated solutions. The technique relies upon the evaporation of solvent from a solution in contact with a hydrophobic membrane to form a super-saturated solution. The operational principle of MCr is essentially the same as that of membrane distillation, which is based on the vapor pressure gradient created across a microporous hydrophobic membrane. The technical feasibility of the MCr process for recovery of salts from seawater brine is well acknowledged in the literature [71–73]. MCr also possesses the potential to realize the objective of zero liquid discharge in industry.

In MCr, the control of the crystallization process is important in order to minimize deposition and growth of crystals on the membrane and to maximize the crystal production and removal in the crystallizer. Depending on the chemical–physical properties of the membrane and on the process parameters (temperature, concentration, flow rate, etc.), the solvent evaporation rate, and hence supersaturation degree and supersaturation rate, might be regulated very precisely. The effect is the control of the nucleation and growth rate by choosing a broad set of available kinetic trajectories in the thermodynamic phase diagram, that are not readily achievable in conventional crystallization methods, and which would lead to the production of specific crystalline morphologies and structures. Furthermore, the surface structure of the membrane, with its cavities, might entrap solute molecules, thus leading to localized supersaturation that promotes the nucleation and the formation of crystals at conditions of supersaturation that would have been inadequate or insufficient for homogeneous nucleation [74].

17.3.2 MEMBRANE CRYSTALLIZATION CONFIGURATION

Principally, MCr can be operated in all four main MD configurations (direct contact, vacuum, sweep gas, and air gap MD) [20].

In its current conception, membrane crystallizer is a system in which a solution containing a non-volatile solute that is likely to be crystallized (defined as the crystallizing solution or feed or retentate), is in contact with, by means of a microporous membrane, a solution on the distillate side. The membrane might be made of polymeric or inorganic materials or a combination of both in a hybrid or composite configuration. Hollow fibers as well as flat-sheet membranes can be employed in a similar manner.

When the membrane is prevented from becoming wet due to the adjacent solutions, no mass transfer through its porous structure is observed directly in liquid phase, but the two subsystems, which are in contact, are subjected to mass interexchange in the vapor phase. The gradient of vapor pressure between the two subsystems induces the evaporation of the volatile component from feed solution, migration through the porous membrane, and, finally, the re-condensation at the distillate side (Figure 17.3a). The continuous removal of solvent from the feed solutions in a membrane crystallizer increases solute concentration thus generating supersaturation.

From a process design point of view, the membrane can be applied for a membrane assisted operation (i.e., on a mixture recirculating loop), or directly for in situ crystallization purposes. The first case can be seen as a typical hybrid process approach and it is shown in Figure 17.4a. The membrane module is here used to generate the supersaturation, or simply to concentrate the solid phase, but the nucleation and the crystal growth takes place in the crystallizer (e.g., [75,76]). In the second case (Figure 17.4b), the crystallization takes place directly in the membrane module where the supersaturation is generated [77–79].

In a recent development of the process, Di Profio et al. [80] proposed a new design of the MCr process in which

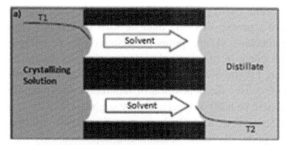
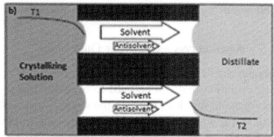
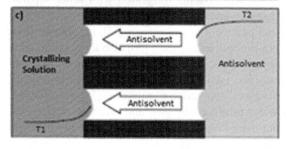

FIGURE 17.3 Basic principle of a membrane crystallizer: (a) solvent removal MCr, where solvent is removed from the crystallizing solution under a temperature gradient (T1 > T2); (b) solvent/antisolvent demixing MCr, in which the preferential evaporation of the solvent induces the increase of the antisolvent volume fraction thus reducing solubility (T1 > T2); (c) antisolvent addition MCr, where an antisolvent is evaporated into the crystallizing solution in vapor phase from the other side of the membrane (T1 < T2). From [82].

crystallization is induced by using antisolvent. This new approach operates in two configurations: first, solvent/antisolvent demixing (Figure 17.3b), and second, antisolvent addition (Figure 17.3c). In both cases, solvent/antisolvent migration occurs in the vapor phase, according to the

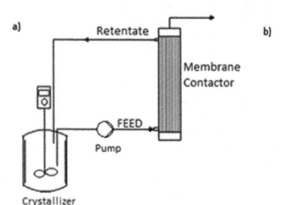
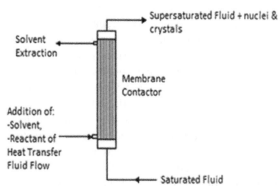

FIGURE 17.4 Schematic representation of the two process designs: (a) Hybrid membrane crystallization process. (b) The crystallization takes place directly in the membrane module.

general concept of membrane crystallization, and, unlike the above-mentioned configuration, not by forcing it in liquid phase through the membrane.

The selective and precise dosing of the antisolvent, controlled by the porous membrane, allows a finer control of the solution composition during the process and at the nucleation point, with consequent improvement of the final crystal characteristics.

According to the considerations above, crystallization by using membranes can be classified depending on the different working principles such as [81]:

1. membrane distillation/osmotic distillation-based processes where diffusion of solvent molecules in vapor phase through a porous membrane, under the action of a gradient of chemical potential as driving force, generates supersaturation in the crystallizing solution;
2. membrane-assisted crystallization in which pressure-driven membrane operations (MF, NF, and RO) are used to concentrate a solution by solvent removal in liquid phase, while crystals are recovered in a separate tank, often operated at lower temperature and with seeding;
3. solid (nonporous) hollow fibers used as heat exchanger to generate supersaturation by cooling;
4. antisolvent (or crystallizing solution) forced directly in the liquid state into the crystallizing solution (or into the antisolvent) through the pores of a membrane under a pressure gradient; and
5. antisolvent membrane crystallization, where dosing of the antisolvent in the crystallizing solution is carried out by means of a membrane, according to the working principle of point 1, in the two solvent/antisolvent demixing and antisolvent addition configurations.

17.3.3 THE NUCLEATION AND CRYSTAL GROWTH PROCESS IN THE CRYSTALLIZER

Crystals' nucleation and growth in the feed solution is induced by generating supersaturation. This is realized by removing the solvent from the crystallizing solution, thus increasing solute concentration up to the overcoming of its solubility limit. Detailed relations and models for heat and mass transport through the membrane in MCr can be described by using the same concepts developed for membrane distillation [83–87]. As a general description, heat and mass transport through membranes occurs only if the overall system is not in thermodynamic equilibrium. For what concerns the mass transport, it can be separated into three steps: mass transfer in feed boundary layer, mass transfer *through the membrane pores*, and mass transfer in permeate boundary layer. The mass transfer in permeate boundary layer is not considered since the mole fraction of the transporting

species in the permeate stream is approximately equal to one. The mass transfer in boundary layers is analysed by film theory, whereas dusty gas model (DGM) is usually employed to describe the mass transfer across the membrane. Dusty gas model elucidates mass transfer in porous media by four possible mechanisms: viscous flow, Knudsen diffusion, molecular diffusion, and surface diffusion. It is general for membrane crystallization in direct contact configuration to neglect surface diffusion and viscous flow, and to employ a Knudsen-molecular diffusion transition model [83,88]:

$$N' = \frac{\varepsilon P D_{ij}}{\tau \delta RT} \ln \left(\frac{2 \frac{2r}{3} \left(\frac{8RT}{\pi M_i} \right)^{1/2}}{p_a^2 + P D_{ij}} \right) \quad (17.9)$$

where N' is the molar trans-membrane flux, ε is the porosity, P is the total pressure, D_{ij} is the diffusivity, τ is membrane tortuosity, δ is membrane thickness, R is ideal gas constant, T is temperature, r is pore size, M_i is molecular weight, and p_a^1 and p_a^2 are the partial pressure of air at feed and membrane surface, respectively. The model can be further simplified in the specific cases. The Knudsen diffusion model is suitable for the system that the collision between molecule and pore wall dominates the mass transport. On the other hand, molecular diffusion model is preferred when the collision between the molecules plays the main role in the mass transfer across the membrane. Nevertheless, if both molecule-pore wall and molecular-molecular collisions occur frequently, the Knudsen-molecular diffusion transition model must be employed. Both molecular diffusion limit and Knudsen-molecular diffusion transition model were successfully applied to describe the flux in DCMD system [22,84,86,89]. In both cases the trans-membrane flux is proportional to membrane porosity ε whereas it is inversely proportional to membrane thickness δ. Therefore, membrane structural properties will strongly affect membrane crystallization performance both in terms of solvent evaporation rate and in terms of crystals nucleation and growth. As a matter of fact, a crystallizing solution can be imagined as a certain number of solute molecules moving among the molecules of solvent and colliding with each other, so that a number of them converge forming clusters. According to Volmer [90], the critical size n^*, which an assembly of molecules must have in order to be stabilized by further growth, is given by:

$$n^* = \frac{32 \pi v_0 \gamma^3}{3 (k_B T)^2 \ln^3 S} \quad (17.10)$$

where v_0 is the molecular volume, γ is the interfacial energy, K_B is Boltzmann's constant, S is the super saturation. **Equation 17.10** shows how critical size n^* depends on supersaturation S: the higher the operating level of supersaturation, the smaller is the size (typically a few tens of molecules). Therefore, the proper choice of membrane chemical-physical properties and process parameters (temperature, concentration, etc.), allow regulating flow rate, supersaturation degree, supersaturation rate, nucleation, and growth.

17.3.4 APPLICATION AND FUTURE DIRECTIONS

Membrane crystallization arises from the need to produce substances in solid-crystalline state required in various sectors of industry, technology, and scientific research. Many products used daily as additives for cosmetics, hygiene and personal care, pharmaceuticals, fine chemicals, pigments, and many other foods are formulated as crystalline powders. Other applications are the fabrication of organic semiconductor single-crystal based devices and photonic crystals, or medical advancement by rationale design of new structure-based drugs. This is because the solid state makes these products more stable for storage and more functional to be managed by users. In some technological fields, such as in microelectronics, in nonlinear optics and in sensing, the solid state is the basis of the realization of single semiconductor crystal devices; furthermore, crystalline materials experience a wide diffusion in heterogeneous catalysis, in the controlled release of active substances and, more generally, in the nanotechnology. As regards research in the medical field, single crystals or crystalline powders are required through a rational strategy of drug design and the development of materials based on the structure. Generally, the crystalline properties have a significant impact compared to their effective use. In the case of crystalline powders, the morphology of the crystals is the dominant feature. During production, preferred crystal forms are required, for example for more efficient filtration of particle recovery from parent solutions by filtration for improved compressibility when producing tablets, and for optical behavior in the case of pigments. In the heterogeneous catalysis, the reduced, very mono dispersed and uniformly shaped crystals are more suitable to obtain the highest surface–volume ratio and greater catalytic efficiency [91,92]. In the pharmaceutical industry, the different polymorphic forms of the same substance are considered different materials, with their characteristic physical, chemical, and biological properties, making each of them a different and patentable drug. In the case of proteins, large crystals (100 mm at least in two dimensions) are required with high order in their crystal lattice for the determination of the structure at the atomic level by X-ray diffraction analysis [93,94]. Other, but not less important, applications have been developed for the treatment of waste water for the recovery of high purity silver [95] or sodium sulphate [96], CO_2 capture [97–99], nanotechnology, such as the synthesis of $BaSO_4$ and $CaCO_3$ particles [100,101], or the recovery of antibiotics [102] or

of polyestyrene microparticles [103]. Membrane crystallization is a process with the potential to achieve the goal of having desalination plants with zero discharge liquids. In principle, MCr could overcome not only the limits of thermal systems but also those of traditional membrane systems such as RO. In fact, the polarization of the concentration does not significantly affect the driving force of the process, and therefore it is possible to reach high recovery factors and high concentrations. Thanks to its characteristics, the integration of MC on RO brine offers the possibility to produce high quality solid materials and controlled properties, with important added values, transforming the traditional problem of disposal of the brines into a potential new profitable market [104]. The potential of membrane crystallization is addressed from an intensification and integration point of view, with particular attention to present and future applications, i.e., green economy and value-added production.

However, the development and application on an industrial scale of membrane crystallization technology requires facing and overcoming the following problems:

- *Development of Proper Membranes for MCr Application*: as for MD, development of new material/membrane preparation methods with extended life, with stable hydrophobic character, with properties tailored for this technology (in terms of pore size, porosity, thickness, thermal conductivity, surface roughness, etc.).
- *Technology Transfer to Industry:* application and testing of the technology on industrial-scale and valuation of benefits/advantages/drawback with respect to conventional process.

17.4 MEMBRANE CONDENSER

17.4.1 MEMBRANE CONDENSER PROCESS

In the last years a new membrane process called "membrane condenser" was developed by Drioli and co-workers [15,17,105–107].

The working principle of membrane condenser consists in condensing and recovering the water contained in the saturate gas on the retentate side of the membrane module, by exploiting the hydrophobic nature of the membrane, whereas the dehydrated gases pass through the membrane in the permeate side. Figure 17.5 schematizes the membrane condenser principle.

In particular, the gaseous stream (e.g., flue gas) exiting from the power plant, cooling towers, stacks, etc., at a certain temperature and, in most of the cases, water saturated, is fed to the membrane condenser kept at a lower temperature for cooling the gas up to a super-saturation state. The water condenses in the membrane module and,

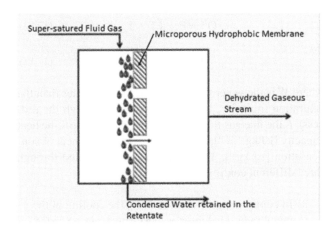

Super-satured Fluid Gas — Microporous Hydrophobic Membrane

Dehydrated Gaseous Stream

Condensed Water retained in the Retentate

FIGURE 17.5 Scheme of the membrane condenser process for the recovery of evaporated "waste" water from a gaseous stream. Reprinted with permission from [107]. Copyright 2013 American Chemical Society.

once this stream is brought into contact with the retentate side of the microporous membranes, their hydrophobic nature prevents the penetration of the liquid into the pores letting pass the dehydrated gases through the membrane. Therefore, the liquid water is recovered at the retentate

side, whereas the other gases remain at the permeate side of the membrane unit.

In comparison to the traditional technologies, membrane condensers are characterized by various aspects that differentiate them from the others (Table 17.1).

First at all, membrane condensers offer higher water recovery with respect to all the other technologies. Despite the adsorption units where problems related to desiccant losses can occur, the membrane condensers can be considered as a clean unit operation. In addition, they do not undergo corrosion phenomena that usually occur in traditional condensers or desiccant units. However, whether the gaseous stream contains hashes or particles, a pre-treatment stage could be required before feeding the membrane condenser. Compared with the dense membrane technology, the main difference can be found in the operating conditions and in the water quality. To promote the permeation of water vapor through dense membranes, a high-pressure difference between the two membrane sides is required, implying investment and operational costs related to the presence of compressors or vacuum pumps. On the contrary, the purity of the water recovered in membrane condensers can be affected by the possible condensation of contaminants – if present in the gaseous stream – but it is

TABLE 17.1
Comparison between Membrane Condenser and Traditional Technologies

	Water Recovery (%)	Water Purity (%)	Maintenance and Durability	Environmental Aspects	Investments Costs	Economic Viability (Euro m⁻³)
Liquid and solid sorption [109]	22-62	>95	Corrosion and salt crystals formation owing to salt desiccants presence and O_2 in the flue gas	Increase of CO_2 emissions Reduction of SO_x emission $CaCl_2$ losses	$5.8*10^6$ \$ (2006) + 200.000 \$/ year (2006) as operational costs	4.4 \$ m⁻³
Cooling with condensation [110]	<70	Sufficient for cooling tower make up. Contaminants in the water	Corrosion owing to the formation of a thin liquid layer of diluted acids and fly ashes forming deposits	Co-capture of SO_x and NO_x could results in an environmental profit reducing the DENOx and FGD systems	$6.4*10^6$ Euro (2011)	1.5–2
Dense membranes [111]	20–40	>95	Ashes removal and FGD necessary to avoid membrane damaging	Clean operation	N/A	1.5 forwet regions; 10 for dry regions
Membrane condensers [108]	>70	Sufficient for cooling tower make up. Contaminants in the water	Ashes removal to avoid membrane damaging	Clean operation	N/A	1.5–2.5 a)

a) Considering only costs related to energy requirements and membrane modules.

Source: Reprinted From CLEAN - Soil, Air, Water, 42, Brunetti, A., S. Santoro, F. Macedonio, A. Figoli, E. Drioli, and G. Barbieri. 2014. Waste gaseous streams: from environmental tssue to source of water by using membrane condensers, 1145–53, Copyright (2014) with permission from Elsevier [108].

sufficient for cooling tower or boiler make-up. However, further purifications would need to make it drinkable. In this context, membrane condensers can be considered also as a proper solution for pre-treating the flue gas streams that have to be fed to another membrane unit for CO_2 separation. In most cases, in fact, the performance of these units is strongly depleted, or the membranes are still irreversibly damaged, by the presence of such contaminants as SO_x, NH_3, etc. The possibility of controlling, by opportunely tuning the operating conditions, the condensation of contaminants in the liquid water recovered in the retentate side of the membrane condenser could lead to two different options for its use: as a unit for water recovery, minimizing the contaminants content, or as the pre-treatment stage in post-combustion capture, forcing most of the contaminants to be retained.

As for the well-known membrane contactors such as "CO_2 adsorbers", many are the challenges that still limit their scaling up from laboratory to industry level. First of all, the fundamental property that the membrane used in membrane condensers has to show is the high hydrophobicity that avoids the penetration of condensed water into the membrane pores and thus favors the separation of liquid phase from the gaseous one recovered on the permeate side. The hydrophobic character of these membranes is influenced by the feed conditions. As aforementioned, in most of the cases, the waste gaseous streams coming out from chimneys of industrial plants contain variable amount of contaminants such as SO_x, NO_x, NH_3, HF, HCl, etc., that on a long-term basis could affect the performance of the membrane condenser, depleting the hydrophobic character of the membrane. The challenge is, thus, to identify materials for making membranes that exhibit excellent resistance and stability toward these components. Actually, these are well represented by PVDF or other superhydrophobic materials such as Hyflon, Teflon, etc., that have the only disadvantage of being quite expensive. The synthesis of new composite membranes where the hydrophobic layer is deposited on a cheaper support can lead to the long-term stability together with a reasonable cost of the unit.

17.4.2 MEMBRANE CONDENSER CONFIGURATIONS

In a membrane condenser (MCo), the selective recovery of evaporated waste water from a gaseous stream is carried out by its cooling down to an appropriate temperature [105,107]. The condenser must cool the entire flow of the combustion gas from its inlet temperature T_i to the temperature T_{wr}, which is the temperature necessary to obtain the necessary water recovery. The total thermal energy Q required to condense the water vapor (Q^I) and to cool the exhaust gas (Q^{II}), has been calculated with the equations:

$$Q^I = W\lambda_v \qquad (17.11)$$

$$Q^{II} = Fc_{pv}\left(T_i - T_{wr}\right) \qquad (17.12)$$

$$Q = Q^I + Q^{II}W \qquad (17.13)$$

where W is the water vapor flow rate [kg/h] higher than the saturation limit (i.e., the water recovered through the process), F the flue gas flow rate [kg h⁻¹], c_{pv} is the specific heat capacity [kJ kg⁻¹ K⁻¹], and λ_v is the water latent heat of condensation [kJ kg⁻¹]. The waste gas can be cooled through three different configurations [5]:

1. In configuration 1 (Figure 17.6), the cooling of the membrane condenser is performed via an external coolant medium before the gas stream enters the membrane module [15]. The supersaturated stream is then put in contact with membrane surfaces where liquid water is retained whereas dehydrated gases are allowed passing through the membrane itself. It has been proved that, utilizing this configuration, the amount of recovered water increases more than proportionally with the growing of ΔT (temperature difference between the feed and the membrane module) and T_{Feed} owing to the exponential dependence of partial pressure of water on temperature.

2. In configuration 2 (Figure 17.7) the feed is cooled inside the membrane module through a cold sweeping gas. In this configuration, energy consumption of the entire system has to take into account also the power required to push the sweeping gas thus overcoming the pressure losses along pipes and membrane module [15]. Cooling via a sweeping gas has the advantage that the condenser heat duty will only depend on the flow rate and temperature of the sweeping gas and not on an external source thus considerably reducing the energy consumption of the entire process.

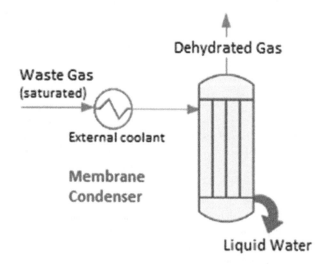

FIGURE 17.6 Scheme of membrane condenser in configuration 1. Reprinted with permission from [15].

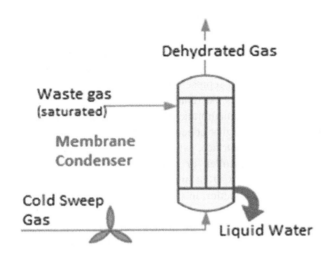

FIGURE 17.7 Scheme of membrane condenser in configuration 2. Reprinted with permission from [15].

In this configuration, the ratio between the cold sweeping gas flow rate and the feed flow rate is defined as the sweeping factor, I:

$$I = \frac{\text{cold sweeping gas flow rate}}{\text{feed waste gas flow rate}}. \quad (17.14)$$

In [15] it was proved that water recovery increases with growing sweeping factor I due to the improving of cooling effect.

3. Configuration 3 (Figure 17.8) is in between the two previous configurations: the fed waste gas is first partially cooled via an external medium and then a sweeping gas is used for the final cooling of the stream [15]. This makes it possible to increase the amount of water recovered as much as possible, avoiding the use of expensive cooling systems. In particular, the cooling water is used for external

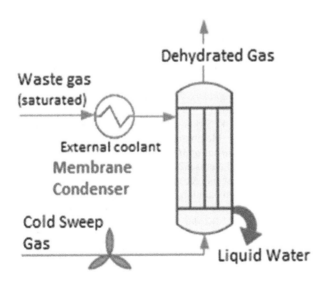

FIGURE 17.8 Scheme of membrane condenser in configuration 3. Reprinted with permission from [15].

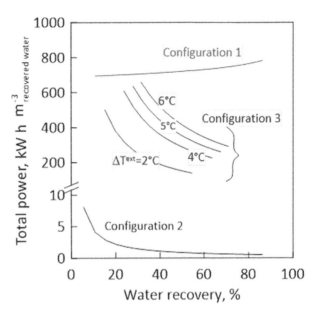

FIGURE 17.9 Energy consumption vs recovered water when the cooling is performed with or without a sweep gas. (Q^{Feed} =0.03 m^3/h, RH=100%, 55°C; sweep gas at 20°C Q^{Sweep}=0.09 m^3/h). Reprinted with permission from [15].

cooling and the air at 20° C as sweeping gas. In configuration 3, as in configuration 2, the maximum amount of recoverable liquid water depends on the temperature of the sweeping gas at the exit of the module.

It was proved [15] that considering waste gas at 55°C and RH=100%, and air at 20°C as cold sweeping gas (for configuration 2 and 3), the highest energy consumption is achieved utilizing configuration 1 (Figure 17.9). On the contrary, configuration 2 is, among those proposed, the one with the lowest energy consumption. For what concerns water recovery, excluding configuration 2 with I=1 and I=10 (the first one because recovers too low an amount (5.42%) of water, the second one because it requires too high a flow rate of cold sweeping gas), it increases going from configuration 1 to 2 at growing sweeping factor I (from around 28 to 39%) (Figure 17.10). The highest maximum amount of liquid water can be obtained utilizing configuration 3 whose energy consumption is in between configuration 1 and 2. In particular, it was found that the maximum amount of recoverable liquid water is of about 46.8% for an external cooling ΔT_{ext} of 6°C (Figure 17.10).

17.4.3 Membrane Characteristics

The membranes to be used in the membrane condenser must possess certain characteristics such as: suitable hydrophobicity, pore size in the micron range (from 0.01 to 1.0 μm, with ideal case between 0.2–0.3 μm), thickness from 20 to 100 μm, resistant to alcohols, surfactant, and chemicals, such as acids and bases, porosity between 70 and 80%, and high thermal conductivity [13]. Of particular importance in

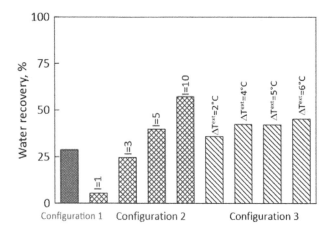

FIGURE 17.10 Maximum amount of recoverable water from the three different proposed configurations (Q^{Feed} =0.03 m³/h, RH=100%, 55°C; sweep gas at 20°C). Reprinted with permission from [15].

the choice of materials will be the values of thermal conductivity, porosity, and thickness of the membrane, fundamental parameters in the calculation of thermal efficiency, and, therefore, of energy consumption [107].

17.4.4 MEMBRANE CONDENSER APPLICATIONS AND FUTURE DIRECTIONS

Membrane condenser was initially developed in the framework of the FP7 project "CAPWA" (capture of evaporated water with novel membranes), and it is now further technologically advanced and scaled up also in the EU H2020 project "MATChING" (Materials & Technologies for Performance Improvement of Cooling Systems performance in Power Plants). While in CAPWA and in MATChING, membrane condenser was and is utilized only for water recovery from waste gaseous streams, a new application can be as pre-treatment stage to further separation units for minimizing the contaminants content of various gaseous streams, removing, in the meantime, part of the water vapor contained in gaseous streams. This aspect is sometimes essential when the gas has to undergo further separation processes, like, for example, in carbon dioxide separation. In fact, current constraints and regulations on CO_2 emissions from power plants have forced manufacturers and researchers to focus on the separation of CO_2 from flue gas streams and to develop specific CO_2 capture technologies that can be retrofitted to existing power plants as well as designed into new plants with the goal to achieve 90% of CO2 capture limiting the increase in cost of electricity to no more than 35%.

Currently, the main strategies for the carbon dioxide capture in a fossil fuel combustion process are: oxy-fuel combustion, pre-combustion capture, and post-combustion capture.

The main technical problems are related to the fact that polymeric membranes cannot withstand high temperatures and/or chemically harsh conditions. Refinery gas streams contain impurities such as water vapor, acid gases, olefins, aromatics, and other organics. Heavy hydrocarbons can be present in the feed also in petrochemical plants and natural gas treatment, representing a problem, mainly in hollow fiber modules. At relatively low concentrations, these impurities cause membrane plasticization and loss of selectivity, while at higher concentrations they can condense on the membrane surface, which could be damaged. Many polymers are swollen or plasticized in the presence of hydrocarbons or CO_2 at high partial pressure: the result is a significant reduction in their separation performance, or their damage. Another issue is physical aging, which negatively affects the properties of interesting polymers (such as poly[1-(trimethylsilyl)-1-propyne] (PTMSP), polymers of intrinsic microporosity (PIMs), etc.) and limits their applicability. The solution for a successful operation of polymeric modules is a careful selection of feed pre-treatment. In this field, membrane condensers can be considered as a proper solution for pre-treating the flue gas streams that have to be fed to another membrane unit for CO_2 separation. In this optic, membrane condensers can be a suitable pre-treatment stage placed before other separation technologies. These latter will be significantly reduced in size and their lifetime will be prolonged with respect to that foreseen without using condensers.

REFERENCES

1. Criscuoli, A., M. C. Carnevale, H. Mahmoudi, S. Gaeta, F. Lentini, and E. Drioli. 2011. "Membrane Contactors for the Oxygen and PH Control in Desalination." *Journal of Membrane Science* 376 (1–2): 207–13. doi: 10.1016/j.memsci.2011.04.018

2. Abdel-Aal, E. A., M. H. H. Mahmoud, M. M. S. Sanad, A. Criscuoli, A. Figoli, and E. Drioli. 2010. "Membrane Contactor as a Novel Technique for Separation of Iron Ions from Ilmenite Leachant." *International Journal of Mineral Processing* 96 (1–4): 62–69. doi: 10.1016/j.minpro.2010.05.002

3. No author. 2006. "Membrane Contactors Improve Production Yields," *Membrane Technology* 4: 5.

4. Fakharian, H., H. Ganji, and A. Naderifar. 2017. "Desalination of High Salinity Produced Water Using Natural Gas Hydrate." *Journal of the Taiwan Institute of Chemical Engineers* 72 (March): 157–62. doi: 10.1016/j.jtice.2017.01.025

5. Li, B., and K. K. Sirkar. 2005. "Novel Membrane and Device for Vacuum Membrane Distillation-Based Desalination Process." *Journal of Membrane Science* 257 (1–2): 60–75. doi: 10.1016/j.memsci.2004.08.040

6. Trusek-Holownia, A., and A. Noworyta. 2002. "Catalytic Membrane Preparation for Enzymatic Hydrolysis Reactions Carried out in the Membrane Phase Contactor." *Desalination* 144 (1–3): 427–32. doi: 10.1016/S0011-9164(02)00356-9

7. Charcosset, C. 2006. "Membrane Processes in Biotechnology: An Overview." *Biotechnology Advances* 24 (5): 482–92. doi: 10.1016/j.biotechadv.2006.03.002

8. Brunel, L., S. Tingry, K. Servat, M. Cretin, C. Innocent, C. Jolivalt, and M. Rolland. 2006. "Membrane Contactors for Glucose/O₂ Biofuel Cell." *Desalination* 199: 426.

9. Bey, S., A. Criscuoli, A. Figoli, A. Leopold, S. Simone, M. Benamor, and E. Drioli. 2010. "Removal of As(V) by PVDF Hollow Fiber Membrane Contactors Using Aliquat-336 as Extractant." *Desalination* 264 (3): 193–200. doi: 10.1016/j.desal.2010.09.027

10. Drioli, E., E. Curcio, and G. di Profio. 2005. "State of the Art and Recent Progresses in Membrane Contactors." *Chemical Engineering Research and Design* 83 (3): 223–33. doi: 10.1205/cherd.04203

11. Quist-Jensen, C. A., A. Ali, S. Mondal, F. Macedonio, and E. Drioli. 2016. "A Study of Membrane Distillation and Crystallization for Lithium Recovery from High-Concentrated Aqueous Solutions." *Journal of Membrane Science* 505 (May): 167–73. doi: 10.1016/j.memsci.2016.01.033

12. Ko, C. C., A. Ali, E, Drioli, K.-L. Tung, C.-H. Chen, Y.-R. Chen, and F. Macedonio. 2018. "Performance of Ceramic Membrane in Vacuum Membrane Distillation and in Vacuum Membrane Crystallization." *Desalination* 440 (March): 48–58. doi: 10.1016/j.desal.2018.03.011

13. Ali, A., J.-H. Tsai, K.-L. Tung, E. Drioli, and F. Macedonio. 2018. "Designing and Optimization of Continuous Direct Contact Membrane Distillation Process." *Desalination* 426 (January): 97–107. doi: 10.1016/j.desal.2017.10.041

14. Drioli, E., A. Ali, and F. Macedonio. 2015. "Membrane Distillation: Recent Developments and Perspectives." *Desalination* 356 (January): 56–84. doi: 10.1016/j.desal.2014.10.028

15. Macedonio, F., A. Brunetti, G. Barbieri, and E. Drioli. 2017. "Membrane Condenser Configurations for Water Recovery from Waste Gases." *Separation and Purification Technology* 181 (June): 60–68. doi: 10.1016/j.seppur.2017.03.009

16. Macedonio, F., and E. Drioli. 2017. "Membrane Engineering for Green Process Engineering." *Engineering* 3 (3): 290–98. doi: 10.1016/J.ENG.2017.03.026

17. Drioli, E., S. Santoro, S. Simone, G. Barbieri, A. Brunetti, F. Macedonio, and A. Figoli. 2014. "ECTFE Membrane Preparation for Recovery of Humidified Gas Streams Using Membrane Condenser." *Reactive and Functional Polymers* 79 (1): 1–7. doi: 10.1016/j.reactfunctpolym.2014.03.003

18. Macedonio, F., M. Cersosimo, A. Brunetti, G. Barbieri, and E. Drioli. 2014. "Water Recovery from Humidified Waste Gas Streams: Quality Control Using Membrane Condenser Technology." *Chemical Engineering and Processing: Process Intensification* 86 (December): 196–203. doi: 10.1016/j.cep.2014.08.008

19. Bey, S., A. Criscuoli, S. Simone, A. Figoli, M. Benamor, and E. Drioli. 2011. "Hydrophilic PEEK-WC Hollow Fibre Membrane Contactors for Chromium (Vi) Removal." *Desalination* 283 (December): 16–24. doi: 10.1016/j.desal.2011.04.038

20. Ruiz Salmón, I., and P. Luis. 2018. "Membrane Crystallization via Membrane Distillation." *Chemical Engineering and Processing: Process Intensification* 123 (November): 258–71. doi: 10.1016/j.cep.2017.11.017

21. Lawson, K. W., and D. R. Lloyd. 1997a. "Membrane Distillation." *Journal of Membrane Science* 124 (1): 1–25. doi: 10.1016/S0376-7388(96)00236-0

22. Macedonio, F., A. Ali, and E. Drioli. 2017. "3.10 Membrane Distillation and Osmotic Distillation." *Comprehensive Membrane Science and Engineering* 3, 282–96. doi: 10.1016/B978-0-12-409547-2.12221-8

23. Baghbanzadeh, M., C. Q. Lan, D. Rana, and T. Matsuura. 2016. "Membrane Distillation." In *Nanostructured Polymer Membranes*, 419–55. Hoboken, NJ: John Wiley & Sons, Inc. doi: 10.1002/9781118831779.ch11

24. Khayet, M., and C. Cojocaru. 2012. "Artificial Neural Network Modeling and Optimization of Desalination by Air Gap Membrane Distillation." *Separation and Purification Technology* 86 (February): 171–82. doi: 10.1016/j.seppur.2011.11.001

25. Meindersma, G. W., C. M. Guijt, and A. B. de Haan. 2006. "Desalination and Water Recycling by Air Gap Membrane Distillation." *Desalination* 187 (1–3): 291–301. doi: 10.1016/j.desal.2005.04.088

26. Xu, J., Y. B. Singh, G. L. Amy, and N. Ghaffour. 2016. "Effect of Operating Parameters and Membrane Characteristics on Air Gap Membrane Distillation Performance for the Treatment of Highly Saline Water." *Journal of Membrane Science* 512: 73–82. doi: 10.1016/j.memsci.2016.04.010.

27. Kiefer, F., M. Spinnler, and T. Sattelmayer. 2018. "Multi-Effect Vacuum Membrane Distillation Systems: Model Derivation and Calibration." *Desalination* 438 (March): 97–111. doi: 10.1016/j.desal.2018.03.024

28. Khayet, M., and T. Matsuura. 2004. "Pervaporation and Vacuum Membrane Distillation Processes: Modeling and Experiments." *AIChE Journal* 50 (8): 1697–712. doi: 10.1002/aic.10161

29. Ma, Q., A. Ahmadi, and C. Cabassud. 2018. "Direct Integration of a Vacuum Membrane Distillation Module within a Solar Collector for Small-Scale Units Adapted to Seawater Desalination in Remote Places: Design, Modeling & Evaluation of a Flat-Plate Equipment." *Journal of Membrane Science* 564 (March): 617–33. doi: 10.1016/j.memsci.2018.07.067

30. Xie, Z., T. Duong, M. Hoang, C. Nguyen, and B. Bolto. 2009. "Ammonia Removal by Sweep Gas Membrane Distillation." *Water Research* 43 (6): 1693–699. doi: 10.1016/j.watres.2008.12.052

31. Bernardo, P., E. Drioli, and G. Golemme. 2009. "Membrane Gas Separation: A Review/State of the Art." *Industrial & Engineering Chemistry Research* 48 (10): 4638–663. doi: 10.1021/ie8019032

32. Gasconsviladomat, F., I. Souchon, V. Athes, and M. Marin. 2006. "Membrane Air-Stripping of Aroma Compounds." *Journal of Membrane Science* 277 (1–2): 129–36. doi: 10.1016/j.memsci.2005.10.023

33. Khalifa, A. E. 2018. "Performances of Air Gap and Water Gap MD Desalination Modules." *Water Practice and Technology* 13 (1): 200–209. doi: 10.2166/wpt.2018.034

34. García-Payo, M. C., C. A. Rivier, I. W. Marison, and U. von Stockar. 2002. "Separation of Binary Mixtures by Thermostatic Sweeping Gas Membrane Distillation." *Journal of Membrane Science* 198 (2): 197–210. doi: 10.1016/S0376-7388(01)00649-4

35. Francis, L., N. Ghaffour, A. A. Alsaadi, and G. L. Amy. 2013. "Material Gap Membrane Distillation: A New Design for Water Vapor Flux Enhancement." *Journal of Membrane Science* 448 (December): 240–47. doi: 10.1016/j.memsci.2013.08.013

36. Alkhudhiri, A., N. Darwish, and N. Hilal. 2012. "Membrane Distillation: A Comprehensive Review." *Desalination* 287 (February): 2–18. doi: 10.1016/j.desal.2011.08.027

37. Feng, C. 2004. "Preparation and Properties of Microporous Membrane from Poly(Vinylidene Fluoride-Co-Tetrafluoroethylene) (F2.4) for Membrane Distillation." *Journal of Membrane Science* 237 (1–2): 15–24. doi: 10.1016/j.memsci.2004.02.007

38. Cui, Z., E. Drioli, and Y. M. Lee. 2014. "Recent Progress in Fluoropolymers for Membranes." *Progress in Polymer Science* 39 (1): 164–98. doi: 10.1016/j.progpolymsci.2013.07.008

39. Tuteja, A., W. Choi, M. Ma, J. M. Mabry, S. A. Mazzella, G. C. Rutledge, G. H. McKinley, and R. E. Cohen. 2007. "Designing Superoleophobic Surfaces." *Science* 318(5856): 1618–1622. doi: 10.1126/science.1148326

40. Chen, L.-H., A. Huang, Y.-R. Chen, C.-H. Chen, C.-C. Hsu, F.-Y. Tsai, and K.-L. Tung. 2018. "Omniphobic Membranes for Direct Contact Membrane Distillation: Effective Deposition of Zinc Oxide Nanoparticles." *Desalination* 428 (February): 255–63. doi: 10.1016/j.desal.2017.11.029

41. Ahuja, A., J. A. Taylor, V. Lifton, A. A. Sidorenko, T. R. Salamon, E. J. Lobaton, P. Kolodner, and T. N. Krupenkin. 2008. "Nanonails: A Simple Geometrical Approach to Electrically Tunable Superlyophobic Surfaces." *Langmuir* 24(1): 9–14. doi: 10.1021/la702327z

42. Brown, P. S., and B. Bhushan. 2016. "Durable, Superoleophobic Polymer-Nanoparticle Composite Surfaces with Re-Entrant Geometry via Solvent-Induced Phase Transformation." *Scientific Reports* 6:21048. doi: 10.1038/srep21048

43. Dufour, R., G. Perry, M. Harnois, Y. Coffinier, V. Thomy, V. Senez, and R. Boukherroub. 2013. "From Micro to Nano Reentrant Structures: Hysteresis on Superomniphobic Surfaces." *Colloid and Polymer Science* 291(2): 409–415. doi: 10.1007/s00396-012-2750-7

44. Lin, S., S. Nejati, C. Boo, Y. Hu, C. O. Osuji, and Menachem Elimelech. 2014. "Omniphobic Membrane for Robust Membrane Distillation." *Environmental Science and Technology Letters* 1(11): 443–447. doi: 10.1021/ez500267p

45. Lee, J., C. Boo, W. H. Ryu, A. D. Taylor, and M. Elimelech. 2016. "Development of Omniphobic Desalination Membranes Using a Charged Electrospun Nanofiber Scaffold." *ACS Applied Materials and Interfaces* 8(17): 11154–11161. doi: 10.1021/acsami.6b02419

46. Boo, C., L. Jongho, and E. Menachem 2016a. "Engineering Surface Energy and Nanostructure of Microporous Films for Expanded Membrane Distillation Applications." *Environmental Science and Technology* 50(15): 8112–8119. doi: 10.1021/acs.est.6b02316

47. Boo, C., L. Jongho, and E. Menachem. 2016b. "Omniphobic Polyvinylidene Fluoride (PVDF) Membrane for Desalination of Shale Gas Produced Water by Membrane Distillation." *Environmental Science and Technology* 50(22): 12275–12282. doi: 10.1021/acs.est.6b03882

48. Rezaei, M., David M. Warsinger, J. H. Lienhard V, Mikel C. Duke, T. Matsuura, and Wolfgang M. Samhaber. 2018. "Wetting Phenomena in Membrane Distillation: Mechanisms, Reversal, and Prevention." *Water Research* 139: 329–52. doi: 10.1016/j.watres.2018.03.058

49. Guillen-Burrieza E., M. O. Mavukkandy, M. R. Bilad, and H. A. Arafat. 2016. "Understanding Wetting Phenomena in Membrane Distillation and How Operational Parameters Can Affect It." *Journal of Membrane Science* 515 (October): 163–74. doi: 10.1016/j.memsci.2016.05.051

50. Wenzel, R. N. 1936. "Resistance of Solid Surfaces to Wetting by Water." *Industrial & Engineering Chemistry* 28 (8): 988–94. doi: 10.1021/ie50320a024

51. Rezaei, M., D. M. Warsinger, J. H. Lienhard V, and W. M. Samhaber. 2017. "Wetting Prevention in Membrane Distillation through Superhydrophobicity and Recharging

an Air Layer on the Membrane Surface." *Journal of Membrane Science* 530 (December 2016): 42–52. doi: 10.1016/j.memsci.2017.02.013

52. Rezaei, M., and W. M. Samhaber. 2016. "Wetting Behaviour of Superhydrophobic Membranes Coated with Nanoparticles in Membrane Distillation." *Chemical Engineering Transactions* 47: 373–78. doi: 10.3303/CET1647063

53. An, A. K., J. Guo, S. Jeong, E.-J. Lee, S. A. A. Tabatabai, and T. Leiknes. 2016. "High Flux and Antifouling Properties of Negatively Charged Membrane for Dyeing Wastewater Treatment by Membrane Distillation." *Water Research* 103 (October): 362–71. doi: 10.1016/j.watres.2016.07.060

54. Kim, H. C., J. Shin, S. Won, J. Y. Lee, S. K. Maeng, and K. G. Song. 2015. "Membrane Distillation Combined with an Anaerobic Moving Bed Biofilm Reactor for Treating Municipal Wastewater." *Water Research* 71 (March): 97–106. doi: 10.1016/j.watres.2014.12.048

55. Curcio, E., and E. Drioli. 2005. "Membrane Distillation and Related Operations—A Review." *Separation & Purification Reviews* 34 (1): 35–86. doi: 10.1081/SPM-200054951

56. Thomas, N., M. O. Mavukkandy, S. Loutatidou, and H. A. Arafat. 2017. "Membrane Distillation Research & Implementation: Lessons from the Past Five Decades." *Separation and Purification Technology* 189 (December): 108–27. doi: 10.1016/j.seppur.2017.07.069

57. Onsekizoglu, P. 2012. "Membrane Distillation: Principle, Advances, Limitations and Future Prospects in Food Industry." In *Distillation – Advances from Modeling to Applications*. InTech.

58. Eykens, L., K. De Sitter, C. Dotremont, L. Pinoy, and B. Van der Bruggen. 2017. "Membrane Synthesis for Membrane Distillation: A Review." *Separation and Purification Technology* 182 (July): 36–51. doi: 10.1016/j.seppur.2017.03.035

59. Attia, H., A. Shirin, C. J. Wright, and N. Hilal. 2017. "Superhydrophobic Electrospun Membrane for Heavy Metals Removal by Air Gap Membrane Distillation (AGMD)." *Desalination* 420 (October): 318–29. doi: 10.1016/j.desal.2017.07.022

60. Essalhi, M., and M. Khayet. 2015. "Membrane Distillation (MD)." In S. Tarleton (Ed.), *Progress in Filtration and Separation*, 61–99. Elsevier.

61. Essalhi, M., and M. Khayet. 2012. "Surface Segregation of Fluorinated Modifying Macromolecule for Hydrophobic/Hydrophilic Membrane Preparation and Application in Air Gap and Direct Contact Membrane Distillation." *Journal of Membrane Science* 417–418 (November): 163–73. doi: 10.1016/j.memsci.2012.06.028

62. Puppolo, M. M., J. R. Hughey, B. Weber, T. Dillon, D. Storey, E. Cerkez, and S. Jansen-Varnum. 2017. "Plasma Modification of Microporous Polymer Membranes for Application in Biomimetic Dissolution Studies." *AAPS Open* 3 (1): 9. doi: 10.1186/s41120-017-0019-4

63. Wei, X., B. Zhao, X. M. Li, Z. Wang, B.-Q. He, T. He, and B. Jiang. 2012. "CF4 Plasma Surface Modification of Asymmetric Hydrophilic Polyethersulfone Membranes for Direct Contact Membrane Distillation." *Journal of Membrane Science* 407–408 (July): 164–75. doi: 10.1016/j.memsci.2012.03.031

64. Kim, H., K. Noh, C. Choi, J. Khamwannah, D. Villwock, and S, Jin. 2011. "Extreme Superomniphobicity of Multiwalled 8 Nm TiO₂ Nanotubes." *Langmuir* 27 (16): 10191–196. doi: 10.1021/la2014978

65. Perreault, F., A, Fonseca de Faria, and M. Elimelech. 2015. "Environmental Applications of Graphene-Based Nanomaterials." *Chemical Society Reviews* 44 (16): 5861–896. doi: 10.1039/C5CS00021A

66. Bhadra, M., S. Roy, and S. Mitra. 2016. "Desalination across a Graphene Oxide Membrane via Direct Contact Membrane Distillation." *Desalination* 378 (January): 37–43. doi: 10.1016/j.desal.2015.09.026

67. Celebi, K., J. Buchheim, R. M. Wyss, A. Droudian, P. Gasser, I. Shorubalko, J.-I. Kye, C. Lee, and H. G. Park. 2014. "Ultimate Permeation Across Atomically Thin Porous Graphene." *Science* 344 (6181): 289–92. doi: 10.1126/science.1249097

68. Moradi, R., J. Karimi-Sabet, M. Shariaty-Niassar, and M. Koochaki. 2015. "Preparation and Characterization of Polyvinylidene Fluoride/Graphene Superhydrophobic Fibrous Films." *Polymers* 7 (8): 1444–463. doi: 10.3390/polym7081444

69. Polat, E. O., O. Balci, N. Kakenov, H. Burkay Uzlu, C. Kocabas, and R. Dahiya. 2015. "Synthesis of Large Area Graphene for High Performance in Flexible Optoelectronic Devices." *Scientific Reports* 5 (1): 16744. doi: 10.1038/srep16744

70. Liu, G., W. Jin, and N. Xu. 2016. "Two-Dimensional-Material Membranes: A New Family of High-Performance Separation Membranes." *Angewandte Chemie International Edition* 55 (43): 13384–3397. doi: 10.1002/anie.201600438

71. Macedonio, F., and E. Drioli. 2010. "Hydrophobic Membranes for Salts Recovery from Desalination Plants." *Desalination and Water Treatment* 18 (1–3): 224–34. doi: 10.5004/dwt.2010.1775

72. Macedonio, F., L. Katzir, N. Geisma, S. Simone, E. Drioli, and J. Gilron. 2011. "Wind-Aided Intensified EVaporation (WAIV) and Membrane Crystallizer (MCr) Integrated Brackish Water Desalination Process: Advantages and Drawbacks." *Desalination* 273 (1): 127–35. doi: 10.1016/j.desal.2010.12.002

73. Cui, Z., X. Li, Y. Zhang, Z. Wang, A. Gugliuzza, F. Militano, E. Drioli, and F. Macedonio. 2018. "Testing of Three Different PVDF Membranes in Membrane Assisted-Crystallization Process: Influence of Membrane Structural-Properties on Process Performance." *Desalination* 440 (August): 68–77. doi: 10.1016/j.desal.2017.12.038

74. Curcio, E., G. Di Profio, and E. Drioli. 2003. "A New Membrane-Based Crystallization Technique: Tests on Lysozyme." *Journal of Crystal Growth* 247 (1–2): 166–76. doi: 10.1016/S0022-0248(02)01794-3

75. Curcio, E., A. Criscuoli, and E. Drioli. 2001. "Membrane Crystallizers." *Industrial & Engineering Chemistry Research* 40 (12): 2679–684. doi: 10.1021/ie000906d

76. Edwie, F., and Tai-Shung Chung. 2013. "Development of Simultaneous Membrane Distillation–Crystallization (SMDC) Technology for Treatment of Saturated Brine." *Chemical Engineering Science* 98 (July): 160–72. doi: 10.1016/j.ces.2013.05.008

77. Kieffer, R., D. Mangin, F. Puel, and C. Charcosset. 2009a. "Precipitation of Barium Sulphate in a Hollow Fiber Membrane Contactor, Part I: Investigation of Particulate Fouling." *Chemical Engineering Science* 64 (8): 1759–767. doi: 10.1016/j.ces.2009.01.011

78. Kieffer, R., D. Mangin, F. Puel, and C. Charcosset. 2009b. "Precipitation of Barium Sulphate in a Hollow Fiber Membrane Contactor: Part II The Influence of Process Parameters." *Chemical Engineering Science* 64 (8): 1885–891. doi: 10.1016/j.ces.2009.01.013

79. Curcio, E., S. Simone, G. Di Profio, E. Drioli, A. Cassetta, and D. Lamba. 2005. "Membrane Crystallization of Lysozyme under Forced Solution Flow." *Journal of Membrane Science* 257 (1–2): 134–43. doi: 10.1016/j.memsci.2004.07.037

80. Di Profio, G., C. Stabile, A. Caridi, E. Curcio, and E. Drioli. 2009. "Antisolvent Membrane Crystallization of Pharmaceutical Compounds." *Journal of Pharmaceutical Sciences* 98 (12): 4902–13. doi: 10.1002/jps.21785

81. Di Profio, G., E. Curcio, and E. Drioli. 2010. "Membrane Crystallization Technology." In E. Drioli and L. Giorno (Eds.), *Comprehensive Membrane Science and Engineering*, 21–46. Elsevier.

82. Drioli, E., G. Di Profio, and E. Curcio. 2012. "Progress in Membrane Crystallization." *Current Opinion in Chemical Engineering* 1 (2): 178–82. doi: 10.1016/j.coche.2012.03.005

83. Srisurichan, S, R. Jiraratananon, and A. Fane. 2006. "Mass Transfer Mechanisms and Transport Resistances in Direct Contact Membrane Distillation Process." *Journal of Membrane Science* 277 (1–2): 186–94. doi: 10.1016/j.memsci.2005.10.028.

84. Schofield, R. W., A. G. Fane, and C. J. D. Fell. 1987. "Heat and Mass Transfer in Membrane Distillation." *Journal of Membrane Science* 33 (3): 299–313. doi: 10.1016/S0376-7388(00)80287-2

85. Laganà, F., G. Barbieri, and E. Drioli. 2000. "Direct Contact Membrane Distillation: Modelling and Concentration Experiments." *Journal of Membrane Science* 166 (1): 1–11. doi: 10.1016/S0376-7388(99)00234-3

86. Phattaranawik, J., R. Jiraratananon, and A. G. Fane. 2003. "Heat Transport and Membrane Distillation Coefficients in Direct Contact Membrane Distillation." *Journal of Membrane Science* 212 (1–2): 177–93. doi: 10.1016/S0376-7388(02)00498-2

87. Martínez, L., and J. M. Rodríguez-Maroto. 2008. "Membrane Thickness Reduction Effects on Direct Contact Membrane Distillation Performance." *Journal of Membrane Science* 312 (1–2): 143–56. doi: 10.1016/j.memsci.2007.12.048

88. Lawson, K. W., and D. R. Lloyd. 1997b. "Membrane Distillation." *Journal of Membrane Science* 124 (1): 1–25. doi: 10.1016/S0376-7388(96)00236-0

89. Bandini, S., C. Gostoli, and G. C. Sarti. 1991. "Role of Heat and Mass Transfer in Membrane Distillation Process, Desalination." *Desalination* 81: 91–106

90. Wagner, C. 1939. "Kinetik Der Phasenbildung. Von Prof. Dr. M. Volmer. (Bd. IV Der Sammlung "Die Chemische Reaktion." Herausgegeben von K. F. Bonhoeffer.) XII Und 220 S. Verlag Th. Steinkopff, Dresden Und Leipzig 1939. Preis Geh. RM. 19,—, Geb. RM. 20,—." *Angewandte Chemie* 52 (30): 503–4. doi: 10.1002/ange.19390523006

91. Margolin, A. L., and M. A. Navia. 2001. "Protein Crystals as Novel Catalytic Materials." *Angewandte Chemie International Edition* 40 (12): 2204–222. doi: 10.1002/1521-3773(20010618)40:12<2204::AID-ANIE2204>3.0.CO;2-J

92. Falkner, J C., A. M. Al-Somali, J. A. Jamison, J. Zhang, S. L. Adrianse, R. L. Simpson, M. K. Calabretta, W. Radding, G. N. Phillips, and V. L. Colvin. 2005. "Generation of Size-Controlled, Submicrometer Protein Crystals." *Chemistry of Materials* 17 (10): 2679–686. doi: 10.1021/cm047924w

93. McPherson, A., and J. A. Gavira. 2014. "Introduction to Protein Crystallization." *Acta Crystallographica Section*

F Structural Biology Communications 70 (1): 2–20. doi: 10.1107/S2053230X13033141

94. Smatanová, I. K. 2002. "Crystallization of Biological Macromolecules." *Materials Structure* 9 (1): 14–15. doi: 063

95. Tang, B., G, Yu, J. Fang, and T, Shi. 2010. "Recovery of High-Purity Silver Directly from Dilute Effluents by an Emulsion Liquid Membrane-Crystallization Process." *Journal of Hazardous Materials* 177 (1–3): 377–83. doi: 10.1016/j.jhazmat.2009.12.042

96. Li, W., B. Van der Bruggen, and P. Luis. 2014. "Integration of Reverse Osmosis and Membrane Crystallization for Sodium Sulphate Recovery." *Chemical Engineering and Processing: Process Intensification* 85 (November): 57–68. doi: 10.1016/j.cep.2014.08.003

97. Ye, W., J. Lin, J. Shen, P. Luis, and B. Van der Bruggen. 2013. "Membrane Crystallization of Sodium Carbonate for Carbon Dioxide Recovery: Effect of Impurities on the Crystal Morphology." *Crystal Growth & Design* 13 (6): 2362–372. doi: 10.1021/cg400072n

98. Ruiz Salmón, I., R. Janssens, and P. Luis. 2017. "Mass and Heat Transfer Study in Osmotic Membrane Distillation-Crystallization for CO_2 Valorization as Sodium Carbonate." *Separation and Purification Technology* 176 (April): 173–83. doi: 10.1016/j.seppur.2016.12.010

99. Luis, P., and B. Van der Bruggen. 2013. "The Role of Membranes in Post-Combustion CO_2 Capture." *Greenhouse Gases: Science and Technology* 3 (5): 318–37. doi: 10.1002/ghg.1365

100. Jia, Z., Z. Liu, and F. He. 2003. "Synthesis of Nanosized $BaSO_4$ and $CaCO_3$ Particles with a Membrane Reactor: Effects of Additives on Particles." *Journal of Colloid and Interface Science* 266 (2): 322–27. doi: 10.1016/S0021-9797(03)00187-5

101. Zhou, J., X. Cao, X. Yong, S.-y. Wang, X. Liu, Y.-l. Chen, T. Zheng, and P.-k. Ouyang. 2014. "Effects of Various Factors on Biogas Purification and Nano-$CaCO_3$ Synthesis in a Membrane Reactor." *Industrial & Engineering Chemistry Research* 53 (4): 1702–706. doi: 10.1021/ie4034939

102. Li, S., X. Li, and D. Wang. 2004. "Membrane (RO-UF) Filtration for Antibiotic Wastewater Treatment and Recovery of Antibiotics." *Separation and Purification Technology* 34 (1–3): 109–14. doi: 10.1016/S1383-5866(03)00184-9

103. Drioli, E., M. C. Carnevale, A. Figoli, and A. Criscuoli. 2014. "Vacuum Membrane Dryer (VMDr) for the Recovery of Solid Microparticles from Aqueous Solutions." *Journal*

of Membrane Science 472 (December): 67–76. doi: 10.1016/j.memsci.2014.08.047

104. Macedonio, F., C. A. Quist-Jensen, O. Al-Harbi, H. Alromaih, S. A. Al-Jlil, F. Al Shabouna, and E. Drioli. 2013. "Thermodynamic Modeling of Brine and Its Use in Membrane Crystallizer." *Desalination* 323 (August): 83–92. doi: 10.1016/j.desal.2013.02.009

105. Macedonio, F., A. Brunetti, G. Barbieri, and E. Drioli. 2012. "Water Recovery from Waste Gaseous Streams: An Application of Hydrophobic Membranes." *Procedia Engineering* 44: 202–3. doi: 10.1016/j.proeng.2012.08.358

106. Macedonio, F., M. Cersosimo, A. Brunetti, G. Barbieri, and E. Drioli. 2014a. "Water Recovery from Humidified Waste Gas Streams: Quality Control Using Membrane Condenser Technology." *Chemical Engineering and Processing: Process Intensification* 86: 196–203. doi: 10.1016/j.cep.2014.08.008

107. Macedonio, F., A. Brunetti, G. Barbieri, and E. Drioli. 2013. "Membrane Condenser as a New Technology for Water Recovery from Humidified 'Waste' Gaseous Streams." *Industrial & Engineering Chemistry Research* 52 (3): 1160–67. doi: 10.1021/ie203031b

108. Brunetti, A., S. Santoro, F. Macedonio, A. Figoli, E. Drioli, and G. Barbieri. 2014. "Waste Gaseous Streams: From Environmental Issue to Source of Water by Using Membrane Condensers." *CLEAN – Soil, Air, Water* 42 (8): 1145–53. doi: 10.1002/clen.201300104

109. Ito, A. 2000. "Dehumidification of Air by a Hygroscopic Liquid Membrane Supported on Surface of a Hydrophobic Microporous Membrane." *Journal of Membrane Science* 175 (1): 35–42. doi: 10.1016/S0376-7388(00)00404-X

110. Bruce, C. F., G. F. Weber, and M. E. Collings. 2006. "Water Extraction from Coal-Fired Power Plant Flue Gas." Final Report: *Cooperative Agreement No.* DE-FC26-*03NT41907*, Pittsburgh, PA and Morgantown, WV.

111. Isetti, C., E. Nannei, and A. Magrini. 1997. "On the Application of a Membrane Air—Liquid Contactor for Air Dehumidification." *Energy and Buildings* 25 (3): 185–93. doi: 10.1016/S0378-7788(96)00993-0

112. Mejia, M, D. Lizeth, C. Castel, C. Lemaitre, and E. Favre. 2018. "Membrane Distillation (MD) Processes for Water Desalination Applications. Can Dense Selfstanding Membranes Compete with Microporous Hydrophobic Materials?" *Chemical Engineering Science* 188 (October): 84–96. doi: 10.1016/j.ces.2018.05.025

18 Role of Hollow Fiber Contactor-Based Technology in Fermentation and Enzymatic Transformation and in Chiral Separations

G. P. Syed Ibrahim, A. M. Isloor, and R. Farnood

CONTENTS

18.1 INTRODUCTION

Increased population and industrial growth demand efficient and low-cost technology for effective separations, adsorption, and gas/liquid and liquid-liquid extraction [1–3]. Among the various technologies available, the membrane separation process is one of the developing technologies. The advantages such as low cost, less time, space and energy consumption, ease of fabrication and operation, high separation efficiency, eco-friendly, and no phase separation make membrane technology attractive for many separation processes [4–6]. There are a few types of membranes which are available for effective separation and being used in various membrane processes like microfiltration (MF), ultrafiltration (UF), nanofiltration (NF), reverse osmosis (RO), gas separation, and dialysis. Among these, UF membranes are most widely used for the effective separation of proteins, heavy metal ions, and dyes with high flux and less pressure. They are also used as a pretreatment membrane for NF and RO membranes [7,8]. The NF membrane can be used for the separation of heavy metal ions, divalent salts, and dye molecules. Moreover, the RO membrane is the ultimate one for the monovalent ions separation [9,10].

The hollow fiber contactors (HFCs) are one of the emerging technologies for gas/liquid and liquid-liquid extraction. In HFCs, the hollow fiber (HF) membrane acts as a selective physical barrier between the two phases, which does not allow the dispersion of one phase within another bypassing the two fluids in the countercurrent manner [1,11,12]. HFC is used as an alternative to the packed towers. When compared with extraction processes, HFC exhibits 500 times better performance for liquid-liquid extraction. The high surface area to volume ratio of the HFC makes

this technology more attractive. Thus, the space required for this method is less compared to conventional methods. Moreover, it is cost-effective and can be easily commercialized. Additionally, these processes tender benefits like no flooding and unloading, the absence of emulsion, and high interfacial areas [13,14]. The processes like RO, UF, and dialysis have been already well established. Still, other separation processes such as adsorption and liquid-liquid extraction using HF contactors are getting consideration due to the above mentioned merits over conventional techniques. The applications of HFC have been extended to many industries such as pharmaceuticals, petrochemicals, and galvanic industries [15,16].

The HFC pores are small and the pressure difference between the fluids is carefully controlled. Thereby, the direct mixing of two fluids can be avoided. In general, microporous membranes with a pore size of 1.0 to 0.05 µm and a thickness of 20 to 100 µm are used [17]. Based on the application, either hydrophilic or hydrophobic membrane material must be chosen. Especially in the case of enzyme immobilization on the HF membrane surface, the HF membrane molecular weight cut-off (MWCO) must be carefully optimized according to the size of the enzyme so that the infiltration of enzyme molecules across the HF membrane can be avoided. The schematic representation of HFC is presented in Figure 18.1. The microporous HF membranes are woven into a fabric and wrapped over the central tube feeder, which delivers the shell side fluid. In addition, the woven fabric bestows higher mass transfer coefficients than the individual fibers by maintaining the uniform spacing. As shown in Figure 18.1, fluid-1 is passing through the lumen surface of the HF. At the same time, fluid-2 flows countercurrently and gets diverted by the baffle across the

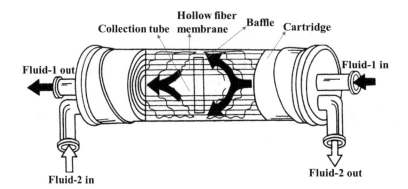

FIGURE 18.1   Schematic representation of hollow fiber contactor [21].

bed of HF. The presence of the baffle enhances efficiency by reducing the shell side sidestepping and it offers a component of velocity normal to the membrane surface, consequently, it delivers a higher mass transfer coefficient than the parallel flow. In addition, the increased surface area, less fouling, and self-supporting character of HF membrane made it more superior to the flat sheet membrane in membrane contactors [18–20]. The importance of HFC can be easily understood from the increased number of publications per year as represented in Figure 18.2.

In general, while the component 'i' is transported from one phase (feed) to another phase (permeate), the following three steps are considered. In the first step, there is a transfer from the feed phase to HF membrane and diffusion across the HF membrane in the second step. Finally, transfer from the HF membrane to the permeate phase. Thus, the flux of the component 'i' can be calculated with regard to a whole mass transfer coefficient [17].

$$j = k \times \Delta c \qquad (18.1)$$

with

$$\frac{1}{k} = \frac{1}{k\,(\text{feed})} + \frac{1}{k\,(\text{membrane})} + \frac{1}{k\,(\text{receiving pahse})} \qquad (18.2)$$

Suppose the mass transfer resistance in the membrane phase, then **Equation 18.2** becomes:

$$j = \frac{Dk}{l} \times \Delta c = \frac{P}{l} \times \Delta c \qquad (18.3)$$

where

- 'k'   signifies the distribution coefficient of component 'i' from feed into membrane phase,
- 'D'   signifies the diffusion coefficient of component 'i' in the membrane,
- 'Δc'   signifies the bulk concentration difference,
- 'l'   signifies the thickness of the membrane.

Over the past few decades, HFC with hydrophobic or hydrophilic microporous membranes is extensively

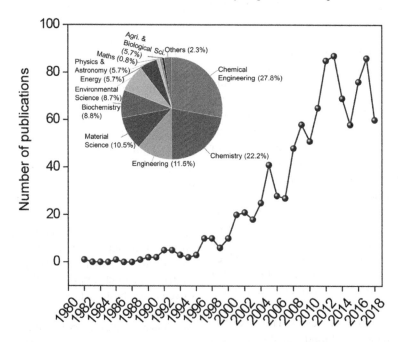

FIGURE 18.2   The number of publications per year and subject area (inset) from the Scopus database using the keyword "hollow fiber contactors" as of December 2018.

explored for the separation processes to overcome the difficulties encountered by conventional methods. As the HFC offers less foaming and higher surface area per unit volume than conventional methods, it is a good candidate for efficient extraction [22,23]. Furthermore, the selection of membrane material affects the chemical stability and separation efficiency of HFC. In the sphere of liquid-liquid extraction, porous HF membranes are highly recommended as the mass transfer resistance is less. However, the nonporous HF membranes are more suitable for gas absorption. For preparing the porous HF membranes, polymers such as polyvinylidene fluoride (PVDF), polyethylene (PE), polyolefin, polyamide (PA), polypropylene (PP), and polyacrylonitrile (PAN) would be the most suitable polymeric materials. Table 18.1 presents the properties of membrane materials for HFC used in fermentation, enzymatic transformation, and in chiral separation.

The conventional methods such as reactive extraction [33], chromatography [34], solvent extraction [35], and electrodialysis [36] were exploited for the extraction of products from the fermentation broth. However, during reactive extraction, various side products were formed as the fermentation broth contains complex components. In the case of electrodialysis, the high energy requirement and membrane pollution limited the usage in industrial scale. Similarly, in chromatography, the disadvantages such as the low capability of adsorption and frequent regeneration of the resin attributed to less usage [37]. Furthermore, the traditional enzymatic transformations were accomplished in a classical batch reactor at a regulated temperature using animal,

vegetable, or microbial origin enzyme. Nevertheless, before recovering the final products, the enzyme was inactivated. In chiral separation, initial developments were carried out with gas-liquid (GLC) and solid-liquid (HPLC) chromatography for enantiomer separation. As these techniques are discontinuous for chiral separation, simulated moving bed chromatography (SMB) was developed, which was very useful in large-scale chiral separation. Nonetheless, the high cost of stationary phase and complex experimental procedure reduced the popularity of this technique. Thus, there is a high demand for new technology to overcome the above-said problems towards fermentation, enzymatic transformation, and chiral separation.

In this chapter, the fabrication of hollow fiber membranes, potting of hollow fibers to fabricate HFC, and membrane materials used for the fabrication of HFC are discussed. This chapter also describes the challenges and evolving applications of hollow fiber contactors in fermentation, enzymatic transformation, and chiral separation. Finally, the problems associated with hollow fiber contactors, advantages, and disadvantages are reviewed.

18.2 CHALLENGES AND EVOLVING APPLICATIONS OF HOLLOW FIBER CONTACTORS IN FERMENTATION AND ENZYMATIC TRANSFORMATION

The toxic products formed during the fermentation process should be removed *in situ* to avoid the inhibition of

TABLE 18.1

Properties of Membrane Materials for Hollow Fiber Contactor Used in Fermentation, Enzymatic, and Chiral Separation

Membrane	Inner Diameter (μm)	Outer Diameter (μm)	Porosity	Pore Size (μm)	Thickness (μm) /Molecular weight cut-off (kDa)	Process	Ref.
Polyolefin	350	430	0.7	0.5	40/-	Extraction of lactic acid	[22]
Polypropylene	600	1020	75%	0.2	-	Extraction of aqueous ethanol and acetone	[24]
Polypropylene	240	300	40%	0.05	-	Extraction of penicillin G	[25]
Polyvinylidene fluoride	1400	2200	-	0.2	40/-	Extraction of surfactin	[26]
Polypropylene	260	-	-	0.04	30/-	Extraction of acetic acid	[27]
Polyamide (For schungsinstitut Berghof, Germany)	1200	2400	-	-	-/50	Conversion of racemic ester of naproxen into S-naproxen acid	[28]
Polyamide (ForschungsinstitutBerghof, Germany)	1000	2000	-	-	-/50	Hydrolysis of olive oil	[29]
Cellulose (LD OC 02, Microdyn, Wuppertal, Germany)	200	-	-	-	-	Enantioselective extraction of N-acetyl-α-amino acids	[30]
Polyacrlonitrile	210	310	-	-	-	Enantioselective extraction of amino acids	[31]
Polypropylene	220	300	40	-	-	Synthesis of chiral amine	[32]

the product formation. The conventional reactors were not energy efficient as the more diluted products recovered from the fermentation broth. Consequently, the large reactors were needed to process the huge broth volume, which led to a capital-intensive process. In recent times, an increased amount of attention has been paid to the progress of effectual extractive fermentation technology, as it improves the productivity and recovers the fermented products *in situ* [38,39]. However, the following prerequisite should be fulfilled by an effective extraction system. The system pH, which may affect the fermenting microbe, has less affinity to form a stable emulsion and water immiscibility [22]. Thereby, an effective system must be constructed for the extraction of fermentation products.

Liquid-liquid extraction of succinic acid from fermentation broth using a cellulose diacetate HFC was demonstrated by Moraes et al. [40]. The extraction of an organic acid such as succinic acid gaining more importance as there is a huge demand to produce biodegradable polymers. In this study, liquid-liquid extraction by HFC exhibited several advantages compared to conventional liquid-liquid extraction. The HF membrane acts as a barrier, where the dispersion of the liquid phase was avoided with the improved contact area. More importantly, the usage of HFC is modular and consumes less energy as claimed by the authors. In general, porous HF membranes afforded less mass transfer resistance as compared to dense membranes. The representative system consisted of two coupled HF modules, which allowed the extraction of succinic acid from the aqueous phase to the organic phase and again re-extraction from the organic phase to the aqueous phase. It was identified that Primene JM-T® amine was the more suitable extractant, where 86% of succinic acid was removed.

A hybrid system was proposed by a fusion of an internal loop airlift reactor (ALR) with a submerged polypropylene HF membrane and a regeneration unit for 2-phenylethanol (PEA) production [41]. PEA produces a rose-like aroma and is used in the food, perfumery, and cosmetic industries. PEA can be prepared by the synthetic and biological pathways. However, for preparing food-grade PEA, a biological method is preferred. In the proposed method, PEA was prepared from L-phenylalanine (Phe) using the yeast *Saccharomyces cerevisiae*. The PEA concentration of around 4 g/L was enough to inhibit the yeast activity. Thus, the *in situ* removal of the product continuously from the fermentation broth was proposed. The fermentation medium was prepared by mixing the aqueous solution of dry corn-steep liquor, vital micro and macro components, glucose (Glu) at the concentration of 10 g/L, and 85 g of lyophilized baker's yeasts. The bioreactor was packed with 13.8 L of fermentation medium and 135 g of L-Phe after 30 min of cultivation. The reaction temperature and pH were maintained at 26 to 28°C and 5.5. The membrane inlet and outlet were connected from the ALR to the regeneration tank. 13.5 L of distilled water was filled in ALR and 5 L/h air was babbled at 25°C. Meanwhile, the organic solvent pentane was flowed across the membrane module at the rate

of 0.15 L/min and distilled continuously in the regeneration unit. Then, the enzymatic transformation was performed by different feeding and aeration approaches. In both cases, a high yield of PEA was obtained. In addition, the experimental results were also compared with simulation results, which were well aligned. The schematic representation of the hybrid system is presented in Figure 18.3.

Kang et al. demonstrated the usage of microporous HF membrane-based extractive fermentor with immobilized yeast (*Saccharomyces cerevisiae*) for the fermentation of glucose to ethanol [42]. The yeast was immobilized on the shell side, and extracting solvent such as oleyl alcohol was flowing on the lumen side of the HFC. It was noted that with oleyl alcohol as extracting solvent the outlet glucose concentration was declined significantly from 0 to 3 at 9 mL/h of substrate flow rate. Furthermore, the increased efficiency of ethanol production by 28% while increasing the solvent flow rate from 0.5 to 0.7 mL/min was also reported [43].

The emulsion formation due to intimate mixing in conventional devices such as mixer-settlers and packed towers is inevitable, which increases the processing time by delaying the phase separation [44]. However, the drawbacks associated with this system can be diminished by using non-dispersive solvent extraction in HFC [23,45]. Extractive fermentation using HFC is one of the growing techniques for the extraction of fermentation products. Chen et al. [26] demonstrated the extraction of surfactin from fermentation broth using n-hexane in PVDF HFC. Surfactin is one of the effective biosurfactants, which is less toxic, biocompatible, and biodegradable. The less efficient extracting solvent n-hexane was used over ethyl acetate (EA), as the EA led to the formation of the crook of hollow fibers. The microorganism *Bacillus subtilis* was used to produce surfactin. In recycling mode, both the organic and aqueous feed solutions were circulated in the shell and lumen side of the HFC at 25°C. The solution flow was stabilized by adjusting the pressure valve, and the flow rate was 2.5 mL/min. The schematic representation

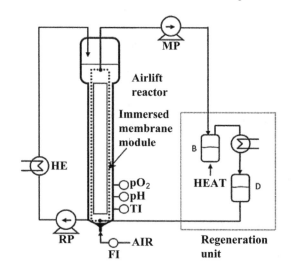

FIGURE 18.3 Schematic representation of the hybrid system [41].

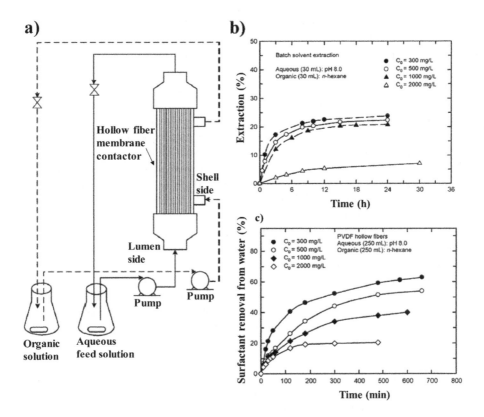

FIGURE 18.4 (a) The schematic representation of surfactant extraction using hollow fiber contactor, (b) solvent extraction in batch mode, and (c) solvent extraction in non-dispersive hollow fiber membrane at a different concentration with respect to time [26].

of surfactin extraction in HFC is represented in Figure 18.4a. The solvent extraction experiment was performed both in batch and non-dispersive mode. It was observed that, in batch mode, only 21% of surfactin was extracted with an initial concentration of less than 1000 mg/L in 12 h (Figure 18.4b). However, in non-dispersive mode, it was possible to extract 40% of surfactin in 8 h of equilibrium time with same initial concentration (Figure 18.4c). The reduced extraction capacity of the HFC was ascribed to the adsorption of surfactin on the PVDF HF membrane surface. Tong et al. investigated the non-dispersive solvent extraction using an anion exchange reaction with tri-n-octylmethylammonium chloride (TOMAC) dissolved in oleyl alcohol for the effective extraction of lactic acid from the fermentation broth [22]. The organic and aqueous feed solutions were circulated through the shell and lumen sides of the hydrophilic HFC by using two microfeeders. The fermentation was carried out using *Lactobacillus rhamnosus* IFO3863 at pH 6 and 42°C. The effect of influent TOMAC concentration on the average flux of extraction was also examined. It was found that the average extraction flux was increased with the concentration of TOMAC and lactic acid. The obtained experimental results were well aligned with the theoretical model.

Penicillin G is the raw material for semi-synthetic penicillin and produced using the strains of *Penicilliumchryogenum* in submerged aerobic fermentation. Conventionally, penicillin G is physically extracted using butyl acetate at the pH of 2.0–2.5 and 0–5°C. Still,

penicillin G decomposes during the extraction process [46]. In addition, excess usage of butyl acetate led to an increase in power consumption for extractant recovery. Therefore, reactive extraction exhibited greater selectivity and yield. Reactive extraction of penicillin G from the aqueous fermentation medium using HFC was explained by Hossain et al. [25]. In the report, solvent extraction and reactive extraction were performed. In solvent extraction, the extraction efficiency was determined based on the polarity of the solvent. The solvent such as tributyl phosphate exhibited higher extraction efficiency as it is more highly polar than other solvents such as butyl acetate. Nevertheless, the cost of this solvent is higher than other solvents, which limits the usage of the solvent extraction. In reactive extraction, Amberlite LA-2 (N-lauryl-N-trialkylmethyl amine) was used as an amine to form a complex with penicillin G. The primary amine is soluble in water and the distribution coefficient of tertiary amine complex is less than the complex formed from a secondary amine. Thus, the usage of secondary amine was recommended over the primary and tertiary amine. The amine is soluble in the organic phase and the penicillin G is soluble in the aqueous phase. The organic and aqueous phase is circulated through the shell and lumen sides of the HFC respectively. Meanwhile, the complexation reaction occurs in the interface between feed and membrane pore mouth. However, the as-formed complex is soluble in the organic phase. Thereby, penicillin G is continuously extracted from the aqueous phase to the organic phase across the HF membranes. At a pH of less than 4.5, extraction of 98% was attained in an hour with a circulation flow rate of 4.17 L/s in HFC.

In another report, reactive extraction of carboxylic acids from the fermentation broth using HFC is reported [47]. In the typical reactive extraction process, the extractants such as tri-octylamine, tri-butyl phosphate, N-methyldioctylamine, and tri-oxtylphosphine oxide were tried using 1-octanol as diluent. In off-line experiments, the extraction efficiency for acetic acid was 15% and almost 100% for caproic acid. However, in the case of in-line experiments, the extraction rate was less, which was attributed to the fouling tendency of the fresh acid-rich effluent of the fermentor. Similarly, Lee et al. described the reactive extraction of acetic acid using different amines such as tri-n-octylamine, tri-ethylamine, and di-ethylamine at the various temperatures in HFC [27]. It was found that, with increasing temperature, the acetic acid removal efficiency was also increased. Furthermore, with methyl isobutyl ketone/tri-n-octylamine and chloroform/tri-n-octylamine the system manifested the acetic acid removal efficiency of 64wt% and 53wt% under 40wt% of tri-n-octylamine at 25°C. The decreasing order of extraction was found to be tri-n-octylamine> tri-ethylamine > di-ethylamine, respectively. Reactive extraction of propionic acid from lactose using amine was stated by Jin and Yang [48]. Propionic acid is an important material to produce pharmaceuticals, food preservatives, herbicides, and so on. The low product concentration, reactor productivity, and product yield in the conventional batch fermentation for the propionic acid production were overcome by the reactive extraction in HFC. As reported, the extractive fermentation exhibited the increased productivity by five-fold, yield to more than 20%, final product concentration of 75 g/L with ~ 90% purity. The increased performance was ascribed to the reduced product inhibition by the HFC. The problem associated with raising the pH of the fermentation and extraction was reduced by the addition of a small amount of propionate, which increased the selectivity towards the propionic acid extraction over acetic acid.

The toxicity and high-power consumption for the organic solvent recovery emphasize the importance of compressed gases. The compressed gas solvent strength and selectivity can be easily altered by temperature and pressure. Further,

it can be removed by simple depressurization, which leaves the solute and aqueous phase free from the solvent and avoids the further purification of the aqueous phase. Even though much research work has shown that CO_2 is a good choice for extraction of fermentation products, compressed CO_2 is not biocompatible with most of the microorganisms, which makes it unsuitable for the *in situ* extraction of fermentation broth. Therefore, the possibility of using compressed gases such as CO_2, propane, and ethane for extracting ethanol and acetone from the fermentation broth using HFC was attempted by Bothun et al. [24]. In the typical work, the ethanol production was carried out using the yeast *Saccharomyces cerevisiae*. The hydrophobic PP HF with a porosity of 75% and a pore radius of 0.2 μm was used. The SEM cross-sectional image of a PP HF membrane is depicted in Figure 18.5. The aqueous phase and compressed solvent were circulating on the lumen and shell side of the HFC in a countercurrent manner. The interface between the solvent and the aqueous phase was maintained by adjusting the trans-membrane pressure. Finally, the compressed solvent containing the fermented product could expand in the cooling water bath to take out the extracted solute. The 20wt% binary mixture of aqueous ethanol and acetone was extracted with compressed propane in the range of 6.4 to 14.3% and 21.8 to 90.6%. Likewise, with compressed CO_2 the acetone and ethanol extraction range were 4.7 to 31.9% and 67.9 to 96.1%.

The increased productivity and selectivity of the enzymes under mild conditions attracted many of the scientists' attention for the further improvement of the system. The enzymes have a range of applications in industries such as food, pharmaceuticals, and so on. At the same time, the high cost of enzymes directed it to utilize its maximum catalytic activity. Therefore, enzymes were immobilized on the HF membrane surface to overcome difficulties such as flooding, emulsion formation, and unloading at low flow rate faced in the conventional systems. The enzymes are immobilized on the HF membrane surface by adsorption or covalent immobilization. The HFC offers not only a biochemical reaction but also the liquid/liquid mass transfer without the dispersion of one phase to another phase. In addition, the stability and resistance towards the organic

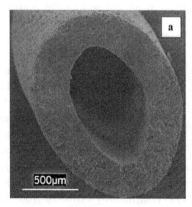

FIGURE 18.5 SEM cross-sectional images of polypropylene hollow fiber membrane (a) 60 ×, and (b) 4000 × magnification [24].

solvent were improved after immobilizing on the HF membrane surface [49]. The molecular weight of most of the enzymes falls in the range of 10 to 80 kDa; thus membrane with MWCO of 1 to 100 kDa is often employed. It is also reported that the electrostatic and hydrophobic interaction of the enzymes and membrane also influence the process performance [50,51]. Furthermore, the enzymes coupled with HF membrane always bestow the advantage of acting as catalyst support and a selective barrier simultaneously while comparing to enzyme reactors [52].

The bioconversion of less water-soluble olive oil and 5,7-diacetoxyflavone using lipase immobilized HF membrane was reported by Giorno et al. [29]. The lipase used in that study was obtained from *Candida rugosa* and *Pseudomonas cepacea*. In the membrane immobilized with lipase from *Candida rugosa*, hydrophilic HF UF membrane with MWCO of 50 kDa, and an internal thin dense layer on the lumen side and external sponge layer on the shell side was used as a support. Further, the organic and aqueous phases were circulated through the shell and lumen sides. The hydrolysis of olive oil and extraction of the product in the aqueous layer was simultaneously performed at basic pH. Moreover, the catalytic performance was influenced by the fluid dynamics conditions and lipase immobilization site on the membrane. The lipase immobilized on the shell side exhibited higher activity (~45%) and stability than the lipase immobilized on the lumen side. It was also identified that the separation of product and reaction rate was increased with respect to trans-membrane pressure from organic phase and aqueous phase flow rate. At the same time, it was decreased with an increase in the organic phase flow rate. Nevertheless, non-aqueous monophasic membrane immobilized with lipase from *Pseudomonas*

cepacea in the wall of the tubular inorganic membrane (MWCO of 10 kDa) demonstrated the ~ 95% of catalytic activity than the free suspended lipase.

In another study, enantioselective hydrolysis of cis-cyclohex-4-ene-1, 2-dicarboxylate (cis-DE) to (1S, 2R)-cyclohex-4-ene-1, 2-dicarboxylic acid monomethyl ester ((1S,2R)-ME) was performed using pig liver esterase (PLE) immobilized HF membrane biphasic contactor [53]. The stability, lower cost, and capacity to hydrolyze large verities of esters made PLE a potential candidate. The reaction scheme for the enantioselective hydrolysis is given in Figure 18.6. The enzyme was immobilized on the HF membrane both by covalent bonding and physical entrapment. In the covalent bonding, the terminal amino group of the nylon microporous membrane was reacted with a carboxylic acid group of the PLE enzyme. In the case of physical immobilization, the enzymes were filtered from outside to the inside of the polysulfone HF UF membrane (MWCO 30 kDa) in dead-end mode. The typical schematic representation of the experimental set-up is represented in Figure 18.7. In summary, the enantiomeric excess (ee) of the final product was in the range of 94 to 97%. In addition, the enzymatic activity was decreased with time in the case of covalent immobilization, which was attributed to the longer immobilization process. However, the physical immobilization of the enzyme on the membrane surface exhibited a moderate decrease in the enzyme activity.

Recently, Sotenko et al. demonstrated the tandem transformation of glycerol to esters using HFC immobilized with *Rhizomucormiehei* lipase [55]. In the typical process, the glycerol from biodiesel waste was converted into ester via tandem fermentation and continuous esterification. Where the HFC improved the stability of the enzyme, contact area

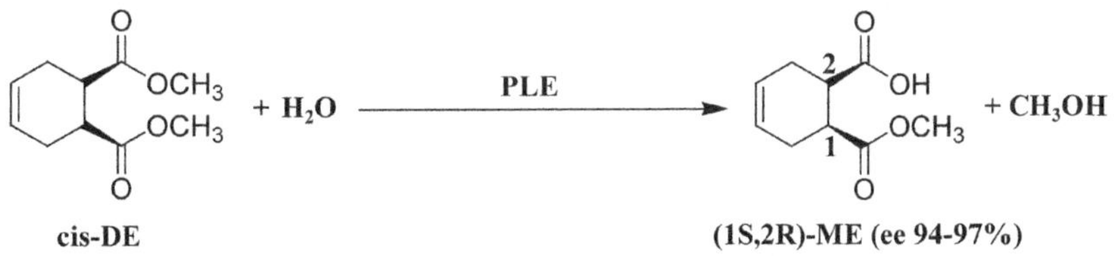

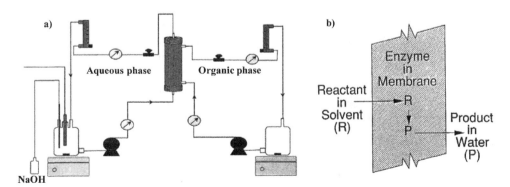

FIGURE 18.6 Reaction scheme for the enantioselective hydrolysis of cis-cyclohex-4-ene-1, 2-dicarboxylate [53].

FIGURE 18.7 Schematic representation of (a) experimental set-up of biphasic hollow fiber membrane contactor, and (b) flux of reactant and product in enzyme immobilized hollow fiber membrane contactor [53, 54].

within the reaction system, and avoided catalyst recovery. PP HF membrane with an inner and outer diameter of 40 µm and 50 µm was used. The lipase was adsorbed on the lumen side of the HF membrane by pumping the 2 mL of enzyme solution for 4 h. Then the membrane was washed with 0.5M potassium phosphate buffer (pH 5.6) for 1 h. The membrane materials such as polyethersulfone (PES), mesoporous synthetic carbon, and PP were screened for the immobilization of enzymes by adsorption. However, the membrane material PP demonstrated the highest enzyme adsorption of 15,000 U. The aqueous phase with a flow rate of 0.8 mL/min and the organic phase with a flow rate of 0.1 mL/min was circulated on the lumen and shell sides of the HF membrane. Thereby the penetration of the organic layer into the aqueous layer was circumvented. The PP membrane exhibited higher stability up to four cycles with a reaction time of 200 h. The first three cycles were performed with phosphate buffer and linoleic acid as an aqueous and organic phase. However, in the fourth cycle, fermentation broth was employed as an aqueous phase. It was identified that the reaction rate and conversion were reduced significantly after the third cycle, where the 1,3-propanediol conversion was 75% up to two cycles in 20 h and the conversion was 50% in the fourth cycle. Moreover, when compared to the batch reactor, HFC exhibited an increased volumetric reaction rate. Nevertheless, in the view of the interfacial area, the HFC showed a reaction rate of 0.67 μmolcm^{-2} h^{-1}, which was less than the batch reactor (1.68 μmol cm^{-2} h^{-1}). The result suggested that in HFC all the contact area is not active for the esterification reaction. Furthermore, this approach reduced the catalytic membrane disadvantages such as stability and incompatibility. The photograph of the biphasic HF membrane contactor is presented in Figure 18.8.

The continuous hydrolysis of penicillin G to 6-aminopenicillanic acid (6-APA) in enzyme membrane reactor (EMR) was stated by Wenten et al. [56]. The HF membrane pores were entrapped with penicillin acylase enzymes. It was reported that the maximum conversion was obtained with less flux rate. At low flux rate, the residence time will be more, which improves the percentage of conversion and reduces the processing cost. More than 90% of the enzyme was immobilized on the HF membrane surface by

dead-end filtration of the phosphate buffer solution containing enzyme from the lumen side to the shell side for 30 min. Thereafter, the HF membrane was washed thoroughly with distilled water until there was no residue of the enzyme. Furthermore, the permeation of hydrolyzed product 6-APA was examined at the TMP of 8 psi. As the pore size of the membrane was bigger than 6-APA, the solute molecules were freely diffused. However, 35% of solute molecules were retained due to the fouling of the HF membrane.

18.3 CHALLENGES AND EVOLVING APPLICATIONS OF HOLLOW FIBER CONTACTORS IN CHIRAL SEPARATION

In the modern world, the usage of racemic drugs for any kind of disease is reduced to a great extent. The demand for the preparation of optically pure and active drug molecules is increasing; thereby the side effects caused by the racemic mixture of drug molecules can be avoided. Therefore, there is a wide scope for the expansion of different strategies for the preparation of enantiomerically pure chiral compounds from the racemic mixtures [57,58]. In all of the biological systems, chirality plays a major role as many of the biological molecules and polymers are chiral. In addition, the biomolecules react more specifically with one enantiomer rather than the other enantiomer and the activity of many enzymes depends on the ee of the reactive species. Still, the chiral separation is a difficult task as the enantiomers have identical physical and chemical properties. According to the Federal Drug Administration (FDA), each enantiomer should be characterized for its purity, pharmacokinetics, and toxicity [13,16,59,60]. For enantiomer separation, Kyba et al. proposed the optical resolution four decades ago, where chiral crown ethers in chloroform were used as selectors [61]. Later, other chiral selectors such as deoxyguanosine derivatives [62], borate complexes [63], and organometallic complexes [64] were reported. Recently, solid-liquid (HPLC) chromatographic method is used widely in preparative scale. However, the cost of stationary phase, chiral selector, and problems associated with implementation in industrial-scale limit this kind of method for further practise. Additionally, the resolution based on the dispersed phase extraction presents limitations, for instance low contact interfacial area, phase coalescence, and phase dispersion in both preparative and continuous mode of operation. Nonetheless, the above said drawbacks can be easily overcome by using HFC [23].

The design of the enantioselective receptor by supramolecular chemistry is one of the major tasks. However, in order to utilize the advantages provided by the chiral receptor, the receptor should be more flexible. The major problem associated with the enantioselective receptor is recycling. Therefore, the chiral receptor becomes more viable only when it is used many times. The enantioselective receptor-mediated membrane overcomes these issues. The receptor can be placed in the membrane covalently or by dissolving

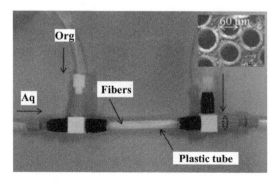

FIGURE 18.8 Photographic image of the biphasic hollow fiber membrane contactor (inset: high magnification image of the hollow fiber membrane) [55].

in the membrane [65,66]. Furthermore, the selectivity of the receptor depends on the diffusion, binding affinity, uptake, and delivery of the two enantiomers [67].

The separation of enantiomers of amino acid derivatives using the HF membrane was demonstrated by Pirkle and Bowen [31]. In that study, (S)-N-(1-naphthyl)leucineoctadecyl ester (Figure 18.9a) was used as a chiral transfer agent. The chiral reagent is insoluble in water and soluble in hexane. It was circulated through the lumen surface of the HF membrane in a closed loop. At the same time, the racemic compound N-(3,5dinitrobenzoyl)leucine is insoluble in hexane and soluble in an aqueous solution of pH 7 was circulated in the shell side of the HF membrane in the closed-loop. However, the enantio selective extraction was carried out by adding phase transfer reagent into the hexane phase. It was also mentioned that with an increase in time the ee was decreased. The schematic representation of enantioselective extraction in a closed-loop is represented in Figure 18.9b.

In another study, steroidal guanidinium was used as a chiral selector, which placed in the cellulose HF membrane, and its performance was also compared with the lab-scale U-tube apparatus [30]. In the typical study, the chiral selector in 2.5% (v/v) 1-octanol in hexane was circulated through the lumen side and 5 g of racemic N-Ac-PheOH in Tris buffer was circulated across the shell side of the one module. At the same time, 0.1 M KBr in Tris buffer was circulated over the shell side of another module. Both concentration and ee were examined over 48 h. Afterward, 60% of the amino acid transported with an ee of 31%. It is also reported that the rate of transport was affected by the buffer. The phosphate buffer exhibited a similar rate of transport like Trisbuffer; however, no enantioselevitiy was observed. Moreover, the usage of receiving phase 0.1 M KBr, without any buffer, revealed an improved enantioselectivity of 33%, whereas the transport of only 12% was observed.

Nowadays, the hollow fiber supported liquid membrane system (HFSLM) for enantiomeric separation is becoming a very fascinating area as it operates effectively by single-unit operation [68,69]. Sunsandee et al. demonstrated the enantiomeric separation of levofloxacin from the racemicmixture of ofloxacin by HFSLM system based on a chiral reagent (O,O'-dibenzoyl-(2R,3R)-tartaric acid ((-)-DBTA) [70]. The

racemic mixture ofloxacin consists of both enantiomers levofloxacin and dextrofloxacin. However, the enantiomer levofloxacin contains the antibacterial effect that is twice as effective as racemic ofloxacin. The chemical structure of ofloxacin and levofloxacin is represented in Figure 18.10a and b. The microporous polypropylene HFC with an inner and outer diameter of 240 µm and 300 µm was used. The porosity and pore size of the HFC was 30% and 0.05 µm. In the typical experiment, (-)-DBTA was dissolved in 0.5 L of 1-decanol and pumped into the HFSLM for 50 min to prepare the liquid membrane. The pumped (-)-DBTA was trapped in the micropores of the HF membrane. The loosely adhered chiral reagent was washed out before the experiment. After that, 5.0 L of feed and 5.0 L of recovery solution were circulated in the lumen and outer phase of the HFSLMcounter-currently. The chiral reagent (-)-DBTA recognizes only levofloxacin as it forms only (-)-DBTA-levofloxacin diastereomeric salt and is transported through the HF membrane phase from feed to recovery phase. Meanwhile, the concentration of levofloxacin in the feed and recovery solution was monitored by HPLC. In this method, the percentage of recovery and extraction was 85.57% and 88.35% with an ee of 80%. Moreover, the Langmuir and Freundlich adsorption isotherms were also studied for HFSLM, where the Langmuir isotherm exhibited a good agreement with the experimental results. The typical transport mechanism of levofloxacin and experimental setup of HFSLM are depicted in Figure 18.10 c and d.

In another study, HFSLM containing O'-dibenzoyl-(2S,3S)-tartaric acid ((+)-DBTA) as a chiral selector for the effective separation of (S)-amlodipine was reported [71]. Amlodipine is a third generation dihydropyridine calcium channel antagonist, which selectively prevents the calcium ion influx through the cell membrane. However, (S)-amlodipine exhibited about 2000 times better calcium channel blocking ability than the (R)-amlodipine. Thereby, separation of (S)-amlodipine enantiomer from the (R,S)-amlodipine is highly encouraged. In addition, the effect of feed pH, amlodipine concentration, chiral selector concentration, stripping phase concentration, and feed solution flow rates were also studied. It was reported that the ee was increased with increasing of pH up to 5 and started decreasing after pH 5. Thus, the optimum pH was identified as pH

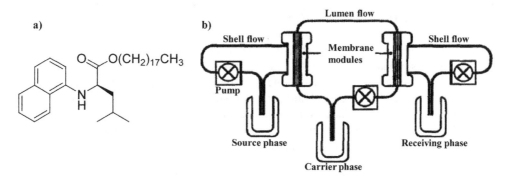

FIGURE 18.9 (a) Chemical structure of (S)-N-(1-naphthyl)leucineoctadecyl ester, and (b) schematic representation of enantioselective extraction of (S)-N-(3,5dinitrobenzoyl)leucine in a closed loop using hollow fiber membrane [31].

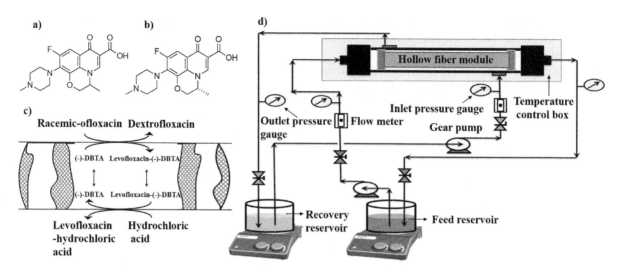

FIGURE 18.10 Chemical structure of (a) ofloxacin, (b) levofloxacin, (c) transport mechanism of levofloxacin, and (d) experimental setup of HFSLM [70].

5 for getting higher ee. Furthermore, it was perceived that higher extraction of (S)-amlodipine was achieved with feed concentration of 4 mmol L^{-1}. The (+)-DBTA concentration also played an important role in the extraction efficiency. The higher extraction efficiency of 77.50% was obtained with (+)-DBTA concentration of 4 mmol L^{-1}. Finally, the stripping phase concentration and flow rate of feed solution were optimized as 4 mmol L^{-1} and 100 mL min^{-1}, where a higher amount of (S)-amlodipine was extracted.

The preparation of enantiomers of both pharmaceutical products and fine chemicals via resolution is one of the widely accepted methodologies. The resolution can be carried out using enzymes, and inorganic or organic catalysts. Among that, enzyme catalyzed reactions are getting more attention for the enantioselective synthesis as they bestow shorter synthetic routes and are safe [72]. The enzyme selectively reacts with any one of the enantiomers to confer the preferred product. The difficulties associated with both classical and dynamic kinetic resolution were overcome by the EMR.

More recently, Yon et al. established the enantiomeric preparation of (S)-ibuprofen acid using EMR, where hydrophilic polyacrylonitrile HF membrane with MWCO of 50 kDa was used [73]. Thereby both enzymatic reaction and product separation were achieved simultaneously. The lipase *Candida rugosa* was immobilized on the HF membrane surface by the ultrafiltration 5g/L lipase solution from the shell side to the lumen side. Subsequently, the membrane was washed with phosphate to remove the free enzymes on the HF membrane surface. After immobilizing the lipase, the organic layer containing racemic ibuprofen ester was circulated through the shell side. The product was continuously extracted into the lumen side. At the same time, the unreacted (R)-ibuprofen ester was transferred into the racemization tank containing OH$^-$ resin, where the racemized product was again transferred to the organic phase. The product concentration was periodically monitored in both the aqueous and organic phases. In addition,

the operational stability of lipase both in EMR and in the free lipase system were compared. It was identified that the lipase activity in the free lipase system was reduced to 88.9% in 120 h, whereas in EMR the activity was reduced only to 53.8%. The increased operational stability of the lipase in EMR was attributed to the reduced mobility of the enzyme in the EMR. Thus, the EMR is more feasible for long time operation. The catalysts such as trioctylamine (TOA) and Amberlyst A26 hydroxide were compared in terms of their racemization efficacy. It was shown that the Amberlyst A26 hydroxide presented the conversion and ee of 96.25% and 98.40% with 38.70 mM of (S)-ibuprofen. However, TOA exhibited the conversion and ee of 86.67% and 94.86% with 34.04 mM of (S)-ibuprofen. The improved activity of Amberlyst A26 hydroxide was ascribed to its sponge-like structure and increased basicity compared to TOA. Furthermore, the conversion and ee were increased with Amberlyst catalyst loading, and the optimal concentration was identified as 6g/column. Other than this concentration, no significant change in the ee was observed. The optimum concentration of enzyme loading was also examined. The product ee was increased from 89.6% to 98.7% when the enzyme concentration was increased from 1 g/L to 5 g/L. The effect of temperature, pH, and flow rate were also intensely investigated. As reported, the conversion was increased with temperature, and at 45°C the system exhibited a higher conversion of 95.9%. However, at an elevated temperature, the conversion was decreased. The reduced conversion was due to the denaturation and deactivation of the enzyme at elevated temperatures. The optimized pH was suggested as pH 8. At that pH, the maximum catalytic activity of 159.7 mmol/h.m^2.g-protein was obtained. The aqueous and organic flow rates were maintained at 200 mL/min and 80 mL/min, respectively. The schematic representation of lipase-catalyzed HF membrane reactor is shown in Figure 18.11.

Sakinah et al. investigated the effect of tapioca starch substrate and enzyme concentration on cyclodextrin (CD)

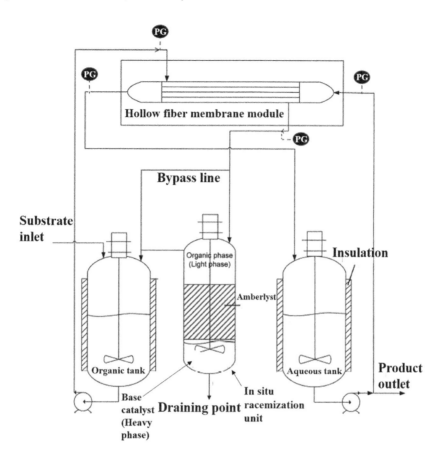

FIGURE 18.11 The schematic representation of the lipase-catalyzed hollow fiber membrane reactor [73].

preparation in EMR [74]. The HF-UF membranes with MWCO of 32 kDa and the enzyme cyclodextrin glycosyltransferase (CGTase) were used in this study. It was identified that the optimum concentration of CGTase was ca. 1%. However, CD production was reduced when the enzyme concentration was above 1%. In addition, at up to 8% of tapioca starch concentration, CD production was increased, which was attributed to the higher amylopectin content. The fouling tendency of the HF membrane was also considered. Thus, at up to the starch feeding rate of 4.41 g/h, CD production was increased. Nevertheless, the CD production decreased with the increase of starch feeding rate. The major fouling mechanism was identified as weak adsorption of starch molecules. Consequently, simple hydraulic cleaning was suggested to improve HF membrane performances.

18.4 CONCLUSION AND FUTURE TRENDS

In conclusion, there is an increased potential for the hollow fiber contactor in the field of fermentation, enzymatic transformation, and chiral separation. The improved surface area and self-supported nature of hollow fiber membranes replaced the flat sheet membranes in many of the fields. The worldwide potential development of the membrane market is ca. 10 to 15% every year. The industrial-scale hollow fiber contactor technology can be easily improved further if the physical, chemical properties, transport mechanism, robustness, and optimal process conditions of hollow fiber

membranes are widely studied. The advanced research and continuous improvement in this field would definitely replace the time and energy-consuming old technologies or blend with them to create some hybrid technology. Furthermore, this kind of new generation membrane has the capacity to perform the reaction and separation simultaneously. Thereby, the equilibrium limitations faced by conventional systems can be easily overcome. The advantages offered by hollow fiber membranes will play a vital part in many areas in the near future. Altogether, the expected industrial scale progress, trustworthy membrane fabrication processes, and the cost of the ultimate solution will be a foremost concern for commercialization.

REFERENCES

1. A.K. Pabby, A.M. Sastre, State-of-the-art review on hollow fiber contactor technology and membrane-based extraction processes, *J. Membr. Sci.*, 430 (2013) 263–303.
2. T. Mulukutla, J. Chau, D. Singh, G. Obuskovic, K.K. Sirkar, Novel membrane contactor for CO₂ removal from flue gas by temperature swing absorption, *J. Membr. Sci.*, 493 (2015) 321–328.
3. G.S. Ibrahim, A.M. Isloor, A.M. Asiri, A. Ismail, R. Kumar, M.I. Ahamed, Performance intensification of the polysulfone ultrafiltration membrane by blending with copolymer encompassing novel derivative of poly (styrene-co-maleic anhydride) for heavy metal removal from wastewater, *Chem. Eng. J.*, 353 (2018) 425–435.

4. G.P.S. Ibrahim, A.M. Isloor, A. Moslehyani, A.F. Ismail, Bio-inspired, fouling resistant, tannic acid functionalized halloysite nanotube reinforced polysulfone loose nanofiltration hollow fiber membranes for efficient dye and salt separation, *J. Water Process Eng.*, 20 (2017) 138–148.

5. R.S. Hebbar, A.M. Isloor, K. Ananda, A. Ismail, Fabrication of polydopamine functionalized halloysite nanotube/polyetherimide membranes for heavy metal removal, *J. Mater. Chem. A*, 4 (2016) 764–774.

6. I. Moideen K, A.M. Isloor, B. Garudachari, A. Ismail, The effect of glycine betaine additive on the PPSU/PSF ultrafiltration membrane performance, *Desalin. Water Treat.*, 57 (2016) 24788–24798.

7. B. Sarkar, S. DasGupta, S. De, Electric field enhanced fractionation of protein mixture using ultrafiltration, *J. Membr. Sci.*, 341 (2009) 11–20.

8. M. Sharma, G. Madras, S. Bose, Unique nanoporous antibacterial membranes derived through crystallization induced phase separation in PVDF/PMMA blends, *J. Mater. Chem. A*, 3 (2015) 5991–6003.

9. H.B. Park, B.D. Freeman, Z.B. Zhang, M. Sankir, J.E. McGrath, Highly chlorine-tolerant polymers for desalination, *Angew. Chem. Int. Ed.*, 47 (2008) 6019–6024.

10. B. Ganesh, A.M. Isloor, A.F. Ismail, Enhanced hydrophilicity and salt rejection study of graphene oxide-polysulfone mixed matrix membrane, *Desalination*, 313 (2013) 199–207.

11. R.W. Baker, Research needs in the membrane separation industry: Looking back, looking forward, *J. Membr. Sci.*, 362 (2010) 134–136.

12. J.G. Crespo, K.W. Böddeker, *Membrane Processes in Separation and Purification*, Springer Science & Business Media (2013).

13. C.A. Afonso, J.G. Crespo, Recent advances in chiral resolution through membrane-based approaches, *Angew. Chem. Int. Ed.*, 43 (2004) 5293–5295.

14. S.K. Gupta, N.S. Rathore, J.V. Sonawane, A.K. Pabby, P. Janardan, R.D. Changrani, P.K. Dey, Dispersion-free solvent extraction of U (VI) in macro amount from nitric acid solutions using hollow fiber contactor, *J. Membr. Sci.*, 300 (2007) 131–136.

15. A. Kiani, R. Bhave, K. Sirkar, Solvent extraction with immobilized interfaces in a microporous hydrophobic membrane, *J. Membr. Sci.*, 20 (1984) 125–145.

16. A.K. Pabby, S.S. Rizvi, A.M. Sastre Requena, *Handbook of Membrane Separations: Chemical, Pharmaceutical, Food, and Biotechnological Applications*, 2nd edition, CRC Press (2015),

17. J. Mulder, *Basic Principles of Membrane Technology*, Springer Science & Business Media (2012).

18. I.M. Kolangare, A.M. Isloor, Z.A. Karim, A. Kulal, A.F. Ismail, A.M. Asiri, Antibiofouling hollow-fiber membranes for dye rejection by embedding chitosan and silver-loaded chitosan nanoparticles, *Environ. Chem. Lett.* 17 (2019) 581–587.

19. V.R. Pereira, A.M. Isloor, A. Zulhairun, M. Subramaniam, W. Lau, A. Ismail, Preparation of polysulfone-based PANI–TiO$_2$ nanocomposite hollow fiber membranes for industrial dye rejection applications, *RSC Adv.*, 6 (2016) 99764–99773.

20. J.V. Sonawane, A.K. Pabby, A.M. Sastre, Au (I) extraction by LIX-79/n-heptane using the pseudo-emulsion-based hollow-fiber strip dispersion (PEHFSD) technique, *J. Membr. Sci.*, 300 (2007) 147–155.

21. E.L. Cussler, Hollow fiber contactors, In J.G. Crespo, K.W. Böddeker (Eds.) *Membrane Processes in Separation and Purification*, Springer Netherlands (1994).

22. Y. Tong, M. Hirata, H. Takanashi, T. Hano, F. Kubota, M. Goto, F. Nakashio, M. Matsumoto, Extraction of lactic acid from fermented broth with microporous hollow fiber membranes, *J. Membr. Sci.*, 143 (1998) 81–91.

23. A. Gabelman, S.-T. Hwang, Hollow fiber membrane contactors, *J. Membr. Sci.*, 159 (1999) 61–106.

24. G.D. Bothun, B.L. Knutson, H.J. Strobel, S.E. Nokes, E.A. Brignole, S. Díaz, Compressed solvents for the extraction of fermentation products within a hollow fiber membrane contactor, *J. Supercrit. Fluids*, 25 (2003) 119–134.

25. M.M. Hossain, J. Dean, Extraction of penicillin G from aqueous solutions: Analysis of reaction equilibrium and mass transfer, *Sep. Purif. Technol.*, 62 (2008) 437–443.

26. H.-L. Chen, R.-S. Juang, Extraction of surfactin from fermentation broth with n-hexane in microporous PVDF hollow fibers: Significance of membrane adsorption, *J. Membr. Sci.*, 325 (2008) 599–604.

27. Y.M. Lee, J.S. Kang, S.Y. Nam, C.H. Choi, Removal of acetic acid with amine extractants from fermentation broth using hydrophobic hollow-fiber membrane contactor, *Sep. Sci. Technol.*, 36 (2001) 457–471.

28. L. Giorno, J. Zhang, E. Drioli, Study of mass transfer performance of naproxen acid and ester through a multiphase enzyme-loaded membrane system, *J. Membr. Sci.*, 276 (2006) 59–67.

29. L. Giorno, R. Molinari, M. Natoli, E. Drioli, Hydrolysis and regioselective transesterification catalyzed by immobilized lipases in membrane bioreactors, *J. Membr. Sci.*, 125 (1997) 177–187.

30. B. Baragaña, A.G. Blackburn, P. Breccia, A.P. Davis, J. de Mendoza, J.M. Padrón-Carrillo, P. Prados, J. Riedner, J.G. de Vries, Enantioselective transport by a steroidal guanidinium receptor, *Chem. Eur. J.*, 8 (2002) 2931–2936.

31. W.H. Pirkle, W.E. Bowen, Preparative separation of enantiomers using hollow-fiber membrane technology, *Tetrahedron: Asymmetry*, 5 (1994) 773–776.

32. Y. Satyawali, E. Ehimen, L. Cauwenberghs, M. Maesen, P. Vandezande, W. Dejonghe, Asymmetric synthesis of chiral amine in organic solvent and in-situ product recovery for process intensification: A case study, *Biochem. Eng. J.*, 117 (2017) 97–104.

33. M. Matsumoto, K. Nagai, K. Kondo, Reactive extraction of 1, 3-propanediol with aldehydes in the presence of a hydrophobic acidic ionic liquid as a catalyst, *Solvent Extr. Res. Dev.*, 22 (2015) 209–213.

34. B. Rukowicz, K. Alejski, A biologically-derived 1,3-propanediol recovery from fermentation broth using preparative liquid chromatography, *Sep. Purif. Technol.*, 205 (2018) 196–202.

35. J.J. Malinowski, Evaluation of liquid extraction potentials for downstream separation of 1, 3-propanediol, *Biotechnol. Tech.*, 13 (1999) 127–130.

36. Y. Gong, Y. Tang, X.-l. Wang, L.-x. Yu, D.-h. Liu, The possibility of the desalination of actual 1, 3-propanediol fermentation broth by electrodialysis, *Desalination*, 161 (2004) 169–178.

37. Z. Li, L. Yan, J. Zhou, X. Wang, Y. Sun, Z.-L. Xiu, Two-step salting-out extraction of 1,3-propanediol, butyric acid and acetic acid from fermentation broths, *Sep. Purif. Technol.*, 209 (2019) 246–253.

38. H. Honda, Y. Toyama, H. Takahashi, T. Nakazeko, T. Kobayashi, Effective lactic acid production by two-stage extractive fermentation, *J. Fer. Bioeng.*, 79 (1995) 589–593.

39. V. Yabannavar, D. Wang, Extractive fermentation for lactic acid production, *Biotechnol. Bioeng.*, 37 (1991) 1095–1100.

40. L.D.S. Moraes, F.D.A. Kronemberger, H.C. Ferraz, A.C. Habert, Liquid–liquid extraction of succinic acid using a hollow fiber membrane contactor, *J. Ind. Eng. Chem.*, 21 (2015) 206–211.

41. M. Mihaľ, Ľ. Krištofíková, J. Markoš, Production of 2-phenylethanol in hybrid system using airlift reactor and immersed hollow fiber membrane module, *Chem. Eng. Process. Process Intensif.*, 72 (2013) 144–152.

42. W. Kang, R. Shukla, K. Sirkar, Ethanol production in a microporous hollow-fiber-based extractive fermentor with immobilized yeast, *Biotechnol. Bioeng.*, 36 (1990) 826–833.

43. G. Frank, K. Sirkar, An integrated bioreactor-separator: In situ recovery of fermentation products by a novel membrane-based dispersion-free solvent extraction technique, *Biotechnol. Bioeng. Symp.* 17 (1986) 303–316.

44. T.C. Lo, M.H. Baird, C. Hanson, *Handbook of Solvent Extraction*, Wiley (1983).

45. R. Prasad, K.K. Sirkar, Membrane-based solvent extraction, In W.S. Winston Ho and K.K. Sirkar (Eds.), *Membrane Handbook*, Springer (1992).

46. M. Reschke, K. Schügerl, Reactive extraction of penicillin I: Stability of penicillin G in the presence of carriers and relationships for distribution coefficients and degrees, *Chem. Eng. J.*, 28 (1984) B1–B9.

47. K. De Sitter, L. Garcia-Gonzalez, C. Matassa, L. Bertin, H. De Wever, The use of membrane based reactive extraction for the recovery of carboxylic acids from thin stillage, *Sep. Purif. Technol.*, 206 (2018) 177–185.

48. Z. Jin, S.T. Yang, Extractive fermentation for enhanced propionic acid production from lactose by Propionibacterium acidipropionici, *Biotechnol. Prog.*, 14 (1998) 457–465.

49. G. Rios, M. Belleville, D. Paolucci, J. Sanchez, Progress in enzymatic membrane reactors–A review, *J. Membr. Sci.*, 242 (2004) 189–196.

50. F. Nau, F. Kerhervé, J. Leonil, G. Daufin, Selective separation of tryptic β-casein peptides through ultrafiltration membranes: Influence of ionic interactions, *Biotechnol. Bioeng.*, 46 (1995) 246–253.

51. S. Bouhallab, G. Henry, Enzymatic hydrolysis of sodium caseinate in a continuous ultrafiltration reactor using an inorganic membrane, *Appl. Microbiol. Biotechnol.*, 42 (1995) 692–696.

52. E. Drioli, G. Iorio, G. Catapano, Enzyme membrane reactors and membrane fermenters. In M.C. Porter (Ed.), *Handbook of Industrial Membrane Technology.* Noyes Publications (1989) 401–481.

53. H.A. Sousa, C. Rodrigues, E. Klein, C.A. Afonso, J.G. Crespo, Immobilisation of pig liver esterase in hollow fibre membranes, *Enzyme Microb. Technol.*, 29 (2001) 625–634.

54. J.L. Lopez, S.L. Matson, A multiphase/extractive enzyme membrane reactor for production of diltiazem chiral intermediate, *J. Membr. Sci.*, 125 (1997) 189–211.

55. M.V. Sotenko, M. Rebroš, V.S. Sans, K.N. Loponov, M.G. Davidson, G. Stephens, A.A. Lapkin, Tandem transformation of glycerol to esters, *J. Biotechnol.*, 162 (2012) 390–397.

56. I. Wenten, I. Widiasa, Enzymatic hollow fiber membrane bioreactor for penicilin hydrolysis, *Desalination*, 149 (2002) 279–285.

57. G. Subramanian, *Chiral Separation Techniques: A Practical Approach*, John Wiley & Sons (2008).

58. T.E. Beesley and J.T. Lee. Method development and optimization of enantioseparations using macrocyclic glycopeptide chiral stationary phases. In G. Subramanian, (Ed.) *Chiral Separation Techniques: A Practical Approach, Third Edition* (2001) Wiley Publications, 1–28.

59. M.E. McCarroll, F.H. Billiot, I.M. Warner, Fluorescence anisotropy as a measure of chiral recognition, *J. Am. Chem. Soc.*, 123 (2001) 3173–3174.

60. H. Ding, P. Carr, E. Cussler, Racemic leucine separation by hollow-fiber extraction, *AIChE J.*, 38 (1992) 1493–1498.

61. E.B. Kyba, K. Koga, L.R. Sousa, M.G. Siegel, D.J. Cram, Chiral recognition in molecular complexing, *J. Am. Chem. Soc.*, 95 (1973) 2692–2693.

62. V. Andrisano, G. Gottarelli, S. Masiero, E.H. Heijne, S. Pieraccini, G.P. Spada, Enantioselective extraction of dinitrophenyl amino acids mediated by lipophilic deoxyguanosine derivatives: Chiral discrimination by self-assembly, *Angew. Chem. Int. Ed.*, 38 (1999) 2386–2388.

63. J.A. Riggs, R.K. Litchfield, B.D. Smith, Molecular recognition and membrane transport with mixed-ligand borates, *J. Org. Chem.*, 61 (1996) 1148–1150.

64. J. Lacour, C. Goujon-Ginglinger, S. Torche-Haldimann, J.J. Jodry, Effiziente enantioselektive Extraktion von Tris (diimin) ruthenium (II)-Komplexen durch chirale, lipophile TRISPHAT-Anionen, *Angew. Chem.*, 112 (2000) 3830–3832.

65. W.H. Pirkle, E.M. Doherty, Enantioselective transport through a silicone-supported liquid membrane, *J. Am. Chem. Soc.*, 111 (1989) 4113–4114.

66. K. Inoue, A. Miyahara, T. Itaya, Enantioselective permeation of amino acids across membranes prepared from 3α-helix bundle polyglutamates with oxyethylene chains, *J. Am. Chem. Soc.*, 119 (1997) 6191–6192.

67. M. Newcomb, J.L. Toner, R.C. Helgeson, D.J. Cram, Host-guest complexation. 20. Chiral recognition in transport as a molecular basis for a catalytic resolving machine, *J. Am. Chem. Soc.*, 101 (1979) 4941–4947.

68. N.M. Kocherginsky, Q. Yang, L. Seelam, Recent advances in supported liquid membrane technology, *Sep. Purif. Technol.*, 53 (2007) 171–177.

69. D. Huang, K. Huang, S. Chen, S. Liu, J. Yu, Rapid reaction-diffusion model for the enantioseparation of phenylalanine across hollow fiber supported liquid membrane, *Sep. Sci. Technol.*, 43 (2008) 259–272.

70. N. Sunsandee, S. Phatanasri, P. Ramakul, U. Pancharoen, Thermodynamic parameters and isotherm application on enantiomeric separation of levofloxacin using hollow fiber supported liquid membrane system, *Sep. Purif. Technol.*, 195 (2018) 377–387.

71. N. Sunsandee, N. Leepipatpiboon, P. Ramakul, U. Pancharoen, The selective separation of (S)-amlodipine via a hollow fiber supported liquid membrane: Modeling and experimental verification, *Chem. Eng. J.*, 180 (2012) 299–308.

72. N. Ran, L. Zhao, Z. Chen, J. Tao, Recent applications of biocatalysis in developing green chemistry for chemical synthesis at the industrial scale, *Green Chem.*, 10 (2008) 361–372.

73. L. Sie Yon, F.N. Gonawan, A.H. Kamaruddin, M.H. Uzir, Enzymatic deracemization of (R, S)-ibuprofen ester via lipase-catalyzed membrane reactor, *Ind. Eng. Chem. Res.*, 52 (2013) 9441–9453.

74. A.M. Sakinah, A. Ismail, R.M. Illias, A. Zularisam, O. Hassan, T. Matsuura, Effect of substrate and enzyme concentration on cyclodextrin production in a hollow fibre membrane reactor system, *Sep. Purif. Technol.*, 124 (2014) 61–67.

19 Advances in Hollow Fiber Contactor Technology in Food, Pharmaceutical, and Biotechnological Separation

T. Patra and S. Ranil Wickramasinghe

CONTENTS

19.1 INTRODUCTION

In recent years, significant progress in terms of research work has been accomplished in the field of design and evolution of hollow fiber (HF) contactors with their vast application in different industrial sectors [1]. Inherent properties like high mass transfer rates and high packing density have made hollow fiber (HF) membrane contactors an important component for the food, pharmaceutical, and biotechnological separation processes. Remarkable progress has been achieved in detailed characterization of HF contactor properties, fundamental aspects, and design considerations for industrial scale applications, which are important for their future application in membrane science and technology. In particular, the concentration-driven separation process of HF contactors has revolutionized the industrial processes for highly sensitive food grade, pharmaceutical, and biomolecule-based components by diminishing the required volume of separation to multiple orders of magnitude. In recent years, there is a surge in basic and applied research for development of new HF contactors with improved properties and new membrane processes for large-scale applications [2]. Several food industries are producing large amounts of effluents with volatile aroma compounds which

can be efficiently removed using these HF contactors based on fluid-fluid contact separation processes. HF contactors with their stable dispersion-free operation have impacted the separation processes involved in the pharmaceutical industry by tackling inherent problems like emulsion formation. Numerous membrane adsorbers have evolved in last few years based on HF contactors for application in biotechnological separation processes. In general, the introduction of a new technology for industrial applications has always been difficult. However, there is always a huge demand for cost-efficient newly developed, improved processes in the separation industry to meet the energy currency demands and product quality requirements.

Further, with increasing environmental legislations, there is a huge resurgence and demand for development of such advanced functional contactors that will be useful in separation processes for future applications. Several HF contactor technologies have found industrial scale applications in the food, pharmaceutical, and biotechnological separation processes. In most of the cases, these contactors are termed by their commercial names like membrane gas adsorbers, membrane extractors, blood oxygenators, etc. With the advent of new HF membrane contactors there has been a

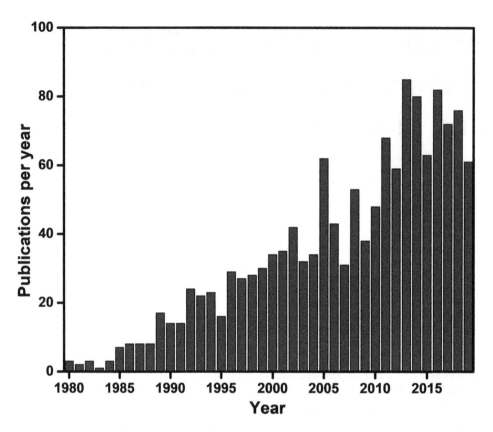

FIGURE 19.1 Publications per year on application of HF membrane contactors.

huge resurgence in terms of the number of publications on HF membrane contactors in recent years as shown in Figure 19.1. This chapter will focus on the recent advances with specific examples, advantages, and disadvantages for future consideration in the field of HF contactors for application in food, pharmaceutical, and biotechnological separations.

19.2 ADVANCES IN HF CONTACTORS FOR APPLICATION IN THE FOOD INDUSTRY

Despite of growing interest in HF contactors, food industries which suffer with several separation and purification issues were largely dependent on conventional processes like gas permeation, electrodialysis, microfiltration, ultrafiltration, nanofiltration, reverse osmosis (RO), etc [3]. Many of these processes are used in industry even today even though in most of the cases these energy intensive processes are not able to meet the customer requirement of fresh and natural qualities for food grade products. HF contactor-based processes are considered as attractive and flexible alternatives with high quality products, which is an essential criterion for food industry. Considering numerous industrial applications based on HF contactors, the most significant processes in terms of the recent advances in food processing are discussed in this section. Several trending processes are emerging in the transformation for the pretreatment and conversion of agricultural raw products to quality food products or beverages as per the requirement and demand of the present-day consumer. Use of HF contactor material

in food industries depends on the specific objectives, like dissolved oxygen removal from feed water, aroma and volatile organic compound removal from effluents, etc.

19.2.1 Dissolved Oxygen Removal

Removal of dissolved oxygen is a necessary process in several industries like food, pharmaceutical, power, semiconductor, etc. [4]. Dissolved oxygen in feed water is one of the major issues in the food processing industry where a highly pure feed water stream is a default requirement to get rid of auto oxidation issues. In general, a typical dissolved oxygen removal process uses hydrophobic and microporous membranes or hollow fibers [5]. Such characteristic membrane contactors require application of vacuum to one side of the microporous membrane, and dissolved oxygen in the feed water is removed through the porous channels. Moreover, non-porous membranes could also be used for such applications with additional process advantages like arbitrary permeate side pressure since the feed water cannot penetrate through the other side of the membrane. Tan et al. reported HF membrane modules for enhanced dissolved oxygen removal using a combined experimental and simulation approach [4]. Shorter length and lumen-side operation was recommended in this study for enhanced mass transfer of dissolved oxygen through membrane modules. Vacuum degassing was used for enhanced performance of the HF membrane modules with an optimized length of 1.6 m. Peng et al. reported woven fabric HF membrane contactors with an RO based membrane degassing system for dissolved oxygen removal from water in pilot scale [6]. The application of a vacuum to the lumen side of the woven fabric HF contactor resulted in excellent degassing performance with enhanced mass transfer as compared to the ones packed with fiber bundles because of the transverse flow of liquid. More than 97.5% oxygen removal was obtained in this process. Bhaumik et al. reported the commercially available small cross-flow blood oxygenator modules, made of hydrophobic polypropylene HF, for dissolved oxygen and carbon dioxide removal from water [7]. Pressure was reduced on the lumen side of the fiber by applying vacuum, and water was fed in cross-flow on the shell side. Under an optimized feed flow rate and other operating flow conditions, almost complete removal was achieved. Shao et al. reported HF membrane contactors for dissolved oxygen removal in a pilot scale process using a vacuum degassing method [8]. Interestingly, feed water was injected on the shell side and pressure was reduced on the lumen side of the HF membrane contactor. In the scope of this study, the effect of different operational parameters like water flow rate, temperature, and vacuum level on the performance of the contactors in terms of dissolved oxygen removal efficiency, and the mass transfer coefficient were investigated. The HF membrane module proved to be a highly efficient mass transfer device. Excellent mass transfer coefficient for the shell side liquid phase of the dissolved oxygen was observed in the HF membrane module, which varied with 0.71 power of the liquid phase flow rate. Organic impurities and the aluminum silicate content of the feed water inflicted fouling to the HF membrane contactors. Figure 19.2 depicts a schematic diagram of the experimental set up for dissolved oxygen removal. Characterization techniques like SEM, EDX, and ATR/FTIR were used to confirm the membrane fouling which resulted in lower efficiency of the membrane contactors. Recovery of the removed oxygen was efficiently performed by an air backwash cleaning method. As a non-chemical method for dissolved oxygen removal, this method is cost-effective and recommended as the regular method for the food industry. Overall, this cleaning method solved the fouling problem encountered with HF membrane contactors for application in dissolved oxygen removal.

Controlling the length of the module is an important factor in terms of application of HF contactors for dissolved oxygen removal efficiency. As the length of the module increases, the efficiency would decrease significantly due to water vapor permeation. Significant challenges remain today in terms of using techniques like vacuum degassing, which could improve the efficiency of dissolved oxygen removal by restriction of the length of the module.

19.2.2 Aroma Industry

Effluents produced by different divisions of the food industry like distilleries, fat, oil, and seafood processing contain large amounts of aroma and other volatile organic compounds [9]. Important fractions of several odorous compounds are of huge importance for food industries as recovery of these peculiar molecules could lead to

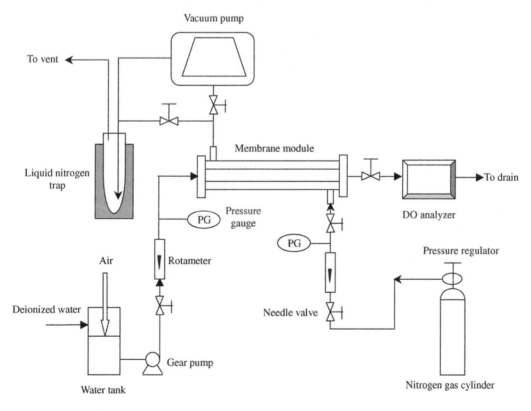

FIGURE 19.2 Schematic diagram of the experimental set up for dissolved oxygen removal (adapted with permission from [4]).

their further use in flavored food products. Bocquet et al. reported a combined experimental and simulation-based approach for a comprehensive analysis of such aroma extraction using HF contactors [10]. Four different odorous molecules like dimethyl trisulfide (DMTS), 2-phenylethanol (PE), benzaldehyde (BA), and ethyl butyrate (EB) with different chemical natures and physicochemical properties were chosen for the study. These four compounds differ in terms of distribution coefficient in aqueous layer and commercially used solvent, hexane, and water. A commercial HF contactor module was used to validate the simulation results. Simulation results were used for a better understanding of the process and improved module configuration. The combined analysis of simulation and experimental results revealed that the extraction was significantly improved for solutes possessing higher distribution coefficient in case of the flow of the feed towards the shell side. Similar results were obtained regardless of the configuration for solutes with lower partition coefficient. Two Liqui-Cel hollow fiber modules fabricated from PP Celgard hollow fibers of varying porosity and fiber i.d. are used in general for separation of such impurities in aroma industry. Hollow fiber module X-30 with ~40% porosity and fiber i.d. of 240 nm performs better for solutes with low partition coefficient values as compared to X-40 with ~25% porosity and fiber i.d. of 200 nm. Decrease in membrane resistance in case of X-30 modules with higher porosity and larger fiber i.d. plays significant role in such cases. This effect is negligible in the case of solutes with higher partition coefficients. However, for a range of solutes with different partition coefficients, the best choice is the X-30 module.

Phenol removal using HF membrane contactors with solvent extraction has been studied widely in recent years [11–13]. Shen et al. reported the extraction of phenol from wastewater and flue gas streams obtained from the lignin and biomass industry [14]. The effect of hydrodynamics using tri-n-butyl phosphate as the extractant was assessed on the overall mass transfer coefficient in the aqueous tube side and the organic shell side for both extraction and stripping. Juang and Wu reported a process based on microporous polypropylene HF membrane contactors for degradation of phenol in aqueous solution using PseudomonasputidaBCRC14365 at 30 °C [15]. Complete phenol degradation was achieved within 28 to 60 h depending on the different concentration of NaCl in the solution. Chung et al. reported a polysulfone HF contactor-based process for removal of phenol with feed concentration as high as 3500 ppm within 250 h using inorganic salt free substrate [16]. In recent years, several studies have explored similar strategy using HF contactors under different operating conditions and advanced supported ionic liquid-based membranes [17,18].

Apart from these advancements in terms of application of HF contactor technology for the aroma industry, several challenges remain even today in this field leaving a significant scope for improvement. For large-scale applications in phenol removal, real time analysis must be performed

to determine the time frame required for phenol degradation. The different process parameters involved in such processes must be optimized for commercialization of such processes. Moreover, cost minimization should also be considered for applications involving ionic liquids as an important component of the technology involved.

19.2.3 Volatile Organic Compounds (VOCs)

Food grade products obtained from industrial processes comprised of VOCs are one of the major hazards in the food industry. These VOCs are mostly originated from chemical processing and can impose a significant threat to the environment and pose health concerns. Recent years have seen increased concerns among the scientific community, and increased government regulatory efforts to get rid of such compounds prior to their release in the air [19,20]. Various physicochemical separation techniques like oxidative and adsorptive methods as well as biologically pretreatment-based processes are available for removal of VOCs [21,22]. All processes suffer with inherent issues like expensive overall cost and limited efficiency in terms of the separation performance. Membranes were applied to address such issues following the gaseous phase adsorptive methods and thereafter used extensively for various removal processes of VOCs [3,23–25]. Rui et al. reported a membrane-based absorptive process to get rid of such components using a mixture of C_6H_6/N_2 as feed [26]. Hydrophobic polypropylene HF contactor-based absorption of C_6H_6 from a mixture of C_6H_6/N_2 using an absorbent comprised of a solution of N-formyl-morpholine in water. In recent years, with increasing environmental norms, significant progress has been made in terms of application of HF contactor-based methods for removal of VOCs. McLeod et al. reported a micro-porous contactor-based method for recovery and concentration of dissolved methane gas [27]. Cookney et al. reported the use of HF membrane contactors for the desorption of dissolved methane gas from analog and anaerobic effluents resulting in 99.8% removal [28].

Vanillin is an intrinsic part of natural vanilla which is extensively used as a flavor component in cosmetics, food, beverages, and pharmaceutical sector, and market size has grown from 12,000 tons yr[1] to 18,700 tons yr[1] in between 2000 and 2016 [29]. The global market of vanillin is expected to grow at CAGR of 8.4% in next five years. However, naturally occurring vanillin contributes <1% of the overall industrial requirement. Hence, advanced cheaper chemical production in large scale was accomplished from biomass-based resources like guaiacol or lignin, accounting for low quality grade vanillin [30]. Natural vanillin production from ferulic acid using microbial processes is considered as the other prominent alternative. Recently, Solvay increased manufacturing capacity of European natural vanillin by 60 metric tons. Bioconversion of ferulic acid to vanillin using microorganisms like actinomycetes, Pseudomonas spp., Bacillus spp., and Aspergillus spp. are used in general [31]. The pervaporation process has been proposed for separation of product removal in such cases. However, the low volatility of vanillin led to a very low flux at the temperature of bioconversion processes [32]. Sciubba et al. reported a membrane-based simultaneous bioconversion and extractive method for large-scale production of vanillin from ferulic acid [33]. For enhanced separation of vanillin, the partition coefficients of vanillin and ferulic acid were measured in different solvents. Among the different solvents, n-butyl acetate and ethyl acetate exhibited higher affinity towards vanillin at pH 7. Interestingly, vanillin was found to be more soluble in water at higher pH, which led to the counter extraction of vanillin from such solvents. Hollow fiber modules were employed for such extraction or counter-extraction processes. The feed was circulated through the lumen or shell side, and depending on time course vanillin was extracted effectively in aqueous phase. Partition coefficient value inflicted significantly lower module efficiency of around ~0.9% due to enhanced mass transfer resistance.

In terms of industrial scale applications for the separation of VOCs using HF contactor technology, there are several advantages like lower specific area requirement and high removal efficiency. However, significant challenges remain still in terms of cost optimization of the processes involving contactor-based technology for such large-scale applications. Recent studies have revealed that the poor shell side dispersion is a major challenge for such applications of HF contactors, which should be considered as an important factor during the design.

19.2.4   Dairy and Beverage Industry

Average growth in the food industry has been 7.5% per year with 30% revenue generation for the overall membrane industry worldwide. Dairy, beverages like wine, beer, fruit and vegetable juices and concentrates, carbonated water, etc., and egg products are among the main components of the food industry where membranes are used to generate almost 100 million-dollar turnovers. Large numbers of applications of membrane operations in separation processes for the food industry are part of the dairy industry. Historically the application of membrane separation has been part of the dairy industry from asymmetric organic membrane, composite inorganic membranes, multi-channel ceramic membranes, to present-day UF membranes. From RO membranes to MF, as well as NF, membranes have found numerous applications in whey concentration due to their selective capabilities of taking care of impurities like large fat globules, microorganisms, casein molecules, etc. More than two out of three of the membrane applications in the dairy industry are devoted to treatment of whey components. All of the different type of membranes ranging from MF to UF as well as NF membranes are used for either whey concentration or valorization [34]. Serum proteins concentrated from whey, commonly known as whey proteins, are a nutritious protein source and are

used in the food industry after demineralization. The supernatant liquid obtained during cheese production by coagulating milk is known as whey and is comprised of ~25% original milk protein. Earlier the whey proteins were discharged into sewers due to high salt and lactose content, but now, with the advent of advanced membrane and membrane contactor-based techniques, additional food value is obtained from such components of the dairy industry. Pouliot et al. reported application of an external electrical field of strength (0–3333 V m^{-1}) during processing of lactoferrin (LF) and whey protein solutions using PVDF membranes [35]. High separation factors of around 3.0 and 9.1 were obtained between LF and the two main whey proteins β-LG and α-LA, respectively. Several other membrane and membrane contactor-based applications in the dairy industry have been reported in recent times leading to products like whey butter, whey buttermilk, etc. [36].

The beverage industry has completely different requirements in terms of production of mineral water with desired characteristics. Removal of oxygenated and sulfidic species obtained in well water is the prerequisite for further carbonation to obtain sparkling water. Earlier studies reported focused on two-step membrane-based processes to achieve such grades acceptable by the beverage industry. However, with the advent of membrane contactors, coupled removal of the impurities present in water and carbonation could be achieved simultaneously [34]. The use of HF membrane contactors for carbonation of water can efficiently decrease the size of the process assembly and reduce the overall cost by reduced CO_2 consumption. Drioli et al. reported the application of commercial HF membrane modules for carbonation of water under different operating conditions like temperature, pressure, flow rate, etc. [37]. Membrane contactors in general present significant advantages in terms of lower CO_2 consumption and release to the environment as compared to conventional methods. Moretti et al. reported a scale-up based study for controlling water gas consumption for application of membrane contactors in the beverage industry [38]. A schematic diagram of the process used in the beverage industry is depicted in Figure 19.3.

In the dairy industry significant advancement has been achieved in terms of application of the contactor-based methods, and several new food grade products are made available today as a result of such efforts. Interestingly, several research works are under process to optimize the process parameter for industrial scale applications using advanced external effects, which could potentially lead to retention of the overall protein content with minimal wastage in the process. Significant efforts have been made in terms of model development for cost minimization for the beverage industry. However, a significant amount of research work in terms of process intensification with improved efficiency under different operating conditions is still under different stages, which could impact the large-scale applications.

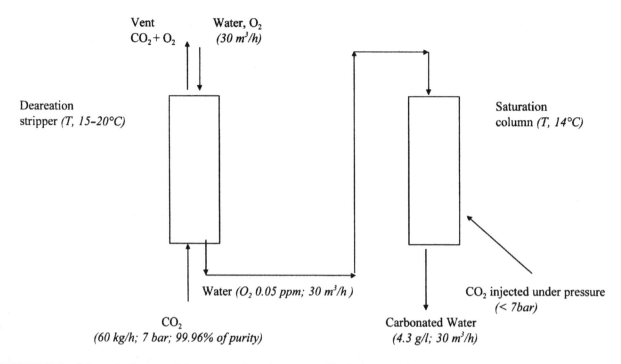

FIGURE 19.3   Schematic diagram of the contactor-based system used in the beverage industry for carbonation (adapted with permission from [38]).

19.2.5 FUTURE SCOPE

The projected market growth of food and beverage-based products like vanillin clearly indicates the huge potential of advancement in improved HF contactors during coming years. Moreover, with increasing industrial norms, the potential use of HF contactors is going to expand to be used in removal of aroma as well as other VOCs. Further, the stepwise advancement in HF membrane contactor technologies could potentially lead to the cost minimization of production units for industrially important products like vanillin. Overall, advanced HF membrane module-based processes have been accepted widely in the food industry in last few decades, and hence, spectacular advancement has been observed in terms of the development of a growing preference for using these modules over the conventional separation processes.

19.3 ADVANCES IN PHARMACEUTICAL APPLICATIONS

HF contactors have found widespread application in different sectors of the pharmaceutical processes. Inherent separation problems associated with different pharmaceutical products increase the overall cost of large-scale production. With intrinsic properties like moderate to high packing density and an excellent mass transfer rate, HF contactors have found their application to be useful in different pharmaceutical industrial processes. From production of pharmaceutical grade water to extraction of antibiotics like penicillin G, drug carriers like Liposomes, HF contactors have contributed immensely in the pharmaceutical production units. This section provides a comprehensive overview

of some of those HF membrane contactors useful in present day pharmaceutical processes in terms of their constituent materials, separation performance, and process parameters.

19.3.1 PURE STEAM

The concentration of dissolved gasses in water used in the pharmaceutical industry is an important parameter. For specific applications, the dissolved CO_2 level in raw water is required to be reduced to obtain water with conductivity <1μS/cm at 25 °C. HF membrane contactors were developed as a long term sustainable pilot-scale and industrial tests to achieve such ultra-pure water [39]. HF membrane contactors are used in the pharmaceutical industry to obtain a highly efficient CO_2 removal process by maximizing surface contact through an efficient chemical free process. Pure steam is an important component in the pharmaceutical industry, which is used in sterilization chambers for sterilization of the instruments, equipment parts, containers, and materials for sterile environments. The recently developed product 3M Liqui-Cell membrane contactors can help meet such requirements of the pharmaceutical industry based on either vacuum operation or sweep operation, and in some cases a combination of both as a result of long-term efforts in this field [40]. Figure 19.4 depicts a schematic diagram of dissolved gas removal strategy using "transverse flow" schematic design. The feed water enters the HF membrane contactors from the shell side of the microporous hollow fiber. Hollow fibers are hydrophobic in nature; hence, water cannot pass through the small pores. On the other hand, the dissolved gasses can permeate through the lumen side of the HF and exit from the lumen side port. A vacuum is

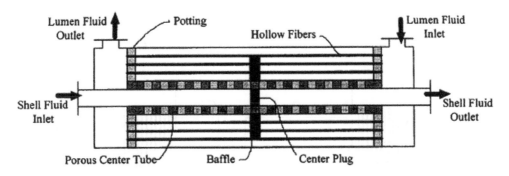

FIGURE 19.4 Dissolved gas removal using "transverse flow" membrane contactor design (adapted with permission from [40]).

applied on the lumen side of the HF, which lowers the gas pressure and results in enhanced gas separation from water by generating a driving force.

In the pharmaceutical industry, this application of HF contactors for generating pure water is an extremely important component. However, the scaling up of this process has peculiar issues like controlling the optimal design of the module length, which significantly contributes to the overall cost. Moreover, the different parameters involved in such processes are critical and need to be optimized during design consideration for commercial applications.

19.3.2 ANTIBIOTIC EXTRACTION

Different types of membrane configurations were explored for the application of HF contactors in antibiotic extraction processes including the hollow fiber supported liquid membrane (HFSLM) device and hollow fiber contained liquid membrane module (HFCLM) [41]. HFSLM devices are comprised of one single fiber set where the organic solution is soaked and supported in the pores of the hydrophobic hollow fibers while the feed and stripping aqueous solutions are circulated through the lumen and shell side separately. The HFCLM device contains two sets of microporous hollow fibers closely packed in a single module: an alkaline feed aqueous solution flows through the lumen side of one fiber set whereas the stripping solution occupies the second fiber set. The extraction of antibiotics produced by fermentation has been one of the major applications of HF membrane contactors for years [42]. Fermentation beer consists of microorganisms and antibiotic. After the filtration of microorganisms, the filtrate obtained from the beer is extracted with organic solvents like alkyl acetate, methylene chloride, higher alcohols, etc. The concentrated antibiotic is further subjected to purification using adsorption and crystallization methods, which increase the overall cost of separation. HF contactors impact the separation process with multiple advantages like high surface area per volume, complete loading, and no flooding. Antibiotic extraction from the beer is accomplished by diffusion across the membrane module and towards the bulk phase resulting in a higher concentration of antibiotic. Rindfleisch et al. first reported a two-step extraction of Penicillin G using Hoechst–Celanese HF membrane module for the

electrodialysis-based synthesis of ampicillin [43]. In this process, Penicillin G was continuously extracted using isodecanol kerosene solution at pH 5 from an aqueous buffer across the membrane of Amberlite LA-2 in a Hoechst–Celanese HF membrane module, followed byre-extraction from the organic phase across the membrane using phosphate buffer solution at pH 8 in a second Hoechst–Celanese HF module. Further, Lazarova et al. reported the use of HF contactors for large-scale application in the extractive removal and stripping of Penicillin G [41]. The kinetics of complex mass transfer processes between penicillin in the aqueous phase and amine-extractant in the organic phase were reported in detail in simultaneous reactive extraction and stripping of Penicillin G. A schematic diagram of such simultaneous extractive removal and stripping of penicillin using HF modules is presented in Figure 19.5.

With the advent of advanced pharmaceutical components, new challenges await with the corresponding extraction processes involving advanced separation complications. In recent years, the multiplicative operational costs involved with the advanced HF contactor design make their application challenging for such applications, and hence, a significant amount of research work in terms of scale-up processes must be associated with such advancements in the pharmaceutical sector.

19.3.3 DRUG CARRIER

With the advent of new drugs, the pharmaceutical industry has realized that its simultaneous coupling with the invention of new drug carrier systems is equally necessary. In a recent report, Laouini et al. reported a low cost and robust method for the preparation of novel liposome preparation using a HF membrane contactor [44]. Commercial Liquicel® Mini Module X50 consisted of 2300 uniformly distributed polypropylene hollow fibers supplied by Alting (Hoerdt, France) was used for this application as shown in the schematic diagram in Figure 19.6. The hollow fiber module had an inner diameter of 220 m and outer diameter of 300 m with 40% porosity and pore sizes of 40 nm with a fiber length of 0.115 m and total active surface area of 0.18 m². Herein, the hydrophobic HF module was properly pretreated with an aqueous ethanolic solution at a low pressure of 0.2 bar for 10 min. Finally, the permeated

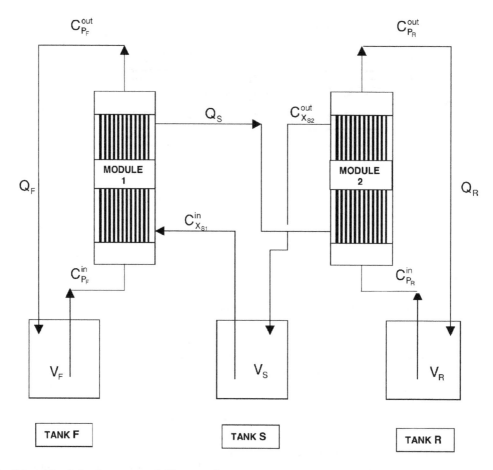

FIGURE 19.5 Schematic of simultaneous penicilin extractive removal and stripping using two HF modules (adapted with permission from [41]).

water drops were visible after 10 min on the filtrate side rather than the outlet side of the membrane device after 10 min. The use of HF membrane contactors provided multiple advantages in this process for liposomes preparation in terms of repeatability and large-scale industrial applications. Nano sized liposome suspension was produced in a simple and fast process at a large scale. Hence, in terms of process intensification, the HF contactors provide new opportunities in development of an optimized industrial processes and a promising strategy for successful applications.

Even with commercial applications available, there are significant challenges remaining in this field in terms of production of suitable HF contactors with the optimized porosity and length required for such specific applications. Moreover, large-scale applications require stepwise model development and process intensification studies for cost efficient technology development useful for present day industrial applications.

19.3.4 FUTURE SCOPE

The reactive extraction of antibiotics and advanced application of HF membrane contactors in the development of drug carriers have embarked on their tremendous potential in pharmaceutical applications. Significant efforts have been made in terms of understanding the overall kinetics of these processes, which could help in improving the design of advanced HF contactors for such application in extractive processes. A reduction of cost by combining multiple steps into one can be achieved by gradual sophistication in terms of design of the contactors and opens another fascinating scope for their application in pharmaceutical pure steam production. Commercial HF contactors have evolved in last few decades leading to significant cost minimization with efficient extraction processes for several pharmaceutical products opening the scope for future applications.

19.4 ADVANCES IN BIOTECHNOLOGICAL SEPARATION PROCESSES

Membrane-based processes have become an important component of the biotechnology industry in last few decades due to their tremendous potential for small as well as large scale biocompatible application. Initial membrane processes were adopted from technology developed for other industrial applications [45]. Bioseparation processes were considered to be a small but extremely important component of the membrane industry by providing new advanced technologies to produce biological drugs until early 2000. However, it has grown immensely in last two decades with the gradual advancement in the design of HF contactors.

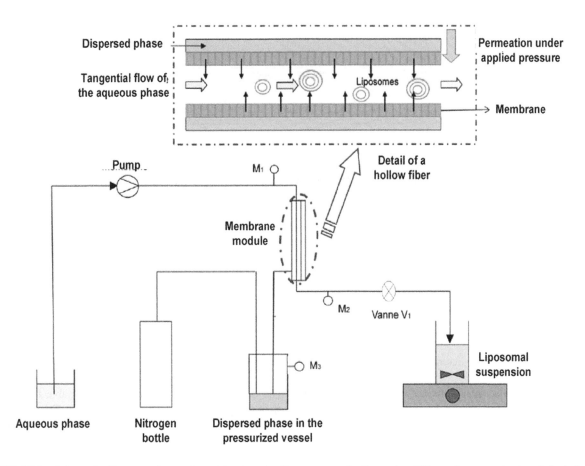

FIGURE 19.6 Schematic diagram of experimental set up for liposome production using HF membrane contactors (adapted with permission from [44]).

In the present day, these processes have found a place on today's priority list in membrane-based processes [2,45]. Other applications of HF contactor based processes are in sterile filtration, and protein concentration, as well as purification. Biotechnological separation processes involved in downstream bioprocessing are in general comprised of four steps: removal of insoluble fraction, isolation, purification of impurities, and polishing. Removal of insoluble fraction and polishing steps are easily accomplished as compared to the other two complicated steps, which pose the maximum challenges. Isolation processes are increasing the cost of the downstream processes involving the generation of a small volume of concentrate as compared to the large volume of the feed. HF contactors are often used in these two complicated steps for extraction and chromatographic application to reduce the overall cost. Further, HF contactors are also widely used in artificial organs, e.g., artificial lungs, kidneys, and livers, as well as some pharmaceutical procedures such as the separation and purification of biological materials. In tissue engineering, using such membranes is of great importance for membrane bioreactors and the culturing of cells sensitive to stress and those which require high mass transfer rates. Among other applications of HF membrane contactors in biotechnological separation processes are separation of waste materials, side product separation for

recovery and recycling of blood plasma, and active fluids like brines, cleaning-in-place solutions, etc.

19.4.1 Blood Oxygenators

Blood oxygenators are used for supplying blood during open-heart surgeries [46]. Skiens et al. invented the earliest use of HF membrane contactors fabricated using poly-4-methylpentene-1 and the thermoplastic polydimethylsiloxane copolymer as blood oxygenators [47]. HF membrane modules are used to separate blood and gas phases. A schematic representation of HF membrane-based blood oxygenators is presented in Figure 19.7. HF modules are fabricated mimicking the pulmonary capillary bed with closely packed microporous fibers in coiled design. Oxygen diffuses through the gas filled micropores to the blood. Gas transferred to the blood oxygenator can be wielded by the perfusionist to optimize oxygenation and discharge of carbon dioxide. Anesthetic vapors are in general transferred through the blood oxygenator with a specific variation for each type of vapor and oxygenator. Mimicking the biological functionality, the surfaces which are exposed to the blood are specifically coated. Wickramasinghe et al. developed the mass transfer and friction factor correlations for flat sheet and HF membrane-based blood oxygenators

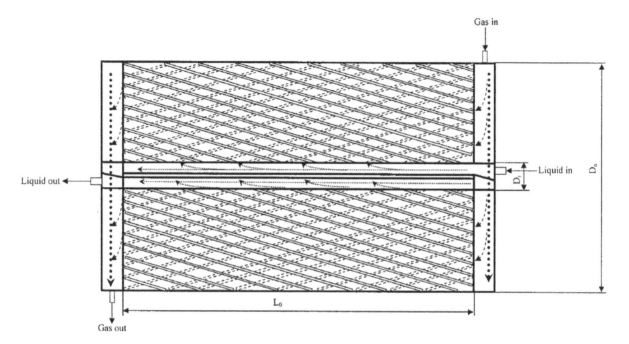

FIGURE 19.7 Schematic representation of HF membrane-based blood oxygenators (adapted with permission from [46]).

[46]. Various specific treatments are available with common specific goals to achieve biocompatibility of the circuit, increased blood compatibility, and increased resistance towards inflammatory response.

Blood oxygenators are well developed in terms of commercial application and have become important components of biotechnological applications of HF membrane contactors. However, there is significant scope left for further cost minimization of such applications for commercial large-scale production.

19.4.2 TISSUE ENGINEERING AND STEM CELL THERAPY

The use of HF contactors is essential for tissue engineering processes involving membrane bioreactors and the culturing of sensitive cells which requires high mass transfer rates. A recent report on the development of artificial lungs for patients suffering from different lung diseases is considered a revolutionary step in terms of technology development based on HF contactors [48]. Figure 19.8 illustrates the schematic representation of a HF contactor-based intravenous lung assisted device (IVLAD). Patients suffering from acute respiratory distress syndrome (ARDS), along with several lung diseases, can use the artificial implantable lungs comprised of microporous fibers of polypropylene with 380 mm inner diameter and 50 mm membrane thickness. The gap between the different fibers in the array was kept constant for any number of tied HFs. Membrane vibrations were used to enhance the performance of the contactors which helped in additional advantage by minimizing fouling. Moreover, the appliance of membrane vibrations was susceptible to improve the efficiency of the blood oxygenator because of ease of implementation, and proved to be less traumatic than use of direct pulsations of fluid flow.

The dimensions of the artificial implantable lungs developed using HF contactors are described in Table 19.1.

Even with the advancements achieved in this field, there is scope left for further research to improve the biocompatibility of the HF membrane contactors for such applications in the case of patients with different types of complex lung diseases.

19.4.3 EXTRACTION OF BIOMOLECULES

HF contactors have become an essentially useful tool for their application in the biotechnology or bioseparation sector recently with their excellent efficiency in terms of extraction of biomolecules. Biomolecules are extracted at low concentration using a complex fluid in biotechnological separation processes involved in downstream processing. The essential requirement is the separation of the biomolecule of interest from a complex mixture of a cell debris, complex substrate materials, and side products. Moreover, the biomolecules of interest are often unstable, and hence, the separation process must be rapid with mild process conditions. In most of these cases, the separation process must be performed in a closed system when operating pathogenic organism. Production steps involved in biotechnological processes like fermentation often involve different module configurations and multistep pretreatment in downstream processing. Hence, the separation process is chosen considering the sensitivity of the product and the viscosity of the fluids involved. In such cases, optimization of the membrane-based separation processes also depends on the adjusting of the membrane module with respect to the fermentation process. Several studies have reported the use of HF membrane contactors for recovery of expensive bi-product from fermentation broth. Bothun et al. reported the use of HF membrane contactors for the extraction of

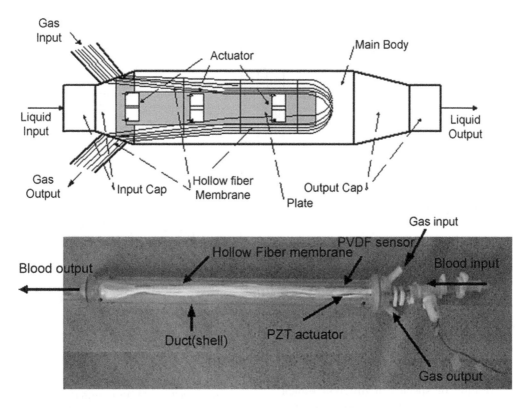

FIGURE 19.8   Schematic representation of HF contactor-based intravenous lung assisted device (IVLAD) (adapted with permission from [48]).

TABLE 19.1

Dimension and Materials of the Artificial Implantable Intravenous Lung Assisted Devices

Parameter	Membrane	Material	Cylinder Duct Inner Diameter (mm)	Number of Hollow Fibers
IVLAD 30-675	Hollow fiber	Polypropylene	30	675
IVLAD 30-450	Hollow fiber	Polypropylene	30	450
IVLAD 30-100	Hollow fiber	Polypropylene	30	100
IVLAD 30-200	Hollow fiber	Polypropylene	30	200
IVLAD 30-300	Hollow fiber	Polypropylene	30	300

ethanol and acetone as "model" fermentation products for extractive fermentation processes [49]. The productivity of fermentation processes is largely affected by product inhibition which requires the use of biocompatible *in situ* extractive methods to provide efficient separation process. Aqueous and compressed solvents like CO_2, ethane, and propane streams were contacted within isotactic polypropylene hydrophobic membrane fiber. The reported process further suggested that by combining the tunability of compressed fluid and high surface area to volume ratio of the HF contactors could be used to replace solvents like hexane with compressed fluids for extraction of numerous natural products like essential fat in biotechnological separation processes. Burge et al. reported the reactive extraction of carboxylic acids using HF membrane contactors from glycerol bioconversion broth by Lactobacillus reuteri [50]. The highly selective advanced process was demonstrated

by extraction of 3-hydroxypropanoic (HP) acid as shown in Figure 19.9. Commercial HF contactor, Liqui-Cel® (2.5 × 8 X50) containing hydrophobic polypropylene fibers (Membrana, USA).

Reactive extraction processes always involve a potential toxic impact on the bacterial physiological state. Hence, further screening experiments are required to design a biocompatible and sustainable HF membrane contactor-based technology. Moreover, large-scale production of such biomolecules via *in situ* product recovery by reactive extraction using HF membrane contactors requires further process optimization prior to large-scale applications.

19.4.4   FUTURE SCOPE

With advanced applications like artificial organs, HF contactors have opened a huge market in last few years for

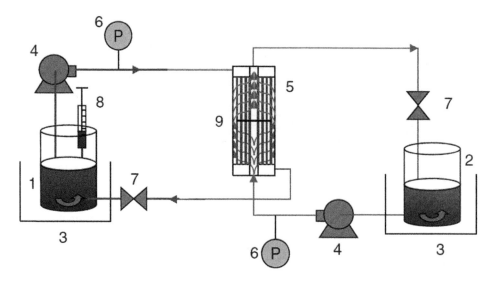

FIGURE 19.9 Experimental setup for the reactive extraction of 3-HP using a HF membrane contactor: (1) Aqueous phase reservoir; (2) Organic phase reservoir; (3) Heated baths; (4) Volumetric pumps; (5) Hollow fiber membrane contactor; (6) Manometers; (7) Pressure regulating valves; (8) Sampling device; (9) Baffle (adapted with permission from [50]).

application in the biotechnology industry. Membranes technologies found applications in the biotechnological industry for separation of species of different dimensions like cellular proteins, important biomolecules from fermentation broths, cell debris, and separation of miscellaneous low molecular weight components from proteins. The fast and aggressive development in the field of membrane chromatography, HPTFF, and electrophoretic membrane contactor could be explored for purification and separation of complex proteins. Commercial processes could be implemented using the results of small-scale comparable yield and the excellent separation performance of HF contactors for extraction of expensive biomolecules. Continuous efforts towards fabrication of advanced materials, modules, and process design could lead to development of advanced biotechnology processes.

Overall, the HF contactor-based processes have found application in all different dimensions of food, pharmaceutical, and biotechnology industry. Several new applications of such processes are in different stages of development, and continuous efforts could lead to the development of long-term sustainable processes. Further studies in terms of detailed understanding of the physical and chemical nature of interaction at the interfaces under the different process conditions would help to improve the available processes in the biotechnology industry. Future applications of membrane contactors could be directed towards decreasing the operating unit size, overall cost reduction, and enhanced membrane throughput. Future developments in the field of HF contactors could thrive to develop advanced technologies in the food, beverage, pharmaceutical, and biotechnology industries by keeping up with ever increasing demand for high purity, quality, yield, and high throughput, with low operating cost-based processes.

REFERENCES

1. Yang, M. C. & Cussler, E. L. Designing hollow-fiber contactors. *AIChE J.* **32**, 1910–1916 (1986).
2. Baker, R. W. Research needs in the membrane separation industry: Looking back, looking forward. *J. Membr. Sci.* **362**, 134–136 (2010).
3. *Handbook of Membrane Separations: Chemical, Pharmaceutical, Food, and Biotechnological Applications* (eds. A. K. Pabby, Rizvi, S. H. H, & Sastre, A. M.) (Boca Raton, FL: CRC Press, Taylor & Francis Group, 2009).
4. Tan, X., Capar, G. & Li, K. Analysis of dissolved oxygen removal in hollow fiber membrane modules: Effect of water vapor. *J. Membr. Sci.* **251**, 111–119 (2005).
5. Ito, A., Yamagiwa, K., Tamura, M. & Furusawa, M. Removal of dissolved oxygen using non-porous HF membranes. *J. Membr. Sci.* **145**, 111–117 (1998).
6. Peng, Z. G., Lee, S. H., Zhou, T., Shieh, J. J. & Chung, T. S. A study on pilot-scale degassing by polypropylene (PP) hollow fiber membrane contactors. *Desalination* **234**, 316–322 (2008).
7. Bhaumik, D., Majumdar, S., Fan, Q. & Sirkar, K. K. Hollow fiber membrane degassing in ultrapure water and microbio-contamination. *J. Membr. Sci.* **235**, 31–41 (2004).
8. Shao, J., Liu, H. & He, Y. Boiler feed water deoxygenation using hollow fiber membrane contactor. *Desalination* **234**, 370–377 (2008).
9. Pabby, A. K. & Sastre, A. M. State-of-the-art review on hollow fiber contactor technology and membrane-based extraction processes. *J. Membr. Sci.* **430**, 263–303 (2013).
10. Bocquet, S., Gascons Viladomat, F., Muvdi Nova, C., Sanchez, J., Athes, V. & Souchon, I. Membrane-based solvent extraction of aroma compounds: Choice of configurations of hollow fiber modules based on experiments and simulation. *J. Membr. Sci.* **281**, 358–368 (2006).
11. Cichy, W., Schlosser, S. & Szymanowski, J. Recovery of phenol with cyanex® 923 in membrane extraction-stripping systems. *Solvent Extr. Ion Exch.* **19**, 905–923 (2001).
12. Cichy, W. & Szymanowski, J. Recovery of phenol from aqueous streams in hollow fiber modules. *Environ. Sci. Technol.* **36**, 2088–2093 (2002).

13. Lazarova, Z. & Boyadzhieva, S. Treatment of phenol-containing aqueous solutions by membrane-based solvent extraction in coupled ultrafiltration modules. *Chem. Eng. J.* **100**, 129–138 (2004).

14. Shen, S. F., Smith, K. H., Cook, S., Kentish, S. E., Perera, J. M., Bowser, T. & Stevens, G. W. Phenol recovery with tributyl phosphate in a hollow fiber membrane contactor: Experimental and model analysis. *Sep. Purif. Technol.* **69**, 48–56 (2009).

15. Juang, R. S. & Wu, C. Y. Microbial degradation of phenol in high-salinity solutions in suspensions and hollow fiber membrane contactors. *Chemosphere* **66**, 191–198 (2007).

16. Chung, T. P., Wu, P. C. & Juang, R. S. Use of microporous hollow fibers for improved biodegradation of high-strength phenol solutions. *J. Membr. Sci.* **258**, 55–63 (2005).

17. Shen, S., Kentish, S. E. & Stevens, G. W. Effects of operational conditions on the removal of phenols from wastewater by a hollow-fiber membrane contactor. *Sep. Purif. Technol.* **95**, 80–88 (2012).

18. Nosrati, S., Jayakumar, N. S. & Hashim, M. A. Performance evaluation of supported ionic liquid membrane for removal of phenol. *J. Hazard. Mater.* **192**, 1283–1290 (2011).

19. Liu, Y., Feng, X. & Lawless, D. Separation of gasoline vapor from nitrogen by hollow fiber composite membranes for VOC emission control. *J. Membr. Sci.* **271**, 114–124 (2006).

20. Sohn, W. I., Ryu, D. H., Oh, S. J. & Koo, J. K. A study on the development of composite membranes for the separation of organic vapors. *J. Membr. Sci.* **175**, 163–170 (2000).

21. Kim, D. J. & Kim, H. Degradation of toluene vapor in a hydrophobic polyethylene hollow fiber membrane bioreactor with Pseudomonas putida. *Process Biochem.* **40**, 2015–2020 (2005).

22. Guizard, C., Boutevin, B., Guida, F., Ratsimihety, A., Amblard, P., Lasserre, J.-C. & Naiglin, S. VOC vapor transport properties of new membranes based on cross-linked flourinated elastomers. *Sep. Purif. Technol.* **22–23**, 23–30 (2001).

23. Sirkar, K. K. Membranes, phase interfaces, and separations: Novel techniques and membranes-an overview. *Ind. Eng. Chem. Res.* **47**, 5250–5266 (2008).

24. Kiani, A., Bhave, R.R., and Sirkar, K.K. Solvent extraction with immobilized interfaces in a microporous hydrophobic membrane. *J Membr Sci.* **20**, 125–145 (1984).

25. Sengupta, A. & Pittman, R. Application of membrane contactors as mass transfer devices. In *Handbook of Membrane Separations* 7–24 (2008) Boca Raton, FL: CRC Press.

26. Li, R., Xu, J., Wang, L., Li, J. & Sun, X. Reduction of VOC emissions by a membrane-based gas absorption process. *J. Environ. Sci.* **21**, 1096–1102 (2009).

27. Mcleod, A., Jefferson, B. & Mcadam, E. J. Toward gas-phase controlled mass transfer in micro-porous membrane contactors for recovery and concentration of dissolved methane in the gas phase. *J. Membr. Sci.* **510**, 466–471 (2016).

28. Cookney, J., Mcleod, A., Mathioudakis, V., Ncube, P., Soares, A., Jefferson, B. & McAdam, E. J. Dissolved methane recovery from anaerobic effluents using hollow fiber membrane contactors. *J. Membr. Sci.* **502**, 141–150 (2016).

29. Walton, N. J., Narbad, A., Faulds, C. B. & Williamson, G. Novel approaches to the biosynthesis of vanillin. *Curr. Opin. Biotechnol.* **11**, 490–496 (2000).

30. Ravishankar, G. A. & Ramachandra Rao, S. Vanilla flavor: Production by conventional and biotechnological routes. *J. Sci. Food Agric.* **80**, 289–304 (2000).

31. Priefert, M. H. Biotechnological production of vanillin. *Appl. Microbiol. Biotechnol.* **56**, 296–314 (2001).

32. Boddeker, K. W., Gatfield, I. L., Jahnig, J. & Schorm, C. Pervaporation at the vapor pressure limit: Vanillin. *J. Membr. Sci.* **137**, 155–158 (1997).

33. Sciubba, L., Gioia, D. Di, Fava, F. & Gostoli, C. Membrane-based solvent extraction of vanillin in hollow fiber contactors. *Desalination* **241**, 357–364 (2009).

34. Saxena, A., Tripathi, B. P., Kumar, M. & Shahi, V. K. Membrane-based techniques for the separation and purification of proteins: An overview. *Adv. Colloid Interface Sci.* **145**, 1–22 (2009).

35. Brisson, G., Britten, M. & Pouliot, Y. Electrically-enhanced crossflow microfiltration for separation of lactoferrin from whey protein mixtures. *J. Membr. Sci.* **297**, 206–216 (2007).

36. Fox, P. F. *Fundamentals of Cheese Science* (Aspen Pub, 2000).

37. Drioli, E., Curcio, E., Criscuoli, A. & Di Profio, G. Di. Integrated system for recovery of $CaCO_3$, NaCl and $MgSO_4 \cdot 7H_2O$ from nanofiltration retentate. *J. Membr. Sci.* **239**, 27–38 (2004).

38. Criscuoli, A., Drioli, E. & Moretti, U. Membrane contactors in the beverage industry for controlling the water gas composition. *Ann. N. Y. Acad. Sci.* **984**, 1–16 (2003).

39. Bazhenov, S. D., Bildyukevich, A. V. & Volkov, A. V. Gas-liquid hollow fiber membrane contactors for different applications. *Fibers* **6**, 1–50 (2018).

40. Sengupta, A., Peterson, P. A., Miller, B. D., Schneider, J. & Fulk, C. W. Large-scale application of membrane contactors for gas transfer from or to ultrapure water. *Sep. Purif. Technol.* **14**, 189–200 (1998).

41. Lazarova, Z., Syska, B. & Schügerl, K. Application of large-scale hollow fiber membrane contactors for simultaneous extractive removal and stripping of penicillin G. *J. Membr. Sci.* **202**, 151–164 (2002).

42. Reed, B. W., Semmens, M. J. & Cussler, E. L. Membrane contactors. In *Membrane Separations Technology. Principles and Applications* (eds. Noble, R. D. & Stern, S. A.) (Elsevier B.V., 1995), p. 467.

43. Rindfleisch, D., Syska, B., Lazarova, Z. & Schügerl, K. Integrated membrane extraction, enzymic conversion and electrodialysis for the synthesis of ampicillin from penicillin G. *Process Biochem.* **32**, 605–616 (1997).

44. Laouini, A., Jaafar-maalej, C., Sfar, S., Charcosset, C. & Fessi, H. Liposome preparation using a hollow fiber membrane contactor—Application to spironolactone encapsulation. *Int. J. Pharm.* **415**, 53–61 (2011).

45. Van Reis, R. & Zydney, A. Bioprocess membrane technology. *J. Membr. Sci.* **297**, 16–50 (2007).

46. Wickramasinghe, S. R., Han, B., Garcia, J. D. & Specht, R. Microporous membrane blood oxygenators. *AIChE J.* **51**, 656–670 (2005).

47. Skiens, W. E., Lipps, B. J., Clark, M. E. & McLain, E. A. Hollow fiber membrane oxygenator. *J. Biomed. Mater. Res.* **5**, 135–148 (1971).

48. Kim, G. B., Kim, S. J., Kim, M. H., Hong, C. U. & Kang, H. S. Development of a hollow fiber membrane module for using implantable artificial lung. *J. Membr. Sci.* **326**, 130–136 (2009).

49. Bothun, G. D., Knutson, B. L., Strobel, H. J., Nokes, S. E., Brignole, E. A., Díaz, S. Compressed solvents for the extraction of fermentation products within a hollow fiber membrane contactor. *J. Supercrit. Fluids* **25**, 119–134 (2003).

50. Burgé, G., Chemarin, F., Moussa, M., Saulou-Bérion, C., Allais, F., Spinnler, H.-É. & Athès, V. Reactive extraction of bio-based 3-hydroxypropionic acid assisted by hollow-fiber membrane contactor using TOA and Aliquat 336 in n-decanol. *J. Chem. Technol. Biotechnol.* **91**, 2705–2712 (2016).

Index

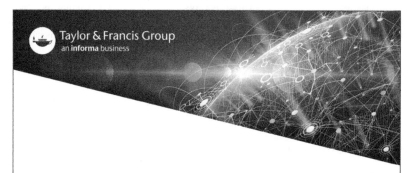

Printed and bound by CPI Group (UK) Ltd, Croydon, CR0 4YY

24/10/2024

01778295-0018